Lecture Notes in Artificial Intelligence 10564

Subseries of Lecture Notes in Computer Science

LNAI Series Editors

Randy Goebel
University of Alberta, Edmonton, Canada
Yuzuru Tanaka
Hokkaido University, Sapporo, Japan
Wolfgang Wahlster
DFKI and Saarland University, Saarbrücken, Germany

LNAI Founding Series Editor

Joerg Siekmann
DFKI and Saarland University, Saarbrücken, Germany

T0212868

More information about this series at http://www.springer.com/series/1244

Serafín Moral · Olivier Pivert
Daniel Sánchez · Nicolás Marín (Eds.)

Scalable Uncertainty Management

11th International Conference, SUM 2017
Granada, Spain, October 4–6, 2017
Proceedings

 Springer

Editors
Serafín Moral 🆔
University of Granada
Granada
Spain

Olivier Pivert
University of Rennes I
Lannion
France

Daniel Sánchez 🆔
University of Granada
Granada
Spain

Nicolás Marín 🆔
University of Granada
Granada
Spain

ISSN 0302-9743   ISSN 1611-3349   (electronic)
Lecture Notes in Artificial Intelligence
ISBN 978-3-319-67581-7   ISBN 978-3-319-67582-4   (eBook)
DOI 10.1007/978-3-319-67582-4

Library of Congress Control Number: 2017953397

LNCS Sublibrary: SL7 – Artificial Intelligence

Printed on acid-free paper

This Springer imprint is published by Springer Nature
The registered company is Springer International Publishing AG
The registered company address is: Gewerbestrasse 11, 6330 Cham, Switzerland

Preface

Managing uncertainty and inconsistency has been extensively explored in the field of artificial intelligence over a number of years. Now, with the advent of massive amounts of data and knowledge from distributed, heterogeneous, and potentially conflicting sources, there is interest in developing and applying formalisms for uncertainty and inconsistency in systems that need to better manage these data and knowledge. To meet the challenge of representing and manipulating large amounts of uncertain information, researchers are drawing from a wide range of different methodologies and uncertainty models. While Bayesian methods remain the default choice in most disciplines, sometimes there is a need for more cautious approaches, relying for instance on imprecise probabilities, ordinal uncertainty representations, or even purely qualitative models.

The International Conference on Scalable Uncertainty (SUM) aims to provide a forum for researchers who are working on uncertainty management, in different communities and with different uncertainty models, to meet and exchange ideas. Previous SUM conferences have been held in Washington DC (2007), Naples (2008), Washington DC (2009), Toulouse (2010), Dayton (2011), Marburg (2012), Washington DC (2013), Oxford (2014), Québec City (2015), and Nice (2016).

This volume contains contributions from the 11th SUM conference, which was held in Granada, Spain, during October 4–6, 2017. The conference attracted 35 submissions, of which 30 were accepted for publication and presentation at the conference, based on three rigorous reviews by the members of the Program Committee or external reviewers.

In addition, the conference greatly benefited from invited lectures by three world-leading researchers: Alberto Bugarín Diz, Fabio Gagliardi Cozman, and Martin Theobald. To further embrace the aim of facilitating interdisciplinary collaboration and cross-fertilization of ideas, and building on the tradition of invited speakers at SUM, the conference featured eight tutorials, covering a broad set of topics related to uncertainty management. We thank Olivier Cailloux, Inés Couso, Francisco Herrera, Rafael Peñaloza Nyssen, Régis Sabbadin, Antonio Salmerón, Laurent Vercouter, and Nic Wilson for preparing and presenting these tutorials. A companion paper for several of them can be found in this volume.

We would like to thank all the authors and invited speakers for their valuable contributions, and the members of the Program Committee and external reviewers for their detailed and critical assessment of the submissions. We are also very grateful to the University of Granada for hosting the conference.

October 2017

Serafín Moral
Olivier Pivert
Daniel Sánchez
Nicolás Marín

Organization

Conference Chair

Daniel Sánchez   University of Granada, Spain

Program Chairs

Serafín Moral   University of Granada, Spain
Olivier Pivert   IRISA, University of Rennes, France

Proceedings Chair

Nicolás Marín   University of Granada, Spain

Local Organization

Alexis G. Arroyo   University of Granada, Spain
Nicolás Marín   University of Granada, Spain
Gustavo Rivas-Gervilla   University of Granada, Spain
M. Dolores Ruiz   University of Cádiz, Spain
Daniel Sánchez   University of Granada, Spain
José M. Soto-Hidalgo   University of Córdoba, Spain

Steering Committee

Didier Dubois   IRIT-CNRS, France
Lluis Godo   IIIA-CSIC, Spain
Eyke Hüllermeier   Universität Paderborn, Germany
Anthony Hunter   University College London, UK
Henri Prade   IRIT-CNRS, France
Steven Schockaert   Cardiff University, UK
V.S. Subrahmanian   University of Maryland, USA

Program Committee

Leila Amgoud   IRIT-CNRS, Toulouse, France
Christoph Beierle   University of Hagen, Germany
Salem Benferhat   Université d'Artois, France
Leopoldo Bertossi   Carleton University, Canada
Fernando Bobillo   University of Zaragoza, Spain
Stéphane Bressan   National University of Singapore, Singapore
Andrea Cali   University of London, Birkbeck College, UK

Federico Cerutti	Cardiff University, UK
Reynold Cheng	The University of Hong Kong, SAR China
Olivier Colot	Université Lille 1, France
Fabio Cozman	Universidade de Sao Paulo, Brazil
Alfredo Cuzzocrea	University of Calabria, Italy
Luis M. de Campos	University of Granada, Spain
Thierry Denoeux	Université de Technologie de Compiègne, France
Sébastien Destercke	CNRS-Université de Technologie de Compiègne, France
Zied Elouedi	ISG Tunis, Tunisia
Alberto Fernández	University of Granada, Spain
Manuel Gómez-Olmedo	University of Granada, Spain
John Grant	Towson University, USA
Gabriele Kern-Isberner	TU Dortmund University, Germany
Kristian Kersting	TU Dortmund University, Germany
Evgeny Kharlamov	University of Oxford, UK
Anne Laurent	University of Montpellier 2, LIRMM, France
Sebastian Link	University of Auckland, New Zealand
Weiru Liu	Queen's University Belfast, UK
Thomas Lukasiewicz	University of Oxford, UK
Silviu Maniu	Université Paris-Sud, France
Thomas D. Nielsen	Aalborg University, Denmark
Francesco Parisi	University of Calabria, Cosenza, Italy
Simon Parsons	King's College London, UK
Gabriella Pasi	Università degli Studi di Milano Bicocca, Italy
Rafael Peñaloza	Free University of Bozen-Bolzano, Italy
Andrea Pugliese	University of Calabria, Italy
Thomas Roelleke	Queen Mary University of London, UK
Sebastian Rudolph	Technische Universität Dresden, Germany
Antonio Salmerón	University of Almería, Spain
Pierre Senellart	Telecom ParisTech, France and NUS, Singapore
Guillermo Simari	Universidad Nacional del Sur, Argentina
Umberto Straccia	ISTI-CNR, Italy
Maurice Van Keulen	University of Twente, The Netherlands
Barbara Vantaggi	La Sapienza University of Rome, Italy
María Amparo Vila	University of Granada, Spain
Jef Wijsen	University of Mons, Belgium
Sławomir Zadrożny	Systems Research Institute PAS, Poland
Marco Zaffalon	Istituto Dalle Molle di Studi sull'Intelligenza Artificiale (IDSIA), Switzerland
Andreas Züfle	Ludwig Maximilian University of Munich, Germany

Additional Reviewers

Zhipeng Huang
Xiadong Li

Sponsoring Institutions

University of Granada (UGR)

UNIVERSIDAD
DE GRANADA

High Technical School of Computer Science and Telecommunications – UGR

Department of Computer Science and Artificial Intelligence – UGR

Administrative Organization

Fundación General UGR-Empresa

Abstracts of Invited Talks

What Uncertainty Models Do We Need for Natural Language in Data-To-Text Systems?

Alberto Bugarín

Centro de Investigación en Tecnoloxías da Información (CiTIUS),
Universidade de Santiago de Compostela
alberto.bugarin.diz@usc.es

Abstract. Automatic data-to-text (D2T) systems are increasingly being used to remove the barrier between data stored in information systems and the people that demand information from them. D2T are a part of the wide field of Natural Language Generation, which automatically generate high-quality narratives which summarize in natural language the most relevant information hidden in the data, which are typically numerical (as in time series), symbolic or both. The produced narratives are provided in a form that can be directly consumed and understood by human users. D2T systems are usually built as a way to directly information to users or as a complement of other means (graphical representations, numeric tables, …). In this talk we will present and describe the general architecture of D2T systems (the so called D2T pipeline). We will also carry out a review of real D2T systems of high impact, in areas such as robot-journalism, meteorological forecasting, and clinical and industrial supervision systems, among others. We will present the models used in D2T for managing uncertainty in the language realizations. Special attention will be devoted to discuss the role that fuzzy logic and Computing with words play in the D2T field as well as to present potential areas of convergence between these two paradigms. Finally, we will present some open issues in D2T related to uncertainty management in natural language and to scalability of this approach within the contex provided by large volumes of data.

Keywords: Data-To-Text systems • Computing with words • Natural language generation

Scalable RDF Data Management
With a Touch of Uncertainty

Martin Theobald

Faculty of Science, Technology & Communication, University of Luxembourg
martin.theobald@uni.lu

Abstract. The invited talk provides an overview of our recent research activities
and also highlights a number of research challenges in the context of extracting,
indexing, and querying large collections of RDF data. A core part of our work
focuses on handling uncertain facts obtained from various information-
extraction techniques, where we aim to develop efficient algorithms for querying the resulting uncertain RDF knowledge base with the help of a probabilistic
database. A further, very recent research focus lies on scaling out these
approaches to a distributed setting. Here, we aim to process declarative queries,
posed in either SQL or logical query languages such as Datalog, via a proprietary,
asynchronous communication protocol based on the Message Passing Interface.
Our current RDF engine, coined "TriAD", has proven to be one of the fasted such
engines over a number of RDF benchmarks with up to 1.8 billion triples.

1   Joint Named-Entity Recognition & Disambiguation

Virtually all of the recently proposed approaches for Named-Entity Recognition and
Disambiguation, henceforth coined "NERD", proceed in two strictly separated stages.
At the NER stage, text spans containing entity mentions are detected and tagged with
coarse-grained types like *Person*, *Organization*, *Location*, etc. This is typically performed by a trained Conditional Random Field (CRF) over word sequences. At the
NED stage, mentions are then mapped to entities in an underlying Knowledge Base
(KB) based on contextual similarity measures and their semantic coherence (see, e.g.,
[6, 9, 11, 13]). This two-stage, pipelined approach has several limitations. First, NER
may produce false positives that can misguide NED. Second, NER may miss out on
some of the entity mentions, and NED has no chance to compensate for these false
negatives. Third, NED is not able to help NER, for example, by disambiguating "easy"
mentions (e.g., of prominent entities with more or less unique names) and then using
the entities and contextual knowledge about them as enriched features for NER. Our
method, called *J*-NERD [14], is based on a supervised, non-linear graphical model that
combines multiple per sentence models into an entity-coherence-aware global model.
The global model detects mention spans, tags them with coarse-grained types, and
maps them to entities in a single joint-inference step based on either the Viterbi
algorithm (for exact inference) or Gibbs sampling (for approximate inference). Our
ongoing work also focuses on the extraction of RDF triples (and even higher-arity
facts) by clustering patterns of verbal phrases into a canonical set of relationships
among entities.

2   Uncertain RDF Data & Inference in Probabilistic Databases

Managing uncertain RDF data obtained from the aforedescribed extraction techniques provides a perfect showcase for Probabilistic Databases (PDBs). PDBs encompass a plethora of applications, ranging from scientific data management, sensor networks, data integration, to information extraction and knowledge management [15]. While classical database approaches benefit from a mature and scalable infrastructure for the management of relational data, probabilistic databases aim to further combine these well-established data management strategies with efficient algorithms for probabilistic inference by exploiting given independence assumptions among database tuples whenever possible. Moreover, PDBs adopt powerful query languages from relational databases, including Relational Algebra, the Structured Query Language (SQL), and logical query languages such as Datalog. The Trio probabilistic database system [12], which we developed at Stanford University back in 2006, was the first such system that explicitly addressed the integration of data management (using SQL as query language), lineage (aka. "provenance") management via Boolean formulas, and probabilistic inference based on the lineage of query answers. The Trio data model, coined "Uncertainty and Lineage Databases" (ULDBs) [1], provides a closed and complete probabilistic extension to the relational data model under all of the common relational (i.e., SQL-based) operations. Our recent research contributions in the domain of PDBs comprise lifted top-k queries over non-materialized database views [3], learning of tuple probabilities from user feedback [5], as well as temporal-probabilistic extensions [2, 4].

3   Distributed RDF Indexing & SPARQL Processing

The last part of the talk takes an in-depth look at the architecture of our TriAD (for "Triple-Asynchronous-Distributed") [7] engine, which provides an end-to-end system for the distributed indexing of large RDF collections and the processing of queries formulated in the SPARQL 1.0 standard. TriAD combines a novel form of sharded, main-memory-based index structures with an asynchronous communication protocol based on the Message Passing Interface (MPI). It thus aims to bridge the gap between shared-nothing MapReduce engines [10], on the one hand, and shared-everything graph engines [16], on the other hand. TriAD is designed to achieve higher parallelism and less synchronization overhead during query executions than MapReduce engines by adding an additional layer of multi-threading for entire execution paths within a query plan that can be executed in parallel. TriAD is the first RDF engine that employs asynchronous join executions, which are coupled with a lightweight join-ahead pruning technique based on graph summarization. Our current work also considers the processing of distributed set-reachability [8] queries, as they may occur, for example, in the recent "Property Paths" extension of SPARQL 1.1.

References

1. Benjelloun, O., Sarma, A.D., Halevy, A.Y., Theobald, M., Widom, J.: Databases with uncertainty and lineage. VLDB J. **17**(2), 243–264 (2008)
2. Dylla, M., Miliaraki, I., Theobald, M.: A temporal-probabilistic database model for information extraction. PVLDB **6**(14), 1810–1821 (2013)
3. Dylla, M., Miliaraki, I., Theobald, M.: Top-k query processing in probabilistic databases with non-materialized views. In: ICDE, pp. 122–133 (2013)
4. Dylla, M., Sozio, M., Theobald, M.: Resolving temporal conflicts in inconsistent RDF knowledge bases. In: BTW, pp. 474–493 (2011)
5. Dylla, M., Theobald, M., Miliaraki, I.: Querying and learning in probabilistic databases. In: Koubarakis, M., et al. (eds) Reasoning Web. Reasoning on the Web in the Big Data Era. Reasoning Web 2014. LNCS, vol. 8714. Springer, Cham (2014)
6. Ferragina, P., Scaiella, U.: TAGME: On-the-fly annotation of short text fragments (by wikipedia entities). In: CIKM, pp. 1625–1628 (2010)
7. Gurajada, S., Seufert, S., Miliaraki, I., Theobald, M.: TriAD: a distributed shared-nothing RDF engine based on asynchronous message passing. In: SIGMOD, pp. 289–300 (2014)
8. Gurajada, S., Theobald, M.: Distributed set reachability. In: SIGMOD, pp. 1247–1261 (2016)
9. Hoffart, J., Yosef, M.A., Bordino, I., Fürstenau, H., Pinkal, M., Spaniol, M., Taneva, B., Thater, S., Weikum, G.: Robust disambiguation of named entities in text. In: EMNLP, pp. 782–792 (2011)
10. Huang, J., Abadi, D.J., Ren, K.: Scalable SPARQL querying of large RDF graphs. PVLDB **4**(11), 1123–1134 (2011)
11. Mendes, P.N., Jakob, M., García-Silva, A., Bizer, C.: DBpedia spotlight: shedding light on the web of documents. In: I-SEMANTICS, pp. 1–8 (2011)
12. Mutsuzaki, M., Theobald, M., de Keijzer, A., Widom, J., Agrawal, P., Benjelloun, O., Sarma, A.D., Murthy, R., Sugihara, T.: Trio-one: layering uncertainty and lineage on a conventional DBMS. In: CIDR, pp. 269–274 (2007)
13. Nguyen, D.B., Hoffart, J., Theobald, M., Weikum, G.: AIDA-light: high-throughput named-entity disambiguation. In: LDOW. CEUR-WS.org (2014)
14. Nguyen, D.B., Theobald, M., Weikum, G.: J-NERD: joint named entity recognition and disambiguation with rich linguistic features. TACL **4**, 215–229 (2016)
15. Suciu, D., Olteanu, D., Ré, C., Koch, C.: Probabilistic Databases. Synthesis Lectures on Data Management, Morgan & Claypool Publishers (2011)
16. Zeng, K., Yang, J., Wang, H., Shao, B., Wang, Z.: A distributed graph engine for web scale RDF data. PVLDB **6**(4), 265–276 (2013)

Contents

Short Papers

Invited Papers

Maximum Likelihood Estimation
and Coarse Data

Inés Couso[1([⊠])], Didier Dubois[2], and Eyke Hüllermeier[3]

[1] Department of Statistics and O.R., Universidad de Oviedo, Oviedo, Spain
couso@uniovi.es
[2] IRIT, CNRS Université Paul Sabatier, Toulouse, France
dubois@irit.fr
[3] Department of Computer Science, Universität Paderborn, Paderborn, Germany
eyke@upb.de

Abstract. The term coarse data encompasses different types of incomplete data where the (partial) information about the outcomes of a random experiment can be expressed in terms of subsets of the sample space. We consider situations where the coarsening process is stochastic, and illustrate with examples how ignoring this process may produce misleading estimations.

Keywords: Coarse data · Grouped data · Coarsening at random · Maximum likelihood · Visible likelihood · Face likelihood

1 Introduction

The term "coarse data" [15] covers a number of situations treated in the literature such as rounded, heaped, censored or partially missing data. It refers to those situations where we do not get access to the exact value of the data, but only to some subset of the sample space that contains it. Thus, formally speaking, the observations are not assumed to belong to the sample space, but to its power set (see [4,8] for further discussions on set-valued data).

One key problem consists in estimating the distribution of the underlying random variable on the basis of the available incomplete sample data. During the last two decades, different authors have independently studied the way to adapt maximum likelihood estimation (MLE) to this case [5,6,11,13,15,17,20]. In fact, the maximum likelihood procedure results in a consistent estimator of the parameter under some regularity conditions [16], and therefore it is one of the most usual approaches in a variety of machine learning problems. One may adapt the MLE method to incomplete data by considering the collection of possible completions of data, which would lead to a set-valued likelihood function.

The first author thanks the Program Committee Chairs for their kind invitation to participate in the conference. The research in this work has been supported by TIN2014-56967-R (Spanish Ministry of Science and Innovation) and FC-15-GRUPIN14-073 (Regional Ministry of the Principality of Asturias).

S. Moral et al. (Eds.): SUM 2017, LNAI 10564, pp. 3–16, 2017.
DOI: 10.1007/978-3-319-67582-4_1

Thus, maximizing the likelihood function for each of the feasible samples would lead to a set-valued counterpart of the MLE. But this does not seem to be the most reasonable procedure (see comments about extension principle -based approaches in [17], for further details). Two dual alternative procedures have been recently explored [13,14,17]. They consist in replacing the set-valued likelihood either by its upper [17] or its lower bound [13], and seeking for the arg max of the corresponding real-valued mappings. They are respectively referred to as the maximax and the maximin estimators. Some properties of both of them have been recently studied in [14].

A third approach focusses on the observations rather than on the underlying (ill-known) outcomes represented by them. The so-called "visible" likelihood function [6,7] represents the probability of observing the actual observed (set-valued) sample, as a function of a vector of parameters. In order to determine such a function, we do not only need to parametrize the underlying experiment, but also of the coarsening process. The joint distribution over the collection of pairs constituted by the outcomes and their corresponding (incomplete) observations is univocally determined by the marginal distribution over the sample space plus a transition probability from the sample space to its power set, representing the coarsening process. The "visible" likelihood function is nothing else but the likelihood of the marginal distribution over the power set, expressed as a function of the vector of parameters. Different aspects of the arg max of this function have been recently studied in the literature [1,3,5–7,20]. This paper surveys those advances.

2   What Has Occurred and What Do We Know About It?

2.1   Preliminaries and Notation

Let a random variable $X : \Omega \to \mathcal{X}$ represent the outcome of a certain random experiment. For the sake of simplicity, let us assume that its range $\mathcal{X} = \{a_1, \ldots, a_m\}$ is finite. Suppose that instead of directly observing X, one observes a coarse version of it, $Y \ni X$. Let $\mathcal{Y} = \{b_1, \ldots, b_r\}$ denote the (finite) set of possible observations, with $b_j = A_j \subseteq \mathcal{X}$, $\forall j = 1, \ldots, r$. Let us introduce the following notation:

- $p_{kj} = P(X = a_k, Y = b_j)$ denotes the joint probability of getting the precise outcome $X = a_k$ and observing $b_j = A_j$,
- $p_{k.} = P(X = a_k)$ denotes the probability that the precise outcome is a_k,
- $p_{.j} = P(Y = b_j)$ denotes the probability that the generation plus the imprecisiation processes lead us to observe $b_j = A_j$.
- $p_{.j|k.} = P(Y = A_j | X = a_k)$ denotes the (conditional) probability of observing $b_j = A_j$ if the precise outcome is a_k,
- $p_{k.|.j} = P(X = a_k | Y = A_j)$ denotes the (conditional) probability that the value of X is a_k if we have been reported that it belongs to $b_j = A_j$.

We may represent the joint distribution of (X, Y) by means of the matrix $(M|\mathbf{p})$:

$$\begin{pmatrix} p_{.1|1.} & \cdots & p_{.r|1.} & p_{1.} \\ \cdots & \cdots & \cdots & \cdots \\ p_{.1|m.} & \cdots & p_{.r|m.} & p_{m.} \end{pmatrix}$$

where $(p_{1.}, \ldots, p_{m.})^T$ characterizes the distribution of the underlying generating process, while $M = (p_{.j|k.})_{k=1,\ldots,m; j=1,\ldots,r}$ represents the coarsening process. M is the so-called mixing matrix [25]. We can alternatively characterise it by means of $(M'|\mathbf{p}')$:

$$\begin{pmatrix} p_{1.|.1} & \cdots & p_{m.|.1} & p_{.1} \\ \cdots & \cdots & \cdots & \cdots \\ p_{1.|.r} & \cdots & p_{m.|.r} & p_{.r} \end{pmatrix}$$

where the vector $(p_{.1}, \ldots, p_{.r})^T$ characterises the probability distribution of the observation process, and $M' = (p_{k.|.j})_{k=1,\ldots,m; j=1,\ldots,r}$ represents the conditional probability of X (precise outcome) given Y (observation).

Now, let us assume that the above joint distribution (or equivalently, each of the matrices $(M|\mathbf{p})$ and $(M'|\mathbf{p}')$) is characterized by means of a (vector of) parameter(s) $\theta \in \Theta$. We naturally assume that the dimension of θ is less than or equal to the number of elements in both matrices, i.e., it is less than or equal to $\min\{m \times (r+1), r(m+1)\}$. We also assume that X cannot be written as a function of Y, because such a situation would involve a trivial coarsening process, were Y is just some kind of "encoding" of X. Similarly, we can assume without much loss of generality that X and Y are not independent. Otherwise, the restriction $X \in Y$ would imply that Y is constant, and its image includes all the possible outcomes for X. Furthermore, the parameter is said to be separable [6] wrt $(M|\mathbf{p})$ if it can be "separated" into two (maybe multidimensional) components $\theta_1 \in \Theta_1$, $\theta_2 \in \Theta_2$ such that $\Theta = \Theta_1 \times \Theta_2$, where $p_{.j|k.}^\theta$ and p_k^θ can be respectively written as functions of θ_1 and θ_2. This definition corresponds to an earlier notion of "distinctness" of the parameters [15]. Alternatively, θ is said to be separable wrt $(M'|\mathbf{p}')$ if it can be "separated" into two (maybe multidimensional) components $\theta_3 \in \Theta_3$, $\theta_4 \in \Theta_4$ such that $\Theta = \Theta_3 \times \Theta_4$ and $p_{k.|.j}^\theta$ and $p_{.j}^\theta$ can be respectively written as functions of θ_3 and θ_4. One may think that the notion of "separability" implies some kind of "independence" between X (outcome) and Y (imprecise observation), but this is not the case, as we illustrate below.

We will provide three examples illustrating three different situations: In the first case, Y can be expressed as a function of X, and therefore it determines a partition on $\mathcal{X}$, but their joint distribution depend on a single one-dimensional parameter. In the second case, Y can also be written as a function of X, but the parameters are separable. In the third case, Y is not a function of X, and in fact, it represents a "coarsening at random process" [15] and the joint distribution of (X, Y) depends on a one-dimensional parameter.

Example 1 (Taken from [7]). Let us consider the following example by Dempster et al. in [10] under the light of our analysis. It is based on a former example by Rao. There is a sample of 197 animals distributed into four categories, so that the observed data consist of:

$$n_{.1} = 125, n_{.2} = 18, n_{.3} = 20, n_{.4} = 34.$$

Suppose that the first category is in fact a mixture of two sub-categories, but we do not have information about the number of individuals observed from each of them. On the other hand, a genetic model for the population specifies the following restrictions about the five categories: $p_{11} = 0.5$, $p_{12} = p_{.4}$, $p_{.2} = p_{.3}$. If we use the notation: $p_{12} = 0.25\pi = p_{.4}$ and $p_{.2} = 0.25(1 - \pi) = p_{.3}$, the corresponding matrix $(M'|\mathbf{p}')$ is given as

$$\begin{pmatrix} \frac{0.5}{0.5+0.25\pi} & \frac{0.25\pi}{0.5+0.25\pi} & 0 & 0 & 0 & 0.5 + 0.25\pi \\ 0 & 0 & 1 & 0 & 0 & 0.25(1-\pi) \\ 0 & 0 & 0 & 1 & 0 & 0.25(1-\pi) \\ 0 & 0 & 0 & 0 & 1 & 0.25\pi \end{pmatrix}$$

and only depends on a single parameter.

Example 2. Let X be the random variable that represents the score shown on the top face of a die. Let $(p_1, \ldots, p_6)$ characterize the probability distribution over the set of possible outcomes. Let us suppose that we are just told whether X takes an even or an odd value. We identify the possible observations (values of Y) respectively with $b_1 = \{1, 3, 5\}$ and $b_2 = \{2, 4, 6\}$. This example is formally equivalent to the case of grouping data, where Y can be expressed as a function of X. In other words, the coarsening process is a deterministic procedure where all the values in the mixing matrix M are either 0 or 1. Thus, the distribution of Y only depends on $\theta_1 = p_1 + p_2 + p_3$. Let us now consider the matrix $M' = (m'_{ij})_{i,j}$ where the two-dimensional variable (X, Y) can take six different values, and its joint distribution can be expressed in terms of $(p_1, \ldots, p_5)$. It can be also written as a function of $\theta_1 = p_1 + p_3 + p_5$ (determining the marginal distribution of Y) and the four-dimensional vector $\boldsymbol{\theta}_2 = (\frac{p_1}{\theta_1}, \frac{p_3}{\theta_1}, \frac{p_2}{1-\theta_1}, \frac{p_4}{1-\theta_1})$ (that characterizes the disambiguation process). Thus, the joint distribution is separable wrt M' and p'.

Example 3. Suppose a coin is flipped and let X be the binary random variable that takes the value 1= "heads", and 0="tails". Suppose that half of the times, we are not informed about the result (regardless what the result is). The coarsening process is therefore characterised as follows:

$$P(Y = \{0, 1\}|X = 0) = P(Y = \{0, 1\}|X = 1) = 0.5,$$

$$P(Y = \{0\}|X = 0) = P(Y = \{1\}|X = 1) = 0.5.$$

This process agrees with the notion of coarsening at random (CAR) introduced by Heitjan and Rubin, to be discussed later on, since the fact of being informed of the result does not depend on the result itself. Furthermore, it satisfies a stronger property called "superset assumption" [18], since we are informed half of the times, on average, about the result of the coin, whatever it is. Notwithstanding, the joint distribution of (X, Y) can be expressed in terms of the one-dimensional parameter $p \in (0, 1)$ denoting the probability of heads. In fact, under the above assumptions, we have:

$$P(X = 1, Y = \{1\}) = P(X = 1, Y = \{0, 1\}) = 0.5p,$$

$$P(X = 0, Y = \{0\}) = P(X = 0, Y = \{0, 1\}) = 0.5(1 - p).$$

As a conclusion, the above example satisfies the so-called property of "missing at random" (MAR), but the joint distribution of (X, Y) is completely characterised by marginal distribution of Y, since both of them depend on the same -single- parameter.

As a matter of fact, the parameter of the joint distribution can be written as a function of the parameter of the distribution of Y when the distribution about the instantiation process is known, given the marginal distribution of Y. This does not seem to be related to the degree of dependence between X and Y. In this case, the problem of identifiability of the parameter of the marginal distribution of Y reduces to the problem of identifiability of the parameter of the joint likelihood function, and therefore, a MLE procedure based on the "visible" likelihood function seems a good option in order to estimate the parameter.

2.2   The Outcomes of an Experiment and Their Incomplete Observations

According to the framework developed in the last subsection, we can easily observe that, given some $b_j = A_j \in \mathcal{Y}$, the two events $X \in A_j$ and $Y = A_j$ do not coincide in general. In fact, it is generally assumed that the latter implies the former, but the equivalence does not hold in general: For, suppose that $X \in A_j$ implies $Y = A_j$. Therefore, for every $a_k \in A_j$, we can derive that $X = a_k$ implies $X \in A_j$ and therefore $Y = A_j$. Thus, we can deduce that $P(Y = A_j | X = a_k) = 1, \forall a_k \in A_j$. Thus, the above implication entails a deterministic coarsening process, inducing a partition over the set of outcomes $\mathcal{X}$.

Let us illustrate the difference between the events $X \in A_j$ and $Y = A_j$ and their corresponding probabilities with an example:

Example 4 (Taken from [7]). Consider the random experiment that consists on rolling a dice. We do not know whether the dice is fair or not. Take a sample of N tosses of the dice and assume that the reporter has told us n_1 of the times that the result was less than or equal to 3 and the remaining $n_2 = N - n_1$ tosses, he told us that it was greater than or equal to 3. After each toss, when the actual result (X) is 3, the reporter needs to make a decision. Let us assume that the conditional probability $P(Y = \{1, 2, 3\} | X = 3)$ is a fixed number $\alpha \in [0, 1]$. The joint distribution of (X, Y) can be written as a function of $(p_1, \ldots, p_6)$ and α, since it is determined by the following matrix: $(M|\mathbf{p})$:

$$\begin{pmatrix} 1 & 0 & p_1 \\ 1 & 0 & p_2 \\ \alpha & 1-\alpha & p_3 \\ 0 & 1 & p_4 \\ 0 & 1 & p_5 \\ 0 & 1 & p_6 \end{pmatrix}$$

corresponding to the joint probability

Y, X	1	2	3	4	5	6
y_1	p_1	p_2	$\alpha\, p_3$	0	0	0
y_2	0	0	$(1-\alpha)\, p_3$	p_4	p_5	p_6

We can easily make the distinction between the two events $X \in \{1,2,3\}$ (the result is less than or equal to 3) and $Y = \{1,2,3\}$ (we are told that the result is less than or equal to 3) and their corresponding probabilities. According to the above notation, the probability of the first event is

$$P(X \in \{1,2,3\}) = p_1 + p_2 + p_3,$$

while the probability of the latter is:

$$P(Y = \{1,2,3\}) =$$
$$P(X=1, Y=\{1,2,3\}) + P(X=2, Y=\{1,2,3\}) + P(X=3, Y=\{1,2,3\})$$
$$= p_1 + p_2 + \alpha\, p_3.$$

3   The Optimization Problem: What Should We Maximise?

Let us consider a sequence $\mathbf{Z} = ((X_1, Y_1), \ldots; (X_N, Y_N))$ of N iid copies of $Z = (X, Y)$. We will use the nomenclature $\mathbf{z} = ((x_1, y_1), \ldots, (x_N, y_N)) \in (\mathcal{X} \times \mathcal{Y})^N$ to represent a specific sample of the vector (X, Y). Thus, $\mathbf{y} = (y_1, \ldots, y_N)$ will denote the observed sample (an observation of the vector $\mathbf{Y} = (Y_1, \ldots, Y_N)$), and $\mathbf{x} = (x_1, \ldots, x_N)$ will denote an arbitrary artificial sample from $\mathcal{X}$ for the unobservable (latent) variable X, that we shall vary in $\mathcal{X}^N$. We can describe any sample $\mathbf{z}$ in frequentist terms assuming exchangeability:

- $n_{kj} = \sum_{i=1}^{N} 1_{\{(a_k, b_j)\}}(x_i, y_i)$ is the number of repetitions of (a_k, b_j) in the sample $\mathbf{z}$;
- $\sum_{k=1}^{m} n_{kj} = n_{\cdot j}$ be the number of observations of $b_j = A_j$ in $\mathbf{y}$;
- $\sum_{j=1}^{r} n_{kj} = n_{k\cdot}$ be the number of appearances of a_j in $\mathbf{x}$.

Clearly, $\sum_{k=1}^{m} n_{k\cdot} = \sum_{j=1}^{r} n_{\cdot j} = N$. Let the reader notice that, once a specific sample $\mathbf{y} = (y_1, \ldots, y_N) \in \mathcal{Y}^N$ has been observed, the number of n_{kj} repetitions of each pair $(a_k, b_j) \in \mathcal{X} \times \mathcal{Y}$ in the sample, can be expressed as a function of $\mathbf{x} = (x_1, \ldots, x_N)$.

3.1 Different Generalizations of the Notion of the Likelihood Function

We may consider the following two generalizations of the likelihood function (and their respective logarithms), depending on whether our sequence of observations $\mathbf{y} = (y_1, \ldots, y_N)$ is interpreted either as a singleton in $\mathcal{Y}^N$ or as a non-trivial subset of $\mathcal{X}^N$:

- $\mathbf{p}(\mathbf{y}; \theta) = \prod_{i=1}^{N} p(y_i; \theta)$ denotes the probability of observing $\mathbf{y} \in \mathcal{Y}^N$, assuming that the value of the parameter is θ. It can be alternatively expressed as $\mathbf{p}(\mathbf{y}; \theta) = \prod_{j=1}^{r} (p_{\cdot j}^{\theta})^{n_{\cdot j}}$, where $n_{\cdot j}$ denotes the number of repetitions of $b_j = A_j$ in the sample of size N (the number of times that the reporter says that the outcome of the experiment belongs to A_j.) The logarithm of this likelihood function will be denoted by

$$L^{\mathbf{y}}(\theta) = \log \mathbf{p}(\mathbf{y}; \theta) = \sum_{i=1}^{N} \log p(y_i; \theta) = \sum_{j=1}^{r} n_{\cdot j} \log p_{\cdot j}^{\theta}.$$

  We call $\mathbf{p}(\mathbf{y}; \theta)$ the *visible likelihood function* [7], because we can compute it based on the available data only, that is the observed sample $\mathbf{y}$. It is also sometimes called the *marginal likelihood of the observed data* in the EM literature, not to be confused with the *marginal likelihood* in a Bayesian context (see [2], for instance).
- Alternatively,

$$\lambda(\mathbf{y}; \theta) = \prod_{j=1}^{r} P(X \in A_j; \theta)^{n_{\cdot j}},$$

  called the "face likelihood" in [9, 20] does not refer to the observation process, and replaces the probability of reporting A_j as the result of an observation (i.e. $P(Y = A_j)$) by the probability that the precise outcome falls inside the set A_j, $P(X \in A_j)$. As we have previously noticed, the occurrence of event "$X \in A_j$" is a consequence of, but does not necessarily coincide with the outcome "$Y = A_j$". In our context, $\mathbf{p}(\mathbf{y}; \theta)$ represents the probability of occurrence of the result "$(Y_1, \ldots, Y_N) = \mathbf{y}$", given the hypothesis θ. Therefore given two arbitrary different samples $\mathbf{y} \neq \mathbf{y}'$ the respective events $(Y_1, \ldots, Y_N) = \mathbf{y}$ and "$(Y_1, \ldots, Y_N) = \mathbf{y}'$" are mutually exclusive. In contrast, $\lambda(\mathbf{y}; \theta)$ denotes the probability of occurrence of the event $(X_1, \ldots, X_N) \in y_1 \times \ldots \times y_N$. Events of this form may overlap, in the sense that, given two different samples $\mathbf{y} \neq \mathbf{y}'$, the corresponding events $(X_1, \ldots, X_N) \in y_1 \times \ldots \times y_N$ and $(X_1, \ldots, X_N) \in y_1' \times \ldots \times y_N'$ are not necessarily mutually exclusive. Therefore $\lambda(\mathbf{y}; \theta)$ can not be regarded as a likelihood in the sense of Edwards [12]. This criterion has been generalized to uncertain data and exploited in the Evidential EM algorithm of Denœux [11]. This extension of EM has been successfully used in some applications (see [23, 24] and references therein).

  The above functions $\mathbf{p}$ and λ do coincide if and only if the coarsening process is deterministic, and therefore, the collection of sets $\{A_1, \ldots, A_r\}$ forms a partition of $\mathcal{X}$. In fact, $P(Y = A_j) \leq P(X \in A_j)$ for every $j = 1, \ldots, r$ and the equalities hold when the coarsening is deterministic. Otherwise, if there exists a pair (k, j) with $a_k \in A_j$ and $P(Y = A_j | X = a_k) < 1$ then we easily derive that $P(Y = A_j)$ is strictly smaller than $P(X \in A_j)$ and therefore, we deduce that $\mathbf{p}(\mathbf{y}, \theta)$ is strictly less than $\lambda(\mathbf{y}, \theta)$. But we may ask ourselves whether the maximization of each of those functions leads or not to the same pair of maximizers, even in those cases where they do not coincide. The next example illustrates a situation where both methods lead to completely different estimators.

Example 5. Consider again the situation described in Example 4. Furthermore, suppose that the reporter provides us with the following additional information: when the result is $X = 3$, he will flip a coin. If it lands heads, he will tells us that the result is less than or equal to 3. Otherwise, he will tell us that it is greater than or equal to 3. Mathematically, $\alpha = P(Y = \{1, 2, 3\} | X = 3) = 0.5$. Under these conditions, the visible likelihood is

$$\mathbf{p}(\mathbf{y}, \theta) = (p_1 + p_2 + 0.5\, p_3)^{300} + (0.5\, p_3 + p_4 + p_5 + p_6)^{700}.$$

It attains its maximum value for every $(\hat{p}_1, \ldots, \hat{p}_6)$ satisfying the restrictions: $\hat{p}_1 + \hat{p}_2 + 0.5\, \hat{p}_3 = 0.3$ and $0.5\, \hat{p}_3 + \hat{p}_4 + \hat{p}_5 + \hat{p}_6 = 0.7$ (The set of solutions is not a singleton). Alternatively, the face likelihood function is calculated as follows:

$$\lambda(\mathbf{y}, \theta) = (p_1 + p_2 + p_3)^{300} + (p_3 + p_4 + p_5 + p_6)^{700}.$$

It attains the maximum value for $(\hat{p}_1, \ldots, \hat{p}_6) = (0, 0, 1, 0, 0, 0)$. In other words, according to this maximization procedure, the experiment is assumed to be deterministic.

Both optimization procedures lead to completely different solutions. In fact, according to the first set of solutions, p_3 is upper bounded by 0.6, while in the second case it is assumed to be equal to 1. Furthermore, according to the Weak Law of Large Numbers, the relative frequencies $\frac{n_{.1}}{N}$ and $\frac{n_{.2}}{N}$ respectively converge in probability to $p_{.1}$ and $p_{.2}$ that, according to the information, respectively coincide with $p_1 + p_2 + 0.5\, p_3$ and $0.5\, p_3 + p_4 + p_5 + p_6$. Thus, the first procedure (the one based on the visible likelihood) satisfies the following consistency property:

$$\lim_{n \to \infty} \hat{p}_1 + \hat{p}_2 + 0.5\, \hat{p}_3 = p_1 + p_2 + 0.5\, p_3$$

and

$$\lim_{n \to \infty} 0.5\, \hat{p}_3 + \hat{p}_4 + \hat{p}_5 + \hat{p}_6 = 0.5\, p_3 + p_4 + p_5 + p_6.$$

In contrast, the estimation based on the face likelihood does not satisfy the above consistency property unless the underlying probability satisfies the following equality:

$$p_1 + p_2 + 0.5\, p_3 = 0.5\, p_3 + p_4 + p_5 + p_6 = 0.5.$$

The differences between the visible and the face likelihood functions have been studied in practice in relation with incomplete ranked data in [1,3]. In fact, incomplete rankings are viewed there as coarse observations of ranked data. Let $\mathcal{X} = \mathbb{S}_3$ denote the collection of rankings (permutations) over a set

$U = \{a_1, a_2, a_3\}$ of 3 items. We denote by $\pi : \{1, 2, 3\} \Rightarrow \{1, 2, 3\}$ a complete ranking (a generic element of $\mathbb{S}_3$), where $\pi(k)$ denotes the position of the kth item a_k in the ranking. An incomplete ranking τ can be associated with the collection of complete rankings that are in agreement with it denoted $E(\tau)$. An important special case is an incomplete ranking τ in the form of a pairwise comparison $a_i \succ a_j$, which is associated with the set of extensions

$$E(\tau) = E(a_i \succ a_j) = \{\pi \in \mathbb{S}_K : \pi(i) < \pi(j)\}.$$

For every pair (i, j), let n_i denote the number of times that the incomplete ranking τ_i is observed in a sample of size N. Let us furthermore assume that the marginal distribution of X on $\mathbb{S}_3$ belongs to a family of distributions parametrized by some vector of parameters θ, while the coarsening process is determined by some λ. The face likelihood based on the above sample is calculated as follow:

$$\lambda(\mathbf{y}; \theta) = \prod_{i=1}^{3} \prod_{j \neq i} P_\theta(X \in E(\tau_i))^{n_i},$$

while the visible likelihood function is calculated as

$$p(\mathbf{y}; \theta) = \prod_{i=1}^{3} \prod_{j \neq i} P_{(\theta, \lambda)}(Y = \tau_i)^{n_i}.$$

They do not coincide in general. Let us consider, for instance, the top-2 setting, in which always the two items on the top of the ranking are observed. The corresponding mixing matrix denotes a one-to-one correspondence between π_i and τ_i, for $i = 1, \ldots, 6$, where:

$$\pi_1(1) = 1, \pi_1(2) = 2, \pi_1(3) = 3 \qquad \tau_1 = a_1 \succ a_2$$
$$\pi_2(1) = 1, \pi_2(2) = 3, \pi_2(3) = 2 \qquad \tau_2 = a_1 \succ a_3$$
$$\pi_3(1) = 2, \pi_3(2) = 1, \pi_3(3) = 3 \qquad \tau_3 = a_2 \succ a_1$$
$$\pi_4(1) = 2, \pi_4(2) = 3, \pi_4(3) = 1 \qquad \text{and} \qquad \tau_4 = a_3 \succ a_1$$
$$\pi_5(1) = 3, \pi_5(2) = 1, \pi_5(3) = 2 \qquad \tau_5 = a_2 \succ a_3$$
$$\pi_6(1) = 3, \pi_6(2) = 2, \pi_6(3) = 1 \qquad \tau_6 = a_3 \succ a_2$$

Thus Y takes the "value" τ_i if and only if $X = \pi_i$, for all $i = 1, \ldots, 6$. Let us furthermore notice that each partial ranking τ_i represents a collection of three different complete rankings:

$$E(\tau_1) = \{\pi_1, \pi_2, \pi_4\}$$
$$E(\tau_2) = \{\pi_1, \pi_2, \pi_3\}$$
$$E(\tau_3) = \{\pi_1, \pi_2, \pi_4\}$$
$$E(\tau_4) = \{\pi_1, \pi_2, \pi_4\}$$
$$E(\tau_5) = \{\pi_1, \pi_2, \pi_4\}$$
$$E(\tau_6) = \{\pi_1, \pi_2, \pi_4\}$$

Thus, the face and the visible likelihood functions are respectively calculated as follows:

$$\lambda(\mathbf{y}; \theta) = \prod_{i=1}^{6} P_\theta(X \in E(\tau_i))^{n_i}$$

while

$$p(\mathbf{y}, \theta) = \prod_{i=1}^{6} \prod P_\theta(X = \pi_i)^{n_i}.$$

They do not lead in general to the same estimations, as it is checked in [3]. In fact, under some general assumptions about the underlying generating process, the visible likelihood-based estimator is consistent, while the face likelihood-based estimator is not. Some additional formal studies about the consistency of both estimators under different assumptions about the coarsening process are performed in [1].

3.2   Different Assumptions About the Coarsening and the Disambiguation Processes

Different assumptions about the coarsening and the disambiguation processes have been investigated in the literature [1,15,18,21,22]. The purpose in some of those cases was to establish simple conditions under which the stochastic nature of the coarsening process could be ignored when drawing inferences from data. This subsection reviews two assumptions, one about the coarsening process and the other one about the disambiguation process, both of them commonly considered in the literature.

Coarsening at Random. One common assumption about the coarsening process is the so-called "coarsening at random" assumption (CAR). It was introduced by Heitjan and Rubin [15]. According to it, the underlying data do not affect the observations. Mathematically,

$$P(Y = A_j | X = a_k) = P(Y_j = A_j | X = a_{k'}), \ \forall a_k, a'_k \in A_j.$$

Two remarkable particular cases of CAR are:

- Grouping. We speak about grouped data [15] when the coarsening process is deterministic, and therefore $P(Y = A_j | X = a_k)$ is either 1 (if $a_k \in A_j$) or 0 (otherwise). In this case, the set $\{A_1, \ldots, A_r\}$ forms a partition of the collection of possible outcomes $\mathcal{X}$.
- Missing at random (MAR).- It particularizes the CAR assumption to the case where data are either completely observed or missing, and therefore, the collection of possible observations is $\mathcal{Y} = \{\{a_1\}, \ldots, \{a_m\}, \mathcal{X}\}$. The MAR assumption means that missingness is not affected by the underlying outcome.

   The first one illustrates the partition case.

Example 6 (Taken from [19]). Let $X = (X_1, X_2)$ with $\mathcal{X} = \mathcal{X}_1 \times \mathcal{X}_2 = \{p, n\} \times \{p, n\}$ We interpret X_1, X_2 as two medical tests with possible outcomes positive or negative. Suppose that test X_1 always is performed first on a patient, and that test X_2 is performed if and only if X_1 comes out positive. Possible observations that can be made then are $b_1 = \{(n, n), (n, p)\}$, $b_2 = \{(p, n)\}$ and $b_3 = \{(p, p)\}$.

These three outcomes determine a partition of $\mathcal{X}$. Therefore, the matrix M is determined by the following 0–1 conditional probabilities, and CAR is trivially satisfied. In fact:

$$P(Y = b_1|X = (n, n)) = P(Y = b_1|X = (n, p)) = 1,$$

$$P(Y = b_2|X = \{(p, n)\}) = 1,$$

$$P(Y = b_3|X = \{(p, p)\}) = 1.$$

The following example illustrates the missing at random assumption:

Example 7 (Taken from [7]). A coin is tossed. The random variable $X : \Omega \to \mathcal{X}$, where $\mathcal{X} = \{h, t\}$, represents the result of the toss. We do not directly observe the outcome, that is reported by someone else, who sometimes decides not to tell us the result. The rest of the time, the information he provides about the outcome is faithful. Let Y denote the information provided by this person about the result. It takes the "values" $\{h\}$, $\{t\}$ and $\{h, t\}$.

This example corresponds to the following matrix $(M|\mathbf{p})$ where $a_{kj} = p_{.j|k.}$, $k = 1, 2; j = 1, 2, 3$:

$$\begin{pmatrix} 1 - \alpha & 0 & \alpha & p \\ 0 & 1 - \beta & \beta & 1 - p \end{pmatrix}$$

The marginal distribution of X (outcome of the experiment) is given as

- $p_{1.} = P(X = h) = p,$
- $p_{2.} = P(X = t) = 1 - p.$

The joint probability distribution of (X, Y) is therefore determined by:

$$\begin{pmatrix} X \backslash Y & \{h\} & \{t\} & \{h, t\} \\ \hline h & (1 - \alpha)p & 0 & \alpha p \\ t & 0 & (1 - \beta)(1 - p) & \beta(1 - p) \end{pmatrix}$$

Under the MAR assumption, we have that $\alpha = \beta$, i.e.,

$$P(Y = \{h, t\}|X = h) = P(Y = \{h, t\}|X = t).$$

When furthermore the model is separable with respect to the matrix $(M|p)$, the coarsening process can be ignored, in the sense that both the visible and the face likelihood lead to the same estimator of the parameter. This has been proved by Heitjan and Rubin (see [15, 20]). Additional conditions under which the stochastic nature of the coarsening process can be ignored in some practical problems have been recently studied in [1, 3].

Uniform Disambiguation Process. We can alternatively make assumptions about the disambiguation process. When dealing with noisy observations, it is not unusual to assume that all the possible outcomes compatible with an observation $Y = A_j$ (i.e., all the elements in A_j) are equally probable, and therefore $P(X = a_k | Y = A_j) = 1_{A_j}(a_k) \cdot \frac{1}{\#A_j}$, $\forall a_k \in A_j$. According to this assumption, the probability induced by X on $\mathcal{X}$ corresponds to the pignistic transform [26] of the mass function derived from the marginal distribution of Y as follows:

$$m(A_j) = P(Y = A_j), \; j = 1, \ldots, r.$$

Contrarily to what happens with the CAR assumption, under this alternative assumption, the face and the visible likelihood do not necessarily lead to the same estimator.

Example 8. Consider once more the situation described in Example 4, and assume a uniform disambiguation process. Let p denote the probability of the event $Y = \{1, 2, 3\}$. The visible likelihood can be written as a function of p as:

$$\mathbf{p}(\mathbf{y}, p) = p^{n.1}(1 - p)^{n.2}.$$

The marginal probability over $\mathcal{X}$ can be written as a function of p as follows:

$$P(X = 1) = P(X = 2) = \frac{p}{3}, \; P(X = 3) = \frac{p}{3} + \frac{1 - p}{4},$$

$$P(X = 4) = P(X = 5) = P(X = 6) = \frac{1 - p}{4}.$$

Therefore, the face likelihood is different from the visible likelihood:

$$\lambda(\mathbf{y}, p) = \left(p + \frac{1 - p}{4} \right)^{n.1} \left(\frac{p}{3} + (1 - p) \right)^{n.2}.$$

4   Concluding Remarks

We have provided an overview of the maximization procedures based on the so-called visible and face likelihood functions. The face likelihood depends on the marginal distribution of X, while the visible likelihood depends on the marginal distribution of Y. Both, the face and the visible likelihoods have their advantages and their caveats. When the parameter is separable with respect to the matrix $(M|p)$ (distinctness in the context of Heitjan and Rubin), the first one only depends on θ_3 while the second one depends on both, θ_3 and θ_4. The MLE based on the visible likelihood is therefore not unique in this case, unless the parameter set Θ_4 is a singleton. But, although the arg max of the face likelihood may be unique in those cases, it is not a consistent estimator in general, as we have observed. The visible likelihood involves the probability of observing the different outcomes $Y = A_j$ (as a function of the parameter) and the proportion of times each of them is observed in the sample. Such a proportion converges in probability to the (true) probability of the event, and therefore, under some regularity conditions, the arg max of the visible function is consistent. Alternatively, the face likelihood replaces the probability of observing $Y = A_j$ by the probability of occurrence of $X \in A_j$. The vector $(q_1, \ldots, q_r)$, where $q_i = P(X \in A_j)$ for all j^1 is not proportional in general to the vector $(p_{.1}, \ldots, p_{.r})$ and therefore, the arg max of the face likelihood is not consistent in general.

Some recent studies compare the maximization of the visible likelihood function with other strategies such as the maximax and the maximin approaches mentioned at the beginning of this paper. In this line, the face likelihood can be regarded as a max-average approach, in the sense that it maximizes the average of the likelihoods of all the feasible samples on $\mathcal{X}^N$ (all the samples of the form $\mathbf{x} = (x_1, \ldots, x_n)$ satisfying the restriction $x_i \in y_i,\ \forall i$) (see [17] for further details.) Further theoretical and empirical studies are needed in order to determine what is the best strategy in each practical situation.

References

1. Ahmadi, M., Hüllermeier, E., Couso I.:, Statistical inference for incomplete ranking data: the case of rank-dependent coarsening. In: Proceedings of the 34th International Conference on Machine Learning (2017 ICML), Sydney (Australia)
2. Chib, S.: Marginal likelihood from the Gibbs output. J. Am. Stat. Assoc. **90**, 1313–1321 (1995)
3. Couso, I., Ahmadi, M., Hüllermeier, E.: Statistical inference for incomplete ranking data: a comparison of two likelihood-based estimators. In: Proceedings of DA2PL 2016 (From Multiple Criteria Decision Aid to Preference Learning), Paderborn (Germany) (2016)
4. Couso, I., Dubois, D.: Statistical reasoning with set-valued information: ontic vs. epistemic views. Int. J. Approximate Reasoning **55**, 1502–1518 (2014)
5. Couso, I., Dubois, D.: Belief revision and the EM algorithm. In: Carvalho, J.P., Lesot, M.-J., Kaymak, U., Vieira, S., Bouchon-Meunier, B., Yager, R.R. (eds.) IPMU 2016. CCIS, vol. 611, pp. 279–290. Springer, Cham (2016). doi:10.1007/978-3-319-40581-0_23
6. Couso, I., Dubois, D.: Maximum likelihood under incomplete information: toward a comparison of criteria. In: Ferraro, M.B., Giordani, P., Vantaggi, B., Gagolewski, M., Gil, M.Á., Grzegorzewski, P., Hryniewicz, O. (eds.) Soft Methods for Data Science. AISC, vol. 456, pp. 141–148. Springer, Cham (2017). doi:10.1007/978-3-319-42972-4_18
7. Couso, I., Dubois, D.: A general framework for maximizing likelihood under incomplete data, under review
8. Couso, I., Dubois, D., Sánchez, L.: Random Sets and Random Fuzzy Sets as Ill-Perceived Random Variables. SAST. Springer, Cham (2014). doi:10.1007/978-3-319-08611-8
9. Dawid, A.P., Dickey, J.M.: Likelihood and Bayesian inference from selectively reported data. J. Amer. Statist. Assoc. **72**, 845–850 (1977)

---

[1] Let the reader notice that this vector does not necessarily represent a probability distribution. In fact, the sum $\sum_{j=1}^r q_j$ is strictly greater than 1, unless the collection of A_j forms a partition of $\mathcal{X}$.

10. Dempster, A.P., Laird, N.M., Rubin, D.B.: Maximum likelihood from incomplete data via the EM algorithm (with discussion). J. Roy. Statist. Soc. B **39**, 1–38 (1977)
11. Denœux, T.: Maximum likelihood estimation from uncertain data in the belief function framework. IEEE Trans. Knowl. Data Eng. **26**, 119–130 (2013)
12. Edwards, A.W.F.: Likelihood. Cambridge University Press, Cambridge (1972)
13. Guillaume, R., Dubois, D.: Robust parameter estimation of density functions under fuzzy interval observations. In: 9th ISIPTA Symposium, Pescara, Italy, pp. 147–156 (2015)
14. Guillaume, R., Couso, I., Dubois, D.: Maximum likelihood and robust optimisation on coarse data. In: 10th ISIPTA Symposium, Lugano, Switzerland, pp. 147–156 (2017)
15. Heitjan, D.F., Rubin, D.B.: Ignorability and coarse data. Ann. Stat. **19**, 2244–2253 (1991)
16. Huber, P.J.: The behavior of maximum likelihood estimates under nonstandard conditions. In: Proceedings of the Fifth Berkeley Symposium on Mathematical Statistics and Probability, vol. 1, pp. 221–233. University of California Press (1967)
17. Hüllermeier, E.: Learning from imprecise and fuzzy observations: data disambiguation through generalized loss minimization. Int. J. Approximate Reasoning **55**, 1519–1534 (2014)
18. Hüllermeier, E., Cheng, W.: Superset learning based on generalized loss minimization. In: Appice, A., Rodrigues, P.P., Santos Costa, V., Gama, J., Jorge, A., Soares, C. (eds.) ECML PKDD 2015. LNCS, vol. 9285, pp. 260–275. Springer, Cham (2015). doi:10.1007/978-3-319-23525-7_16
19. Jaeger, M.: Ignorability in statistical and probabilistic inference. J. Artif. Intell. Res. (JAIR) **24**, 889–917 (2005)
20. Jaeger, M.: The AI&M procedure for learning from incomplete data. In: Proceedings of Uncertainty in Artificial Intelligence Conference (UAI-06), pp. 225–232 (2006)
21. Plass, J., Augustin, T., Cattaneo, M., Schollmeyer, G.: Statistical modelling under epistemic data imprecision: some results on estimating multinomial distributions and logistic regression for coarse categorical data. In: Proceedings of the 9th International Symposium on Imprecise Probability: Theories and Applications (ISIPTA 2015), Pescara (Italy) (2015)
22. Plass, J., Cattaneo, M.E.G.V., Schollmeyer, G., Augustin, T.: Testing of coarsening mechanisms: coarsening at random versus subgroup independence. In: Ferraro, M.B., Giordani, P., Vantaggi, B., Gagolewski, M., Gil, M.Á., Grzegorzewski, P., Hryniewicz, O. (eds.) Soft Methods for Data Science. AISC, vol. 456, pp. 415–422. Springer, Cham (2017). doi:10.1007/978-3-319-42972-4_51
23. Quost, B., Denœux, T.: Clustering and classification of fuzzy data using the fuzzy EM algorithm. Fuzzy Sets Syst. **286**, 134–156 (2016)
24. Ramasso, E., Denœux, T.: Making use of partial knowledge about hidden states in HMMs: an approach based on belief functions. IEEE Trans. Fuzzy Syst. **22**(2), 395–405 (2014)
25. Sid-Sueiro, J.: Proper losses for learning from partial labels. In: Proceedings of Neural Information Processing Systems Conference (NIPS 2012), Lake Tahoe, Nevada, USA (2012)
26. Smets, P.: Constructing the pignistic probability function in a context of uncertainty. In: Henrion M., et al. (eds.) Uncertainty in Artificial Intelligence 5, pp. 29–39. North-Holland, Amsterdam (1990)

Reasons and Means to Model Preferences as Incomplete

Olivier Cailloux[1](✉) and Sébastien Destercke[2]

[1] Université Paris-Dauphine, PSL Research University,
CNRS, LAMSADE, 75016 Paris, France
olivier.cailloux@dauphine.fr
[2] Sorbonne Université, UMR CNRS 7253 Heudiasyc, Université de Technologie
de Compiègne, CS 60319, 60203 Compiègne cedex, France

Abstract. Literature involving preferences of artificial agents or human beings often assume their preferences can be represented using a complete transitive binary relation. Much has been written however on different models of preferences. We review some of the reasons that have been put forward to justify more complex modeling, and review some of the techniques that have been proposed to obtain models of such preferences.

1 Introduction

Preferences of agents are usually assumed to be representable with a weak order (a complete and transitive binary relation). We are interested in discussing the completeness assumption.

Preference models are especially important in two fields: choosing an alternative when it is evaluated according to different aspects (multi-criteria decision making, or MCDM), and picking an alternative whose quality depends on states of the world that are uncertainly known (decision making under uncertainty, or DMU). In MCDM, the common assumption is that the alternatives, i.e., the state of the world, is known without ambiguity, and the difficulty is to determine the structure of the user's preferences over these well-defined alternatives. In DMU, the alternatives are usually not described over several criteria, but the problem is to recommend an alternative given our uncertainty about the world.

In this paper, we review some reasons to relax preference completeness and modeling approaches (either in MCDM or DMU) that support this relaxation. We discuss in particular reasons to consider that the assumption of completeness is empirically falsified. Although these reasons are not new, we think it is interesting to discuss this question here and now because of (as we perceive it) a relative ignorance of these discussions in research fields that use preference models but are not specialized in preference modeling per se, and because of recent and ongoing advances in analysis of incomplete preferences. We try to cover a wide

This work has received support under the program "LABEX MS2T" launched by the French Government and implemented by ANR with the references ANR-11-IDEX-0004-02.

© Springer International Publishing AG 2017
S. Moral et al. (Eds.): SUM 2017, LNAI 10564, pp. 17–30, 2017.
DOI: 10.1007/978-3-319-67582-4_2

scope by discussing some of the goals, assumptions and basic definitions related to preference modeling and reviewing a wide range of techniques for obtaining such models. In counterpart, this review does not claim to be comprehensive and does not provide technical details. To further simplify the discussion, we pretend that the MCDM and DMU contexts are sharply separated. (In reality, it is often possible to cover MCDM contexts while taking uncertainty into account (Keeney and Raiffa 1993).) We also do not discuss transitivity.

We briefly present the MCDM and DMU settings considering completeness in Sect. 2. Section 3 discusses completeness in descriptive and normative approaches (recalling their difference at the same time). Finally, we review models that departs from completeness in Sect. 4.

2   Assuming Completeness

In this section, we are going to recall the main models that consider completeness and transitivity of preferences as a consequence of natural requirements, if not as pre-requisite of any preference modeling. We will also recall normative views and descriptive views of these concepts.

2.1   MCDM

We consider a simple and classical setting in MCDM. We assume that the alternatives are evaluated using a set of criteria G, each having an evaluation scale X_g. The set of all possible alternatives is $\mathcal{X} = \prod_{g \in G} X_g$, that is, every combination of evaluations are considered possible. We are interested in a preference relation $\succeq$ defined as a binary relation over $\mathcal{X}$.

Example 1. Say the Decision Maker (DM) must choose what to plant in her garden. The set of alternatives $\mathcal{X}$ are all possible vegetables, the criteria $G = \{g_1, g_2, g_3\}$ measure the taste, quantity, and price of each vegetable. The scales are $X_{g_1} = \{A, B, C, D\}$, a set of labels, with x_1 representing the taste of the vegetable $x \in \mathcal{X}$ as considered by the DM (A is the worst taste, D the best), $X_{g_2} = [0, 100]$, with x_2 representing the number of meals that the DM would enjoy if deciding to plant x, and $X_{g_3} = \mathbb{R}$, thus x_3 indicates the price to pay for planting x.

Typical approaches in MCDM assume that there is some real-valued function $v : \mathcal{X} \to \mathbb{R}$ mapping alternatives to their values, and that $x \succeq y$ iff $v(x) \geq v(y)$.

2.2   DMU

In the simplest form of DMU considered here (SDMR, for Simple Decision Making under Risk), we consider a set S of possible states of the world, a finite set of consequences C, and each act $x : S \to C$ is modeled as a function where $x(s)$ is the consequence of performing x when s is the actual state of the world.

Define $\mathscr{X}$, for simplicity, as all possible or imaginary acts (thus $\mathscr{X} = C^S$). In SDMR, uncertainty is modeled by a probability measure p over the power set of S, $\mathscr{P}(S)$, thus with $p(s) \in [0,1]$ indicating the probability of occurence of s (with $s \subseteq S$), and $p(S) = 1$. We consider a preference relation $\succeq$ defined as a binary relation over $\mathscr{X}$. Given an act x and a probability measure p, it is usually convenient to view x as p_x, a probability mass over the consequences: define $p_x : C \to [0,1]$ as $p_x(c) = p(x^{-1}(c))$, where $x^{-1}(c)$ designate the set of states in which x leads to the consequence c. Such a p_x is usually called a lottery.

Example 2. Assume you want to go out and wonder about taking or leaving your umbrella. You consider relevant weather state to be A = "shiny" and B = "raining", with $A, B \subseteq S$. We assume $A \cup B = S$ for simplicity. Two simple actions are x_1: "take the umbrella" and x_2: "leave the umbrella home", and the consequences are c_1: "encumbered" (when taking your umbrella, irrelevant of the weather), c_2: "free" (when leaving your umbrella and weather is A), and c_3: "wet" (when leaving your umbrella and weather is B). Assume the probabilities of the states A and B are 0.2 and 0.8. Then the constant act x_1 can also be described as p_{x_1} with $p_{x_1}(c_1) = 1$ and p_{x_1} being zero everywhere else, and similarly the act x_2 can be associated to p_{x_2} where $p_{x_2}(c_1) = 0$, $p_{x_2}(c_2) = 0.2$, $p_{x_2}(c_3) = 0.8$.

In most DMU frameworks, consequences can be mapped to a real-valued reward or utility through a function $u_1 : C \to \mathbb{R}$, in which case $u_1(x(s))$ denotes the utility of performing x in state s, and acts can be evaluated using a utility function $u : \mathscr{X} \to \mathbb{R}$ defined as $u(x) = \sum_{s \in S} p(s)u_1(x(s))$, such that $u(x) \geq u(y)$ iff $x \succeq y$. It follows from this definition that u and u_1 are coherent, in the following sense: given an act x that brings a consequence c with probability one, $u(x) = u_1(c)$.

Expected utility has been justified axiomatically by different authors, the main ones being Savage (1972), de Finetti (2017) and von Neumann and Morgenstern (2004) (hereafter, vNM). In the de Finetti setting, utilities are given as random variables, and a precise price can be associated to each random variable. That reasoning should be probabilistic and choices made according to expected utility follow from two axioms: linearity and boundedness of those prices. vNM postulate conditions on $\succeq$ ensuring that utility functions u and u_1 satisfying the above conditions exist. The axioms assume completeness of the preferences, and the probabilities are assumed to be given. In the Savage setting, both probabilities and expected utility follow from axioms about preferences between acts. In particular, his first axiom (P1) is that any pair of act should be comparable. Completeness is therefore postulated in the axioms, and expected utility and probabilistic reasoning follow from the axioms.

While these theoretical constructs have set very strong foundations for the use of probabilities, in practice experiments such as the Ellsberg (1961) urn (contradicting Savage sure-thing principle) suggest that people do not always act according to expected utility (MacCrimmon and Larsson 1979).

Since then, many different extensions have been proposed (Wakker 2010, Quiggin 2012). Others propose to relax the probabilistic assumption, for instance by considering a possibilistic setting (e.g., Dubois et al. (2003) discuss Savage-like axioms), by considering sets of probabilities such as in decision under ambiguity (Gajdos et al. 2008), or by simply considering completely missing information, such as Wald's (1992) celebrated maximin criterion.

All models presented thus far assume that $\succeq$ is complete (by which we mean that if $\succeq$ is incomplete, then no suitable function exists in the class of functions admitted by models presented thus far).

3   Questioning Completeness

Before discussing the reasonableness of restrictions about $\succeq$, we need to say a word about what those preferences really represent and what the goal of modeling those may be. Indeed, the meaning of completeness depend on whether a descriptive or a normative approach is adopted. In particular, we will later discuss "how much" descriptive one must accept to be in order for the completeness hypothesis to stand.

3.1   Descriptive and Normative Approaches

In the descriptive approach to preferences, the goal of the model is to reflect the observed behavior of a DM. Typically, a set of sample choices of the DM is first collected, say, of choices of food products in his favorite store, and we would then try to obtain the model that best reflects his choice attitude. Or, we would query an individual's preference about pairs of objects, and then try to build a predictive model on the whole set of possible pairs of alternatives (a method called active learning in the machine learning community). Such a model may be used to predict his behavior, e.g. for marketing or regulation purposes.

Under the normative approach, the goal is to model the way the DM ought to choose rationally. Rationality may corresponds to accepted external norms, or to rules accepted by the DM after careful thinking. (In the second case, the term prescriptive or constructive may be used instead of normative, but different authors use these terms differently (Roy 1993; Tsoukiàs 2007); we will stick to the term "normative" as an umbrella.) In both cases, the decision outcome using such approach may differ from empirically observed decisions. Consider as an example a recruiter in an enterprise who wants to model the recruitment procedure. After having collected data, it may appear that for some (possibly unconscious) reason, the recruitment is biased against some particular socio-economic category. The DM may then want to find a recruitment strategy that avoids such biases, therefore actively trying to build a model contradicting empirical observations.

McClennen (1990), Guala (2000) discuss philosophical grounds for accepting a normative model. Anand (1987), Mandler (2001) discuss normative grounds for usual axioms about preferences, including completeness.

Choosing between normative or descriptive approaches is not always easy. For instance recommender systems often adopt a descriptive approach. But descriptive approaches will, by design, reflect our cognitive limitations. Those limitations are numerous and sometimes obviously not in agreement with what the DM himself would do when thinking more carefully, as will be illustrated in Sect. 3.3. Providing (more) normative-based automatic recommendations might help provide sound advices, help increase serendipity, and possibly build trust (or avoid mistrust) in the recommender system. For example, the DM might appreciate that the recommender system's advices protect him from exploitations of the DM's cognitive limitations by merchants. (As an old but known example, "the credit card lobby is said to insist that any price difference between cash and card purchases should be labeled a cash discount rather than a credit surcharge" (Tversky and Kahneman 1986).)

3.2 Defining and Testing Incompleteness

Defining and testing incompleteness in preferences requires to define "preference" (and thus $\succeq$), as its everyday usage can be ambiguous: Frankfurt (1971) gives seven interpretations of "to want to", and this exercice transposes, *mutatis mutandis*, to the notion of preference.

Here is what vNM say about the preference relation (we have taken this from the very insightful presentation of the vNM approach by Fishburn (1989)): "It is clear that every measurement – or rather every claim of measurability – must ultimately be based on some immediate sensation, which possibly cannot and certainly need not be analyzed any futher. In the case of utility the immediate sensation of preference – of one object or aggregate of objects as against another – provides this basis" (3.1.2); "Let us for the moment accept the picture of an individual whose system of preferences is all-embracing and complete, i.e. who, for any two objects or rather for any two imagined events, possesses a clear intuition of preference. More precisely we expect him, for any two alternative events which are put before him as possibilities, to be able to tell which of the two he prefers." (3.3.2) (The "events" correspond to our alternatives.)

Expanding on vNM, we define that the DM *prefers* a to b when expressing an intuitive attraction towards a when presented with a and b, or an equal attraction towards a and b; and this attraction does not change along a reasonable time span and as well as when irrelevant changes in the context happen. Here, we assume that a, b are alternatives in $\mathscr{X}$ described by their evaluations on the criteria (in MCDM) or by the relevant probability distributions and consequences (in SDMR), and consider as irrelevant changes anything that does not change those descriptions.

Under this definition, postulating completeness of $\succeq$ amounts to say that choices of the DM will not change along time or when irrelevant changes happen. While this is not the only possible definition (others will be mentioned), it appears reasonable and sufficiently formal to make the condition empirically testable.

A first, immediate argument against completeness is that preferences are not stable over even very short period of time, a well-accepted fact in experimental psychology. Quoting Tversky (1969), individuals "are not perfectly consistent in their choices. When faced with repeated choices between x and y, people often choose x in some instances and y in others. Furthermore, such inconsistencies are observed even in the absence of systematic changes in the decision maker's taste which might be due to learning or sequential effects. It seems, therefore, that the observed inconsistencies reflect inherent variability or momentary fluctuation in the evaluative process".

This argument may not be strong enough however. In absence of other arguments, one might agree that preferences are in reality incomplete but claim that they may appropriately be *modeled* as complete: a model of complete preferences would simply deviate from time to time from what individuals declare because of (perhaps rare) random fluctuations in their expressions of preferences. In order to discuss this hypothesis, we turn to the second (and much more interesting) reason for failure of completeness, which is also brought by the literature in empirical psychology. It appears that preferences change may not be attributed solely to random fluctuations: they change in systematic ways according to changes in the presentation of the alternatives or the context that should have no impact from a normative point of view.

3.3 Empirical Evidence of Incompleteness

In multicriteria contexts, psychologists have shown systematic differences between the so-called choice and matching elicitation procedures (Tversky et al. 1988). Assume you want to know which of two alternatives x, y the DM prefers, in a problem involving two criteria. You can present both and directly ask for a choice. Alternatively, with the matching procedure, you present x with its two evaluations $g_1(x), g_2(x)$, and y' with only $g_1(y') = g_1(y)$, and ask the DM for which value $g_2(y')$ y' would be indifferent to x. Assuming $\succeq$ satisfies dominance and transitivity, you then know that $x \succeq y$ iff $g_2(y') \geq g_2(y)$. Although the two elicitation procedures should be equivalent, the authors confirm the prominence hypothesis stating that the more prominent criterion has more importance in choice than in matching. One of their study confront the subject to a hypothetical choice between two programs for control of a polluted beach. Program x completely cleans up the beach at a yearly cost of \$750 000; program y partially cleans it up for a yearly cost of \$250 000. They assume that pollution is the more prominent criterion here, hence expect that x will be chosen more often in choice than in matching. Indeed, 48% out of the 104 subjects confronted with a choice procedure selected x, whereas only 12% out of the 170 subjects selected it in a matching procedure. Similar effects apply to lotteries in SDMR (Luce 2000).

This phenomenon is known as preference reversal due to a breach of procedure invariance. Another reversal is the one due to description invariance (or framing effect), showing that preferences can change by changing the descriptions of alternatives. In Tversky and Kahneman (1981), two groups have to choose a program to prepare against an epidemic outspring that would result otherwise in 600 deaths. The two groups are presented with the same numeric alternatives x and y, but on the first group the alternatives are presented in terms of numbers of life saved, while in the second they are presented in terms of death counts. The experiment shows that preferences differ predictably in the two groups.

It is indeed well-known that results are perceived differently depending on their descriptions as losses or gains Thaler (1980).

Numerous other studies exist that show and discuss preference reversal effects (Deparis 2012, Chap. 2 (from which we took the two studies described here above), Lictenstein and Slovic 2006a, Tversky et al. 1990, Kahneman et al. 1981, Kahneman and Tversky 2000). How to best account for and predict preference reversals is still debated, but their existence is consensual (Wakker 2010, Birnbaum 2017). Some skeptics did try to show that preference reversals could be attributed to deficiencies in the design of the studies, but finally came around (Slovic and Lichtentstein 1983).

This shows that the $\succeq$ relation cannot be expected to be complete given our definition. For some alternatives, individuals may be led to declare different preferences, denoting an absence of a clear, intuitive preference for each pairs of alternatives. When thinking more about the comparison and presented with different views of the same problem, individuals may in some cases change their preference. This has been studied empirically (Slovic and Tversky 1974, Mac-Crimmon and Larsson 1979, Lichtenstein and Slovic 2006b) and Savage (1972, pp. 101–103) famously reported that it happened to him.

One may of course want to preserve completeness of preferences, for example to preserve mathematical and computational simplicity. One way to do so, common in experimental psychology, is to restrict further the frame in which preferences are considered. For instance, Luce (2000) indicates clearly that he studies preferences in terms of choice, not judgment; MacCrimmon et al. (1980) exclude some kind of loteries from the scope of the model. In such cases, completeness may well be justified. In other settings, such as normative approaches or recommender systems, it is unclear that such reductions should be enforced, as they may be hard to impose in practice or lead to behavior that the user may not desire.

We also mention two related interesting articles: Deparis et al. (2012) study the behavior of individuals when they are allowed to make explicit statements of incomparability; Danan and Ziegelmeyer (2006) propose to consider that an incomparability is observed whenever the DM is ready to pay a small price to postpone the decision.

Another, more evident, reason to be interested in models allowing incompleteness is that it may well be that provided information is insufficient to obtain a fully precise models.

The next section describes approaches that allow incomplete preference representations.

4   Dropping Completeness

4.1   Incompleteness in MCDM

Some approaches in MCDM in the family of outranking methods (Roy 1996, Greco et al. 2016, Bouyssou et al. 2000, 2006, Bouyssou and Pirlot 2015) can represent incomparabilities. A much used idea is to take into account two points of view, leading to weak-orders $\succeq^1$ and $\succeq^2$, then define $\succeq = \succeq^1 \cap \succeq^2$. Thus, when the two weak-orders strongly disagree about some pair of objects, the result can declare them incomparable. As an example, consider (a simplification of) the ELECTRE III method (our much simplified description only consider the aspects sufficient to obtain incomparabilities). It builds a concordance relation C that determines whether alternative x is sufficiently better than y, by accounting only for the criteria in favor of x; and a discordance relation D that determines whether x is so much worst than y on some criterion that x cannot possibly be considered better than y (thus implementing a veto effect). Precise definitions of C and D depend on parameters to be fixed when implementing the method. Then, the model declares that $x \succeq y$ iff xCy and not xDy.

Example 3. Consider $\mathscr{X} = \mathbb{R}^3$, each criteria to be maximized, and a model according to which xCy iff x is better than or equal to y for at least two criteria, and xDy iff for some g, $y_g - x_g \geq 2$. Such a model would consider the two alternatives $x = (0,0,2)$ and $y = (1,1,0)$ as incomparable: neither $x \succeq y$ nor $y \succeq x$ hold.

Such approaches tend to consider incomparabilities as intrinsic to the preferences, since even a completely specified preference could lead to incomparabilities.

Robust methods in MCDM exist that distinguish conclusions about preferences that hold for sure, given limited preferential information from the DM, from conclusions that possibly hold. Such methods typically start from a class M of possible models (similar to hypothesis space in machine learning) assumed to be candidate representative models of the DM preferences. A robust method, given a class M and a set of constraints C reducing the set of possibles models (typically preference statements given by the DM), will consider that a is necessarily preferred to b, $a \succeq^N b$, whenever $a \succeq b$ for all relations $\succeq$ in M that satisfy C (Greco et al. 2008).

Example 4. Assume that the only thing you know about the DM is that she prefers $x = (0,0,2)$ to $y = (0,4,0)$, and you assume that $\succeq$ satisfies preferencial independence, meaning that the way two alternatives compare does not change when changing equal values on a given criterion. Thus, M contains all relations that satisfy preferencial independence, and C is the constraint $x \succ y$. You may then conclude that $a = (3,0,2)$ is preferred to $b = (3,4,0)$, thus, $a \succ^N b$, but you ignore whether $c = (1,1,1)$ is preferred to $d = (0,2,2)$, thus, $\neg(c \succeq^N d)$ and $\neg(d \succeq^N c)$.

In such approach, the relation $\succeq^N$ is able to represent incomparabilites. Incomparabilities stem here from a lack of knowledge, and are not intrinsic to the modeled preference relation, as in principle one could collect enough constraints C about M to identify a unique compatible relation $\succ$ on a set of alternatives.

It is of course also possible to include in M some models that allow for incomparabilities (Greco et al. 2011).

4.2   Incompleteness in DMU

As recalled in Sect. 1, probability theory and expected utility are the most widely used tools when having to decide under uncertainty, and naturally induce completeness of preferences. It should however be noted early scholars were critical about the fact that completeness could hold in practice. von Neumann and Morgenstern (1953, p. 630) for example themselves considered completeness as a strong condition: "it is very dubious, whether the idealization of reality which treats this postulate as a valid one, is appropriate or even convenient".

Many attempts to relax the completeness axioms does so by considerings axioms leading to deal with sets of utilities and sets of probabilities (Aumann 1962), entangling together aspects about decision and about information modeling.

Keeping Precise Probabilities but Not Expected Utility. Even when having precise probabilities, there are alternatives to expected utility that induce incomplete preferences. One of them that is particularly interesting is the notion of stochastic dominance (Levy 1992). Assuming that the set of consequences is completely ordered by preference, which we denote by $C = \{c_1, \cdots, c_n\}$ where c_{i-1} is preferred to c_i, then a lottery p_x is said to stochastically dominate p_y iff, for all $1 \leq i \leq n$:

$$p_x(\{c_1, \ldots, c_i\}) = \sum_{j=1}^{i} p_x(c_j) \geq p_y(\{c_1, \ldots, c_i\}) = \sum_{j=1}^{i} p_y(c_j). \tag{1}$$

Since Inequality (1) can be satisfied for some i and not for others, possible incomparabilities immediately follow.

Example 5. Consider the set of consequences $C = \{c_1, c_2, c_3\}$ and the following lotteries (induced by different acts x_1, x_2, x_3), given in vectorial forms: $p_1 = (0.5, 0.3, 0.2)$, $p_2 = (0.6, 0.3, 0.1)$ and $p_3 = (0.7, 0, 0.3)$. Then x_2 stochastically dominates x_1, while x_3 is incomparable to both x_1 and x_2, according to stochastic dominance.

The notion of stochastic dominance has some very attractive properties, as:

1. it does not necessitate to define utilities over consequences, and merely requires them to be linearly ordered;
2. it can be perceived as a criterion allowing for utilities to be ill-defined, as p_x stochastically dominates p_y if and only if x has a higher expected utility than y for any increasing utility function u defined over C.

Incompleteness from Non-precise Probabilities. In the past few decades, different scholars have challenged the need for precise probabilities associated to classical axiomatics, advocating the use of imprecisely defined prices (expected values) or of imprecisely defined probabilities. To mention but a few:

- Levi (1983) advocates the uses of sets of probabilities within a logical interpretation of probabilities;
- Walley (1991) extends the de Finetti axioms by assuming that an agent would give different buying and selling prices for an act, therefore allowing indecision if the price is between these bounds;
- Shafer and Vovk (2005) explores a probabilistic setting centered on the notion of Martingale.

Such theories can most of the time be associated to the use of convex sets of probabilities, and give rise to decision rules that extend expected utility but do allow incomparabilities. Once we accept that a convex set $\mathcal{P}$ of probabilities (or a formally equivalent representation) can represent our knowledge, incompleteness may ensue.

A prototypical way to induce incompleteness between acts from incompleteness in probabilities is to adapt expected utility criterion, and among rules doing so, maximality is a popular one (it is championed by Walley, but is considered as early as the 60's (Aumann 1962)). Given acts x_1, x_2, maximality says that

$$x_1 \succeq x_2 \text{ iff } u(x_1) \geq u(x_2) \text{ for all } p \in \mathcal{P}.$$

Maximality reduces to expected utility when $\mathcal{P}$ is a singleton.

Example 6. Going back to Example 2, imagine that x_1 is indifferent to x_2 exactly when $p(A) = p(\text{"shiny"}) = 1/3$. Thus, $u_1(\text{"encumbered"}) = 1/3 u_1(\text{"free"}) + 2/3 u_1(\text{"wet"})$. Then, x_1 and x_2 will be incomparable according to maximality as soon as $\mathcal{P}$ contains at least one mass where $p(A) < 1/3$, and another where $p(A) > 1/3$.

It should be noted that other authors have proposed different rules: for instance Levi (1983) recommends to use a decision rule, often called E-admissibility, that does not give rise to an incomplete order between acts, but rather selects all the acts that are Bayes optimal according to at least one probability $p \in \mathcal{P}$. In terms of order, this comes down to consider a set of possible linear ordering, and to retain only those elements that are maximal for at least one of them.

Working with Sets of Probabilities and Utilities. Sets of probabilities are helpful to represent incomplete beliefs or lack of information, yet it is natural to also consider cases where the DM cannot provide a fully accurate estimation of utilities associated to consequences, or even to completely order them. In some sense, stochastic dominance is an extreme view of such a case, where consequences are ordered but the utility function is left totally unspecified.

Other works have dealt with partially specified utilities.

- Dubra et al. (2004) represents preferences over lotteries by a set of utility functions. Preference holds whenever the expected utility for the preferred alternative is higher for all utility functions. This idea has been applied in other contexts (Ok 2002).

- Dubra and Ok (2002) propose to view the preference relation as a completion of an intuitive partial preference relation: the DM knows intuitively the result of some comparisons, and compute the other ones by applying some reasoning process. They also obtain a preference relation that is representable using a set of utility functions. This approach directly tackles some of the shortcomings described in Sect. 3.2.
- Manzini and Mariotti (2008) use a utility function and a vagueness function, representing the preference using intervals of utilities rather than real valued utilities. (Beyond DMU, also using this representation, Masatlioglu and Ok (2005) assume that a specific alternative called the status quo alternative is prominently chosen whenever the DM faces a choice about which incomparability occur.)

There exist a few works where both requirements of precise probabilities and utilities are relaxed. This can be traced back at least to Aumann (1962) whose axioms do not require uniqueness of utilities: $x \succ y \Rightarrow u(x) > u(y)$, without requiring the reverse. More recently, Galaabaatar and Karni (2013) are interested in the Savage-like context where probabilities are unknown and represent an incomplete preference relation in uncertaintly using a set of pairs of probabilities and utilities.

5   Incompleteness: Absence of Knowledge or Knowledge of Absence?

We have tried to browse a general picture of reasons why preference modeling should accommodate for incompleteness, and how it can do so in multi-criteria problems and uncertainty modeling.

One issue that transpired in most of the paper is whether incompleteness should be considered as an intrinsic, or ontic property of the preferences, in which case incomparability express a knowledge of absence of relation, or if incompleteness should be considered as an incomplete, epistemic description of a complete order, in which case it expresses an absence of knowledge. This mirrors different views about probability sets (Walley's consider that they model belief, without assuming an existing precise unknown distribution, while robust Bayesians consider the opposite).

Our opinion is that both views can be legitimate in different settings, and also that beyond the philosophical interest of distinguishing the two, this can have an important practical impact: knowing that incomparabilities are observable facts may influence strongly our information collection protocol; also, a same piece of information will be interpreted differently. If a DM pick act a among three acts $\{a, b, c\}$, in the espistemic interpretation, we woud deduce $a \succ \{b, c\}$, but in the ontic one we could only deduce that a is a maximal element ($\neg(b \succ a)$ and $\neg(c \succ a)$)

References

Anand, P.: Are the preference axioms really rational? Theory Decis. **23**(2), 189–214 (1987). doi:10.1007/BF00126305

Aumann, R.J.: Utility theory without the completeness axiom. Econometrica **30**(3), 445–462 (1962). doi:10.2307/1909888. ISSN: 0012–9682

Birnbaum, M.H.: Empirical evaluation of third-generation prospect theory. Theory Decis. 1–17 (2017). doi:10.1007/s11238-017-9607-y. ISSN 0040–5833, 1573-7187

Bouyssou, D., Pirlot, M.: A consolidated approach to the axiomatization of outranking relations: a survey and new results. Ann. Oper. Res. **229**(1), 159–212 (2015). doi:10.1007/s10479-015-1803-y. ISSN: 0254–5330, 1572-9338

Bouyssou, D., Marchant, T., Pirlot, M., Perny, P., Tsoukiàs, A., Vincke, P.: Evaluation and decision models: a critical perspective. In: International Series in Operations Research & Management Science, vol. 32. Kluwer Academic, Dordrecht (2000). http://doi.org/10.1007/978-1-4615-1593-7

Bouyssou, D., Marchant, T., Pirlot, M., Tsoukiàs, A., Vincke, P.: Evaluation and Decision Models with Multiple Criteria: Stepping Stones for the Analyst. International Series in Operations Research and Management Science, vol. 86, 1st edn. Springer, Boston (2006). ISBN: 0-387-31098-3. http://doi.org/10.1007/0-387-31009-1

Danan, E., Ziegelmeyer, A.: Are preferences complete? An experimental measurement of indecisiveness under risk (2006). http://www.ericdanan.net/research. Working paper

de Finetti, B.: Theory of Probability: A Critical Introductory Treatment, 6th edn. Wiley, New York (2017)

Deparis, S.: Etude de l'Effet du Conflit Multicritère sur l'Expression des Préférences : une Approche Empirique. Ph.D. thesis, École Centrale Paris, Châtenay-Malabry, France, June 2012

Deparis, S., Mousseau, V., Öztürk, M., Pallier, C., Huron, C.: When conflict induces the expression of incomplete preferences. Eur. J. Oper. Res. **221**(3), 593–602 (2012). doi:10.1016/j.ejor.2012.03.041. ISSN: 03772217

Dubois, D., Fargier, H., Perny, P.: Qualitative decision theory with preference relations and comparative uncertainty: an axiomatic approach. Artif. Intell. **148**(1–2), 219–260 (2003)

Dubra, J., Ok, E.A.: A model of procedural decision making in the presence of risk*. Int. Econ. Rev. **43**(4), 1053–1080 (2002). doi:10.1111/1468-2354.t01-1-00048. ISSN: 1468–2354

Dubra, J., Maccheroni, F., Ok, E.A.: Expected utility theory without the completeness axiom. J. Econ. Theory **115**(1), 118–133 (2004). doi:10.1016/S0022-0531(03)00166-2. ISSN: 0022–0531

Ellsberg, D.: Risk, ambiguity, and the savage axioms. Q. J. Econ. **75**, 643–669 (1961)

Fishburn, P.C.: Retrospective on the utility theory of von Neumann and Morgenstern. J. Risk Uncertainty **2**(2), 127–157 (1989). doi:10.1007/BF00056134

Frankfurt, H.G.: Freedom of the will and the concept of a person. J. Philos. **68**(1), 5–20 (1971). doi:10.2307/2024717

Gajdos, T., Hayashi, T., Tallon, J.-M., Vergnaud, J.-C.: Attitude toward imprecise information. J. Econ. Theory **140**(1), 27–65 (2008)

Galaabaatar, T., Karni, E.: Subjective expected utility with incomplete preferences. Econometrica **81**(1), 255–284 (2013). doi:10.3982/ECTA9621. ISSN: 1468–0262

Greco, S., Mousseau, V., Słowiński, R.: Ordinal regression revisited: multiple criteria ranking using a set of additive value functions. Eur. J. Oper. Res. **191**(2), 415–435 (2008). doi:10.1016/j.ejor.2007.08.013

Greco, S., Kadziński, M., Mousseau, V., Słowiński, R.: ELECTREGKMS: robust ordinal regression for outranking methods. Eur. J. Oper. Res. **214**(1), 118–135 (2011). doi:10.1016/j.ejor.2011.03.045

Greco, S., Ehrgott, M., Figueira, J.R. (eds.): Multiple Criteria Decision Analysis - State of the Art Surveys. Springer, New York (2016). http://doi.org/10.1007/978-1-4939-3094-4

Guala, F.: The logic of normative falsification: rationality and experiments in decision theory. J. Econ. Methodol. **7**(1), 59–93 (2000). doi:10.1080/135017800362248. ISSN: 1350–178X

Kahneman, D., Tversky, A. (eds.): Choices, Values, and Frames. Cambridge University Press, Cambridge (2000). ISBN: 978-0-521-62749-8

Kahneman, D., Slovic, P., Tversky, A.: Judgement Under Uncertainty - Heuristics and Biases. Cambridge University Press, Cambridge (1981)

Keeney, R.L., Raiffa, H.: Decisions with Multiple Objectives: Preferences and Value Tradeoffs, 2nd edn. Cambridge University Press, Cambridge (1993). ISBN: 978-0-521-43883-4

Levi, I.: The Enterprise of Knowledge: An Essay on Knowledge, Credal Probability, and Chance. MIT press, Cambridge (1983)

Levy, H.: Stochastic dominance and expected utility: survey and analysis. Manage. Sci. **38**(4), 555–593 (1992)

Lichtenstein, S., Slovic, P. (eds.): The Construction of Preference. Cambridge University Press, New York (2006a). ISBN: 0-521-83428-7

Lichtenstein, S., Slovic, ;P.: Reversals of preferences between bids and choices gambling decisions. In: Lichtenstein, S., Slovic, P. (eds.) The Construction of Preference, pp. 52–68. Cambridge University Press (2006b). ISBN: 0-521-83428-7

Luce, R.D.: Utility of Gains and Losses: Measurement-theoretical and Experimental Approaches. Lawrence Erlbaum Publishers, Mahwah (2000)

MacCrimmon, K.R., Larsson, S.: Utility theory: axioms versus 'Paradoxes'. In: Allais, M., Hagen, O. (eds.) Expected Utility Hypotheses and the Allais Paradox, pp. 333–409. Springer, Netherlands (1979). http://doi.org/10.1007/978-94-015-7629-1_15

MacCrimmon, K.R., Stanbury, W.T., Wehrung, D.A.: Real money lotteries: a study of ideal risk, context effects, and simple processes. In: Wallsten, T.S. (ed.) Cognitive Processes in Choice and Decision Behavior. L. Erlbaum Associates, Hillsdale (1980)

Mandler, M.: A difficult choice in preference theory: rationality implies completeness or transitivity but not both. In: Millgram, E. (ed.) Varieties of Practical Reasoning. MIT Press, Cambridge (2001)

Manzini, P., Mariotti, M.: On the representation of incomplete preferences over risky alternatives. Theory Decis. **65**(4), 303–323 (2008). doi:10.1007/s11238-007-9086-7

Masatlioglu, Y., Ok, E.A.: Rational choice with status quo bias. J. Econ. Theory **121**(1), 1–29 (2005). doi:10.1016/j.jet.2004.03.007. ISSN: 0022–0531

McClennen, E.: Rationality and Dynamic Choice: Foundational Explorations. Cambridge University Press, Cambridge (1990)

Ok, E.A.: Utility representation of an incomplete preference relation. J. Econ. Theory **104**(2), 429–449 (2002). doi:10.1006/jeth.2001.2814. ISSN: 0022–0531

Quiggin, J.: Generalized Expected Utility Theory: The Rank-Dependent Model. Springer, Netherlands (2012)

Roy, B.: Decision science or decision-aid science? Eur. J. Oper. Res. **66**, 184–203 (1993). doi:10.1016/0377-2217(93)90312-B

Roy, B.: Multicriteria Methodology for Decision Aiding. Kluwer Academic, Dordrecht (1996)

Savage, L.J.: The Foundations of Statistics. Dover Publications, New York (1972). Second revised edition. ISBN 978-0-486-62349-8

Shafer, G., Vovk, V.: Probability and Finance: It's Only a Game!, vol. 491. Wiley, New York (2005)

Slovic, P., Lichtentstein, S.: Preference reversals: a broader perspective. Am. Econ. Rev. **73**, 596–605 (1983)

Slovic, P., Tversky, A.: Who accepts Savage's axiom? Behav. Sci. **19**, 368–373 (1974). doi:10.1002/bs.3830190603

Thaler, R.: Toward a positive theory of consumer choice. J. Econ. Behav. Organ. **1**, 39–60 (1980)

Tsoukiàs, A.: On the concept of decision aiding process: an operational perspective. Ann. Oper. Res. **154**(1), 3–27 (2007). doi:10.1007/s10479-007-0187-z. ISSN: 0254-5330

Tversky, A.: Intransitivity of preferences. Psychol. Rev. **76**, 31–48 (1969)

Tversky, A., Kahneman, D.: The framing of decisions and the psychology of choice. Science **211**, 453–458 (1981)

Tversky, A., Kahneman, D.: Rational choice and the framing of decisions. J. Bus. **59**(4), S251–S278 (1986). doi:10.1086/296365. ISSN: 0021-9398

Tversky, A., Sattath, S., Slovic, P.: Contingent weighting in judgment and choice. Psychol. Rev. **95**(3), 371–384 (1988). doi:10.1037/0033-295X.95.3.371

Tversky, A., Slovic, P., Kahneman, D.: The causes of preference reversals. Am. Econ. Rev. **80**(1), 204–217 (1990)

von Neumann, J., Morgenstern, O.: Theory of Games and Economic Behavior, 3rd edn. Princeton University Press, Princeton (1953)

von Neumann, J., Morgenstern, O.: Theory of games and economic behavior. Princeton classic editions. Princeton University Press, Princeton, 60th anniversary edition (2004). ISBN: 978-0-691-13061-3

Wakker, P.P.: Prospect Theory: For Risk and Ambiguity. Cambridge University Press, Cambridge (2010). ISBN: 978-1-139-48910-2

Wald, A.: Statistical decision functions. In: Breakthroughs in Statistics, pp. 342–357. Springer (1992)

Walley, P.: Statistical Reasoning with Imprecise Probabilities. Taylor & Francis, London (1991). ISBN: 9780412286605

Fuzzy Description Logics – A Survey

Stefan Borgwardt[1] and Rafael Peñaloza[2(✉)]

[1] Chair for Automata Theory, Technische Universität Dresden, Dresden, Germany
stefan.borgwardt@tu-dresden.de
[2] KRDB Research Centre, Free University of Bozen-Bolzano, Bolzano, Italy
rafael.penaloza@unibz.it

1 History

Mathematical Fuzzy Logics [51,60] have a long tradition with roots going back to the many-valued logics of Łukasiewicz, Gödel, and Kleene [57,68,73] and the Fuzzy Set Theory of Zadeh [111]. Their purpose is to model vagueness or imprecision in the real world, by introducing new *degrees of truth* as additional shades of gray between the Boolean *true* and *false*. For example, one can express the distinction between a person x having a *high fever* or a *low fever* as the degree of truth of the logical statement $\mathsf{Fever}(x)$. One of the central properties of fuzzy logics is *truth functionality*—the truth degree of a complex logical formula is uniquely determined by the truth degrees of its subformulas. This is a fundamental difference to other quantitative logics like probabilistic or possibilistic logics [56,83]. The semantics of fuzzy logics are thus given by functions interpreting the logical constructors conjunction, implication, and negation. For example, the truth degree of the conjunction $(\mathsf{Fever} \wedge \mathsf{Headache})(x)$ can be computed as a function of the degrees of $\mathsf{Fever}(x)$ and $\mathsf{Headache}(x)$. The functions proposed by Zadeh [111] are a popular choice, because they lead to good computational properties. A different approach uses operators called *triangular norms (t-norms)* and their associated *residua* to interpret conjunction and implication [60,69].

More recently, Description Logics (DLs) were developed as fragments of first-order logic, with a focus on their computational properties [3]. They use *concepts* (unary predicates, such as Fever and $\mathsf{Headache}$) and *roles* (binary predicates like $\mathsf{hasSymptom}$) to describe knowledge about the world. For example, the description logic axiom

$$\exists\mathsf{hasDiagnosis.Flu} \sqsubseteq \exists\mathsf{hasSymptom.Fever} \sqcap \exists\mathsf{hasSymptom.Headache} \tag{1}$$

says that whenever patients are diagnosed with flu, they must have the symptoms fever and headache, i.e. fever and headache are necessary symptoms for a flu diagnosis. Different choices of concept constructors (such as conjunction $\sqcap$ and existential restrictions $\exists$) can be used to tailor the description logic to the needs of a specific domain, ranging from lightweight to very expressive logics, for which nevertheless highly optimized reasoning systems have been developed. The first Fuzzy Description Logics (FDLs) were developed based on Zadeh's fuzzy semantics [95,106,110], and classical DL algorithms were extended to deal with

© Springer International Publishing AG 2017
S. Moral et al. (Eds.): SUM 2017, LNAI 10564, pp. 31–45, 2017.
DOI: 10.1007/978-3-319-67582-4_3

the additional expressivity provided by the truth degrees. Since then, a multitude of combinations of description logics with fuzzy semantics have been investigated, they have been subject of several surveys and monographies [10,24,47,74,102], and many FDL reasoners have been implemented [1,14,23,86,107]. Returning to the example axiom (1), FDLs enable us to grade a flu diagnosis as *mild* or *severe*, based on the severity of its symptoms.

In this survey, we focus on a prototypical FDL based on the classical description logic $\mathcal{ALC}$, and demonstrate the effects of different fuzzy semantics on the complexity of reasoning with general TBoxes. Section 2 introduces the basic syntax, semantics, and reasoning problems of the logic, and the subsequent chapters deal with different kinds of semantics, sorted roughly by their complexity: Zadeh semantics and semantics based on finitely many degrees of truth (Sect. 3), semantics based on the *Gödel t-norm* (Sect. 4), and other t-norms such as the *Łukasiewicz t-norm* or the *product t-norm* (Sect. 5). We conclude with a discussion of related logics and reasoning problems, and some open problems.

2   The Prototypical Fuzzy Description Logic

As a prototypical FDL, we briefly introduce a generic fuzzy extension of $\mathcal{ALC}$, called L-$\mathfrak{N}\mathcal{ALC}$, where L denotes an algebra specifying the fuzzy semantics, and $\mathfrak{N}$ denotes the presence of an additional concept constructor $\boxminus$ called *residual negation*. In the following sections, we instantiate this generic definition with different concrete semantics.

Truth degrees. We consider algebras of the form $\mathsf{L} = (\underline{\mathsf{L}}, *_\mathsf{L}, \Rightarrow_\mathsf{L})$ where $\underline{\mathsf{L}}$ is a totally ordered set of *truth degrees*, including 0 (*false*) and 1 (*true*), respectively; $*_\mathsf{L}$ is a *t-norm*, i.e. an associative, commutative, and monotonic binary operator on $\underline{\mathsf{L}}$ that has unit 1; and $\Rightarrow_\mathsf{L}$ is a binary operator on $\underline{\mathsf{L}}$ called *implication function*. To simplify the notation, we usually denote the set $\underline{\mathsf{L}}$ as L. In most cases, $\Rightarrow_\mathsf{L}$ is a *residuum* of $*_\mathsf{L}$ (formally defined in Sect. 4), and the associated *residual negation* is the function $x \mapsto x \Rightarrow_\mathsf{L} 0$, for $x \in \mathsf{L}$. Another popular fuzzy negation function is the *involutive negation* defined by $x \mapsto 1 - x$. In the following, we consider a logic that uses both involutive and residual negation.

Syntax. Let $\mathsf{N_I}$, $\mathsf{N_C}$, and $\mathsf{N_R}$ be mutually disjoint sets of *individual-*, *concept-*, and *role names*. *Concepts* of L-$\mathfrak{N}\mathcal{ALC}$ are defined by the following grammar rule, where $A \in \mathsf{N_C}$ and $r \in \mathsf{N_R}$:

$$C ::= A \mid \neg C \mid \boxminus C \mid C \sqcap C \mid \exists r.C \mid \forall r.C$$

A *TBox* is a finite set of *general concept inclusions* (GCIs) $\langle C \sqsubseteq D \geq q \rangle$, where C, D are concepts and $q \in [0, 1]$. An *ABox* is a finite set of *concept assertions* $\langle C(a) \bowtie q \rangle$ and *role assertions* $\langle r(a, b) \bowtie q \rangle$ with $a, b \in \mathsf{N_I}$, $r \in \mathsf{N_R}$, C a concept, $\bowtie \in \{\leq, \geq\}$, and $q \in [0, 1]$. An *ontology* is composed of a TBox and an ABox. We refer to GCIs and assertions as *axioms*.

Semantics. *Interpretations* $\mathcal{I} = (\Delta^\mathcal{I}, \cdot^\mathcal{I})$ consist of a non-empty set $\Delta^\mathcal{I}$, called *domain*, and an *interpretation function* $\cdot^\mathcal{I}$ that maps every $a \in \mathsf{N_I}$ to an element

$a^{\mathcal{I}} \in \Delta^{\mathcal{I}}$, every $A \in N_C$ to a function $A^{\mathcal{I}} : \Delta^{\mathcal{I}} \to L$, and every $r \in N_R$ to $r^{\mathcal{I}} : \Delta^{\mathcal{I}} \times \Delta^{\mathcal{I}} \to L$. This function is extended to concepts C by similarly assigning them a truth function $C^{\mathcal{I}} : \Delta^{\mathcal{I}} \to L$ as follows. For every $x \in \Delta^{\mathcal{I}}$,

$$(\neg C)^{\mathcal{I}}(x) := 1 - C^{\mathcal{I}}(x);$$
$$(\boxminus C)^{\mathcal{I}}(x) := C^{\mathcal{I}}(x) \Rightarrow_L 0;$$
$$(C \sqcap D)^{\mathcal{I}}(x) := C^{\mathcal{I}}(x) *_L D^{\mathcal{I}}(x);$$
$$(\exists r.C)^{\mathcal{I}}(x) := \sup_{y \in \Delta^{\mathcal{I}}} r^{\mathcal{I}}(x,y) *_L C^{\mathcal{I}}(y); \text{ and}$$
$$(\forall r.C)^{\mathcal{I}}(x) := \inf_{y \in \Delta^{\mathcal{I}}} r^{\mathcal{I}}(x,y) \Rightarrow_L C^{\mathcal{I}}(y).$$

The interpretation $\mathcal{I}$ *satisfies* (or is a *model* of)

- the GCI $\langle C \sqsubseteq D \geq q \rangle$ iff $C^{\mathcal{I}}(x) \Rightarrow_L D^{\mathcal{I}}(x) \geq q$ holds for all $x \in \Delta^{\mathcal{I}}$;
- the concept assertion $\langle C(a) \bowtie q \rangle$ iff $C^{\mathcal{I}}(a^{\mathcal{I}}) \bowtie q$;
- the role assertion $\langle r(a,b) \bowtie q \rangle$ iff $r^{\mathcal{I}}(a^{\mathcal{I}}, b^{\mathcal{I}}) \bowtie q$; and
- the ontology $\mathcal{O}$ iff it satisfies all axioms in $\mathcal{O}$.

We are interested in deciding *consistency* in L-$\mathfrak{N}\mathcal{ALC}$, i.e. whether a given ontology has a model.

Witnessed models. The semantics of the existential restriction $\exists r.C$ computes the supremum over a potentially infinite set of truth degrees, which may lead to unwanted or unexpected results. For that reason, *witnessed* models have been introduced [61], in which this supremum—and, dually, the infimum from $\forall r.C$—is required to be reached by some domain element, becoming a maximum (or minimum, respectively). After having been introduced for FDLs, witnessed models were also studied in the context of fuzzy predicate logics [62–64]. Following the standard approach in FDLs, in the following we implicitly restrict all models to be witnessed, in particular for Sects. 4 and 5. For work that does not enforce this restriction, we refer the interested reader to [24,29,32,37,40,48,61].

In the following sections, we instantiate the logic L-$\mathfrak{N}\mathcal{ALC}$ with different semantics, represented by various choices for the algebra L, and discuss the effect of these choices on the complexity of deciding (witnessed) consistency.

3 Zadeh and Finitely Valued Semantics

We start with the "traditional" FDL Z-$\mathfrak{N}\mathcal{ALC}$, where $Z = ([0,1], *_Z, \Rightarrow_Z)$ is given by the *Gödel* (or *minimum*) t-norm $x *_Z y = \min\{x,y\}$ and the *Kleene-Dienes implication* $x \Rightarrow_Z y = \max\{1 - x, y\}$ [55,68]. These functions were initially proposed in the context of Fuzzy Set Theory [111] to model the intersection and inclusion of fuzzy sets. This semantics allows various simplifications in Z-$\mathfrak{N}\mathcal{ALC}$. For example, $\neg C$ is equivalent to $\boxminus C$, and—as in classical logic—$\forall r.C$ can be expressed as $\neg \exists r.\neg C$. This is not always the case for other semantics. It was noticed early on that this choice of functions also results in an *effectively* *finitely valued* FDL, i.e. that the set of truth degrees can be restricted without loss of generality to be finite, as long as it contains 0, 0.5, and 1, and is closed under application of the involutive negation $x \mapsto 1 - x$ [97]. For this reason, we see it as a special case of *finitely valued* FDLs, where L is a fixed, finite set, which can be assumed to be of the form $\{0, \frac{1}{n}, \ldots, \frac{n-1}{n}, 1\}$, and $*_L$ and $\Rightarrow_L$ are arbitrary functions as introduced in Sect. 2. In such logics, the restriction to witnessed models is not necessary anymore, since any supremum over a finite set of values is automatically a maximum. Reasoning approaches for Zadeh and finitely valued FDLs can be divided into three classes: tableaux-, crispification-, and automata-based algorithms.

Tableaux algorithms were developed as extensions of classical tableaux techniques for description logics. The basic idea is to iteratively apply *tableaux rules* to decompose complex assertions into simpler ones. For example, $\langle (C \sqcap D)(a) \geq q \rangle$ is split into $\langle C(a) \geq q_1 \rangle$ and $\langle D(a) \geq q_2 \rangle$, where q_1 and q_2 are nondeterministically chosen such that $q = q_1 *_L q_2$. For an assertion $\langle (\exists r.C)(a) \geq q \rangle$ involving an existential restriction, a new individual x has to be introduced, together with assertions $\langle r(a, x) \geq q_1 \rangle$ and $\langle C(x) \geq q_2 \rangle$ as above. In general, there are many choices for q_1 and q_2, which result in a high degree of nondeterminism; however, for some semantics (e.g. Zadeh) the rules can be simplified. Since there are only finitely many possibilities for such assertions over a single individual, appropriate *blocking conditions* can ensure termination of these algorithms. Fuzzy tableaux algorithms have been developed, starting from the initial works in [95, 96, 106], for very expressive extensions of L-$\mathcal{NALC}$ with more concept constructors [38, 43, 88, 89, 92, 93]. They usually do not provide tight complexity bounds for deciding consistency, but they have been successfully implemented in FDL reasoners like *fuzzyDL* [23] and *FiRE* [86].

Crispification algorithms are also among the first reasoning methods to be developed for finitely valued FDLs [97, 100]. The idea is to translate the fuzzy ontology into a classical DL ontology, and then use optimized algorithms for classical reasoning. In this translation, each concept name A is replaced by finitely many classical concept names $A_{\geq q}$ that represent all those individuals that belong to A with a degree of at least q. The order structure on L is then expressed by GCIs $A_{\geq q_2} \sqsubseteq A_{\geq q_1}$ where $q_1 < q_2$. Since the order has to be preserved also for role names, these reductions introduce so-called *role inclusion axioms*, which usually do not increase the complexity of reasoning. Finally, the original fuzzy axioms are recursively translated into classical axioms by using the new concept and role names. The ontology resulting from this translation is consistent in the classical sense iff the original fuzzy ontology is consistent. The first reductions for general finitely valued semantics included an exponential blowup [13, 15, 16, 20, 22, 75], which can however be avoided by preprocessing the ontology [35]; again, this problem does not occur when using the Zadeh semantics [11, 12, 97]. Based on these polynomial translations, it can be shown that deciding consistency in finitely valued FDLs has the same complexity as in the underlying classical DLs. In particular, consistency in L-$\mathcal{NALC}$ with finite L is EXPTIME-complete. It was pointed out recently [35] that some reductions [16, 20, 75] are incorrect for so-called *number restrictions*, which allow to restrict the number of r-successors of a particular type; unfortunately, no alternative reduction has been found so far. The crispification approach has been implemented in the FDL reasoner *DeLorean* [14].

Automata-based algorithms generalize similar techniques from classical DLs. The basic idea behind these algorithms is to use tree automata to decide the existence of a *forest-shaped* model of the ontology [36, 41, 45]. In some cases, e.g. in the presence of *nominals* (concepts that can refer to specific individual names), a forest-shaped model does not need to exist, but the automata-based techniques can be adapted [25]. One disadvantage of this approach is that it cannot handle ABoxes naturally. In fact, a pre-completion step, which is based on the tableau rules mentioned before, is necessary to ensure correctness of the algorithms [38]. On the other hand, automata-based algorithms are useful for finding tight complexity bounds. In particular, using the notions from [4], one can show that consistency in L-$\mathcal{NALC}$ is PSPACE-complete when restricted to so-called *acyclic* TBoxes [41].

In summary, reasoning in finitely valued FDLs (including those using Zadeh semantics) is usually decidable and has the same complexity as in the underlying classical DLs; moreover, efficient implementations of reasoning algorithms are available. The only known exceptions are FDLs that are less expressive than L-$\mathcal{NALC}$, where additional truth degrees can actually increase the complexity of reasoning, e.g. from P to EXPTIME in $\mathcal{EL}$ [26].

4   Gödel Semantics

Hájek initiated a systematic investigation of FDLs with t-norm-based semantics in [61], founded on his work on Mathematical Fuzzy Logic [60, 65]. Such semantics use the infinite set of truth degrees $[0, 1]$ and t-norms $*_L$ as introduced in Sect. 2. However, as implication function they use an associated *residuum* $\Rightarrow_L$ that satisfies, for all $x, y, z \in [0, 1]$, the equivalence $x *_L z \leq y$ iff $z \leq x \Rightarrow_L y$. We consider here only *continuous* t-norms $*_L$, which ensures that their residuum is unique [69]. It is long known [82] that all infinitely many continuous t-norm can be constructed using so-called *ordinal sums* from three fundamental t-norms: the *Gödel t-norm* (which is also used in the Zadeh semantics), the *Łukasiewicz t-norm*, and the *product t-norm*. Without going into the details of this construction, we focus here on these fundamental t-norms themselves; however, in the literature many results for their combinations have been obtained.

In this section, we consider the first of the three fundamental t-norms. As mentioned earlier, the Gödel semantics $G = ([0, 1], *_G, \Rightarrow_G)$ differs from the Zadeh semantics only by using as implication function the *Gödel residuum*, which is defined, for all $x, y \in [0, 1]$, as

$$x \Rightarrow_G y = \begin{cases} 1 & \text{if } x \leq y, \\ y & \text{otherwise.} \end{cases}$$

The first work to deal with Gödel semantics in FDLs was [13]. There, the authors attempted to replicate the ideas successfully employed for the Zadeh semantics, and show that reasoning can be restricted to a finite subset of truth degrees. Unfortunately, this claim turned out to be incorrect. The main reason lies in the fact that the constructors $\boxminus$ and $\forall$, which require the Gödel residuum in their interpretation, can be used to guarantee the existence of truth degrees strictly greater than others, which may, however, be arbitrarily close.

Using this insight, it was later shown in [31] that G-$\mathfrak{N}\mathcal{ALC}$ does not have the finite model property (FMP). That is, there exist consistent G-$\mathfrak{N}\mathcal{ALC}$ ontologies that have only infinite models. Since classical $\mathcal{ALC}$ has the FMP, a direct consequence of this result is that it is impossible to apply the crispification approach from finitely valued FDLs to the Gödel semantics. This result was strengthened further, by showing that some ontologies can only be satisfied by models that use infinitely many different truth degrees [31].

Interestingly, it turns out that satisfiability of a G-$\mathfrak{N}\mathcal{ALC}$ ontology does not depend on the precise truth degrees used by an interpretation, but rather on their relative order. Moreover, only finitely many such orderings are relevant. Hence, a new crispification approach was developed in [44], where the classical concepts represent orderings between the degrees given to different fuzzy concepts. Building on this idea, a tableaux-based method, which decomposes complex concepts into simpler (ordered) concepts, was developed in [46]. From these algorithms, it was possible to obtain complexity bounds for reasoning in FDLs under Gödel semantics that mostly match those of reasoning in their classical variants. In particular, deciding consistency in G-$\mathfrak{N}\mathcal{ALC}$ is again ExpTime-complete.

5   Łukasiewicz and Product Semantics

We now consider the remaining two of the three fundamental continuous t-norms. Following the formalization of [60], the product semantics is given by the algebra $\Pi = ([0,1], *_\Pi, \Rightarrow_\Pi)$, with $x *_\Pi y = x \cdot y$ and

$$x \Rightarrow_\Pi y = \begin{cases} 1 & \text{if } x \leq y, \\ \frac{y}{x} & \text{otherwise.} \end{cases}$$

The Łukasiewicz semantics is given by $Ł = ([0,1], *_Ł, \Rightarrow_Ł)$, where

$$x *_Ł y = \max\{x + y - 1, 0\} \text{ and}$$
$$x \Rightarrow_Ł y = \min\{1 - x + y, 1\}.$$

After t-norm-based semantics were proposed for FDLs in [61], many tableaux algorithms were developed for these logics [17,18,59,87,103,104]. Most of them are based on a novel combination of traditional tableaux algorithms with mixed integer programming solvers. The former decompose complex assertions into smaller ones, while generating a set of constraints; and the latter are then used to find a solution for these constraints. In a simple example, $\langle (C \sqcap D)(a) \geq 0.5 \rangle$ is decomposed into $\langle C(a) \geq x \rangle$ and $\langle D(a) \geq y \rangle$, and the constraint $x \cdot y \geq 0.5$ is added to a set of inequations (under product semantics). At the end, an external solver computes a solution to this inequation, e.g. $\{x \mapsto 0.8, y \mapsto 0.7\}$. Unfortunately, as observed in [5,9], these logics also lack the FMP and, consequently, none of these algorithms can decide consistency in the presence of GCIs. In fact, these algorithms are correct only for so-called *unfoldable TBoxes* (similar to acyclic TBoxes). A detailed discussion of the reasons for restricting the expressivity of the ontologies can be found in [2]. Different algorithms tailored towards FDLs using the product and Łukasiewicz t-norms in the absence of GCIs were used to show decidability in [1,47,48,61].

Surprisingly, many such FDLs were subsequently shown to have an undecidable consistency problem, first for product semantics [5–7], and then for Łukasiewicz semantics [49]. These results are based on reductions from the Post Correspondence Problem. They were later generalized to cover a variety of combinations of t-norms and concept constructors [33,39], including both $\Pi\text{-}\mathcal{NALC}$ and $\text{Ł-}\mathcal{NALC}$, even if all axioms are restricted to be *crisp*, i.e. of the form $\langle \alpha \geq 1 \rangle$. At the same time, it was discovered that these results are quite sensitive to the choice of concept constructors and axioms allowed. For example, consistency in $\Pi\text{-}\mathcal{NALC}$ becomes decidable if the constructor $\neg$ and assertions of the form $\langle \alpha \leq q \rangle$ are disallowed. Indeed, it was shown that, to decide consistency in this restricted logic, it suffices to consider classical interpretations, which use only the truth values 0 and 1; effectively, the logic cannot even be considered to be fuzzy [30,33]. On the other hand, consistency in $\text{Ł-}\mathcal{NALC}$ remains undecidable even without $\neg$ and $\forall$, and with only crisp axioms [39]. Very recently, it was discovered that even $\text{Ł-}\mathcal{EL}$, which extends a logic with polynomial complexity in the classical case, has an undecidable consistency problem [27,28]; however, no such result is known for the variant $\Pi\text{-}\mathcal{EL}$ with product semantics. For a detailed discussion of the border between decidability and undecidability in t-norm-based FDLs, we refer the reader to [33].

6 Related Notions

Much research effort has been devoted to extending FDLs towards even more expressive languages. For example, more complex assertions like $\langle \mathsf{Tall}(a) > \mathsf{Tall}(b) \rangle$ allow to compare fuzzy degrees between different individuals a and b. Usually, such extensions do not affect the complexity of consistency [31,44–46]. Going one step further, one can also allow comparisons inside concepts like $\mathsf{Tall} > \forall \mathsf{friend}.\mathsf{Tall}$, representing the set of all people that are taller than all their friends. These latter extensions have so far been studied only for the Zadeh semantics [67,72].

The papers [21,70,109] propose aggregation operators that generalize concept conjunctions, and for example allow to express weighted sums of truth degrees, like a $\frac{1}{2}\mathsf{Comfortable} + \frac{1}{2}\mathsf{Cheap}$ hotel. Due to their generality, however, one has to be careful not to obtain an undecidable logic, e.g. by restricting to unfoldable TBoxes. Similarly, one can replace the quantifiers in FDLs by more general functions [85], or introduce fuzzy modifiers like very [66,112] that can scale and transform the interpretations of concepts, e.g. very Tall.

A different direction for generalization is to allow the truth degrees to be only partially ordered. In particular, extensions of description logics with lattice-based semantics have been studied in [24,25,36,41,43,58,100]. If the underlying lattice is finite, then methods similar to those described in Sects. 3 and 4 can be used to show tight complexity bounds or provide reductions to classical reasoning. Other (infinite) semantics remain largely unexplored, but they are expected to be undecidable in most cases.

The standard approach to integrate datatypes into description logics is to include *concrete domains*, which provide access to new concrete predicates whose interpretation is fixed, e.g. the total order on the natural numbers. In [80], the classical description logic $\mathcal{ALC}$ is extended with the use of *fuzzy* concrete domains, which provide built-in fuzzy predicates. In the context of FDLs, fuzzy concrete domains have also been studied in [18,98], but are usually restricted to unary concrete predicates.

A different reasoning task often considered in description logics is to answer conjunctive queries (CQs) over an ontology $\mathcal{O}$. More precisely, one is interested in retrieving all the individuals that satisfy some given properties in every model of $\mathcal{O}$. In FDLs this problem becomes more involved, since individuals may satisfy the query to some intermediate truth degree. This has motivated different approaches and solutions to CQ answering over fuzzy ontologies, for example by crispification or adaptation of classical CQ answering techniques such as query rewriting [35,75,78,84,99].

A lot of research has been done also on fuzzy extensions of less expressive description logics like $\mathcal{EL}$ [8,26,42,76,90], $\mathcal{FL}_0$ [34], and rule-based languages [77, 94,108]. Since the semantics of the standard Web Ontology Language OWL 2 is based on classical description logics,[1] several proposals have been made for a fuzzy extension of OWL 2 in order to make FDLs more accessible, and fuzzy plug-ins for ontology editors have been developed [19,88,91].

We have focused here mainly on the theoretical aspects of FDLs and in particular on the complexity of reasoning in ontologies built with these languages. The importance of this study is highlighted by several applications that have been considered. Some of these applications include medicine [81], information retrieval [79,101], recommendation [50,53], and detection [52,54]. Another important aspect is the problem of constructing such ontologies in the first place. Approaches for learning FDL axioms from data have been suggested in [71,105].

7   Conclusions

As it can be seen from this survey, FDLs are a very active research topic. Particularly during the current decade, much effort has been made to understand the computational properties of this family of logics. As a result, the computational complexity of deciding consistency of FDL ontologies and other related reasoning tasks is mostly known. The most notable exceptions are perhaps the

---

[1] https://www.w3.org/TR/owl2-primer/.

cases with restricted expressivity—either by the limiting the concept constructors as in $\mathcal{FL}_0$ or DL-Lite, or by constraining the available axioms like for acyclic TBoxes—with general t-norm semantics.

Apart from the identified undecidable logics, there are several candidates of languages that can be used to effectively represent and reason about imprecise knowledge. Interestingly, these cases are still expressive enough for the needs of some existing applications, and mostly retain the same computational complexity as their underlying classical formalisms. However, the development of efficient, scalable reasoners is still ongoing. Future work should focus on the development and implementation of specialized optimizations. These reasoners can then be used to further promote the use of FDLs in practical applications.

References

1. Alsinet, T., Barroso, D., Béjar, R., Bou, F., Cerami, M., Esteva, F.: On the implementation of a Fuzzy DL solver over infinite-valued product logic with SMT solvers. In: Liu, W., Subrahmanian, V.S., Wijsen, J. (eds.) SUM 2013. LNCS, vol. 8078, pp. 325–330. Springer, Heidelberg (2013). doi:10.1007/978-3-642-40381-1_25
2. Baader, F., Borgwardt, S., Peñaloza, R.: On the decidability status of fuzzy $\mathcal{ALC}$ with general concept inclusions. J. Philos. Logic **44**(2), 117–146 (2015)
3. Baader, F., Calvanese, D., McGuinness, D.L., Nardi, D., Patel-Schneider, P.F. (eds.): The Description Logic Handbook: Theory, Implementation, and Applications, 2nd edn. Cambridge University Press (2007)
4. Baader, F., Hladik, J., Peñaloza, R.: Automata can show PSpace results for description logics. Inf. Comput. **206**(9–10), 1045–1056 (2008)
5. Baader, F., Peñaloza, R.: Are fuzzy description logics with general concept inclusion axioms decidable? In: Proceedings of the 2011 IEEE International Conference on Fuzzy Systems (FUZZ-IEEE 2011), pp. 1735–1742 (2011)
6. Baader, F., Peñaloza, R.: GCIs make reasoning in fuzzy DL with the product T-norm undecidable. In: Proceedings of the 24th International Workshop on Description Logics (DL 2011), pp. 37–47 (2011)
7. Baader, F., Peñaloza, R.: On the undecidability of fuzzy description logics with GCIs and product T-norm. In: Tinelli, C., Sofronie-Stokkermans, V. (eds.) FroCoS 2011. LNCS, vol. 6989, pp. 55–70. Springer, Heidelberg (2011). doi:10.1007/978-3-642-24364-6_5
8. Bobillo, F.: The role of crisp elements in fuzzy ontologies: the case of fuzzy OWL 2 EL. IEEE Trans. Fuzzy Syst. **24**(5), 1193–1209 (2016)
9. Bobillo, F., Bou, F., Straccia, U.: On the failure of the finite model property in some fuzzy description logics. Fuzzy Sets Syst. **172**(1), 1–12 (2011)
10. Bobillo, F., Cerami, M., Esteva, F., García-Cerdaña, À., Peñaloza, R., Straccia, U.: Fuzzy description logics. In: Handbook of Mathematical Fuzzy Logic, vol. 3, 58, chap. XVI, pp. 1109–1188. College Publications (2016)
11. Bobillo, F., Delgado, M., Gómez-Romero, J.: A crisp representation for fuzzy $\mathcal{SHOIN}$ with fuzzy nominals and general concept inclusions. In: da Costa, P.C.G., d'Amato, C., Fanizzi, N., Laskey, K.B., Laskey, K.J., Lukasiewicz, T., Nickles, M., Pool, M. (eds.) URSW 2005-2007. LNCS, vol. 5327, pp. 174–188. Springer, Heidelberg (2008). doi:10.1007/978-3-540-89765-1_11

12. Bobillo, F., Delgado, M., Gómez-Romero, J.: Optimizing the crisp representation of the fuzzy description logic $\mathcal{SROIQ}$. In: da Costa, P.C.G., d'Amato, C., Fanizzi, N., Laskey, K.B., Laskey, K.J., Lukasiewicz, T., Nickles, M., Pool, M. (eds.) URSW 2005-2007. LNCS, vol. 5327, pp. 189–206. Springer, Heidelberg (2008). doi:10.1007/978-3-540-89765-1_12

13. Bobillo, F., Delgado, M., Gómez-Romero, J.: Crisp representations and reasoning for fuzzy ontologies. Int. J. Uncertainty Fuzziness Knowl. Based Syst. **17**(4), 501–530 (2009)

14. Bobillo, F., Delgado, M., Gómez-Romero, J.: Reasoning in fuzzy OWL 2 with DeLorean. In: Bobillo, F., Costa, P.C.G., d'Amato, C., Fanizzi, N., Laskey, K.B., Laskey, K.J., Lukasiewicz, T., Nickles, M., Pool, M. (eds.) UniDL/URSW 2008-2010. LNCS, vol. 7123, pp. 119–138. Springer, Heidelberg (2013). doi:10.1007/978-3-642-35975-0_7

15. Bobillo, F., Delgado, M., Gómez-Romero, J., Straccia, U.: Fuzzy description logics under Gödel semantics. Int. J. Approximate Reasoning **50**(3), 494–514 (2009)

16. Bobillo, F., Delgado, M., Gómez-Romero, J., Straccia, U.: Joining Gödel and Zadeh fuzzy logics in fuzzy description logics. Int. J. Uncertainty Fuzziness Knowl. Based Syst. **20**(4), 475–508 (2012)

17. Bobillo, F., Straccia, U.: A fuzzy description logic with product t-norm. In: Proceedings of the 2007 IEEE International Conference on Fuzzy Systems (FUZZ-IEEE 2007), pp. 1–6 (2007)

18. Bobillo, F., Straccia, U.: Fuzzy description logics with general t-norms and datatypes. Fuzzy Sets Syst. **160**(23), 3382–3402 (2009)

19. Bobillo, F., Straccia, U.: Fuzzy ontology representation using OWL 2. Int. J. Approximate Reasoning **52**(7), 1073–1094 (2011)

20. Bobillo, F., Straccia, U.: Reasoning with the finitely many-valued Łukasiewicz fuzzy description logic $\mathcal{SROIQ}$. Inf. Sci. **181**, 758–778 (2011)

21. Bobillo, F., Straccia, U.: Aggregation operators for fuzzy ontologies. Appl. Soft Comput. **13**(9), 3816–3830 (2013)

22. Bobillo, F., Straccia, U.: Finite fuzzy description logics and crisp representations. In: Bobillo, F., Costa, P.C.G., d'Amato, C., Fanizzi, N., Laskey, K.B., Laskey, K.J., Lukasiewicz, T., Nickles, M., Pool, M. (eds.) UniDL/URSW 2008-2010. LNCS, vol. 7123, pp. 99–118. Springer, Heidelberg (2013). doi:10.1007/978-3-642-35975-0_6

23. Bobillo, F., Straccia, U.: The fuzzy ontology reasoner fuzzyDL. Knowl.-Based Syst. **95**, 12–34 (2016)

24. Borgwardt, S.: Fuzzy Description Logics with General Concept Inclusions. Ph.D. thesis, Technische Universität Dresden, Germany (2014)

25. Borgwardt, S.: Fuzzy DLs over finite lattices with nominals. In: Proceedings of the 27th International Workshop on Description Logics (DL 2014), pp. 58–70 (2014)

26. Borgwardt, S., Cerami, M., Peñaloza, R.: The complexity of subsumption in fuzzy $\mathcal{EL}$. In: Proceedings of the 24th International Joint Conference on Artificial Intelligence (IJCAI 2015), pp. 2812–2818 (2015)

27. Borgwardt, S., Cerami, M., Peñaloza, R.: The complexity of fuzzy $\mathcal{EL}$ under the Łukasiewicz t-norm. Int. J. Approximate Reasoning (2017, submitted)

28. Borgwardt, S., Cerami, M., Peñaloza, R.: Łukasiewicz fuzzy $\mathcal{EL}$ is undecidable. In: Proceedings of the 30th International Workshop on Description Logics (DL 2017) (to appear, 2017)

29. Borgwardt, S., Distel, F., Peñaloza, R.: Gödel negation makes unwitnessed consistency crisp. In: Proceedings of the 25th International Workshop on Description Logics (DL 2012), pp. 103–113 (2012)

30. Borgwardt, S., Distel, F., Peñaloza, R.: How fuzzy is my fuzzy description logic? In: Gramlich, B., Miller, D., Sattler, U. (eds.) IJCAR 2012. LNCS, vol. 7364, pp. 82–96. Springer, Heidelberg (2012). doi:10.1007/978-3-642-31365-3_9

31. Borgwardt, S., Distel, F., Peñaloza, R.: Decidable Gödel description logics without the finitely-valued model property. In: Proceedings of the 14th International Conference on Principles of Knowledge Representation and Reasoning (KR 2014), pp. 228–237 (2014)

32. Borgwardt, S., Distel, F., Peñaloza, R.: Gödel description logics with general models. In: Proceedings of the 27th International Workshop on Description Logics (DL 2014), pp. 391–403 (2014) (poster paper)

33. Borgwardt, S., Distel, F., Peñaloza, R.: The limits of decidability in fuzzy description logics with general concept inclusions. Artif. Intell. **218**, 23–55 (2015)

34. Borgwardt, S., Leyva Galano, J.A., Peñaloza, R.: The fuzzy description logic G-$\mathcal{FL}_0$ with greatest fixed-point semantics. In: Proceedings of the 14th European Conference on Logics in Artificial Intelligence (JELIA 2014), pp. 62–76 (2014)

35. Borgwardt, S., Mailis, T., Peñaloza, R., Turhan, A.Y.: Answering fuzzy conjunctive queries over finitely valued fuzzy ontologies. J. Data Semant. **5**(2), 55–75 (2016)

36. Borgwardt, S., Peñaloza, R.: Description logics over lattices with multi-valued ontologies. In: Proceedings of the 22nd International Joint Conference on Artificial Intelligence (IJCAI 2011), pp. 768–773 (2011)

37. Borgwardt, S., Peñaloza, R.: Non-Gödel negation makes unwitnessed consistency undecidable. In: Proceedings of the 25th International Workshop on Description Logics (DL 2012), pp. 411–421 (2012) (poster paper)

38. Borgwardt, S., Peñaloza, R.: A tableau algorithm for fuzzy description logics over residuated De Morgan lattices. In: Krötzsch, M., Straccia, U. (eds.) RR 2012. LNCS, vol. 7497, pp. 9–24. Springer, Heidelberg (2012). doi:10.1007/978-3-642-33203-6_3

39. Borgwardt, S., Peñaloza, R.: Undecidability of fuzzy description logics. In: Proceedings of the 13th International Conference on Principles of Knowledge Representation and Reasoning (KR 2012), pp. 232–242 (2012)

40. Borgwardt, S., Peñaloza, R.: About subsumption in fuzzy $\mathcal{EL}$. In: Proceedings of the 26th International Workshop on Description Logics (DL 2013) (2013) (poster paper)

41. Borgwardt, S., Peñaloza, R.: The complexity of lattice-based fuzzy description logics. J. Data Semant. **2**(1), 1–19 (2013)

42. Borgwardt, S., Peñaloza, R.: Positive subsumption in fuzzy $\mathcal{EL}$ with general t-norms. In: Proceedings of the 23rd International Joint Conference on Artificial Intelligence (IJCAI 2013), pp. 789–795 (2013)

43. Borgwardt, S., Peñaloza, R.: Consistency reasoning in lattice-based fuzzy description logics. Int. J. Approximate Reasoning **55**(9), 1917–1938 (2014)

44. Borgwardt, S., Peñaloza, R.: Reasoning in expressive description logics under infinitely valued Gödel semantics. In: Lutz, C., Ranise, S. (eds.) FroCoS 2015. LNCS, vol. 9322, pp. 49–65. Springer, Cham (2015). doi:10.1007/978-3-319-24246-0_4

45. Borgwardt, S., Peñaloza, R.: Reasoning in fuzzy description logics using automata. Fuzzy Sets Syst. **298**, 22–43 (2016)

46. Borgwardt, S., Peñaloza, R.: Algorithms for reasoning in very expressive description logics under infinitely valued Gödel semantics. Int. J. Approximate Reasoning **83**, 60–101 (2017)

47. Cerami, M.: Fuzzy Description Logics from a Mathematical Fuzzy Logic Point of View. Ph.D. thesis, Universitat de Barcelona, Spain (2012)

48. Cerami, M., Esteva, F., Bou, F.: Decidability of a description logic over infinite-valued product logic. In: Proceedings of the 12th International Conference on Principles of Knowledge Representation and Reasoning (KR 2010), pp. 203–213 (2010)
49. Cerami, M., Straccia, U.: On the (un)decidability of fuzzy description logics under Łukasiewicz t-norm. Inf. Sci. **227**, 1–21 (2013)
50. Ciaramella, A., Cimino, M.G.C.A., Marcelloni, F., Straccia, U.: Combining fuzzy logic and semantic web to enable situation-awareness in service recommendation. In: Bringas, P.G., Hameurlain, A., Quirchmayr, G. (eds.) DEXA 2010. LNCS, vol. 6261, pp. 31–45. Springer, Heidelberg (2010). doi:10.1007/978-3-642-15364-8_3
51. Cintula, P., Hájek, P., Noguera, C. (eds.): Handbook of Mathematical Fuzzy Logic, vol. 37–38. College Publications (2011)
52. Dasiopoulou, S., Kompatsiaris, I., Strintzis, M.G.: Applying fuzzy DLs in the extraction of image semantics. In: Spaccapietra, S., Delcambre, L. (eds.) Journal on Data Semantics XIV. LNCS, vol. 5880, pp. 105–132. Springer, Heidelberg (2009). doi:10.1007/978-3-642-10562-3_4
53. Di Noia, T., Mongiello, M., Straccia, U.: Fuzzy description logics for component selection in software design. In: Bianculli, D., Calinescu, R., Rumpe, B. (eds.) SEFM 2015. LNCS, vol. 9509, pp. 228–239. Springer, Heidelberg (2015). doi:10.1007/978-3-662-49224-6_19
54. Díaz-Rodríguez, N., Cadahía, O., Cuéllar, M., Lilius, J., Calvo-Flores, M.: Handling real-world context awareness, uncertainty and vagueness in real-time human activity tracking and recognition with a fuzzy ontology-based hybrid method. Sensors **14**(20), 18131–18171 (2014)
55. Dienes, Z.P.: On an implication function in many-valued systems of logic. J. Symbolic Logic **14**(2), 95–97 (1949)
56. Dubois, D., Prade, H.: Possibilistic logic - an overview. In: Computational Logic, Handbook of the History of Logic, vol. 9, pp. 283–342. Elsevier (2014)
57. Gödel, K.: Zum intuitionistischen aussagenkalkül. Anzeiger der Akademie der Wissenschaften in Wien, vol. 69, pp. 286–295 (1932) (reprinted, Gödel 1986)
58. Goguen, J.A.: L-fuzzy sets. J. Math. Anal. Appl. **18**(1), 145–174 (1967)
59. Haarslev, V., Pai, H.I., Shiri, N.: A formal framework for description logics with uncertainty. Int. J. Approximate Reasoning **50**(9), 1399–1415 (2009)
60. Hájek, P.: Metamathematics of Fuzzy Logic (Trends in Logic). Springer, Netherlands (2001)
61. Hájek, P.: Making fuzzy description logic more general. Fuzzy Sets Syst. **154**(1), 1–15 (2005)
62. Hájek, P.: On witnessed models in fuzzy logic. Math. Logic Q. **53**(1), 66–77 (2007)
63. Hájek, P.: On witnessed models in fuzzy logic II. Math. Logic Q. **53**(6), 610–615 (2007)
64. Hájek, P.: On witnessed models in fuzzy logic III - witnessed Gödel logics. Math. Logic Q. **56**(2), 171–174 (2010)
65. Haniková, Z., Godo, L.: Petr Hájek, obituary. Fuzzy Sets Syst. (in Press, 2017)
66. Hölldobler, S., Nga, N.H., Khang, T.D.: The fuzzy description logic $\mathcal{ALC}_{FLH}$. In: Proceedings of the 9th IASTED International Conference on Artificial Intelligence and Soft Computing (ASC 2005), pp. 99–104 (2005)
67. Kang, D., Xu, B., Lu, J., Li, Y.: Reasoning for a fuzzy description logic with comparison expressions. In: Proceedings of the 19th International Workshop on Description Logics (DL 2006), pp. 111–118 (2006)
68. Kleene, S.C.: Introduction to Metamathematics. Van Nostrand, New York (1952)

69. Klement, E.P., Mesiar, R., Pap, E.: Triangular Norms. Springer, Netherlands (2000)

70. Kułacka, A., Pattinson, D., Schröder, L.: Syntactic labelled tableaux for Łukasiewicz fuzzy $\mathcal{ALC}$. In: Proceedings of the 23rd International Joint Conference on Artificial Intelligence (IJCAI 2013), pp. 762–768 (2013)

71. Lisi, F.A., Straccia, U.: A FOIL-like method for learning under incompleteness and vagueness. In: Zaverucha, G., Santos Costa, V., Paes, A. (eds.) ILP 2013. LNCS, vol. 8812, pp. 123–139. Springer, Heidelberg (2014). doi:10.1007/978-3-662-44923-3_9

72. Lu, J., Kang, D., Zhang, Y., Li, Y., Zhou, B.: A family of fuzzy description logics with comparison expressions. In: Wang, G., Li, T., Grzymala-Busse, J.W., Miao, D., Skowron, A., Yao, Y. (eds.) RSKT 2008. LNCS, vol. 5009, pp. 395–402. Springer, Heidelberg (2008). doi:10.1007/978-3-540-79721-0_55

73. Łukasiewicz, J.: O logice trójwartościowej. Ruch filozoficzny 5, 170–171 (1920)

74. Lukasiewicz, T., Straccia, U.: Managing uncertainty and vagueness in description logics for the semantic web. J. Web Semant. 6(4), 291–308 (2008)

75. Mailis, T., Peñaloza, R., Turhan, A.-Y.: Conjunctive query answering in finitely-valued fuzzy description logics. In: Kontchakov, R., Mugnier, M.-L. (eds.) RR 2014. LNCS, vol. 8741, pp. 124–139. Springer, Cham (2014). doi:10.1007/978-3-319-11113-1_9

76. Mailis, T., Stoilos, G., Simou, N., Stamou, G.B., Kollias, S.: Tractable reasoning with vague knowledge using fuzzy $\mathcal{EL}^{++}$. J. Intell. Inf. Syst. 39(2), 399–440 (2012)

77. Mailis, T., Stoilos, G., Stamou, G.: Expressive reasoning with Horn rules and fuzzy description logics. Knowl. Inf. Syst. 25(1), 105–136 (2010)

78. Mailis, T., Turhan, A.Y.: Employing $DL - Lite_R$-reasoners for fuzzy query answering. In: Proceedings of the 4th Joint International Semantic Technology Conference (JIST 2014), pp. 63–78 (2014)

79. Meghini, C., Sebastiani, F., Straccia, U.: A model of multimedia information retrieval. J. ACM 48(5), 909–970 (2001)

80. Merz, D., Peñaloza, R., Turhan, A.: Reasoning in $\mathcal{ALC}$ with fuzzy concrete domains. In: Proceedings of the 37th German Conference on Artificial Intelligence (KI 2014), pp. 171–182 (2014)

81. Molitor, R., Tresp, C.B.: Extending description logics to vague knowledge in medicine. In: Fuzzy Systems in Medicine, pp. 617–635 (2000)

82. Mostert, P.S., Shields, A.L.: On the structure of semigroups on a compact manifold with boundary. Ann. Math. 65(1), 117–143 (1957)

83. Nilsson, N.J.: Probabilistic logic. Artif. Intell. 28(1), 71–88 (1986)

84. Pan, J.Z., Stamou, G.B., Stoilos, G., Taylor, S., Thomas, E.: Scalable querying services over fuzzy ontologies. In: Proceedings of the 17th International World Wide Web Conference (WWW 2008), pp. 575–584 (2008)

85. Sánchez, D., Tettamanzi, A.G.B.: Fuzzy quantification in fuzzy description logics. In: Fuzzy Logic and the Semantic Web, vol. 1, chap. 8, pp. 135–159 (2006)

86. Stoilos, G., Simou, N., Stamou, G., Kollias, S.: Uncertainty and the semantic web. IEEE Intell. Syst. 21(5), 84–87 (2006)

87. Stoilos, G., Stamou, G.B.: A framework for reasoning with expressive continuous fuzzy description logics. In: Proceedings of the 22nd International Workshop on Description Logics (DL 2009) (2009)

88. Stoilos, G., Stamou, G.B.: Reasoning with fuzzy extensions of OWL and OWL 2. Knowl. Inf. Syst. 40(1), 205–242 (2014)

89. Stoilos, G., Stamou, G.B., Kollias, S.D.: Reasoning with qualified cardinality restrictions in fuzzy description logics. In: Proceedings of the 2008 IEEE International Conference on Fuzzy Systems (FUZZ-IEEE 2008), pp. 637–644 (2008)

90. Stoilos, G., Stamou, G.B., Pan, J.Z.: Classifying fuzzy subsumption in fuzzy-$\mathcal{EL}$+. In: Proceedings of the 21st International Workshop on Description Logics (DL 2008) (2008)

91. Stoilos, G., Stamou, G.B., Pan, J.Z.: Fuzzy extensions of OWL: logical properties and reduction to fuzzy description logics. Int. J. Approximate Reasoning **51**(6), 656–679 (2010)

92. Stoilos, G., Stamou, G.B., Pan, J.Z., Tzouvaras, V., Horrocks, I.: Reasoning with very expressive fuzzy description logics. J. Artif. Intell. Res. **30**, 273–320 (2007)

93. Stoilos, G., Straccia, U., Stamou, G.B., Pan, J.Z.: General concept inclusions in fuzzy description logics. In: Proceedings of the 17th European Confernce on Artificial Intelligence (ECAI 2006), pp. 457–461 (2006)

94. Stoilos, G., Venetis, T., Stamou, G.: A fuzzy extension to the OWL 2 RL ontology language. Comput. J. **58**(11), 2956–2971 (2015)

95. Straccia, U.: A fuzzy description logic. In: Proceedings of the 15th National Confernce on Artificial Intelligence (AAAI 1998), pp. 594–599 (1998)

96. Straccia, U.: Reasoning within fuzzy description logics. J. Artif. Intell. Res. **14**, 137–166 (2001)

97. Straccia, U.: Transforming fuzzy description logics into classical description logics. In: Alferes, J.J., Leite, J. (eds.) JELIA 2004. LNCS, vol. 3229, pp. 385–399. Springer, Heidelberg (2004). doi:10.1007/978-3-540-30227-8_33

98. Straccia, U.: Description logics with fuzzy concrete domains. In: Proceedings of the 21st Conference on Uncertainty in Artificial Intelligence (UAI 2005), pp. 559–567 (2005)

99. Straccia, U.: Answering vague queries in fuzzy DL-Lite. In: Proceedings of the 11th International Conference on Information Processing and Management of Uncertainty in Knowledge-Based Systems (IPMU 2006), pp. 2238–2245 (2006)

100. Straccia, U.: Description logics over lattices. Int. J. Uncertainty Fuzziness Knowl. Based Syst. **14**(1), 1–16 (2006)

101. Straccia, U.: An ontology mediated multimedia information retrieval system. In: Proceedings of the 40th IEEE International Symposium on Multiple-Valued Logic (ISMVL 2010), pp. 319–324 (2010)

102. Straccia, U.: Foundations of Fuzzy Logic and Semantic Web Languages. CRC Studies in Informatics, Chapman & Hall (2013)

103. Straccia, U., Bobillo, F.: Mixed integer programming, general concept inclusions and fuzzy description logics. In: Proceedings of the 5th EUSFLAT Conference (EUSFLAT 2007), pp. 213–220 (2007)

104. Straccia, U., Bobillo, F.: Mixed integer programming, general concept inclusions and fuzzy description logics. Mathware & Soft Comput. **14**(3), 247–259 (2007)

105. Straccia, U., Mucci, M.: pFOIL-DL: Learning (fuzzy) EL concept descriptions from crisp OWL data using a probabilistic ensemble estimation. In: Proceedings of the 30th Annual ACM Symposium on Applied Computing (SAC 2015), pp. 345–352 (2015)

106. Tresp, C.B., Molitor, R.: A description logic for vague knowledge. In: Proceedings of the 13th European Conference on Artificial Intelligence (ECAI 1998), pp. 361–365 (1998)

107. Tsatsou, D., Dasiopoulou, S., Kompatsiaris, I., Mezaris, V.: LiFR: a light-weight fuzzy DL reasoner. In: Presutti, V., Blomqvist, E., Troncy, R., Sack, H., Papadakis, I., Tordai, A. (eds.) ESWC 2014. LNCS, vol. 8798, pp. 263–267. Springer, Cham (2014). doi:10.1007/978-3-319-11955-7_32

108. Venetis, T., Stoilos, G., Stamou, G., Kollias, S.: f-DLPs: Extending description logic programs with fuzzy sets and fuzzy logic. In: Proceedings of the 2007 IEEE International Conference on Fuzzy Systems (FUZZ-IEEE 2007), pp. 1–6 (2007)

109. Vojtáš, P.: A fuzzy EL description logic with crisp roles and fuzzy aggregation for web consulting. In: Proceedings of the 11th International Conference on Information Processing and Management of Uncertainty in Knowledge-Based Systems (IPMU 2006), pp. 1834–1841 (2006)

110. Yen, J.: Generalizing term subsumption languages to fuzzy logic. In: Proceedings of the 12th International Joint Conference on Artificial Intelligence (IJCAI 1991), pp. 472–477 (1991)

111. Zadeh, L.A.: Fuzzy sets. Inf. Control **8**(3), 338–353 (1965)

112. Zadeh, L.A.: A fuzzy-set-theoretic interpretation of linguistic hedges. J. Cybern. **2**(3), 4–34 (1972)

Regular Papers

Using k-Specificity for the Management of Count Restrictions in Flexible Querying

Nicolás Marín[(✉)], Gustavo Rivas-Gervilla, and Daniel Sánchez

Department of Computer Science and A.I., University of Granada,
18071 Granada, Spain
{nicm,daniel}@decsai.ugr.es, g.r.gervilla@gmail.com

Abstract. In the field of Fuzzy Set Theory, special attention has been paid to the problem of determining whether a fuzzy set is a singleton, by means of the well-known measures of specificity. This has been done, for example, to be able to measure the level of uncertainty associated with the fuzzy set or, also, to be able to determine the discriminatory power of the property associated with the fuzzy set in a given context. This concept was extended to that of k-specificity in order to determine the difficulty of choosing k objects in a fuzzy set. In this paper we study bounding properties for k-specificity measures, and we introduce their use in flexible querying, analyzing their computation, and comparing the information provided by these measures with the tightly related fuzzy cardinality measures.

1 Introduction

One of the recurrent problems in the field of Fuzzy Sets Theory is the calculation of the cardinality of a fuzzy set [1–7]. The computation of this value can be useful in many problems, for example, in fields like Flexible Querying or Fuzzy Data Mining.

The problem is very complex, among other reasons, because the meaning of the cardinality may respond to very different semantics [1]. For example, it is not the same to determine the mass of objects that fulfill to some extent the property represented by the set, than to determine which natural numbers could be used to describe the number of objects that belongs to the fuzzy set.

In this work we focus on a specific information retrieval problem within Flexible Querying: retrieving those sets within a collection that satisfy the condition that the number of objects in the set that are compatible with a given fuzzy constraint is exactly n, being n a natural number or 0.

To address the problem, we employ a special type of measures based on the notion of specificity. The concept of specificity was introduced by Yager [8] and aims to determine to what extent a fuzzy set is a singleton. Measures of specificity have been widely analyzed in the field of Soft Computing [9–13].

As we will see, the concept of specificity can be extended to analyze to what extent a fuzzy set can be considered to be a crisp set of k elements. This can be achieved by means of measures of k-specificity [14]. In this paper we consider the

© Springer International Publishing AG 2017
S. Moral et al. (Eds.): SUM 2017, LNAI 10564, pp. 49–63, 2017.
DOI: 10.1007/978-3-319-67582-4_4

use of a slight variation of these measures in flexible querying. For such purpose, we introduce and study two types of k-specificity measures, namely, *predicate-like* and *index-like*, and their suitability for the proposed problem in terms of their convergence to 0 as the membership function of the target set evolves.

The work is organized as follows: Sect. 2 presents a description of the problem, defining a formal framework for the rest of the paper. Section 3 recalls specificity and k-specificity measures, as well as existing bounding conditions and families for specificity measures. New results regarding bounding conditions for k-specificity are introduced in Sect. 4, together with some new proposals of measures. The relationship between k-specificity and fuzzy cardinality in the field of fuzzy set theory is studied in Sect. 5. Section 6 discusses some questions about the calculation of this type of measures and Sect. 7 presents some illustrative examples in the context of scene retrieval. Finally, Sect. 8 concludes the work.

2   Formalization of the Problem

When working in an environment with uncertainty, the subset of objects in a given set that meet a fuzzy restriction is a fuzzy set. Each object of the original set belongs to this fuzzy subset of objects with a degree that indicates the matching of the object with the given restriction.

Once the fuzzy subset of objects satisfying the restriction is determined, we might need to compute the number of objects that comply with the constraint.

To cite an example where this problem may arise, consider the case of imposing conditions on a HAVING clause when grouping query results on fuzzy databases, where the tuples belonging to each group accomplish the fuzzy constraint raised in the WHERE clause to a certain degree. The problem of the aggregate functions under fuzziness has been widely studied in the field of flexible querying, although, as indicated in [15,16], the analysis of the functions derived from the conventional COUNT() is of special complexity, among other reasons, because the semantics of the count may vary depending on the purpose of the query.

Another case is found in information retrieval systems, when the user wants to retrieve collections based on the number of objects in the collection that satisfy a given condition. For example, think of image retrieval systems in which the user wants to get those images in which a number of objects appear that fits a given fuzzy pattern.

In this work we will focus on those cases where the count is aimed at determining whether the number of objects compatible with the imposed restriction is a certain natural number n. That is, in a collection of sets of objects, we want to select those sets in which there are exactly n objects that satisfy a certain fuzzy restriction. Although several non-complex approaches can be made to this problem (for example by using a simple threshold), in this work this problem is approached from a broader perspective, defining a new class of operators derived from the notion of specificity introduced by Yager [8] and connecting with the tightly related concept of cardinality [1]. Formally, the problem can be defined as follows.

Let $\mathscr{O}$ be the universe of objects in a given context, and let DB be a collection of sets $\{O_1, ..., O_t\} \subseteq \{0,1\}^{\mathscr{O}}$ (i.e. $\forall i \in \{1..t\}, O_i \subseteq \mathscr{O}$).

Let $\mathscr{P}$ be a set of properties that can be predicated of objects in $\mathscr{O}$ and P be a restriction built using properties of $\mathscr{P}$.

The problem under study in this paper is to determine those sets $O \in DB$ where the number of objects compatible with P is a given $n \in \mathbb{N} \cup \{0\}$.

If, $\forall o \in \mathscr{O}$, $\mu_P(o)$ stands for the accomplishment degree of object o with restriction P, then, $\forall O \in DB$, O_P stands for the fuzzy subsets of objects compatible with P and $O_P(o) = \mu_P(o)$.

Next section is devoted to analyze fuzzy measures of the form $Sp_k : [0,1]^{\mathscr{O}} \to [0,1]$ that permits to assess the degree to which only k objects compliant with P can be found in the set of objects O.

3   Specificity

In order to solve the problem described in the previous section, we shall use measures of k-specificity [14], for any nonnegative integer k.

3.1   Specificity Measures

The idea of k-specificity measure is inspired by the idea of measures of specificity [8–13]. Given a finite set $\mathscr{O}$, specificity measures are functions of the form $[0,1]^{\mathscr{O}} \to [0,1]$ assessing the degree to which fuzzy sets in $[0,1]^{\mathscr{O}}$ are close to be a singleton. For every $A \in [0,1]^{\mathscr{O}}$, let us consider that memberships in A are ranked in nonincreasing order as $a_1 \geq a_2 \geq \cdots \geq a_m$. Let us also consider $a_0 = 1$ and $a_i = 0 \ \forall i > m$.

Measures of specificity are required to satisfy the following properties [17]:

- S1: $Sp(A) = 1$ iff $A = \{o\} \subseteq \mathscr{O}$.
- S2: $Sp(\emptyset) = 0$.
- S3: Let $A, A' \in [0,1]^{\mathscr{O}}$ such that $a_1 \geq a_1'$ and $a_i \leq a_i' \ \forall 2 \leq i \leq m$, with a_i and a_i' being the memberships of A and A', respectively, arranged in nonincreasing order. Then $Sp(A) \geq Sp(A')$.

Specificity measures can be seen as measuring the degree to which the cardinality of the set A is 1, since being a singleton and having cardinality 1 are equivalent.

3.2   Families of Bounded Specificity Measures

In [18], specificity measures were classified into different kinds according to their behaviour. The following definitions were introduced for that purpose:

Definition 1 ([18]) *A specificity measure Sp is ∞-bounded iff $Sp(A) \leq a_1 \ \forall A \in [0,1]^{\mathscr{O}}$.*

Definition 2 ([18]) *Let $z \in \mathbb{N}$, $z > 1$. A specificity measure Sp is z-bounded iff $Sp(A) \leq a_1 - a_z \; \forall A \in [0,1]^{\mathscr{O}}$.*

As shown in [18], if Sp is z-bounded then

- Sp is z'-bounded $\forall z' \geq z > 1$, and
- Sp is ∞-bounded.

These properties lead to the following definition:

Definition 3 ([18]) *A specificity measure Sp is said to be* bounded *iff Sp is z-bounded or ∞-bounded. For every bounded measure Sp, its* bound *is defined as:*

$$bound(Sp) = \begin{cases} min\{z \mid Sp \text{ is } z\text{-bounded }\} & \exists z \; Sp \text{ is } z\text{-bounded} \\ \infty & otherwise \end{cases} \tag{1}$$

On this basis, specificity measures can be classified into the following three groups [18]:

- *Predicate-like* specificity measures are the 2-bounded measures. These measures are the most restrictive ones, since when $a_1 = a_2$ (in the crisp case, when there are at least two elements with degree 1 in the set), they yield a value 0.
- *Index-like* specificity measures are those bounded measures with $bound(Sp) = \infty$ satisfying $Sp(A) > 0 \; \forall A \neq \emptyset$. They never yield a value 0 for nonemtpy sets.
- *Intermediate* specificity measures are the rest. These can be seen as a kind of "index-like measures with a saturation point", being non-decreasing for an amount of elements less or equal to a certain value.

As explained in [18], predicate-like specificity measures are useful for representing the fulfilment of the predicate "to be a singleton", whilst the rest of measures are useful when ranking fuzzy sets in terms of their closeness to be a singleton. Hence, the most appropriate measures for computing specificity in the context introduced in Sect. 2 are predicate-like specificity measures, because of the semantics of the problem.

In [19], predicate-like specificity measures have been employed for solving the problem introduced in Sect. 2 for the particular case of cardinality being exactly 1. In the next section we describe the extension to the case of k-specificity.

3.3 k-Specificity Measures

Let $A \in [0,1]^{\mathscr{O}}$ with $|\mathscr{O}| = m$. Let us again consider that memberships in A are ranked in nonincreasing order as $a_1 \geq a_2 \geq \cdots \geq a_m$, and also $a_0 = 1$ and $a_i = 0$ $\forall i > m$. A measure of k-specificity for $k \in \mathbb{N} \cup \{0\}$ is a function $Sp_k : [0,1]^{\mathscr{O}} \rightarrow [0,1]$ satisfying the following properties for every $A, A' \in [0,1]^{\mathscr{O}}$ [14]:

- NC1: $Sp_k(A) = 1$ iff A is a crisp set and $|A| = k$.

– NC2: $Sp_k(\emptyset) = 0$ iff $k > 0$
– NC3: With respect to a_i, $Sp_k(A)$ is strictly increasing when $1 \leq i \leq k$, and strictly decreasing when $k + 1 \leq i \leq m$.

In this paper we shall use slightly different definitions of the abovementioned properties in order to better accomplish to the properties required for specificity measures, as the latter are a particular case of k-specificity:

– NC1: $Sp_k(A) = 1$ iff $a_k = 1$ and $a_{k+1} = 0$ (equivalent to the property NC1 in [14]).
– NC2: If $a_k = 0$ then $Sp_k(A) = 0$ (note that as a particular case, $a_1 = 0$ implies $A = \emptyset$ and $Sp_1(\emptyset) = 0$).
– NC3: Let A, A' such that $a_i \geq a'_i \ \forall i \leq k$ and $a_i \leq a'_i \ \forall i > k$. Then $Sp_k(A) \geq Sp_k(A')$.

It is easy to show that when $k = 1$, the previous properties coincide with the three properties of specificity measures introduced in Sect. 3.

As the two sets of properties are intended to capture the same semantics that motivates the definitions in [14], in this paper, the measures safisfying the second set of properties will be also called k-specificity measures.

4 New Results on k-Specificity

4.1 Bounding of k-Specificity Measures

The definition of the different kinds of bounding conditions introduced in previous sections for specificity measures can be extended to the case of measures of k-specificity as follows:

Definition 4. *Let* $k \in \mathbb{N} \cup \{0\}$*. A measure of k-specificity Sp_k is ∞-bounded iff* $Sp_k(A) \leq a_k \ \forall A \in [0,1]^{\mathscr{O}}$*.*

Note that since $a_0 = 1$ by definition, all measures of 0-specificity are ∞-bounded. This is not necessarily the case for $k > 0$, as it is shown in [18] for $k = 1$.

Definition 5. *Let* $k, z \in \mathbb{N} \cup \{0\}$*,* $0 \leq k < z$*. A measure of k-specificity Sp_k is z-bounded iff $Sp_k(A) \leq a_k - a_z \ \forall A \in [0,1]^{\mathscr{O}}$*.*

The notion of predicate-like k-specificity measure generalizes that of specificity measure as follows:

Definition 6. *A k-specificity measure is said to be predicate-like iff it is $(k+1)$-bounded.*

Note that when $k = 1$, the different bounding conditions and the notion of predicate-like measure coincide with those of specificity measures. As in the case of specificity measures, we shall consider that predicate-like k-specificity measures are the appropriate measures for solving the problem introduced in Sect. 2.

4.2   Some New Measures of k-Specificity

Let us introduce some new k-specificity measures.

Definition 7. *The fractional measure of k-specificity, Sp_k^f, is defined as*

$$Sp_k^f(A) = \begin{cases} \dfrac{\prod\limits_{i=0}^{k} a_i^2}{\prod\limits_{i=0}^{k} a_i + \sum\limits_{i=k+1}^{m} a_i} & a_k > 0 \\[2em] 0 & otherwise \end{cases} \tag{2}$$

It is easy to show that Sp_k^f satisfies all the properties of k-specificity measures:

Proof. With regard to the three properties of k-specificity measures:

(NC1) $Sp_k^f(A) = \frac{a^2}{a+b}$ where $a = \prod\limits_{i=0}^{k} a_i$ and $b = \sum\limits_{i=k+1}^{m} a_i$. If $Sp_k^f(A) = 1$ then $a > 0$ and it is $\frac{a^2}{a+b} = \frac{a}{1+\frac{b}{a}}$. Since $a \leq 1$ and $\frac{b}{a} \geq 0$ it is:

$$Sp_k^f = 1 \Leftrightarrow \begin{cases} a = 1 \Leftrightarrow a_i = 1 \ \forall i \leq k \\ \wedge \\ 1 + \frac{b}{a} = 1 \Leftrightarrow b = 0 \Leftrightarrow a_i = 0 \ \forall i > k \end{cases}$$

(NC2) Immediate by definition.

(NC3) Under the conditions of this property, a increases and b decreases, and hence $Sp_k^f(A) = \frac{a}{1+\frac{b}{a}}$ increases.

It is easy to show that when $k = 1$, $Sp_k^f(A) = Sp^f(A)$, with $Sp^f(A)$ being the fractional specificity measure introduced by Yager in [17] as

$$Sp^f(A) = \frac{a_1^2}{\sum_{i=1}^{m} a_i} \tag{3}$$

for $A \neq \emptyset$, and 0 otherwise.

In [18] it is shown that this measure is not 2-bounded in the case $k = 1$. This can be easily extended to any other value k: consider a crisp set A with $|A| = k + 1$, that is, $a_1 = a_2 = \cdots = a_{k+1} = 1$. Any predicate-like measure satisfies $Sp_k(A) \leq a_k - a_{k+1} = 0$, whilst $Sp_k^f(A) = 1/2 > 0$. Hence, Sp_k^f is not a predicate-like measure for any k. This result can be extended to any other z-bounded property for any $z > k$, that is, Sp_k^f is not z-bounded for any $z > k$. It is also easy to show that Sp_k^f is a ∞-bounded measure.

The example of a crisp set with $|A| = k + 1$ is useful for illustrating why predicate-like k-specificity measures are required for solving our problem: for any measure it is expected that $Sp_k(A) = 0$ since there are $k + 1$ crisp objects in A, and hence the degree to which we can say that the cardinality of A is exactly k is expected to be 0. This is only guaranteed when $Sp_k(A) \leq a_k - a_{k+1} = 0$.

Let us now introduce two predicate-like k-specificity measures:

Definition 8. *The expression*

$$Sp_{\Lambda,k}(A) = a_k(a_k - a_{k+1}) \tag{4}$$

defines a measure of k-specificity.

It is trivial to show that $Sp_{\Lambda,k}$ satisfies all the properties of k-specificity measures. It is also immediate that $Sp_{\Lambda,k}(A) \leq a_k - a_{k+1}$, and hence $C_{\Lambda,k}$ is a predicate-like k-specificity measure. As a particular case, when $k = 1$, $Sp_{\Lambda,k}(A) = Sp_{\Lambda}(A)$, the specificity measure Sp_{Λ} introduced in [20] as

$$Sp_{\Lambda}(A) = a_1(a_1 - a_2) \tag{5}$$

Definition 9. *The expression*

$$Sp_{L,k}(A) = a_k - a_{k+1} \tag{6}$$

defines a measure of k-specificity.

Again, it is trivial to show that $Sp_{L,k}$ satisfies all the properties of k-specificity measures. It is also trivially $Sp_{\Lambda,k}(A) \leq a_k - a_{k+1}$, and hence $Sp_{L,k}$ is a predicate-like k-specificity measure. We shall discuss further on $Sp_{L,k}$ in the next section.

5   Relation to Fuzzy Cardinalities

It is obvious that the problem introduced in Sect. 2 has to do with the issue of cardinality of fuzzy sets. One way to solve the problem, as a direct extension of the crisp case, is to compute the cardinality of each fuzzy set O_i and to choose those sets for which cardinality is equal to k.

However, both the definition and calculation of the cardinality of fuzzy sets, as well as determining the degree to which the cardinality is k, are not trivial issues. There are many proposals in the literature for computing the cardinality of a fuzzy set, see [1–7] and references therein.

The existing proposals do not provide a *cardinality* in the sense of cardinality of crisp sets, but *information* about the cardinality of the fuzzy set. One of the most important approaches is that of fuzzy cardinality as a fuzzy subset of the nonnegative integers. That is, the information about the cardinality of A is provided as a fuzzy subset $Card(A)$ where $Card(A)(k)$ is the degree to which k can be considered as the cardinality of A. The semantics of such degree varies depending on the way $Card$ is defined.

The relation to the issue of k-specificity measures is obvious: both predicate-like k-specificity measures and fuzzy cardinalities provide a degree to which the cardinality is k, for every nonnegative integer k. However, the contribution of k-specificity measures in this context is that they are required to satisfy certain properties, whilst the only property required for fuzzy cardinalities in the literature is that they provide a crisp singleton as result when the set A is crisp. More specifically, when A is crisp with $|A| = k$, it is expected that $Card(A)(k) = 1$ and $Card(A)(k') = 0 \ \forall k' \neq k$.

The required properties for k-specificity measures, introduced in previous sections, are reasonable and derive directly from the properties required from specificity measures. Hence, they open the door to the definition and study of both new and existing fuzzy cardinality measures satisfying properties similar to those of specificity.

Let us discuss some existing fuzzy cardinalities from the point of view of the properties required for k-specificity measures. Probably the first such cardinality proposal is Zadeh's first fuzzy cardinality, introduced in [3], that can be formulated as:

$$Z(A)(k) = \begin{cases} 0 & a_k = a_{k+1} \\ a_k & otherwise \end{cases} \tag{7}$$

The corresponding measure $Sp_k(A) = Z(A)(k)$ does not satisfy the properties required for k-specificity measures. For instance, for the fuzzy set $A = 1/o_1 + 0.5/o_2 + 0.5/o_3$, being $a_1 = 1$ and $a_2 = a_3 = 0.5$, it is $Sp_1(A) = Z(A)(1) = 1$, whilst $a_2 > 0$, and hence property NC1 is not satisfied.

As another example we can mention the fuzzy cardinality ED introduced in [1] as:

$$ED(A)(k) = a_k - a_{k+1} \tag{8}$$

It is obvious that $ED(A)(k) = Sp_{L,k}(A)$, and hence this fuzzy cardinality is consistent with the properties required for k-specificity measures.

These last two fuzzy cardinalities share the feature of being *nonconvex*. For instance, let again $A = 1/o_1 + 0.5/o_2 + 0.5/o_3$. Then $Z(A) = 0/0 + 1/1 + 0/2 + 0.5/3$, and $ED(A) = 0/0 + 0.5/1 + 0/2 + 0.5/3$. The fact that convexity is not mandatory for fuzzy cardinalities is discussed in [1] on two basis: first, the possible nonnegative integers in the support of a fuzzy cardinality are expected to be the cardinalities of the alpha-cuts of the set A. Second, whilst a concept like "around 2" has to be represented necessarily by a convex fuzzy set, the information about the cardinality of a fuzzy set is not mandatorily a concept of this kind. In the example above, if we are very strict we have only one element in A (o_1), and hence under that restriction, the cardinality of A is 1. If we relax our criterion and we allow that elements with degree 0.5 to be in A, then we have also o_2 and o_3 in A, and the cardinality is 3. But there is no way in which the cardinality can be 2, since the cardinality is at least 1, and if we admit o_2 to be in A, then we have also to admit o_3, and vice versa. Hence, in cases like this, convexity is counterintuitive.

Convexity turns out to be intuitive for fuzzy cardinalities that are not intended to measure the degree to which the cardinality is *exactly* k for every nonnegative integer k, but *at least* k. Such fuzzy cardinalities have no relation to k-specificity measures, since they intend to measure different things. Other fuzzy cardinalities offer a mixed behaviour. For example, the measure introduced in [4], that can be formulated as:

$$DP(A)(k) = \begin{cases} 0 & a_{k+1} = 1 \\ a_k & otherwise \end{cases} \tag{9}$$

measures the degree to which the cardinality is exactly k for objects in the core of A, and the degree to which the cardinality is at least k for the rest of elements. The same counterexample employed for the fuzzy cardinality Z serves to show that DP does not satisfy property NC1.

Let us conclude our discussion about the relation between k-specificity measures and cardinality of fuzzy sets with some brief comments about scalar and gradual cardinality measures.

Scalar measures yield a single number as cardinality of a fuzzy set. The most employed measure of cardinality of a fuzzy set is the sigma-count [2], that can be defined as

$$SC(A) = \sum_{i=1}^{m} a_i \tag{10}$$

The sigma count is in general a real number. However, its semantics is that of a summary of the available information about the cardinality of the fuzzy set, in the form of an expected value, and cannot be interpreted as a crisp cardinality of a fuzzy set. Indeed, taking $a_k - a_{k+1}$ as the probability that the cardinality of an alpha-cut taken at random is the same as that of the alpha-cut of level $\alpha = a_k$ (which happens for values of α in the real interval $(a_{k+1}, a_k]$), the sigma-count can be seen as the expected value of the cardinality of an alpha-cut of A taken at random uniformly. It can be seen as well as a centre of mass of the fuzzy cardinality ED of Eq. (8) [1]. The sigma-count can also be interpreted as a measure of the *energy*, in the sense of *amount of membership*, of A [2].

Many problems are known since long ago when using sigma-count for measuring cardinality, such as the addition of many small memberships adding to a significant value, and the fact that the same value is obtained for very different sets [7]. In our example above, for $A = 1/o_1 + 0.5/o_2 + 0.5/o_3$ we get $SC(A) = 2$, and hence one may think that $Sp_2(A) = 1$, when in fact any predicate-like k-specificity measure yields a value 0 since it is $Sp_2(A) \leq a_2 - a_3 = 0$.

A final comment concerns cardinalities measured by means of *gradual numbers* [21]. Gradual numbers are an assignment of numbers to levels. Gradual cardinality assigns to each level $\alpha \in (0, 1]$ the cardinality of the corresponding alpha-cut of that level. In order to obtain a measure of k-specificity, one should measure to which degree each possible cardinality is representative, or holds, in the gradual cardinality. In [22], the proposal for providing fuzzy summaries of gradual numbers (and gradual sets in general) is to assign to every number k the addition of those differences $a_i - a_{i+1}$ such that the cardinality of the alpha-cut of level a_i is k. This procedure yields exactly $Sp_{L,k}$ as k-specificity measure.

6   Efficient Computation of the Count

One of the advantages of the new class of measures considered in the previous section is that it allows to propose strategies for the efficient computation of Sp_k in big datasets (for a given restriction P).

Let us consider the case of the measure $Sp_{L,k}$ (the discussion below is also valid for the case of $Sp_{\Lambda,k}$). The computation of $Sp_{L,k}$ according to Eq. (6) involves the execution of two processes for each set O of the collection DB:

– Obtaining the ranking of a_i values or, at least, getting the $k+1$ first ones.
– Computing Eq. (6).

Of the two previous processes, it is evident that the one that supposes a higher computational cost is the first one. In the worst case, it implies to apply a ranking technique on the degrees of O_P, admitting, depending on the particular count measure, more efficient solutions as k decreases. The second process does not imply a significant computational cost since it involves the calculation of a simple arithmetic expression with the a_i values.

In any case, once the computation of $Sp_{L,k}$ has been solved for a certain value of k, the computation of any other $Sp_{L,k'}$, with $k' < k$, can reuse the results of the first process. From a general point of view, having pre-calculated the ordered set of degrees for each O_P, allows the direct computation of $Sp_{L,k}$ whatever the chosen k.

This can be used to develop both special index structures and storage strategies for recent queries, allowing the efficient computations of counts in relation to a given constraint, especially in problems that involve large amounts of data.

In addition, a simple incremental updating of these structures can be considered when new objects are incorporated in the considered sets. The update of the structures after the insertion of a new object o in a given set O, only implies adding $O_P(o)$ in the right place of the ordered list $a_1, ..., a_m$.

7   An Illustrative Example

Refer4Learning [19] is an application that helps to teach primary visual concepts such as the size, color, and position of simple geometric objects to children in the early stages of their education.

To do this, it presents scenes composed of various objects to children and proposes exercises of recognition in the image of those objects that match a given textual description.

With this type of applications, the teacher can work on linguistic expressions built on the properties of objects paying attention to both the semantics of the visual concepts and the notion of cardinality. One of the exercises that can be inserted in a tool like this is to locate the n objects of an image that fit with a certain description: for example, locate the five large green triangles in the image. As the visual properties that the system handles are inherently fuzzy, the compliance of the objects with a given description is also a matter of degree. Therefore, this kind of exercises permits children to get acquainted with the usual vagueness of natural language.

Refer4Learning uses a repository of training images, each of them composed by a set of objects. Given a certain expression P, the retrieval of images from

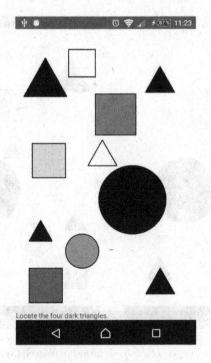

Fig. 1. Example scene with $Sp_{L,4} = 1$

the repository can be made by means of the computation of an appropriate k-specificity measure Sp_k. We shall use $Sp_{L,k}$ in our examples.

For instance, Figs. 1 and 2 show some example scenes with their corresponding value of $Sp_{L,4}$ when the considered expression is: Locate the four dark triangles.

As can be observed, the first scene has $Sp_{L,4} = 1$ (see Fig. 1). Four objects clearly match with the proposed expression and the children will have no problem in locating them. The value of $Sp_{L,4}$ diminishes when either the compatibility of the four target objects decreases or additional objects match the proposed description: Figs. 2(a) and 2(b) show example scenes of this, in which triangles being dark to degree 0.5 are considered.

The lower the value of $Sp_{L,4}$, the higher the difficulty for the children in the task of locating the objects. If the compatibility of the target objects continues decreasing (respectively, the compatibility of additional objects increases), $Sp_{L,4}$ will finally reach the 0 value (Fig. 2(c) and (d) shows example scenes with $Sp_{L,4} = 0$).

To conclude our example, Table 1 shows some of the k-specificity measures introduced in this paper, applied to the images in Figs. 1 and 2 for the fuzzy restriction *dark triangle*. Note that for the case $k = 4$ the different measures provide rather similar results. However, the fact that Sp_k^f is not suitable for solving our problem can be appreciated in the case $k = 2$, for which the predicate-

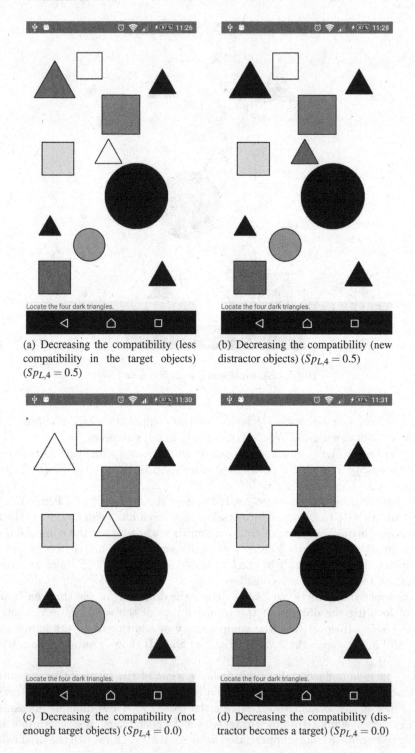

(a) Decreasing the compatibility (less compatibility in the target objects) $(Sp_{L,4} = 0.5)$

(b) Decreasing the compatibility (new distractor objects) $(Sp_{L,4} = 0.5)$

(c) Decreasing the compatibility (not enough target objects) $(Sp_{L,4} = 0.0)$

(d) Decreasing the compatibility (distractor becomes a target) $(Sp_{L,4} = 0.0)$

Fig. 2. Examples scenes with lower values of $Sp_{L,4}$

Table 1. Different k-specificity measures applied to the images in Figs. 1 and 2 for the fuzzy restriction *dark triangle*.

	Sp_4^f	$Sp_{\Lambda,4}$	$Sp_{L,4}$	Sp_2^f	$Sp_{\Lambda,2}$	$Sp_{L,2}$
Figure 1	1	1	1	1/3	0	0
Figure 2a	0.5	0.25	0.5	2/5	0	0
Figure 2b	2/3	0.5	0.5	2/7	0	0
Figure 2c	0	0	0	1/2	0	0
Figure 2d	0	0	0	1/4	0	0

like measures $Sp_{\Lambda,2}$ and $Sp_{L,2}$ provide 0 as result, as expected since there are at least three triangles being dark to degree 1 in all the images. On the contrary, Sp_2^f provides degrees greater than 0 for all the images.

8 Conclusions

In this paper we have employed the concept of k-specificity to determine the extent to which a fuzzy set of objects can be considered to be a crisp set of k objects. For this purpose, we have formally introduced two types of measures, *predicate-like* and *index-like*, depending on the way that they have to converge to the value 0 as the membership function of the fuzzy set evolves. We have also introduced new k-specificity measures.

The context of application that has been considered in this paper is the selection of those sets within a database that fulfill the condition that, in the set, there are k objects compatible with a given fuzzy restriction. As we have seen, the proposed k-specificity measures have the advantage that they are suitable for defining efficient computational strategies to solve the selection process with different values of k, or under the incremental addition of new objects to the sets, which is highly appreciated in the case of large datasets. According to their properties, *predicate-like* k-specificity measures seem to be the most appropriate for the resolution of this type of retrieval problems.

Finally, we have also seen that the concept of k-specificity can be related to the concept of fuzzy cardinality introduced in [1] for a particular case of measure of k-specificity.

Acknowledgments. This work has been partially supported by the Spanish Ministry of Economy and Competitiveness and the European Regional Development Fund - ERDF (Fondo Europeo de Desarrollo Regional - FEDER) under project TIN2014-58227-P *Descripción lingüística de información visual mediante técnicas de minería de datos y computación flexible.*

References

1. Delgado, M., Martín-Bautista, M., Sánchez, D., Vila, M.: A probabilistic definition of a nonconvex fuzzy cardinality. Fuzzy Sets Syst. **126**(2), 41–54 (2002)
2. Luca, A.D., Termini, S.: A definition of a nonprobabilistic entropy in the setting of fuzzy sets theory. Inf. Control **20**, 301–312 (1972)
3. Zadeh, L.A.: A theory of approximate reasoning. Mach. Intell. **9**, 149–194 (1979)
4. Dubois, D., Prade, H.: Fuzzy cardinality and the modeling of imprecise quantification. Fuzzy Sets Syst. **16**, 199–230 (1985)
5. Wygralak, M.: Vaguely Defined Objects Representations. Fuzzy Sets and Nonclassical Cardinality Theory. Kluwer Academic Press, Dordrecht (1996)
6. Wygralak, M.: Cardinalities of Fuzzy Sets. Studies in Fuzziness and Soft Computing, vol. 118. Springer, Heidelberg (2003). doi:10.1007/978-3-540-36382-8
7. Chamorro-Martínez, J., Sánchez, D., Soto-Hidalgo, J., Martínez-Jiménez, P.: A discussion on fuzzy cardinality and quantification. Some applications in image processing. Fuzzy Sets Syst. **257**, 85–101 (2014)
8. Yager, R.: Measuring tranquility and anxiety in decision-making: an application of fuzzy sets. Internat. J. Gen. Syst. **8**, 139–146 (1982)
9. Dubois, D., Prade, H.: Properties of measures of information in evidence and possibility theories. Fuzzy Sets Syst. **24**, 161–182 (1987)
10. Yager, R.R.: Ordinal measures of specificity. Internat. J. Gen. Syst. **17**, 57–72 (1990)
11. Yager, R.R.: On the specificity of a possibility distribution. Fuzzy Sets Syst. **50**, 279–292 (1992)
12. Garmendia, L., Yager, R., Trillas, E., Salvador, A.: On t-norms based specificity measures. Fuzzy Sets Syst. **133**(2), 237–248 (2003)
13. Garmendia, L., Yager, R., Trillas, E., Salvador, A.: Measures of specificity of fuzzy sets under t-indistinguishabilities. IEEE Trans. Fuzzy Syst. **14**(4), 568–572 (2006)
14. González Sánchez, J.L., González del Campo, R., Garmendia, L.: Some new measures of k-specificity. In: Luaces, O., Gámez, J.A., Barrenechea, E., Troncoso, A., Galar, M., Quintián, H., Corchado, E. (eds.) CAEPIA 2016. LNCS, vol. 9868, pp. 489–497. Springer, Cham (2016). doi:10.1007/978-3-319-44636-3_46
15. Marín, N., Sánchez, D., Serrano, J.M., Vila, M.A.: Problems of fuzzy queries involving aggregation functions: the "select count" case. In: Proceedings Joint 9th IFSA World Congress and 20th NAFIPS International Conference, vol. 4, p. 2132–2137 (2001)
16. Pivert, O., Prade, H.: Dealing with aggregate queries in an uncertain database model based on possibilistic certainty. In: Laurent, A., Strauss, O., Bouchon-Meunier, B., Yager, R.R. (eds.) IPMU 2014. CCIS, vol. 444, pp. 150–159. Springer, Cham (2014). doi:10.1007/978-3-319-08852-5_16
17. Yager, R.R.: Expansible measures of specificity. Internat. J. Gen. Syst. **41**(3), 247–263 (2012)
18. Marín, N., Rivas-Gervilla, G., Sánchez, D., Yager, R.: On families of bounded specificity measures. In: IEEE International Conference on Fuzzy Systems, FUZZ-IEEE Naples, Italy (2017)
19. Marín, N., Rivas-Gervilla, G., Sánchez, D.: Scene selection for teaching basic visual concepts in the Refer4Learning app. In: IEEE International Conference on Fuzzy Systems, FUZZ-IEEE Naples, Italy (2017)
20. Marín, N., Rivas-Gervilla, G., Sánchez, D., Yager, R.: Specificity measures and referential success. IEEE Trans. Fuzzy Syst. (in press, 2017). doi:10.1109/TFUZZ.2017.2694803

21. Dubois, D., Prade, H.: Gradual elements in a fuzzy set. Soft. Comput. **12**, 165–175 (2008)
22. Sánchez, D., Delgado, M., Vila, M., Chamorro-Martínez, J.: On a non-nested level-based representation of fuzziness. Fuzzy Sets Syst. **192**(1), 159–175 (2012)

Comparing Machine Learning and Information Retrieval-Based Approaches for Filtering Documents in a Parliamentary Setting

Luis M. de Campos$^{(\boxtimes)}$ (iD), Juan M. Fernández-Luna (iD), Juan F. Huete (iD),
and Luis Redondo-Expósito

Departamento de Ciencias de la Computación e Inteligencia Artificial,
ETSI Informática y de Telecomunicación, CITIC-UGR, Universidad de Granada,
18071 Granada, Spain
{lci,jmfluna,jhg,luisre}@decsai.ugr.es

Abstract. We consider the problem of building a content-based rec-
ommender/filtering system in a parliamentary context which, given a
new document to be recommended, can decide those Members of Parliament who should receive it. We propose and compare two different
approaches to tackle this task, namely a machine learning-based method
using automatic document classification and an information retrieval-
based approach that matches documents and legislators' representations.
The information necessary to build the system is automatically extracted
from the transcriptions of the speeches of the members of parliament
within the parliament debates. Our proposals are experimentally tested
for the case of the regional Andalusian Parliament at Spain.

Keywords: Content-based recommender systems · Information filtering · Information retrieval · Machine learning · Parliamentary documents

1 Introduction

Politicians in general and Members of Parliament (MP) in particular, need to be
concerned about the reality of the territory, region or country where they develop
their activity. This is particularly true in relation to these matters more related
with their specific political interests. For example, an MP who is specialized
in educational issues or the health minister should be specially interested in
receiving information concerning their respective fields of interest. However, at
present, the amount of information that is generated and is available through
the Information and Communication Technologies (ICT) is enormous, so it is
not easy to decide what is interesting and what is not. As Shamin and Neuhold
stated in [21], in the context of the European Parliament, "MPs need to be
selective in their information input".

Let us consider a stream of documents that may be distributed among the
MPs. These documents can be news releases, technical reports or parliamentary
initiatives, for example. We would like to build an automated system able to

© Springer International Publishing AG 2017
S. Moral et al. (Eds.): SUM 2017, LNAI 10564, pp. 64–77, 2017.
DOI: 10.1007/978-3-319-67582-4_5

recommend those MPs who should receive each document, taking into account both its own content and the specific interests and preferences of each MP.

Therefore, our research falls in the context of content-based recommender/filtering systems [10,18], which suggest items to users according to their preferences (represented by a profile or model of some kind), also taking into account some characteristics of the items (their textual content in our case). There are a lot of works addressing the recommendation/filtering problem in many domains and applications (see for example the three survey papers [3,16,17]). However, we are not aware of any such a system in a parliamentary context, except our own previous work [7,20]. Content-based recommender systems can be built using either information retrieval-based (IR) methods, which generate recommendations heuristically [1,2,9,15], or machine learning-based (ML) methods, mainly supervised classification algorithms for learning user models [4,5,12,13,19,22].

The objective of this paper is precisely to study and compare the capabilities of IR-based and ML-based methods in the parliamentary context we are considering. Therefore we propose two relatively simple approaches to create the recommender system, both based on first building a training document collection. One approach uses an Information Retrieval System (IRS) to explore this document collection, whereas the other uses this collection to generate a set of classifiers, one per MP. The training document collection will be obtained from the transcriptions of the speeches of the MPs in the parliamentary debates. The basic assumption is that these documents can provide information about the interests and preferences of the MPs. In order to compare our proposals, we shall perform experiments using a collection of MPs interventions from the regional Parliament of Andalusia at Spain.

The rest of the paper is organized in the following way: Sect. 2 gives details of the proposed IR-based and ML-based approaches to be compared. Section 3 contains the experimental part of the paper. Finally, Sect. 4 includes the concluding remarks and some proposals for future work.

2   Approaches for Recommending

The scenario that we consider is the following: we have a set of MPs $\mathcal{MP} = \{MP_1, \ldots, MP_n\}$. To the parliament documents arrive that must be distributed among the MPs according to their interests and preferences. We want to build a system that, given a new document, automatically selects those MPs that could be interested in reading it. Associated to each MP_i there is a set of documents $\mathcal{D}_i = \{d_{i1}, \ldots, d_{im_i}\}$, each d_{ij} representing the transcription of the speech of MP_i when participating in the discussion of a parliamentary initiative. The complete set of documents is $\mathcal{D} = \cup_{i=1}^n \mathcal{D}_i$. $\mathcal{D}$ is the training document collection that will be used by both the IR-based and the ML-based approaches.

2.1   The ML-Based Approach

The idea is simply to use the transcriptions of the speeches of the MPs in the parliamentary debates, $\mathcal{D}$, as training data to train a binary classifier (relevant/non-relevant) for each MP. Then, given a new document to be filtered/recommended, we use all these classifiers to decide which MPs should receive this document, namely those MPs whose corresponding classifier predicts the relevant class or, alternatively, assuming that the classifiers give a numerical output (a score) instead of a binary value, we could generate a ranking of MPs in decreasing order of score, thus recommending the document to those MPs whose score is greater than a given threshold.

In order to build a standard binary classifier for each MP we need training data (documents in this case), both positive (relevant documents) and negative (irrelevant documents). We shall consider that the own interventions/speeches of an MP are positive training data for building the classifier for this MP. Therefore, for each MP_i the set of positive examples is precisely $\mathcal{D}_i$. We shall also consider that all the interventions which are not from an MP are negative training data for the classifier associated to this MP. Hence the set of negative examples for each MP_i is $\mathcal{D} \setminus \mathcal{D}_i$.

2.2   The IR-Based Approach

In this case we are going to use the documents in $\mathcal{D}$ in two different ways to feed an Information Retrieval System (IRS). This IRS will be used to retrieve the documents that are more similar to the document to be filtered/recommended, which plays the role of a query to the system. The two ways in which $\mathcal{D}$ is transformed into an indexed document collection, which were originally proposed in [7], are the following:

The Collection of MP Interventions. The documents to be indexed by the IRS are just those in $\mathcal{D}$, i.e. all the interventions of all the MPs in the training set. In this case, what we obtain as the output for a query (which is the document to be filtered) is a ranking of documents, each one associated with an MP. Then we replace a document in the ranking by its associated MP. However, this new ranking of MPs may contain duplicate MPs with different scores (corresponding to different interventions of the same MP). In order to get a ranking of non duplicate MPs, we remove all the occurrences of an MP except the one having the maximum score. We call this approach IR-i.

The Collection of MP Profiles. To avoid the previous problem of having to remove duplicates from the ranked list retrieved by the IRS, another option is to group together all the interventions of each MP in only one document, thus obtaining a document collection with as many documents as MPs. More precisely, from each set $\mathcal{D}_i$ we build the single document $d_i = \cup_{j=1}^{m_i} d_{ij}$ and then use $\cup_{i=1}^{n} d_i$ as the document collection to be indexed by the IRS. In this case the output of the system as response to a query is directly a ranked list of MPs. We call this approach IR-p.

In the two cases considered the system obtains a ranked list of MPs in decreasing order of score. Nevertheless, and due to efficiency considerations, an IR system does not compute the document-length normalization and as consequence the output scores vary with the number of terms in the query. Although these raw scores are valid for obtaining a MP's ranking (not to compute the length normalization does not affect the ranking, the final aim of an IR system) this is not the case for document recommendation purposes. Particularly, in this problem we are looking for a common threshold that should be used to recommend a document to those MPs whose score is greater than this value, independently of the query. In order to be able to determine such threshold, the raw scores are normalized by dividing by the maximum score. Note that in this case the normalized score represents a similarity percentage with respect to the top ranked MP.

3   Experimental Evaluation

The evaluation of our proposals will be carried out using all the 5,258 parliamentary initiatives discussed in the 8th term of office of the Andalusian Parliament at Spain[1], marked up in XML [8].

Each initiative contains, among other things, the transcriptions of all the speeches of the MPs who intervene in the debate, together with their names. There is a total of 12,633 different interventions, but we have only considered the interventions of those MPs who participate in at least 10 different initiatives, a total of 132 MPs. All the initiatives were preprocessed by removing stop words and performing stemming.

Regarding the evaluation methodology, we shall use the repeated holdout method [14]. Concretely, the set of initiatives is randomly partitioned into a training and a test set (containing in our case 80% and 20% of the initiatives, respectively), and the process is repeated (5 times in our case), thus averaging the results of the different rounds.

From the initiatives in the training set, we extract the interventions of all the MPs to form our training document collection $\mathcal{D}$. Then we build a classifier for each MP, following the ML-based approach (described in Sect. 2.1), and also an IRS (in the two ways described in Sect. 2.2) following the IR-based approach. In order to train a binary classifier for each MP_i from $\mathcal{D}_i$ and $\mathcal{D} \setminus \mathcal{D}_i$, we have used Support Vector Machines [6], which is considered as the state-of-the-art technique for document classification (we used the implementations of SVM available in R^2). From the IR perspective, we have used the BM25 information retrieval model (using the implementation in the search engine library Lucene[3]), which is also a state-of-the-art technique in document retrieval [1].

---

[1] http://www.parlamentodeandalucia.es.
[2] https://cran.r-project.org.
[3] https://lucene.apache.org.

The initiatives in the test set are used as the documents to be filtered/recommended (using only the transcriptions of all the speeches within each initiative as the text of the document). We consider that each test initiative is relevant only for those MPs who participate in it. Notice that this is a very conservative assumption, since this initiative could also be relevant to other MPs interested in the same topics discussed in it, but it is the only way to establish a kind of "ground truth".

The evaluation measures used to assess the quality of the filtering/ recommendation system are those typically used in text classification: we compute the precision, recall and the F-measure of the results associated to each MP_i. Precision is the ratio between the number of truly relevant test initiatives for MP_i which are correctly identified by the system (True Positives, TP_i) and the total number of test initiatives identified as relevant for MP_i ($TP_i + FP_i$, being FP_i the False Positives), $p_i = TP_i/(TP_i + FP_i)$. Recall is the ratio between TP_i and the number of test initiatives which are truly relevant for MP_i ($TP_i + FN_i$, being FN_i the False Negatives), $r_i = TP_i/(TP_i + FN_i)$ (see Table 1). Then we can compute the F-measure, as the harmonic mean of precision and recall, $F_i = 2p_i r_i/(p_i + r_i)$. To summarize all the measures, associated to each MP_i, we shall use both macro-averaged (M) and micro-averaged (m) measures [23]:

$$Mp = \frac{1}{n}\sum_{i=1}^{n} p_i \qquad Mr = \frac{1}{n}\sum_{i=1}^{n} r_i \qquad MF = \frac{1}{n}\sum_{i=1}^{n} F_i$$

$$mp = \frac{\sum_{i=1}^{n} TP_i}{\sum_{i=1}^{n}(TP_i + FP_i)} \qquad mr = \frac{\sum_{i=1}^{n} TP_i}{\sum_{i=1}^{n}(TP_i + FN_i)} \qquad mF = \frac{2mp\,mr}{mp + mr}$$

Table 1. Relations between TP_i, FP_i and FN_i with true relevance of the documents to be recommended and the scores.

	Truly relevant	Truly irrelevant
Score $\geq$ threshold	TP_i	FP_i
Score $<$ threshold	FN_i	TN_i

All the previous performance measures heavily depend on the selected threshold used to recommend the document to those MPs whose score is greater than this threshold. We will experiment with different thresholds, ranging from 0.1 to 0.9. It should be noticed that, as the scores obtained by the ML-based and the IR-based approaches represent different things (probability in one case and similarity with the best result in the other), the same happens with the thresholds.

We are going to also use another evaluation measure that does not depend on any threshold but it measures directly the ranking quality. This measure is the well-known in the IR field Normalized Discounted Cumulative Gain (NDCG) [11]. This evaluation metric tries to estimate the cumulative relevance gain obtained by examining the first documents (MPs in our case) in a retrieved list of results. Since users tend to check only the first results, a discounting factor is used to reduce the document effect over the metric value as its position increases within the ranking. The metric value for a given list of MPs, is calculated as follows:

$$NDCG@k = \frac{1}{N} \sum_{i=1}^{k} \frac{2^{rel(d_i)} - 1}{\log(i+1)}, \tag{1}$$

where k is the number of results evaluated (10 in our experiments); i is the ranking position of the MP being evaluated; d_i is the MP at position i; $rel(d_i)$ is the relevance value of d_i (either 0 or 1 in our case); the normalization factor N is the DCG for the ideal ranking, where all the relevant results are located consecutively in the first positions of the ranking. With this normalization, the

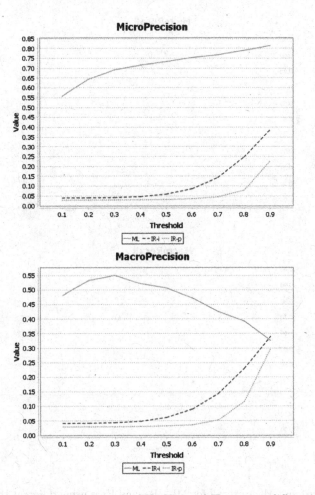

Fig. 1. Micro and Macro precision for ML, IR-i and IR-p using different thresholds.

metric values are always between 0 and 1, making it possible to calculate averages among different documents. This metric is computed for all the documents in the test set and then averaged.

3.1   Results

The results of our experiments for (macro and micro) precision, recall and F, using different thresholds (from 0.1 to 0.9) are displayed in Figs. 1, 2 and 3, respectively.

We can observe that, in general, the lower the threshold, the easier the system assigns the relevant value to documents, which increases the number of false positives and decreases precision. At the same time the number of false negatives decreases, thus increasing recall. When the threshold is high, the opposite

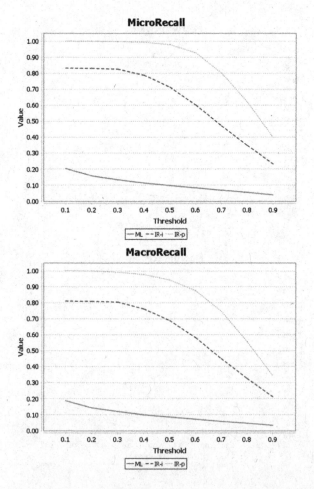

Fig. 2. Micro and Macro recall for ML, IR-i and IR-p using different thresholds.

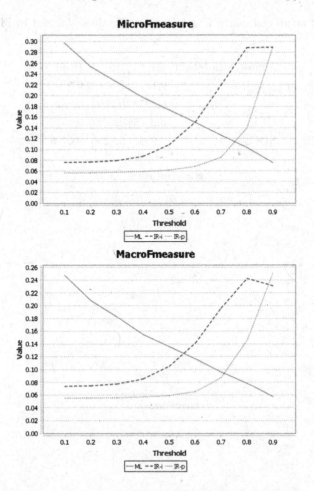

Fig. 3. Micro and Macro F measures for ML, IR-i and IR-p using different thresholds.

situation occurs, increasing precision and decreasing recall. The only anomaly to this general behaviour is with the ML-based approach and macro precision, which tends to decrease as the threshold increases. This may be due to a bad behaviour of this approach with those MPs having a low number of interventions (thus generating a poor training set), where the number of true positives decreases, even more steeply than the number of false positives, as the threshold increases (remember that with the macro measures all the MPs are equally important, independently on their number of interventions). More insights about this question will be given in the next section.

Nevertheless, the behaviour of the two approaches is quite different. The ML-based approach obtains relatively good precision values, much better than those of the IR-based approach. However, the recall values of the ML-based approach are very bad, whereas those of the IR-based approach are quite good.

Table 2. Best micro and macro F and NDCG@10 values obtained by ML, IR-i and IR-p.

Approach threshold	ML	IR-i	IR-p
	0.1	0.8	0.9
mF	0.2978	0.2896	0.2829
MF	0.2475	0.2423	0.2513
NDCG@10	0.6263	0.6246	0.6776

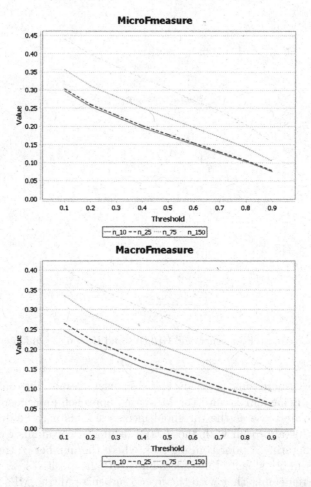

Fig. 4. Micro and Macro F measures for ML, using different thresholds and varying the minimum number of interventions.

The two IR-based approaches are quite similar, although IR-p gets more extreme values than IR-i (better in recall and worse in precision).

The F measure, which establishes a balance between precision and recall, clearly indicates that the ML-based approach works better with low thresholds and the opposite is true for the IR-based approach. However, there is no clear winner. Table 2 contains the best F values obtained by each approach, as well as the corresponding values of the NDCG@10 measure.

The values of mF and MF are very similar for the three methods, ML is slightly better in mF and IR-p is slightly better in MF. In fact a t-test (using the results of the five random partitions, and a confidence level of 99%) does not report any statistically significant differences between these methods. Concerning NDCG, a t-test indicates that IR-p is significantly better than both ML and IR-i, although there is no significant difference between ML and IR-i.

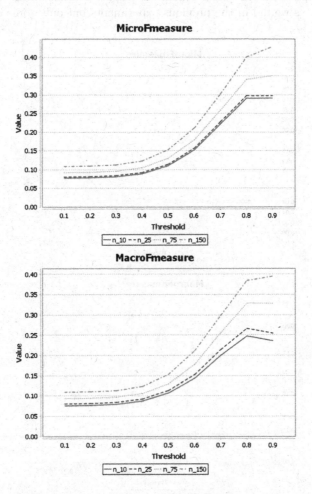

Fig. 5. Micro and Macro F measures for IR-i, using different thresholds and varying the minimum number of interventions.

3.2   Results When Varying the Number of Initiatives

As we said at the beginning of Sect. 3, the MPs being considered in this study are those who participate in at least 10 initiatives. This includes both MPs scarcely participating in the debates and other much more active (taking part in hundreds of initiatives). We want to evaluate the quality of the results depending on the number of initiatives where the MPs intervene.

To this end we have repeated our previous experiments, but fixing the minimum number of interventions of an MP which are necessary to include him in the study to greater values, concretely to 25, 75, and 150. Our goal is to evaluate whether a greater number of interventions of an MP translates into a better training set and hence to better results. For space reasons we do not include all the figures as we did in the previous experiments but only some of them for

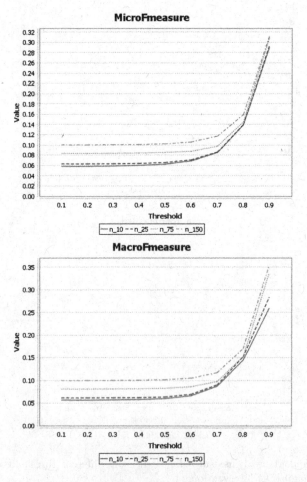

Fig. 6. Micro and Macro F measures for IR-p, using different thresholds and varying the minimum number of interventions.

illustrative purposes (micro and macro F for ML, IR-i and IR-p, see Figs. 4, 5 and 6, respectively).

As we can observe the trends are the same as in the previous experiments: in the ML-based approach the F measures decrease as the threshold increases, whereas the opposite is true for the IR-based approach. Moreover, the results are consistently better as the number of interventions required increases. Therefore, the two approaches could potentially reach better results if more training documents for each MP were available.

In Table 3 we show the best F values obtained for the different numbers of interventions, as well as the NDCG@10 values. For the F measures, the t-tests indicate that there are not significant differences between ML and IR-i in any case, whereas both ML and IR-i are significantly better than IR-p for micro F with sizes 75 and 150. For NDCG, again the differences between ML and IR-i are not significant but IR-p is significantly better than ML and IR-i with all the sizes.

Table 3. Best micro and macro F and NDCG@10 values obtained by ML, IR-i and IR-p, using different minimum numbers of interventions.

Approach	mF			MF			NDCG@10		
	ML	IR-i	IR-p	ML	IR-i	IR-p	ML	IR-i	IR-p
10	0.2978	0.2896	0.2829	0.2475	0.2423	0.2513	0.6263	0.6246	0.6776
25	0.3037	0.2971	0.2939	0.2658	0.2661	0.2829	0.6267	0.6242	0.6806
75	0.3568	0.3509	0.3085	0.3355	0.3288	0.3368	0.6132	0.6192	0.7086
150	0.4408	0.4282	0.3120	0.4039	0.3948	0.3532	0.5622	0.5744	0.6782

4   Concluding Remarks

In this paper we have proposed and compared two different approaches to build a system able to recommend/filter documents to the Members of Parliament. One approach is based on machine learning techniques, namely automatic document classification, whereas the other is based on information retrieval methods. The two approaches start from a collection of training documents composed of the interventions of the MPs in the parliamentary debates, which is assumed contains information about the interests and preferences of MPs. While the ML-based approach uses this collection to train a binary classifier for each MP, the IR-based approach uses an information retrieval system to index this collection and then retrieves the MPs which are more similar to the document to be recommended/filtered. In the two cases the output of the system is a ranked list (in decreasing order of score) of MPs. Then, given a fixed threshold, the system recommends the target document to those MPs whose score is above the threshold.

The two studied approaches behave quite differently in terms of recall and precision, and their best performance is attached using very different thresholds.

However, in terms of the F measures, both macro and micro (and to a lesser extent in terms of NDCG@10), the best results with both approaches are quite similar. Therefore, there is not a clear reason to prefer one approach to the other.

A possible weakness of the ML-based approach is that all the interventions which are not from an MP are considered as negative training data for the classifier associated to this MP. This is questionable: the interventions of other MPs which are about the same topics considered of interest for a given MP may be also relevant for him. For example an MP whose main area of interest is health could find interesting the interventions of other MPs also dealing with health. In this way the negative training data being used could contain positive data and this can limit the capacity of the classifier to discriminate between relevant and irrelevant documents. Therefore, we are interested for future research in using the so-called positive unlabeled learning techniques [24], which only assume the existence of a set of positive training data and a (usually larger) set of unlabeled data, but there is no negative training data.

Acknowledgements. This work has been funded by the Spanish "Ministerio de Economía y Competitividad" under projects TIN2013-42741-P and TIN2016-77902-C3-2-P, and the European Regional Development Fund (ERDF-FEDER).

References

1. Baeza-Yates, R., Ribeiro-Neto, B.: Modern Information Retrieval. Addison-Wesley, Boston (2011)
2. Belkin, N.J., Croft, W.B.: Information filtering and information retrieval: two sides of the same coin? Commun. ACM **35**, 29–38 (1992)
3. Bobadilla, J., Hernando, A., Fernando, O., Gutiérrez, A.: Recommender systems survey. Knowl. Based Syst. **46**, 109–132 (2013)
4. Billsus, D., Pazzani, M., Chen, J.: A learning agent for wireless news access. In: Proceedings of the International Conference on Intelligent User Interfaces, pp. 33–36 (2002)
5. Cohen, W.: Learning rules that classify e-mail. In: Papers from the AAAI Spring Symposium on Machine Learning in Information Access, pp. 18–25 (1996)
6. Cristianini, N., Shawe-Taylor, J.: An Introduction to Support Vector Machines and Other Kernel-Based Learning Methods. Cambridge University Press, New York (2000)
7. de Campos, L.M., Fernández-Luna, J.M., Huete, J.F.: A lazy approach for filtering parliamentary documents. In: Kő, A., Francesconi, E. (eds.) EGOVIS 2015. LNCS, vol. 9265, pp. 364–378. Springer, Cham (2015). doi:10.1007/978-3-319-22389-6_26
8. de Campos, L.M., Fernández-Luna, J.M., Huete, J.F., Martin-Dancausa, C.J., Tur-Vigil, C., Tagua, A.: An integrated system for managing the andalusian parliament's digital library. Program Electron. Libr. Inf. Syst. **43**, 121–139 (2009)
9. Foltz, P., Dumais, S.: Personalized information delivery: an analysis of information filtering methods. Commun. ACM **35**, 51–60 (1992)
10. Hanani, U., Shapira, B., Shoval, P.: Information filtering: overview of issues, research and systems. User Model. User Adapt. Interact. **11**, 203–259 (2001)
11. Jarvelin, K., Kekalainen, J.: Cumulative gain-based evaluation of IR techniques. ACM Trans. Inf. Syst. **20**, 422–446 (2002)

12. Kim, J., Lee, B., Shaw, M., Chang, H., Nelson, W.: Application of decision-tree induction techniques to personalized advertisements on internet storefronts. Int. J. Electron. Commerce **5**, 45–62 (2001)
13. Jennings, A., Higuchi, H.: A user model neural network for a personal news service. User Model. User Adapt. Interact. **3**, 1–25 (1993)
14. Lantz, B.: Machine Learning with R. Packt Publishing Ltd., Birmingham (2013)
15. Loeb, S.: Architecting personal delivery of multimedia information. Commun. ACM **35**, 39–48 (1992)
16. Lops, P., de Gemmis, M., Semeraro, G.: Content-based recommender systems: state of the art and trends. In: Ricci, F., Rokach, L., Shapira, B., Kantor, P. (eds.) Recommender Systems Handbook. Springer, Boston (2011)
17. Lu, J., Wu, D., Mao, M., Wang, W., Zhang, G.: Recommender system application developments: a survey. Decis. Support Syst. **74**, 12–32 (2015)
18. Pazzani, M.J., Billsus, D.: Content-based recommendation systems. In: Brusilovsky, P., Kobsa, A., Nejdl, W. (eds.) The Adaptive Web. LNCS, vol. 4321, pp. 325–341. Springer, Heidelberg (2007). doi:10.1007/978-3-540-72079-9_10
19. Pazzani, M., Billsus, D.: Learning and revising user profiles: the identification of interesting web sites. Mach. Learn. **27**, 313–331 (1997)
20. Ribadas, F.J., de Campos, L.M., Fernández-Luna, J.M., Huete, J.F.: Concept profiles for filtering parliamentary documents. In: Proceedings of the 7th International Joint Conference on Knowledge Discovery, Knowledge Engineering and Knowledge Management, vol. 1, pp. 409–416 (2015)
21. Shamin, J., Neuhold, C.: 'Connecting Europe': the use of 'new' information and communication technologies within European parliament standing committees. J. Legislative Stud. **13**, 388–402 (2007)
22. Tjoa, A.M., Hofferer, M., Ehrentraut, G., Untersmeyer, P.: Applying evolutionary algorithms to the problem of information filtering. In: Proceedings of the 8th International Workshop on Database and Expert Systems Applications, pp. 450–458 (1997)
23. Tsoumakas, G., Katakis, I., Vlahavas, I.: Mining multi-label data. In: Maimon, O., Rokach, L. (eds.) Data Mining and Knowledge Discovery Handbook. Springer, Boston (2009)
24. Zhang, B., Zuo, W.: Learning from positive and unlabeled examples: a survey. In: International Symposiums on Information Processing, pp. 650–654 (2008)

Eliciting Implicit Evocations Using Word Embeddings and Knowledge Representation

Sébastien Harispe[(✉)], Massissilia Medjkoune, and Jacky Montmain

LGI2P Research Center, École des Mines d'Alès,
Parc Scientifique G. Besse, 30035 Nîmes Cedex 1, France
{Sebastien.Harispe,Massissilia.Medjkoune,Jacky.Montmain}@mines-ales.fr

Abstract. Automatic elicitation of implicit evocations - i.e. indirect references to entities (e.g. objects, persons, locations) - is central for the development of intelligent agents able of understanding the meaning of written or spoken natural language. This paper focuses on the definition and evaluation of models that can be used to summarize a set of words into a unique unambiguous entity identifier selected from a given ontology; the ability to accurately perform this task being a prerequisite for the detection and elicitation of implicit evocations on spoken and written contents. Among the several strategies explored in this contribution, we propose to compare hybrid approaches taking advantages of knowledge bases (symbolic representations) and word embeddings defined from large text corpora analysis. The results we obtain highlight the relative benefits of mixing symbolic representations with classic word embeddings for this task.

1 Introduction

Developing automatic approaches enabling human spoken and written productions to be deeply understood is central for the development of artificial agents capable of complex human-machine interactions and collaborations. This broad challenge, largely studied by the Artificial Intelligence community[1], aims at developing approaches capable of *capturing* the meaning conveyed by units of language (from word utterances to sequences of phrases); this is central for numerous processes widely studied in the literature: Question Answering, Information Extraction and Information Retrieval, among others.

In this paper, we focus on studying aspects tightly related to the development of approaches for understanding the meaning and semantics of large units of languages such as sentences or paragraphs. The positioning of our work is, broadly speaking, closely related to Named-Entity Recognition (NER), i.e. detection of explicit entity mentions in texts [12]. We are more particularly interested in fine-grained entity recognition, not only aiming at detecting classes of entities as it is classically done in NER by detecting references to persons or locations for instance. We are here rather interested in entity linking, i.e. at linking specific unambiguous identifiers provided by knowledge bases - such as DBpedia

---

[1] e.g. in the Natural Language Processing and Computational Linguistics domains.

© Springer International Publishing AG 2017
S. Moral et al. (Eds.): SUM 2017, LNAI 10564, pp. 78–92, 2017.
DOI: 10.1007/978-3-319-67582-4_6

and Yago Uniform Resource Identifiers (URIs) or WordNet synset identifiers [1,11] - to entities mentioned into texts [10]. We study in particular the problem of eliciting implicit evocations, i.e. references to unambiguous entities that are not mentioned by lexical forms of those entities. As an example, in the sentence *"I've visited the capital of Spain as well as Picasso birthplace city last summer"*, the utterances *'capital of Spain'* refers to Madrid, *'Picasso birthplace city'* to Málaga. Similarly, telling you that *"this morning I've eaten a yellow tropical fruit very much liked by monkeys"* should give you a good idea of the kind of fruit I've eaten (i.e. a Banana).

The aim of this paper is to study automatic approaches that are able to elicit implicit evocations; similarly to the way humans are most often able to understand them. Developing approaches enabling such a process is important for capturing the meaning of units of languages; their direct applications for semantic indexing and information retrieval, as well as their indirect potential applications to question answering, information extraction, topic identification or sentiment analysis to cite a few, are numerous. Due to the breadth and complexity of the task, we here focus on eliciting implicit evocations considering the words mentioning the entity evocation to be given, e.g. considering a bag of words extracted from the initial sentence {*'yellow', 'tropical fruit', 'monkeys'*} we expect the approach to identify the entity Banana.[2] We also consider that there is no need to take into consideration contextual information for detecting the implicit evocation - otherwise stated, the knowledge base that is required to answer the question is therefore considered to be static and not contextual.

The paper is organized as follows; Sect. 2 presents related works as well as the fundamental notions on which will be based our contributions. Section 3 introduces the different models that can be used to detect implicit entity evocations from bags of words. Section 4 presents the evaluation protocol as well as the results obtained during the empirical evaluation. Section 5 summarizes the main results and concludes this work.

2   Related Works and Problem Setting

This section introduces related works, formalizes the problem setting and presents the fundamental notions on which the models that will be introduced afterwards are based; notations are also defined hereafter.

Eliciting implicit evocations is closely related to well-known problems studied in Artificial Intelligence, in particular the reversed dictionary task, Topic Modeling, Language Model and text summarization. In the reversed dictionary or word access task, a word has to be found considering a given description; a problem closely related to the one considered in this paper - related recent work also refer to phrase embedding [6,15]; these approaches consider known term descriptions which is not considered hereafter. Topic Modeling techniques,

---

[2] We do not consider in this paper the complex problem of detecting implicit evocations. Note also the special syntax used to refer to the non-ambiguous entity reference Banana compared to its ambiguous lexical form *banana*.

for instance, can be used to analyze large corpora in order to generate topics by detecting frequent word collocations [14]. The aim of these approaches is slightly different since generated topics have to be extracted from large corpora by analyzing word usage statistics – topics are also *per definition* always abstract notions and cannot therefore be used straightforwardly for eliciting potentially specific entities. Considering our setting, it could be tempting for example to define a probabilistic model based on a conditional probability estimation enabling to compute $p(\text{Banana} \mid$ *'yellow', 'tropical fruit', 'monkeys'*$)$. More generally, the problem could be studied by considering an approach based on language models - i.e. models largely used in machine translation, speech recognition or text summarization to cite a few. However, this study does not consider such models due to the curse of dimensionality [9] hampering their use for eliciting implicit evocations - indeed, despite the use of existing smoothing techniques [3], computing language models taking into account potentially large contexts (e.g. 5 to 10 words) is not possible. Other techniques based on neural probabilistic language models could also be considered to answer this limit [2]; more recent techniques based on sequence learning, e.g. based on Long Short-Term Memory neural network architectures, could also be worth studying [7]. Such techniques will only be partially and indirectly considered through the use of word embeddings techniques.

2.1   Explicit and Implicit Evocations

In this contribution we are interested by detecting implicit evocations; we introduce this notion by providing some illustrations as well as elements of definition – relationships with state-of-the-art notions such as topic identification or NER have been mentioned above.

First of all, entity evocations are here defined as strongly supported references to non-ambiguous notions or entities. As an example, several evocations could be detected from the following sentences "*I went to Paris last week, the Eiffel Tower is amazing... I love France!*". It is relatively easy to detect that it is highly probable that a reference to **Paris**, the capital of **France** called *Paris* is made. Note however that due to the ambiguous nature of words, it could not be the case; the word, i.e. surface form, *Paris* could indeed refer to other cities, e.g. **Paris_(Tennessee)**, or even other entities that are not locations. Nevertheless, considering the context which is defined by the sentence meaning, and in particular the utterrances of the words *France*, and *Eiffel Tower*, most people would understand the utterance of the string *Paris* as a will of the speaker to explicitly refer to the capital of France; here we consider that an *explicit* evocation has been made since a word corresponding to a lexical form of the entity, despite being ambiguous, explicitly refers to it. It is important to understand that evocations are here not necessarily understood as the intended speaker evocation; they are rather considered to be the consensual agreement towards understood evocations, i.e. the disambiguation most people would consider based on the context of utterance of words - e.g. as an example, nothing restrict the speaker of aforementioned sentences to say that he is referring to

Paris_(Tennessee); even if most person would agree that discussing with such a speaker would thus be quite challenging. We therefore consider that, in most cases, intended evocations correspond to evocations most target recipients of a spoken/written message would consider. Explicit entity evocations could also be more refined than single word utterances, e.g. *The City of Lights* could be used to mention Paris. All the examples provided so far were referring to the notion of explicit entity evocations since all of them could have been linked to a unique lexical/surface form of the entity.

Implicit entity evocations refer to entity evocations that cannot be directly associated to a word utterance, i.e. a surface form. As an example, the sentence *"Bob bought an expensive red sport car of a famous italian brand"* is most likely to refer to the fact that Bob bought a car from the italian car brand, Ferrari - otherwise stated, most of us would understand *Bob bought a* Ferrari. Additional examples are provided in the introduction section. Note that we could discuss in details the technical differences that we consider between surface forms of an entity and implicit references. Indeed, in some cases, judgement aiming at distinguishing if an evocation is explicit or implicit may depend on subjective evaluations. As an example, considering that *The City of Lights* is an explicit reference to the city Paris could be surprising considering that mentioning *the capital of France* would be considered as an implicit reference to the same city. We therefore stress that we consider explicit references to be lexical/surface forms of a concept. We thus consider that the utterance '*The City of Lights*', contrary to the utterance '*the capital of France*', is a lexical entry linked to the concept Paris in an index (e.g. a dictionary). Thus, considering that the lexical entry '*the capital of France*' is no linked to Paris in any index - no dictionary will give you such a lexical form -, more refined techniques have to be used to elicit the reference to Paris. As an example, this implicit evocation could be detected by taking advantage of a database or a knowledge base for answering the question *What's the capital of France?* - a process which is highly more complex than searching for a specific entry into a lookup table index to further resolve any ambiguity associated to word utterances.

Note that, independently to any context, implicit entity evocations can also be considered from a set of words (Table 1). In that case, the problem setting is close to a simplified form of the Pyramid game[3] (considering no interaction and no word ordering): a set of words is provided and a unique implicit evocation has to be provided by considering semantic relationships between the words. This is the setting we consider in this paper.

2.2   Problem Setting and Global Strategies Evaluated

Formal Definition. Considering a vocabulary T and a set of entities $\mathcal{E}$ partially ordered into a taxonomy $O = (\preceq, \mathcal{E})$, we are looking for a function:

$$f : \mathcal{P}(T) \to \mathcal{E} \tag{1}$$

---

[3] https://en.wikipedia.org/wiki/Pyramid_(game_show).

Table 1. Examples of evaluation entries

Given words	Expected evocation
place, study, teacher	School
food, italy, round, tomato	Pizza
yellow, fruits, monkeys	Banana
city, UK, capital	London

The function f therefore aims at reducing a set of terms into a unique entity reference corresponding to the implicit mentioned entity. More generally, we are looking for a total order $\preceq_{\mathcal{E}}$ among the entities w.r.t. their relevancy for summarizing a given set of terms $T' \subset T$. To this aim we are looking for a scoring function evaluating the relevancy to associate a specific evocation to a given set of terms:

$$s : \mathcal{P}(T) \times \mathcal{E} \to \mathbb{R} \tag{2}$$

We will focus on the definition of the scoring function s in this paper. We therefore consider the following definitions: $f(T') := \arg\max_{e \in \mathcal{E}} s(T', e)$; the considered total order $\preceq_{\mathcal{E}}$ is thus defined such as $s(T', e') \le s(T', e) \to e' \preceq_{\mathcal{E}} e$.

Evaluated Strategies. Different types of knowledge have to be taken into account for detecting implicit evocations. Only considering our simplified problem setting in which a set of words is evaluated, two types of information seems important for answering the task; (i) abstract restriction and enumerations, as well as (ii) salient properties definitions. Examples are provided:

- *Abstract restriction/enumeration* - need for a partial ordering of entities. An implicit evocation often refers to a general class to which the target implicit evocation refers to, e.g. *Paris* refers to a specific `Capital`, *expensive red sport car* refers to a `Car`. In those cases, it's important to know what are the instances of a specific class in order to be able to consider potentially relevant restrictions - i.e. group of entities in which candidates will be evaluated. In a similar manner, by mentioning *Krakatoa, Etna, Mont St. Helens* or *Eyjafjallajokull* the concept `Volcano` is clearly implicitly mentioned by providing explicit references of specific instances of volcano. Detecting such implicit evocations requires taking advantage of knowledge representations that will be used to identify a set of evocations referring to an abstract class.
- *Salient property*: Most of us would link the evocations {*green, monster, angry, muscle*} to the concept `Hulk`; this is because `Hulk` has a *green* skin, has a *muscular* type, and refers to a famous *angry monster*. Detecting mentions of such an implicit evocation requires linking provided evocations to salient properties of an entity of interest. To this aim an approach enabling to link properties values to specific entities has to be defined.

In this paper we consider that exhaustive formalized knowledge bases answering our needs, i.e. defining extensive properties values for a large number of entities, are not available.[4] Indeed, despite the large efforts made for defining extensive knowledge bases [1,11], the properties to be analyzed for detecting implicit evocations are too broad, e.g. despite an URI exists for the concept Hulk, no property defines its skin color in DBpedia. We however consider that large text corpora are freely available (as it is the case today - Wikipedia for instance), and that it could be an interesting strategy to try mixing large scale text analysis (e.g. for capturing word relatedness enabling to detect a 'sort of' link between the words *green* and *hulk*), as well as large taxonomical ordering of entity provided by existing knowledge bases.

As it has been mentioned in the previous section, implicit entity evocations are tightly linked to the notion of context. An evocation is indeed often explained by utterances of words that could be linked to entities that are members of the same conceptual neighborhood. As an example, the implicit evocation of Ferrari mentioned earlier could have been explained by its narrow relationships with the concepts car, Italy and brand. Interestingly, the strength of a relationship between words or entities can therefore be discussed through the notions of semantic similarities/proximities [5].

In this context, we therefore propose to define and to compare different strategies taking advantage of (i) terms relationships extracted from large corpora analysis - through term semantic relatedness estimations -, as well as (ii) conceptual relationships defined by a partial ordering of entities provided by a knowledge base. The models discussed in this paper consider this postulate. Considering the type of strategies we will evaluate, two notions are of major importance: semantic relatedness of terms and semantic similarity of concepts/entities; both are briefly introduced in the following subsection.

2.3   Estimating Similarities and Relatedness of Words and Entities

Both word relatedness/proximity and entity similarity estimations from text and knowledge base analysis have been, and are still, extensively studied in particular by the NLP community. Word relatedness and entity similarity are extensively used in information retrieval, question answering, among others. A short introduction to these notions is provided hereafter - the reader can refer to the extensive literature and surveys for additional information, e.g. [5].

Estimating Word Relatedness. Considering a vocabulary T, word relatedness estimations aim at defining a function $\sigma_{TT} : T \times T \rightarrow [0,1]$ such as σ_{TT} enables capturing the intuitive (but weakly defined) notion of relatedness – generally defined as the strength of the semantic link established between units of language, here a pair of words [5]; once again most people will agree that the two words (*banana, monkey*) are more related than the two words (*banana, lion*).

Among the various approaches defined for comparing a pair of words, most recent strategies aims at (i) building a vector representation of words

---

[4] and that expecting such bases to exist in the near future is just illusionary.

(called embeddings) that will further be compared using traditional vector comparison metrics, most often the cosine similarity of vector representations. Technical details of most approaches therefore rely on defining the strategy used for building embeddings. Those strategies rely on the consideration that word meaning is defined by its context of use. Embeddings will thus be built by (indirectly) analyzing word collocations. Most recent strategies rely on predictive approaches, e.g. by building word embeddings by using internal representations of words that have been built by a neural network trained to predict a word considering a given context or a context considering given words. Further details related to word embeddings are out of the scope of this paper.

Estimating Entity Similarity. Considering a partial ordering $O = (\preceq, \mathcal{E})$ among a set of entities $\mathcal{E}$ (individuals and concepts of a knowledge base). The similarity of two entities is defined by $\sigma_{\mathcal{E}\mathcal{E}} : \mathcal{E} \times \mathcal{E} \to [0,1]$. An example of similarity measure proposed by Lin's measure is presented [8]:

$$sim(e, e') = \frac{2 \cdot IC(MICA(e, e'))}{IC(e) \times IC(e')} \tag{3}$$

with $MICA(e, e')$ the Most Informative Common Ancestor of entities e and e' with regards to a function evaluating the information content of an entity, with $IC : \mathcal{E} \to [0,1]$, and $x \preceq y \to IC(x) \geq IC(y)$, i.e. an entity is always considered to be more informative than its ancestors, e.g. $IC(\texttt{Paris}) > IC(\texttt{Capital})$.

3   Models for Detecting Implicit Entity Evocations

This section presents the various model proposals that are used to distinguish a ranked list of entity evocations for a provided set of terms (by defining Eq. 2, page 4). These models consider that word vector representations, as well as a labeling function linking terms to entities are provided. The labeling function defines the sets of labels that refer to a specific entity, i.e. $\pi : \mathcal{E} \to \mathcal{P}(T)$, e.g. $\pi(\texttt{Person}) = \{person, human, \ldots\}$.[5]
Two general types of models are presented:[6]

1. *Vector Aggregation Model* (VAM). The aim is to encompass the meaning of a set of terms by aggregating commonly used word embeddings.
2. *Graph-based Model* (GM). The aim is to detect implicit entity evocation by using a pre-built structure mixing both links between entities and terms as well as relationships between terms.

---

[5] Note the ambiguity at terminological level, a given term can refer to several entity. In addition, due to the transitivity induced by the relationship defining the considered partial ordering O, the set of entities that are potentially, implicitly or explicitly, evocated by a term $t \in T$ is defined by the set: $\bigcup_{e \in \mathcal{E}, t \in \pi(e)} \{x | e \preceq x\} \subseteq \mathcal{E}$; considering $\texttt{car} \prec \texttt{vehicule}$, mentioning $\texttt{car}$ makes you implicitly mention $\texttt{vehicule}$.

[6] Nothing excludes that specific models generated by one approach cannot be expressed by the other approach.

Both models are detailed hereafter. They will next be compared by analyzing their performances w.r.t an empirical evaluation.

3.1   Vector Aggregation Model

The Vector Aggregation Model (VAM) relies on a generic three-step strategy for analyzing a set of terms $T' \subset T$:

1. Computation of a conceptual evocation vector $t \in \mathbb{R}^{|\mathcal{E}|}$ for each term $t \in T'$ - the aim of this representation is to encompass all potential explicit and implicit entity evocations that are made by t.
2. Aggregation of the entity evocation vector of the terms composing the set of terms to evaluate. We will evaluate aggregations that generate vector representations of T' into $\mathbb{R}^{|\mathcal{E}|}$.
3. Analysis of aforementioned aggregation product in order to compute the ranked list of entity evocations.

These three steps are detailed.

Establishing the Link Between Words and Entities. We consider that without prior knowledge about context, the degree of evocation of an entity by a term is defined by the function $\sigma_{T\mathcal{E}} : T \times \mathcal{E} \to [0,1]$, defined such as:

$$\sigma_{T\mathcal{E}}(t,e) = \max_{t' \in \pi(e)} \sigma_{TT}(t,t') \tag{4}$$

Otherwise stated, the relationship considered between a term and an entity only depends on the semantic relatedness that can be distinguished at word level. Note that no prior knowledge about word usage is taken into account in this approach. Therefore, considering a term t, every entity $e \in \mathcal{E}$ with $t \in \pi(e)$ will have the same $\sigma_{T\mathcal{E}}(t,e)$ value – which will be maximal if the σ_{TT} function respects the identity of the indiscernible.[7]

Finally, without applying any preprocessing step excluding potential conflicting entity evocations, we consider the evocation of a term $t \in T$ to be defined by the function $\rho_{T\mathcal{E}} : T \to \mathbb{R}^n$ with ($|\mathcal{E}| = n$):

$$\rho_{T\mathcal{E}}(t) = [\sigma_{T\mathcal{E}}(t,e_1),\ldots,\sigma_{T\mathcal{E}}(t,e_n)]^{\intercal} \tag{5}$$

---

[7] Otherwise stated, by observing the word utterance *Paris*, all concepts having this specific string as label, e.g. **Paris** (France), **Paris_Tennesse**, will have the same evocation degree value. This is obviously not how humans process information. Indeed, without context, or only considering poor contextual information, people rely most often on evocation likelihood (considering their body of knowledge). Therefore, to refine the approach, we could also estimate the probability that a given term refers to an entity. Several approaches could be explored, e.g. analyzing usage of Word-Net synsets. This information is however difficult to obtain for entities that are not mentioned into this structured lexicon, which hampers the general aspect of the approach. We therefore consider that no prior knowledge about word-entity evocation is provided by excluding the use of statistics about word-entity usage.

This vector represents the potential entity evocations of a term without distinguishing among potentially conflictual evocations. We however consider that it represents a footprint encompassing all entity evocations a word could refer to.

Aggregation of the Information Provided by Several Words. Several approaches can be considered for aggregating the degrees of evocation of a set of terms $T' \subset T$. To this purpose we consider a general function $P_{\mathcal{E}} : \mathcal{P}(T) \to \mathbb{R}^n$. Two definitions of $P_{\mathcal{E}}$ will further be considered; both of them are based on an element-wise aggregation: (i) $P_{\mathcal{E}}^{min}(T') = \wedge_{t \in T'} \rho_{T\mathcal{E}}(t)$ defining the aggregation to be the minimal evocation value among all terms, and (ii) a less constraining evaluation summing the evocations $P_{\mathcal{E}}^{sum}(T') = \sum_{t \in T'} \rho_{T\mathcal{E}}(t)$.

Ranking Conceptual Evocations. We consider that, because of the nature of the function used to build the vector representations, as well as the aggregation operator, implicitly mentioned entities could be detected by analyzing associated dimension values in $P_{\mathcal{E}}$. More precisely, it is expected that evocation values associated to implicitly mentioned entities will diverge from the values that would be expected if randomly selected terms were used to build the vector representation. We therefore consider that the distribution of the value for a given entity and a given size of set of terms is known. This distribution is estimated by computing associated $P_{\mathcal{E}}$ representations for randomly sampled sets of terms of a specific size. The distribution stores for each entity the number of time a randomly composed set of terms has obtained a specific evocation value. Using this estimated distribution we can compute the probability that the observed value for a given set of terms is an *artefact*, or indeed seems to refer to an implicit evocation. We therefore consider that implicitly evocations are those for which observed values highly diverge from the expected one.

Several approaches have been tested for defining the ranking function; the raw score (a metric taking on the standard deviation[8] σ and the mean μ) is presented. Considering a given set of term T', we denote rs_{e_i} the raw score of T' w.r.t $e_i \in \mathcal{E}$:

$$rs_{e_i}(T') = \frac{P_{\mathcal{E}}(T')_i - \mu_{e_i}}{\sigma_{e_i}} \tag{6}$$

μ_{e_i} and σ_{e_i} respectively denote the median and the standard deviation of evocation values for the entity e_i computed during the sampling process associated to samplings of size $|T'|$.

3.2   Graph-Based Model

The Graph-based Model (GM) approach is based on a graph propagation strategy aiming at distinguishing what are the most relevant entities to be considered given a set of terms. Defined graph data structure aims at modelling relationships: among the terms, among the entities, as well as among terms and entities.

---

[8] Recall standard deviation: $\sigma = \sqrt{E[X^2] - E[X]^2}$.

We first present the graph structure. Next, the propagation approach used for distinguishing entity evocations is introduced.

Graph Model. Formally, let's consider a weighted directed graph $G = (V, E)$ with $V = T \cup \mathcal{E}$ and $E \subseteq V \times V$. Three types of relationships are distinguished:

1. relationships among terms, i.e. from $T \times T$; those relationships are weighted using a σ_{TT} measure capturing the relationships among terms. The weight of a relationship $(t, t') \in E$ is defined by a function $w_{TT} : T \times T \to [0, 1]$:

$$w_{TT}(t, t') = \frac{\sigma_{TT}(t, t')}{\sum_{t'' \in T} \sigma_{TT}(t, t'')} \tag{7}$$

This weighting function definition aims at normalizing the σ_{TT} scores considering that scores distributions may highly differ between terms.

2. relationships among entities, i.e. from $\mathcal{E} \times \mathcal{E}$; those relationships are given by the partial ordering O; the weight of the relationships are provided by a $\sigma_{\mathcal{E}\mathcal{E}}$ measure. More precisely, the relationships between entities are defined as follows: (1) building of a graph $G' = (\mathcal{E}, E_{\mathcal{E}\mathcal{E}})$ from O by considering that $(e, e') \in E_{\mathcal{E}\mathcal{E}}$ iff $e \preceq e'$ or $e' \preceq e$ in O; (2) apply a transitive reduction to G'; (3) weigh the relationships considering a $\sigma_{\mathcal{E}\mathcal{E}}$ measure – the weights are here also defined by normalizing considering all relationships defined in G'.

$$w_{\mathcal{E}\mathcal{E}}(e, e') = \frac{\sigma_{\mathcal{E}\mathcal{E}}(e, e')}{\sum_{e'' \in \mathcal{E} | (e, e'') \in E_{\mathcal{E}\mathcal{E}}} \sigma_{\mathcal{E}\mathcal{E}}(e, e'')} \tag{8}$$

3. relationships between terms and entities, i.e. from $(T \times \mathcal{E}) \cup (\mathcal{E} \times T)$; those relationships are given by the labeling function π. With $e \in \mathcal{E}, t \in T$, we consider that both $(t, e) \in E$ and $(e, t) \in E$ iff $t \in \pi(e)$, i.e. iff the term t is a label (refers) to the entity e.[9]

Propagation Model. Considering a given set of terms $T' \subset T$. The propagation model adopted to distinguish relevant entities is defined in Algorithm 1; the propagation procedure is detailed by Algorithm 2. The proposed approach is discussed hereafter. As it is defined in Algorithm 1, the global strategy aims at:

1. Computing the entity evocation degree for each term composing the query (lines 3–9). This is done by propagating a fixed quantity from each node composing the query (line 7).
2. Aggregating those results in order to compute, for the full set of terms, the entity evocation scores for each entity (lines 10–13).

---

[9] Those relationships could have also been weighted by considering word usage frequency. However, as stated before, we consider that no weighting function is defined here - even if analyzing σ_{TT} scores distributions could have been used.

The details of Algorithm 1 are now provided. At line 1–2, we initialize the map data structures.[10] that will be used to store the (temporary) results. The entity evocation degrees for each query term is stored into *query_term_evocation_map* – for instance, the entity evocation for the term t is stored as a map into *query_term_evocation_map[t]*; *query_term_evocation_map[t][e]* is the evocation degree of entity e by the term t. From line 3 to 9 we compute the entity evocations for each term defining the query (discussed later). From lines 10 to 13 those results are aggregated using a specific strategy. The sum and the median will be considered - intuitively, the median is used to express the fact that we not only want a high score; but we also want the score to be supported by a shared contribution of the terms composing the query.

Algorithm 1. Propagation algorithm

Data: The graph G structuring terms and entities; a set of terms $T' \subset T$, with $|T'| \ll |T|$, ϵ threshold value: stopping criteria.

Result: A data structure storing the relevance of each entity.

1 $query\_term\_evocation\_map \leftarrow map()$
2 $concept\_score \leftarrow map()$;
3 **for** $t \in T'$ **do**
4 | $ev\_map \leftarrow map()$;
5 | $visited\_node \leftarrow \{\}$;
6 | $score \leftarrow 1$;
7 | $propagate(t, visited\_node, 1, ev\_map)$ // cf. Algorithm 2;
8 | $query\_term\_evocation\_map[t] = ev\_map$;
9 **end**
10 **for** $e \in \mathcal{E}$ **do**
11 | $entity\_scores[e] = aggregate(e, query\_term\_evocation\_map)$;
12 | // The aggregate function can just be a sum, min, average...;
13 **end**
14 **return** $entity\_scores$;

Details of the propagation are defined by Algorithm 2.[11] The propagating process is defined using a recursive procedure aiming at propagating values avoiding already processed nodes. Depending of the type of node being processed (term or entity), the propagation aims at extending to other terms or entities. When a term is processed (line 2 to 13) a quantity is propagated to all entities that could be referred by the term (without any *a priori* consideration about term usage). The evocation is next propagated to those entities if the propagated quantity is important enough (line 6). The propagation is also performed to the neighboring terms by taking into account the distance between the terms at terminological level – line 8 to 10. When an entity node is processed (line 13 to 22) the propagation to the terminological level is performed by considering the labels associated to the entity. The propagation is also performed at the entity level, also taking into account the entity similarity that can be computed by analyzing entities' topological ordering (cf. weight definition Eq. 8).

---

[10] A map or dictionary stores a value for a specific key.

[11] We consider to be known G, T and $\mathcal{E}$ the terms and entities, ϵ the threshold value defining when to stop the propagation, *synDecFactor* a decay factor for handling synonyms while propagating, *eSmoothingFactor* a smoothing factor for reducing the impact of excessively considering the taxonomy on the results.

Algorithm 2. *propagate* routine

Data: A given node v of the graph, a set of visited node S, *score*: a score value (to propagate), *qtem* (for *query_term_evocation_map*): a map for storing entity evocation scores.

Result: None - updated evocation vector

1   $S.add(v)$

2   **if** $v \in T$ **then**

3      $\mathcal{E}' = \{e \in \mathcal{E} | v \in \pi(e)\}$

4   **for** $e \in \mathcal{E}'$ **do**

5         $qtem[e] = qtem[e] + score$

6   **if** $score \geq \epsilon$ and $e \notin S$ **then** $propagate(e, score)$;

7   **end**

8   **for** $t \in T$ **do**

9         $p\_value \leftarrow w_{TT}(t, v) \times score$

10   **if** $p\_value \geq \epsilon$ and $t \notin S$ **then**
          $propagate(t, p\_value \times synDecFactor)$;

11   **end**

12   **end**

13   **else**

14   // $v \in \mathcal{E}$

15   **for** $t \in \pi(v)$ **do**

16   **if** $score \geq \epsilon$ and $t \notin S$ **then** $propagate(t, p\_value)$;

17   **end**

18   **for** $e \in \{e \in \mathcal{E} | (v, e) \in E \vee (e, v) \in E\}$ **do**

19         $p\_value \leftarrow w_{\mathcal{E}\mathcal{E}}(v, e) \times score \times eSmoothingFactor$

20   **if** $p\_value \geq \epsilon$ and $e \notin S$ **then** $propagate(e, p\_value)$;

21   **end**

22   **end**

23   $S.remove(v)$

---

4   Evaluation and Results

4.1   Evaluation Protocol

The proposed evaluation is based on a set of expected entity evocations for given sets of words. Table 1 presents some of the 220 entries composing the evaluation set. Expected implicit evocations for each entry have been linked to WordNet 3.1 [11], a widely used lexical database. WordNet defines an ordering among sets of synonyms providing both, (i) the set of entities and their partial ordering (O), as well as (ii) the labeling function - π function.

The performance of the different approaches is evaluated by considering the number of queries for which the expected answer is provided among the top k results. For each approach, six evaluation settings have been compared by evaluating if the expected answer is found among sets composed of 1, 2, 3, 5, 10 or 20 best ranked results. In each setting, the ranked list of entities is computed by considering a set $\mathcal{E}' \subset \mathcal{E}$ corresponding to the expected answers for all evaluated queries ($|\mathcal{E}'| = 198$). Implementations of the $\sigma_{\mathcal{E}\mathcal{E}}$ measure have been made using SML (Semantic Measures Library) [4]. The σ_{TT} function used in the experiments uses Glove word embeddings [13]. Datasets, tested methods Java implementations as well as complete technical details about the evaluation are provided at https://github.com/sharispe/ICE.

Six models have been evaluated:

- Two Vector-based Aggregation Model definitions: VAM_MIN uses an aggregation strategy based on the min, VAM_SUM uses the sum.
- Four Graph-based Model (GM) definitions: two strategies using an aggregation approach based on median, using propagations at entity level or not (GM_MEDIAN_KB and GM_MEDIAN respectively); two strategies using an aggregation approach based on sum, using propagations at entity level or not (GM_SUM_KB and GM_SUM respectively).

4.2   Evaluation Results

Results are presented in Table 2. Considering the performance of evaluated systems setting k to 1 and 2, the results show that the best performance is obtained using a Vector-based Aggregation Model configuration taking advantage of the sum aggregation approach (VAM_SUM). It is interesting to underline that this approach does not take into account any information provided by the ordering of entities - while providing a 0.05 recall improvement over the best results that have been obtained using an approach taking advantage of taxonomic information (GM_MEDIAN_KB). Note also the critical impact of modifying the aggregation strategy using a VAM approach: by using a min aggregation strategy the performance highly decreases (e.g. a 0.11 difference is observed between VAM_MIN and VAM_SUM using $k = 1$). Considering the graph-based approach, the results highlight a large benefit of using taxonomical information for eliciting implicit entity evocations. Indeed, using both median and sum approaches, incorporating information provided by the taxonomy leads to a significant performance increase (cf. comparison of the scores between GM_MEDIAN/GM_MEDIAN_KB, as well as GM_SUM/GM_SUM_KB). It is finally worth noting that by setting k greater than 2, the best performances are achieved using a graph-based model taking advantage of taxonomical information. These results stress that using taxonomical information helps better identifying the semantic neighborhood of expected results, e.g. setting k = 20 GM_SUM_KB achieves a 0.72 recall while the VAM_SUM performance is 0.66.

Table 2. Evaluation results (recall).

Approach k=	1	2	5	10	20	50	100
VAM_MIN	0.17	0.25	0.34	0.41	0.51	0.67	0.81
VAM_SUM	**0.28**	**0.39**	0.51	0.6	0.66	0.8	0.86
GM_MEDIAN	0.20	0.29	0.46	0.54	0.59	0.66	0.74
GM_SUM	0.18	0.29	0.47	0.59	0.65	0.84	0.88
GM_MEDIAN_KB	0.24	0.35	0.49	0.58	0.64	0.74	0.81
GM_SUM_KB	0.23	0.32	**0.55**	**0.63**	**0.72**	**0.86**	**0.9**

5   Conclusion

In this paper, we have introduced the challenge of eliciting implicit entity evocations by stressing (i) its applications for improving automatic approaches enabling human spoken and written productions to be deeply understood, and (ii) its link to existing NLP and AI challenges (e.g. NER, Topic Modelling, Language Model). Several models mixing word embeddings analysis and symbolic representations provided by existing knowledge bases have been proposed. These models can be used to distinguish relevant implicit entities mentioned from a set of terms - they can therefore be used as core elements of more complex systems aiming at providing automatic analysis of the semantics of large units of language. The preliminaries results obtained in the performed experiments highlight the potential benefits of defining an hybrid approach combining word embeddings with symbolic representations for the task - even if additional experiments and configuration settings have further to be proposed and evaluated. To this aim, implementation source code, evaluation dataset and details of the performed experiments are shared to the community (cf. Sect. 4).

References

1. Auer, S., Bizer, C., Kobilarov, G., Lehmann, J., Cyganiak, R., Ives, Z.: Dbpedia: a nucleus for a web of open data. In: The semantic web, pp. 722–735 (2007)
2. Bengio, Y., Ducharme, R., Vincent, P., Jauvin, C.: A neural probabilistic language model. J. Mach. Learn. Res. **3**, 1137–1155 (2003)
3. Chen, S.F., Goodman, J.: An empirical study of smoothing techniques for language modeling. In: Proceedings of the 34th Annual Meeting on Association for Computational Linguistics, pp. 310–318. Association for Computational Linguistics (1996)
4. Harispe, S., Ranwez, S., Janaqi, S., Montmain, J.: The semantic measures library and toolkit: fast computation of semantic similarity and relatedness using biomedical ontologies. Bioinformatics **30**(5), 740–742 (2014)
5. Harispe, S., Ranwez, S., Janaqi, S., Montmain, J.: Semantic Similarity from Natural Language and Ontology Analysis, vol. 8. Morgan & Claypool Publishers, San Rafael (2015)
6. Hill, F., Cho, K., Korhonen, A., Bengio, Y.: Learning to understand phrases by embedding the dictionary. Trans. Assoc. Comput. Linguist. **4**, 17–30 (2015)
7. Jozefowicz, R., Vinyals, O., Schuster, M., Shazeer, N., Wu, Y.: Exploring the limits of language modeling. arXiv preprint arXiv:1602.02410 (2016)
8. Lin, D., et al.: An information-theoretic definition of similarity. ICML **98**, 296–304 (1998)
9. Manning, C.D., Raghavan, P., Schütze, H., et al.: Introduction to information retrieval, vol. 1. Cambridge University Press, Cambridge (2008)
10. Mendes, P.N., Jakob, M., García-Silva, A., Bizer, C.: Dbpedia spotlight: shedding light on the web of documents. In: Proceedings of the 7th International Conference on Semantic Systems, pp. 1–8. ACM (2011)
11. Miller, G.A.: Wordnet: a lexical database for English. Commun. ACM **38**(11), 39–41 (1995)

12. Nadeau, D., Sekine, S.: A survey of named entity recognition and classification. Lingvisticae Investigationes **30**(1), 3–26 (2007)
13. Pennington, J., Socher, R., Manning, C.D.: Glove: global vectors for word representation. EMNLP **14**, 1532–1543 (2014)
14. Steyvers, M., Griffiths, T.: Probabilistic topic models. Handb. Latent Semant. Anal. **427**(7), 424–440 (2007)
15. Thorat, S., Choudhari, V.: Implementing a reverse dictionary, based on word definitions, using a node-graph architecture. In: Proceedings of COLING 2016 (2016)

K-Nearest Neighbour Classification
for Interval-Valued Data

Vu-Linh Nguyen[1]($\boxtimes$), Sébastien Destercke[1], and Marie-Hélène Masson[1,2]

[1] UMR CNRS 7253 Heudiasyc, Sorbonne Universités,
Université de technologie de Compiègne, CS 60319, 60203 Compiègne cedex, France
{linh.nguyen,sebastien.destercke,mylene.masson}@hds.utc.fr
[2] Université de Picardie Jules Verne, Amiens, France

Abstract. This paper studies the problem of providing predictions with a K-nn approach when data have partial features given in the form of intervals. To do so, we adopt an optimistic approach to replace the ill-known values, that requires to compute sets of possible and necessary neighbours of an instance. We provide an easy way to compute such sets, as well as the decision rule that follows from them. Our approach is then compared to a simple imputation method in different scenarios, in order to identify those ones where it is advantageous.

1 Introduction

The K-nearest neighbor method (K-nn) is a simple but efficient classification method [1,6,16]. In classical K-nn, each label is assigned a predicted score and the one with the highest score will be considered as the optimal label of the target instance [1,6,16]. Such procedures usually assume that all training data are precisely specified.

In this paper, we are interested in the case where the features of some training data are imprecisely known, that is are known to lie in an interval. In this case, the notion of nearest neighbour is no longer well-defined, and the learning process has to be modified accordingly. Learning from interval-valued or partial data is not new, but has regained some interest in the last few years [2,13–15]. In practice, such interval-valued data can come from imprecise measurement devices, imperfect knowledge of an expert, or can also be the result of the summary of a huge data set.

In this paper, we intend to apply some generic learning procedures fitted to partial data [8,9] to the specific case of interval-valued K-nn methods. It should be noted that while the problem of modelling the uncertainty or the imprecision of the decision within a K-nn procedure applied to precise data has been well-treated in the literature (see *e.g.* [3,7,10]), few works deal with the problem of applying a K-nn method to interval-valued data [2]. Imputation methods [4] offer a way to solve this problem by replacing imprecise data by precise values, but typically do not aim at improving as much as possible the method accuracy. Instead, maximax or optimistic approaches [8] do intend to improve as much as possible the resulting accuracy.

© Springer International Publishing AG 2017
S. Moral et al. (Eds.): SUM 2017, LNAI 10564, pp. 93–106, 2017.
DOI: 10.1007/978-3-319-67582-4_7

In order to derive our K-nn method, we will adopt an approach similar to the one we previously successfully implemented for partially specified labels [11]. This approach was based on the use of two sets, the sets of possible and necessary predicted labels. These sets correspond to the sets of labels that would be predicted for at least one or all replacement(s) of the partial features, respectively. An important step of our approach will be to determine those sets for the case of interval-valued data from the sets of necessary and possible neighbours of an instance. We deal with this issue in Sect. 3, after having introduced our notations and settings in Sect. 2.

Our adaptation of the K-nn procedure, following the maximax approach put forward by Hüllermeier [8] to build predictive models from partial data, will then be derived in Sect. 4. We then provide some experimental results on several data sets in Sect. 5.

2   Preliminaries

In our setting, we assume that we have an imprecise training set $\mathbf{D} = \{(\mathbf{X}_n, y_n)|n = 1, \ldots, N\}$, used to make predictions, with the imprecise features $\mathbf{X}_n \subset \mathbb{R}^P$ and the precise label $y_n \in \Omega = \{\lambda_1, \ldots, \lambda_M\}$. We assume that X_n^p contains the precise value x_n^p in form of a of closed interval, or in other words, $X_n^p = [a_n^p, b_n^p]$. We are interested into predicting the class of a target instance $\mathbf{t}$, whose features are precisely known.

Let us first remind that in case of precise data, the Euclidean distance between a training instance $\mathbf{x}_n$ and a target instance $\mathbf{t}$ is given by

$$d(\mathbf{x}_n, \mathbf{t}) = \left(\sum_{p=1}^{P} \left(x_n^p - t^p \right)^2 \right)^{1/2}. \tag{1}$$

Then for a given target instance $\mathbf{t}$ and a number of nearest neighbours K, its nearest neighbour set in $\mathbf{D}$ will be denoted by $\mathbf{N_t} = \{\mathbf{x}_k^t | k = 1, \ldots, K\}$ where $\mathbf{x}_k^t$ is its k-th nearest neighbour. In the unweighted version of K-nn, the optimal prediction of $\mathbf{t}$ is

$$h(\mathbf{t}) = \arg \max_{\lambda \in \Omega} \sum_{\mathbf{x}_k^t \in \mathbf{N_t}} \mathbb{1}_{\lambda = y_k^t}, \tag{2}$$

with $\mathbb{1}_A$ is the indicator function of A ($\mathbb{1}_A = 1$ if A is true and 0 otherwise). The idea of the above method is to allow each nearest neighbour to give a vote for its label and the one with the highest number of votes is considered as the optimal prediction.

However, in case of imprecise feature data, there may be some uncertainty about what is the nearest neighbour set $\mathbf{N_t}$ of a target instance $\mathbf{t}$. As a consequence, Eq. (2) is no-longer applicable in order to make a decision on $\mathbf{t}$, the target instance $\mathbf{t}$ is ambiguous, and this ambiguity must be resolved in some way to make a decision. We will first focus on the problem of determining the ambiguity when having to decide the class of $\mathbf{t}$. Denote by $\mathbf{L}$ the set of all possible precise replacements of training set $\mathbf{D}$:

$$L = \left\{ 1 = \{(x_n, y_n) | x_n \in X_n, n = 1, \ldots, N\} \right\}. \tag{3}$$

To each replacement $1 \in L$ corresponds a well-defined nearest neighbour set N_t^1, on which Eq. (2) can be applied to find the optimal prediction(s) as follows

$$h_1(t) = \arg\max_{\lambda \in \Omega} \sum_{x_k^t \in N_t^1} \mathbb{1}_{\lambda = y_k^t}. \tag{4}$$

The sets of possible and necessary predicted labels are then defined as the sets of labels predicted for at least one replacement and for all possible replacements, respectively. Formally, this gives

$$PL_t = \left\{ \lambda \in \Omega | \exists 1 \in L \text{ s.t } \lambda \in h_1(t) \right\} \tag{5}$$

and

$$NL_t = \left\{ \lambda \in \Omega | \forall 1 \in L \text{ s.t } \lambda \in h_1(t) \right\}. \tag{6}$$

A target instance t is said to be ambiguous if and only if $PL_t \neq NL_t$. As we will see in the next section, determining such sets in the case of interval-valued features requires to compute the sets of necessary and possible neighbours. If the instance t is non-ambiguous, then the predictive value is clear and nothing needs to be done. If it is ambiguous, then an additional procedure must be performed to pick a prediction within PL_t. In this paper, we will adopt a maximax approach presented in Sect. 4.

3 Determining Ambiguous Instances

This section focus on determining whether a given instance is ambiguous, and what are the resulting possible and necessary label sets. In order to do so, we will first have to determine the possible and necessary neighbours.

3.1 Determining Interval Ranks

Given an imprecise training data set D and a precise instance t, Groenen *et al.* [5] provides simple formulae to determine the imprecise distance $d(X_n, t) = [\underline{d}(X_n, t), \overline{d}(X_n, t)]$ of $X_n \in D$ with respect to t:

$$\overline{d}(X_n, t) = \left(\sum_{p=1}^{P} \left[|c_n^p - t^p| + r_n^p \right]^2 \right)^{1/2}, \tag{7}$$

and

$$\underline{d}(X_n, t) = \left(\sum_{p=1}^{P} \max \left[0, |c_n^p - t^p| - r_n^p \right]^2 \right)^{1/2}, \tag{8}$$

where $c_n^p = (b_n^p + a_n^p)/2$ and $r_n^p = (b_n^p - a_n^p)/2$, $p = 1, \ldots, P$. Such interval distance allow us to define a partial order on the set $\mathbf{D}$ of training instance as follows

$$\mathbf{X}_i \succeq \mathbf{X}_j \text{ if } \underline{d}(\mathbf{X}_i, \mathbf{t}) \geq \overline{d}(\mathbf{X}_j, \mathbf{t}) \tag{9}$$

where $\mathbf{X}_i \succeq \mathbf{X}_j$ means that $\mathbf{X}_i$ is farther than $\mathbf{X}_j$ from $\mathbf{t}$. As demonstrated by Patil and Taille [12, Sect. 4.1], this partial order then allows us to derive interval rank values as we have that

$$\mathbf{X}_i \succeq \mathbf{X}_j \Rightarrow r(\mathbf{X}_i) \geq r(\mathbf{X}_j),$$

where $r(\mathbf{X}_i)$ is the rank that can be assigned to $\mathbf{X}_i$. Once the relation $\succeq$ is determined, $\mathbf{D}$ is a poset (partially ordered set) and the corresponding relation matrix, denoted by ζ, is a $N \times N$ matrix defined as

$$\zeta_{i,j} = \begin{cases} 1 \text{ if } \mathbf{X}_i \succeq \mathbf{X}_j \\ 0 \text{ otherwise.} \end{cases} \tag{10}$$

The results given by Theorems 1 and 2 in [12, Sect. 4.1] imply that each instance $\mathbf{X}_n$ can be associated to an imprecise rank which measures how close it is to the target instance $\mathbf{t}$ $i.e.$ $\mathbf{r}_n = [\underline{r}_n, \overline{r}_n]$ where

$$\underline{r}_n = \sum_{j=1}^{N} \zeta_{n,j} \text{ and } \overline{r}_n = N + 1 - \sum_{j=1}^{N} \zeta_{j,n}. \tag{11}$$

Example 1. Let us consider an example where $|\mathbf{D}| = 5$ and target instance $\mathbf{t}$ as illustrated in Fig. 1. Using the relation (9), the corresponding ζ matrix is given in Table 1.

By applying (11), we can easily compute the imprecise ranks of the training instances.

$$([\underline{r}_1, \overline{r}_1], [\underline{r}_2, \overline{r}_2], [\underline{r}_3, \overline{r}_3], [\underline{r}_4, \overline{r}_4], [\underline{r}_5, \overline{r}_5]) = ([2,4], [1,1], [2,4], [2,4], [5,5]). \tag{12}$$

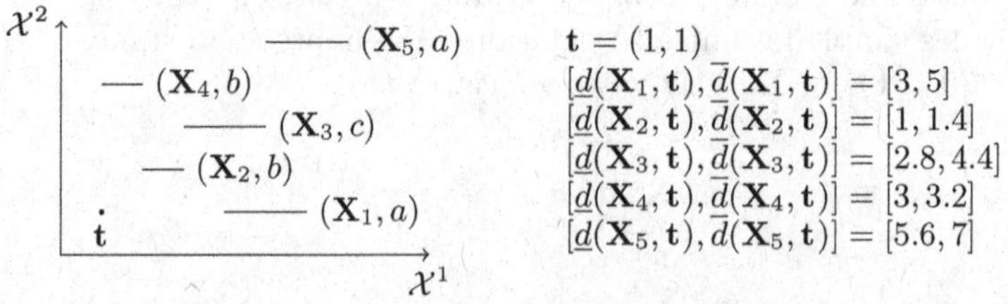

Fig. 1. Example with $|\mathbf{D}| = 5$

Table 1. The corresponding ζ matrix for example in Fig. 1

	$\mathbf{X}_1$	$\mathbf{X}_2$	$\mathbf{X}_3$	$\mathbf{X}_4$	$\mathbf{X}_5$	$\sum_r$
$\mathbf{X}_1$	1	1	0	0	0	2
$\mathbf{X}_2$	0	1	0	0	0	1
$\mathbf{X}_3$	0	1	1	0	0	2
$\mathbf{X}_4$	0	1	0	1	0	2
$\mathbf{X}_5$	1	1	1	1	1	5
$\sum_c$	2	5	2	2	1	

3.2 Determining the Possible Label Set

Let us now focus on the problem of determining whether a given label λ is a possible prediction for $\mathbf{t}$. Denoting by $\mathbf{R_t} = \{r_n = [\underline{r}_n, \overline{r}_n] | n = 1, \ldots, N\}$ the imprecise ranks of the instances in $\mathbf{D}$, we can easily determine the sets of possible and necessary neighbours as

$$\mathbf{PN_t} = \{\mathbf{X}_n | \underline{r}_n \leq K\} \tag{13}$$

and

$$\mathbf{NN_t} = \{\mathbf{X}_n | \overline{r}_n \leq K\}. \tag{14}$$

We have that $\mathbf{X}_n \in \mathbf{NN_t}$ if it is in the set of neighbours $\mathbf{X}_n \in \mathbf{N_t^l}$ for any replacement $\mathbf{l}$, while $\mathbf{X}_n \in \mathbf{PN_t}$ if $\mathbf{X}_n \in \mathbf{N_t^l}$ only for some replacement $\mathbf{l} \in \mathbf{L}$. For each label $\lambda \in \Omega$, we can then compute its minimum number of votes

$$s_\mathbf{t}^{small}(\lambda) = \left| \{\mathbf{X}_n | \mathbf{X}_n \in \mathbf{NN_t}, y_n = \lambda\} \right|, \tag{15}$$

given by its necessary neighbours. From s^{small} can then be deduced the maximal and minimal number of votes it can receive from K neighbours, according to the following formulae

$$s_\mathbf{t}^{max}(\lambda) = \min \left[K - \sum_{\lambda' \neq \lambda} s_\mathbf{t}^{small}(\lambda'), \left| \{\mathbf{X}_n | \mathbf{X}_n \in \mathbf{PN_t}, y_n = \lambda\} \right| \right], \tag{16}$$

and

$$s_\mathbf{t}^{min}(\lambda) = \max \left[s_\mathbf{t}^{small}(\lambda), K - \sum_{\lambda' \neq \lambda} s_\mathbf{t}^{max}(\lambda') \right]. \tag{17}$$

These scores are simply derived from the fact that, among the K neighbours, at least $s^{small}(\lambda)$ among them must give their votes to label λ. This is proved in the next Lemma, where it is shown that $s_\mathbf{t}^{min}(\lambda)$ and $s_\mathbf{t}^{max}(\lambda)$ are the minimum and maximum number of votes that can be given to λ over all replacements $\mathbf{l} \in \mathbf{L}$.

Lemma 1. *Given number of nearest neighbours K, a target instance $\mathbf{t}$, the corresponding maximum and minimum score vectors $(s_\mathbf{t}^{min}(\lambda_1), \ldots, s_\mathbf{t}^{min}(\lambda_M))$ and $(s_\mathbf{t}^{max}(\lambda_1), \ldots, s_\mathbf{t}^{max}(\lambda_M))$, then for any $\lambda \in \Omega$, we have that*

$$s_t^{min}(\lambda) = \min_{l \in \mathbf{L}} s_t^l(\lambda) \; and \; s_t^{max}(\lambda) = \max_{l \in \mathbf{L}} s_t^l(\lambda) \tag{18}$$

and consequently, we have that, for $\forall l \in \mathbf{L}$,

$$s_t^{max}(\lambda) \geq s_t^l(\lambda) \geq s_t^{min}(\lambda), \forall \lambda \in \Omega. \tag{19}$$

Proof. The relation that $s_t^{max}(\lambda) = \max_{l \in \mathbf{L}} s_t^l(\lambda)$ can be simply proved by observing that $K - \sum_{\lambda' \neq \lambda} s_t^{small}(\lambda')$ bounds the number of instance that could be in the set of nearest neighbours and have λ for label, while the value $|\{\mathbf{X}_n | \mathbf{X}_n \in \mathbf{PN_t}, y_n = \lambda\}|$ simply gives the maximal number of such elements that are available within the set of possible neighbours, and that may be chosen freely to be/not be in the neighbour set, as long as they remain lower than the bound $K - \sum_{\lambda' \neq \lambda} s_t^{small}(\lambda')$. So, maximising this number of elements simply provides $s_t^{max}(\lambda)$.

Let us now prove that $s_t^{min}(\lambda) = \min_{l \in \mathbf{L}} s_t^l(\lambda)$, recalling that we just proved that $s_t^{max}(\lambda)$ is reachable for some replacement. We are going to focus on two cases:

1. $s_t^{small}(\lambda) \geq K - \sum_{\lambda' \neq \lambda} s_t^{max}(\lambda')$, meaning that $s_t^{min}(\lambda) = s_t^{small}(\lambda)$, hence for every replacement there is at least $s_t^{small}(\lambda)$ nearest neighbors of label λ. Furthermore, $s_t^{small}(\lambda) \geq K - \sum_{\lambda' \neq \lambda} s_t^{max}(\lambda')$ implies that $\sum_{\lambda' \neq \lambda} s_t^{max}(\lambda') + s_t^{small}(\lambda) \geq K$, meaning that we can choose the remaining $K - s_t^{small}(\lambda)$ neighbours so that they vote for other labels. In other words, we can find a replacement l where $s_t^{small}(\lambda) = s_t^l(\lambda)$, proving that $s_t^{min}(\lambda) = \min_{l \in \mathbf{L}} s_t^l(\lambda)$ in the first case.
2. $s_t^{small}(\lambda) < K - \sum_{\lambda' \neq \lambda} s_t^{max}(\lambda')$, meaning that $s_t^{min}(\lambda) = K - \sum_{\lambda' \neq \lambda} s_t^{max}(\lambda')$. First note that for any replacement we cannot have $s_t^l(\lambda) < K - \sum_{\lambda' \neq \lambda} s_t^{max}(\lambda')$, otherwise the set of nearest neighbour would be necessarily lower than K. $s_t^{min}(\lambda)$ then reaches this lower bound by simply taking the replacement s^l for which we have $s_t^l(\lambda') = s_t^{max}(\lambda')$, proving that $s_t^{min}(\lambda) = \min_{l \in \mathbf{L}} s_t^l(\lambda)$ in the second case.

$\square$

For a label $\lambda_m \in \Omega$, the relations among scores (19) and the definition of the possible label set (5) imply that λ_m is a possible label ($\lambda_m \in \mathbf{PL_t}$) if and only if there is a replacement $l \in \mathbf{L}$ with a score vector $(s_t^l(\lambda_1), \ldots, s_t^l(\lambda_M))$ such that

$$\sum_{i=1}^{M} s_t^l(\lambda_i) = K, \tag{20}$$

and

$$\min \left(s_t^l(\lambda_m), s_t^{max}(\lambda_i) \right) \geq s_t^l(\lambda_i) \geq s_t^{min}(\lambda_i), i = 1, \ldots, M. \tag{21}$$

The condition $\sum_{i=1}^{M} s_t^l(\lambda_i) = K$ simply ensures that l is a legal replacement. The constraint (21) then ensures that all other labels have a score lower than

$s_t^l(\lambda_m)$ for the replacement l (note that $\min(s_t^l(\lambda_m), s_t^{max}(\lambda_m)) = s_t^l(\lambda_m)$), and that their scores are bounded by Eq. (19).

The question is now to know whether we can instantiate such a vector making a winner of λ_m. To achieve this task, we will first maximise its score, such that $s_t^l(\lambda_m) = s_t^{max}(\lambda_m)$. The scores of all other labels λ_i is also lower-bounded by $s_t^{min}(\lambda_i)$, meaning that among the K neighbours we choose in l, only $K - s_t^{max}(\lambda_m) - \sum_{i=1,i\neq m}^{M} s_t^{min}(\lambda_i)$ remain to be fixed in order to specify the score vector. Then we can focus on the relative difference between $s_t^{min}(\lambda_i)$ and the additional number of chosen neighbours voting for λ_i. Solving the problem defined by Eqs. (20) and (21) is equivalent to determine a score vector $(w(\lambda_1), \ldots, w(\lambda_{m-1}), w(\lambda_{m+1}), \ldots, w(\lambda_M))$ with $w(\lambda_i) = s_t^l(\lambda_i) - s_t^{min}(\lambda_i)$, $\forall \lambda_i \neq \lambda_m$, s.t.

$$\sum_{i=1,i\neq m}^{M} w(\lambda_i) = K - s_t^{max}(\lambda_m) - \sum_{i=1,i\neq m}^{M} s_t^{min}(\lambda_i), \tag{22}$$

$$\min\left(s_t^{max}(\lambda_m), s_t^{max}(\lambda_i)\right) - s_t^{min}(\lambda_i) \geq w(\lambda_i) \geq 0, \forall \lambda_i \neq \lambda_m. \tag{23}$$

Equation (22) again ensures that the replacement is a legal one (the number of neighbours sums up to K), and Eq. (23) ensures that λ_m is a winning label. Also note that if $\exists \lambda_i \in \Omega \setminus \{\lambda_m\}$ s.t $s_t^{max}(\lambda_m) < s_t^{min}(\lambda_i)$, then there no chance for λ_m to be a possible label.

We will now give a proposition allowing to determine in an easy way if a label belongs to the set of possible labels.

Proposition 1. *Given the number of nearest neighbours K, a target instance* t, *its corresponding maximum and minimum score vectors* $(s_t^{min}(\lambda_1), \ldots, s_t^{min}(\lambda_M))$ *and* $(s_t^{max}(\lambda_1), \ldots, s_t^{max}(\lambda_M))$. *Assuming that* $s_t^{max}(\lambda_m) \geq s_t^{min}(\lambda_i)$, *for* $\forall \lambda_i \in \Omega \setminus \{\lambda_m\}$, *then* λ_m *is a possible label if and only if*

$$K \leq s_t^{max}(\lambda_m) + \sum_{i=1,i\neq m}^{M} \min\left(s_t^{max}(\lambda_m), s_t^{max}(\lambda_i)\right) \tag{24}$$

Proof. ($\Rightarrow$) Let us prove that λ_m being a possible label implies (24). First, if $\lambda_m \in \mathbf{PL_t}$ and l is a legitimate replacement, we have that

$$w(\lambda_i) \leq \min\left(s_t^{max}(\lambda_m), s_t^{max}(\lambda_i)\right) - s_t^{min}(\lambda_i), \forall i \neq m \tag{25}$$

otherwise λ_m would not be a winner, or we would give a higher score to λ_i than it actually can get (we would have $s^l(\lambda_i) > s_t^{max}(\lambda_i)$). Since for any replacement we have that Eq. (22) must be satisfied, we have necessarily

$$K - s_t^{max}(\lambda_m) - \sum_{i=1,i\neq m}^{M} s_t^{min}(\lambda_i) = \sum_{i=1,i\neq m}^{M} w(\lambda_i).$$

If we replace $w(\lambda_i)$ by its upper bound (25), we get the following inequality

$$K - s_t^{max}(\lambda_m) - \sum_{i=1,i\neq m}^{M} s_t^{min}(\lambda_i) \leq \sum_{i=1,i\neq m}^{M} \min\left(s_t^{max}(\lambda_m)), s_t^{max}(\lambda_i)\right)$$
$$- \sum_{i=1,i\neq m}^{M} s_t^{min}(\lambda_i),$$

that is equivalent to the relation

$$K \leq s_t^{max}(\lambda_m) + \sum_{i=1,i\neq m}^{M} \min\left(s_t^{max}(\lambda_m), s_t^{max}(\lambda_i)\right).$$

($\Leftarrow$) Let us now show that if the conditions given by Eqs. (22) and (23) are satisfied, then $\lambda_m \in \mathbf{PL_t}$. First remark that, once we have assigned the maximal score to λ_m and the minimal ones to the other labels, there remain

$$K - s_t^{max}(\lambda_m) - \sum_{i=1,i\neq m}^{M} s_t^{min}(\lambda_i)$$

neighbours to choose from. We also know from (23) that at most

$$\sum_{i=1,i\neq m}^{M} \left[\min\left(s_t^{max}(\lambda_m), s_t^{max}(\lambda_i)\right) - s_t^{min}(\lambda_i)\right]$$

neighbours can still be affected to other labels than λ_m without making it a loser. Clearly, if

$$K - s_t^{max}(\lambda_m) - \sum_{i=1,i\neq m}^{M} s_t^{min}(\lambda_i) \leq \sum_{i=1,i\neq m}^{M} \left[\min\left(s_t^{max}(\lambda_m), s_t^{max}(\lambda_i)\right) - s_t^{min}(\lambda_i)\right],$$

we can reach the number of K neighbours without making λ_m a loser, or inversely letting λ_m be a winner for the chosen replacement, meaning that $\lambda_m \in \mathbf{PL_t}$. $\square$

Example 2. Let us continue with the data set in Example 1 with value $K = 3$. From Table 1 and the interval ranks (12), we can see that

$$\mathbf{PN_t} = \{(\mathbf{X}_1, a), (\mathbf{X}_2, b), (\mathbf{X}_3, c), (\mathbf{X}_4, b)\}, \mathbf{NN_t} = \{(\mathbf{X}_2, b)\}.$$

Then the maximum and minimum scores for all the labels are

$$(s_t^{min}(a), s_t^{min}(b), s_t^{min}(c)) = (0, 1, 0)$$
$$(s_t^{max}(a), s_t^{max}(b), s_t^{max}(c)) = (1, 2, 1).$$

We will now determine whether a given label in $\Omega = \{a, b, c\}$ is a possible label. For label a, we have that

$$s_t^{max}(a) + \min\left(s_t^{max}(a), s_t^{max}(b)\right) + \min\left(s_t^{max}(a), s_t^{max}(c)\right) = 1 + 1 + 1 = 3 \geq K,$$

hence $a \in \mathbf{PL_t}$. The same procedure applied to b and c gives the result $\mathbf{PL_t} = \{a, b, c\}$.

3.3 Determining Necessary Label Set

Let us now focus on characterizing the set $\mathbf{NL_t}$. The following propositions gives a very easy way to determine it, by simply comparing the minimum score of a given label λ to the maximal scores of the others.

Proposition 2. *Given the maximum and minimum scores* $\left(s_t^{min}(\lambda_1), \ldots, s_t^{min}(\lambda_M)\right)$ *and* $\left(s_t^{max}(\lambda_1), \ldots, s_t^{max}(\lambda_M)\right)$, *then a given label* λ *is a necessary label if and only if*

$$s_t^{min}(\lambda) \geq s_t^{max}(\lambda'), \forall \lambda' \neq \lambda. \tag{26}$$

Proof. ($\Rightarrow$) We proceed by contradiction. Assuming that $\exists \lambda \in \mathbf{NL_t}$ and $\exists \lambda' \in \Omega$ where $s_t^{min}(\lambda) < s_t^{max}(\lambda')$, we show that we can always find a replacement $l \in \mathbf{L}$ s.t $s_t^l(\lambda) < s_t^l(\lambda')$, or in other words, $\exists l \in \mathbf{L}$ s.t $\lambda \notin h_l(\mathbf{t})$, and therefore λ is not necessary. Let us consider the two cases

1. $K - \sum_{\lambda'' \neq \lambda} s_t^{max}(\lambda'') \geq s_t^{small}(\lambda)$, then for $\forall \lambda'' \neq \lambda$, we give its the maximum score s.t $s_t^l(\lambda'') = s_t^{max}(\lambda'')$ and give λ the score $s_t^l(\lambda) = K - \sum_{\lambda'' \neq \lambda} s_t^{max}(\lambda'')$. Then it is clear that

$$s_t^l(\lambda) = K - \sum_{\lambda'' \neq \lambda} s_t^{max}(\lambda'') = s_t^{min}(\lambda) < s_t^{max}(\lambda') = s_t^l(\lambda').$$

2. $K - \sum_{\lambda'' \neq \lambda} s_t^{max}(\lambda'') < s_t^{small}(\lambda)$, then we give λ a score $s_t^l(\lambda) = s_t^{small}(\lambda)$ and give λ' a score $s_t^l(\lambda') = s_t^{smax}(\lambda')$. As we have

$$K < \sum_{\lambda'' \neq \{\lambda, \lambda'\}} s_t^{max}(\lambda'') + s_t^{small}(\lambda) + s_t^{max}(\lambda')$$

by assumption, we can choose $K - s_t^{small}(\lambda) - s_t^{smax}(\lambda')$ nearest neighbours from at most $\sum_{\lambda'' \neq \{\lambda, \lambda'\}} s_t^{max}(\lambda'')$ possible nearest neigbhours whose labels are not λ or λ'. In such a replacement we have $s_t^l(\lambda) < s_t^l(\lambda')$.

($\Leftarrow$) We are going to prove that (26) implies that the label $\lambda \in \mathbf{NL_t}$ is necessary. Let us first note that

$$\min_{l \in \mathbf{L}} s_t^l(\lambda) = s_t^{min}(\lambda) \text{ and } \max_{l \in \mathbf{L}} s_t^l(\lambda') = s_t^{max}(\lambda'), \forall \lambda' \neq \lambda,$$

then (26) ensures that, for any replacement $l \in \mathbf{L}$,

$$s_t^l(\lambda) \geq \min_{l \in \mathbf{L}}(s_t^l(\lambda)) \geq \max_{l \in \mathbf{L}}(s_t^l(\lambda')) \geq s_t^l(\lambda'), \forall \lambda' \neq \lambda,$$

which is sufficient to get the proof. $\qquad\square$

Example 3. Consider the data set given in Example 2 with the maximum and minimum scores of the labels are

$$(s_t^{min}(a), s_t^{min}(b), s_t^{min}(c)) = (0, 1, 0)$$
$$(s_t^{max}(a), s_t^{max}(b), s_t^{max}(c)) = (1, 2, 1).$$

Then (26) implies that the necessary label set $\mathbf{NL_t} = \{b\}$.

4   Learning from Interval-Valued Feature Data

We are now going to present a maximax approach that can be used to make decision on interval-valued feature data. Let us first note that whenever the data is imprecise, the decision rule (2) is no longer well-defined. In case of *set-valued labels*, such a decision rule can be generalized as a maximax rule [9] where the optimal prediction of $\mathbf{t}$ is

$$h(\mathbf{t}) = \arg\max_{\lambda \in \Omega} \sum_{\mathbf{x}_k^t \in \mathbf{N_t}} \mathbb{1}_{\lambda \in \mathbf{y}_k^t}. \tag{27}$$

The idea of the above method is to assign for each label the highest number of votes that it could get. Let call such number of votes by optimal score, then the label with the highest optimal score will be considered as the optimal decision of $\mathbf{t}$. Note that in case of interval-valued feature data, as point out in Lemma 1, the score $s^{max}(\lambda)$ defined in (16) is nothing else but the optimal score of λ. Then the maximax approach can be then generalized for *interval-valued feature data* as follows

$$h(\mathbf{t}) = \arg\max_{\lambda \in \Omega} s_{\mathbf{t}}^{max}(\lambda) \tag{28}$$

$$= \arg\max_{\lambda \in \Omega} \left(\min\left[K - \sum_{\lambda' \neq \lambda} s_{\mathbf{t}}^{small}(\lambda'), \left| \{\mathbf{X}_n | \mathbf{X}_n \in \mathbf{PN_t}, y_n = \lambda\} \right| \right] \right).$$

The procedure to make predictions is summarized in Algorithm 1. It is then clear that if $\lambda \in h(\mathbf{t})$, then λ is the winner in at least one replacement $\mathbf{l} \in \mathbf{L}$. Of course, unless we have $|\mathbf{PL_t}| = |\mathbf{NL_t}| = 1$, we cannot be sure that λ is the prediction that fully precise data would have given us. It merely says that it is the most promising one, in the sense that it is the one with the highest potential score. We may suspect that the higher is $|\mathbf{PL_t}|$, the more likely we are to commit mistakes, as the ambiguity increases. It would then be interesting to wonder if we could reduce $|\mathbf{PL_t}|$ by querying the data and making some of their feature precise, using techniques similar to active learning. Yet, we leave the investigation of such an approach to future research.

It may also happen that Eq. (28) returns multiple instances that have the highest number of votes. We can then follow a different strategy, where we consider the result of the K-nn procedure for a peculiar replacement. Since every label receives its maximal number of votes by considering the lower distance $\underline{d}(\mathbf{X}_n, \mathbf{t})$, a quite simple idea is to consider the result obtained by Eq. (27) when we consider the replacement $\mathbf{l}$ giving $d(\mathbf{X}_n, \mathbf{t}) = \underline{d}(\mathbf{X}_n, \mathbf{t})$ for every $\mathbf{X}_n$.

5   Experiments

We run experiments on a contaminated version of 6 standard benchmark data sets described in Table 2. By contamination, we mean that we introduce artificially imprecision in these precise data sets. These data sets have various numbers

---

Algorithm 1. Maximax approach for interval-valued training data.

Input: **D**-imprecise training data, **T**-test set, K-number of nearest neighbours
Output: $\{p(\mathbf{t})|\mathbf{t} \in \mathbf{T}\}$-predictions

1 **foreach** $\mathbf{t} \in \mathbf{T}$ **do**
2   compute its zeta matrix ζ through (7)–(10);
3   **foreach** $\mathbf{X}_n \in \mathbf{D}$ **do**
4   compute imprecise rank $[\underline{r}_n, \overline{r}_n]$ defined in (11);
5   determine the $\mathbf{PN}_\mathbf{t}$ and $\mathbf{NN}_\mathbf{t}$ defined in (13)–(14);
6   **foreach** $\lambda \in \Omega$ **do**
7   compute $s_\mathbf{t}^{max}(\lambda)$ through (15)–(16);
8   determine $h(\mathbf{t})$ defined in (28);
9   **if** $|h(\mathbf{t})| = 1$ **then**
10         $p(\mathbf{t}) = h(\mathbf{t})$;
11   **else**
12   replace the imprecise distances by $\mathbf{d_t} = \{\underline{d}(\mathbf{X}_n, \mathbf{t})|n = 1, \dots, N\}$;
13   determine $p(\mathbf{t})$ by performing classical K-nn on $\mathbf{d_t}$;

---

of classes and features, but have a relatively small number of instances, for the reason that handling imprecise data is mainly problematic in such situations: when a lot of data are present, we can expect that enough sufficiently precise data will exist to reach an accuracy level similar to the one of fully precise methods.

Table 2. Data sets used in the experiments

Name	# instances	# features	# labels
Iris	150	4	3
Seeds	210	7	3
Glass	214	9	6
Ecoli	336	7	8
Dermatology	385	34	6
Vehicle	846	18	4

Our experimental setting is as follows: given a data set, we randomly chose a training set **D** consisting of 10% of instances and the rest (90%) as a test set **T**, to limit the number of training samples. For each training instance $\mathbf{x}_i \in \mathbf{D}$ and each feature x_i^j, a biased coin is flipped in order to decide whether or not the feature x_i^j will be contaminated; the probability of contamination is p and we have tested different values of it ($\{0.2, 0.4, 0.6, 0.8\}$). In case x_i^j is contaminated, its precise value is transformed into an interval which can be asymmetric with respect to x_i^j. To do that, a pair of widths $\{l_i^j, r_i^j\}$ will be generated from two Beta distributions,

Table 3. Experimental Results: Accuracy of classifiers (%)

		Iris	Seeds	Glass	Ecoli	Derma	Vehicle
$p = 0.2, \epsilon = 0.25$	Precise	91.55	84.88	49.70	75.21	82.26	53.55
	Imputation	88.93	83.79	47.30	74.40	80.20	49.45
	Maximax	**89.39**	**83.80**	**48.37**	**74.57**	**81.19**	**53.21**
$p = 0.2, \epsilon = 0.5$	Precise	91.57	85.15	50.46	74.98	81.76	53.65
	Imputation	89.07	**84.16**	47.41	**74.23**	77.41	50.35
	Maximax	**89.43**	83.92	**48.54**	74.13	**80.55**	**53.19**
$p = 0.2, \epsilon = 1$	Precise	91.35	85.39	50.49	75.11	82.13	53.65
	Imputation	88.80	**84.36**	47.48	**74.52**	75.12	50.76
	Maximax	**89.08**	84.31	**48.73**	74.35	**80.54**	**53.24**
$p = 0.4, \epsilon = 0.25$	Precise	91.44	85.31	50.34	75.33	82.26	53.54
	Imputation	87.70	83.83	46.70	**74.49**	75.87	49.88
	Maximax	**88.59**	**83.88**	**48.06**	74.02	**80.32**	**52.95**
$p = 0.4, \epsilon = 0.5$	Precise	91.14	85.26	50.20	75.47	82.04	53.50
	Imputation	87.00	**83.77**	46.31	**74.60**	75.14	49.70
	Maximax	**87.42**	83.61	**47.69**	73.87	**79.75**	**52.79**
$p = 0.4, \epsilon = 1$	Precise	91.11	85.33	50.18	75.36	82.24	53.52
	Imputation	**86.87**	**83.80**	46.17	**74.62**	73.10	49.77
	Maximax	86.59	83.52	**47.58**	73.57	**79.51**	**52.70**
$p = 0.6, \epsilon = 0.25$	Precise	92.53	84.59	50.82	74.54	81.10	53.25
	Imputation	80.46	**80.88**	43.56	**72.27**	75.38	43.41
	Maximax	**84.86**	80.85	**45.90**	69.48	**77.40**	**50.87**
$p = 0.6, \epsilon = 0.5$	Precise	92.00	85.39	50.97	74.86	81.98	53.38
	Imputation	80.06	**82.51**	44.04	**73.13**	73.28	45.10
	Maximax	**82.43**	82.06	**46.08**	70.24	**77.29**	**50.75**
$p = 0.6, \epsilon = 1$	Precise	91.66	85.57	51.01	74.83	81.97	53.46
	Imputation	80.22	**82.47**	44.37	**73.45**	68.41	46.48
	Maximax	**80.79**	82.16	**46.19**	70.47	**75.84**	**50.59**
$p = 0.8, \epsilon = 0.25$	Precise	91.62	85.46	50.74	74.97	81.91	53.40
	Imputation	79.13	**81.92**	44.34	**73.27**	69.42	44.52
	Maximax	**81.26**	81.86	**45.88**	70.19	**76.04**	**48.88**
$p = 0.8, \epsilon = 0.5$	Precise	91.27	85.29	50.85	74.92	82.08	53.44
	Imputation	78.53	81.95	44.33	**73.34**	69.00	44.18
	Maximax	**80.92**	**82.00**	**45.66**	70.17	**75.71**	**48.32**
$p = 0.8, \epsilon = 1$	Precise	91.16	85.35	50.71	75.00	82.18	53.45
	Imputation	78.58	82.04	44.25	**73.60**	66.67	44.71
	Maximax	**80.38**	**82.47**	**45.48**	70.46	**74.99**	**47.92**

Fixed parameters: $K = 3, \beta = 10$

$Beta(\alpha_l, \beta)$ and $Beta(\alpha_r, \beta)$. To control the skewness of the generated data, we introduce a so called unbalance parameter ϵ and assign $\{\alpha_l, \alpha_r\} = \{\beta * \epsilon, \beta/\epsilon\}$. Then the generated interval valued data is $X_i^j = [x_i^j + l_i^j(\underline{D}^j - x_i^j), x_i^j + r_i^j(\overline{D}^j - x_i^j)]$ where $\underline{D}^j = \min_i(x_i^j)$ and $\overline{D}^j = \max_i(x_i^j)$. As usual when working with Euclidean distance based K-nn, data is normalized. Then, the proposed method is used to make predictions on the test set and its accuracy is compared with the accuracy of two other cases: classical K-nn when fully precise data is given, and a basic imputation method consisting in replacing an interval-valued data X_i^j by its middle value, i.e., $x_i^j = (\underline{X}_i^j + \overline{X}_i^j)/2$. The disambiguated data is used to make predictions under the classical K-nn procedure.

Because the training set is randomly chosen and contaminated, the results maybe affected by random components. Then, for each data set, we repeat the above procedure 100 times and compute the average results. The experimental results on the data sets described in Table 2 with several combinations of parameters (K, p, ϵ, β) are given in the Table 3, with the best results between imputation and the presented method put in bold (the precise case only serves as a reference value of the best accuracy achievable). These first results show that the difference between the two approaches is generally small. Surprisingly, this is true for all explored settings, even for skewed imprecision and high uncertainty ($\epsilon = 0.25$, $p = 0.8$). However, on the two data sets dermatology and vehicle, our approach really provides a significant, consistent increase of accuracy, and this even for low and balanced imprecision ($\epsilon = 1$, $p = 0.2$). In the future, we intend to do more experiments (varying K, increasing the number of data sets) and also try to understand the origin of the witnessed difference.

6   Conclusion

In this paper, we have proposed a maximax approach to deal with K-nn predictions when features are imprecisely specified. Our method mainly relies on identifying possible neighbours in an efficient manner, using the partial orders induced by distance intervals to do so. First experiments suggest that a simple imputation method could often work as well as the presented approach, but for some data sets our approach can bring a real advantage. Compared to imputation methods, our approach also provides us with information about how uncertain our prediction is, by identifying possible and necessary neighbours.

Such information is instrumental in the next step we envision for this work: determining which sample feature should be queried first to improve the overall algorithm accuracy, much like what we did for the case of partial labels [11]. Also, investigating the decision rules and querying procedure when both training and test data can be imprecise is another future direction though defining the partial order (9) is still a challenge.

Acknowledgement. This work was carried out in the framework of Labex MS2T and EVEREST projects, which were funded by the French National Agency for Research (Reference ANR-11-IDEX-0004-02, ANR-12-JS02-0005).

References

1. Cover, T., Hart, P.: Nearest neighbor pattern classification. IEEE Trans. Inf. Theory **13**(1), 21–27 (1967)
2. de Souza, R.M., De Carvalho, F.D.A.: Clustering of interval data based on city-block distances. Pattern Recogn. Lett. **25**(3), 353–365 (2004)
3. Denoeux, T.: A k-nearest neighbor classification rule based on dempster-shafer theory. IEEE Trans. Syst. Man Cybern. **25**, 804–813 (1995)
4. Donders, A.R.T., van der Heijden, G.J., Stijnen, T., Moons, K.G.: Review: a gentle introduction to imputation of missing values. J. Clin. Epidemiol. **59**(10), 1087–1091 (2006)
5. Groenen, P.J., Winsberg, S., Rodriguez, O., Diday, E.: I-scal: multidimensional scaling of interval dissimilarities. Comput. Stat. Data Anal. **51**(1), 360–378 (2006)
6. Hastie, T., Tibshirani, R., Friedman, J., Franklin, J.: The elements of statistical learning: data mining, inference and prediction. Math. Intell. **27**(2), 83–85 (2005)
7. Holmes, C., Adams, N.: A probabilistic nearest neighbour method for statistical pattern recognition. J. Roy. Statist. Soc. Ser. B **64**, 295–306 (2002)
8. Hüllermeier, E.: Learning from imprecise and fuzzy observations: data disambiguation through generalized loss minimization. Int. J. Approximate Reasoning **55**(7), 1519–1534 (2014)
9. Hüllermeier, E., Beringer, J.: Learning from ambiguously labeled examples. Intell. Data Anal. **10**(5), 419–439 (2006)
10. Keller, J., Gray, M., Givens, J.: A fuzzy k-nearest neighbor algorithm. IEEE Trans. Syst. Man Cybern. **15**, 580–585 (1985)
11. Nguyen, V.-L., Destercke, S., Masson, M.-H.: Querying partially labelled data to improve a k-nn classifier. In: AAAI, pp. 2401–2407 (2017)
12. Patil, G., Taillie, C.: Multiple indicators, partially ordered sets, and linear extensions: multi-criterion ranking and prioritization. Environ. Ecol. Stat. **11**(2), 199–228 (2004)
13. Silva, A.P.D., Brito, P.: Linear discriminant analysis for interval data. Comput. Stat. **21**(2), 289–308 (2006)
14. Utkin, L.V., Chekh, A.I., Zhuk, Y.A.: Binary classification SVM-based algorithms with interval-valued training data using triangular and epanechnikov Kernels. Neural Netw. **80**, 53–66 (2016)
15. Utkin, L.V., Coolen, F.P.: Interval-valued regression and classification models in the framework of machine learning. In: ISIPTA, vol. 11, pp. 371–380. Citeseer (2011)
16. Wu, X., Kumar, V., Quinlan, J.R., Ghosh, J., Yang, Q., Motoda, H., McLachlan, G.J., Ng, A., Liu, B., Philip, S.Y., et al.: Top 10 algorithms in data mining. Knowl. Inf. Syst. **14**(1), 1–37 (2008)

Estimating Conditional Probabilities by Mixtures of Low Order Conditional Distributions

Andrés Cano, Manuel Gómez-Olmedo, and Serafín Moral[✉]

Dpto. Ciencias de la Computación e IA, Universidad de Granada,
18071 Granada, Spain
{acu,mgomez,smc}@decsai.ugr.es

Abstract. Estimating probabilities of a multinomial variable conditioned to a large set of variables is an important problem due to the fact that the number of parameters increases in an exponential way with the number of conditional variables. Some models, such as noisy-or gates make assumptions about the relationships between the variables that assume that the number of parameters is linear. However, there are cases in which these hypothesis do not make sense. In this paper, we present a procedure to estimate a large conditional probability distribution by means of an average of low order conditional probabilities. In this way the number of necessary parameters can be reduced to a quantity which can be estimated with available data. Different experiments show that the quality of the estimations can be improved with respect to a direct estimation.

Keywords: Bayesian networks · Parametric estimation · Large dimension conditional probabilities

1 Introduction

Parametric estimation in a Bayesian network consists in the estimation of a conditional probability distribution for each variable given its parents [9]. Maximum likelihood estimation or Bayesian procedures, such as Laplace correction, are the most usual procedures. However, there is an important problem when the number of variables to which we are conditioning is high: for each configuration of values of the parents we have to estimate a distribution for the variable. As the number of configurations is exponential as a function of the number of parents and each element of the sample can be used only for one of these estimation, the sample used to estimate each of these distributions can be very short or even null.

In some situations, it is possible to assume a model of interaction between a variable and its parents which can reduce the number of parameters, as the case of noisy-or gates [8,12]. Even if in [15], it is reported that noisy-or gates can provide a reasonable estimation in about 50% of the conditional probability

© Springer International Publishing AG 2017
S. Moral et al. (Eds.): SUM 2017, LNAI 10564, pp. 107–118, 2017.
DOI: 10.1007/978-3-319-67582-4_8

tables of some well known networks, the experimental setting is very limited and these results can be strongly dependent on the particular case we are considering. So, in many practical situations we may have large conditional probabilities in which this model does not provide a reasonable fit.

In this paper, we propose a general estimation procedure that consists in estimating a set of low dimension conditional distributions for the variable given different small subsets of the original set of parents. The final estimation is an average of these smaller conditional probability distributions. This average can be uniform or weighted by assuming the existence of a hidden variable and using an **EM** algorithm to estimate the weights [5].

It is true that some learning procedures do not induce a too complex network for a given sample size: for example, when using Bayesian scores [11], in general the complexity of the network depends on the sample size, in such a way that with small sample sizes we have many missing links with respect to the true network. But this is only a tendency and there are many other learning procedures such as the **PC** algorithm [14] in which this is not always true: we can learn very complex networks from small datasets. For example, if we have a network with only one variable Y dependent on a large set of variables and this set of variables are independent (there are only links from the variables in this set to Y), then this structure has a high number of parameters while **PC** algorithm can identify it, using only order 0 conditional independence sets which can be quite reliable with low sample sizes. Furthermore, the presence of links can be known from experts in the field and non induced by data. In this case, if the network is complex, then a normal dataset can be insufficient to estimate the parameters of the network.

This paper is organized as follows: Sect. 2 introduces the problem and the notation; Sect. 3 describes the proposed estimation procedure; Sect. 4 provides some insights about the data structures that can be used to efficiently compute with these conditional distributions; Sect. 5 presents the experimental work and Sect. 6 is devoted to the conclusions.

2   Notation and Problem Formulation

Assume that we have a variable Y taking values on a finite set U_Y and a set of variables $\mathbf{X} = \{X_1, \ldots, X_n\}$, where each variable X_i takes values on the set U_{X_i}. We also have a set $\mathcal{D}$ of N vectors of observations for variables $(Y, X_1, \ldots, X_n)$. We want to estimate the conditional probability $P(Y|X_1, \ldots, X_n)$. If U_Y has r values and each U_{X_i} has r_i values, this implies to estimate $r \cdot \prod_{i=1}^{n} r_i$ parameters. A subset of $\{X_1, \ldots, X_n\}$ will be denoted in boldface $\mathbf{Z}$. A generic value of set $\mathbf{Z}$ will be denoted as $\mathbf{z}$ and called a configuration of $\mathbf{Z}$. The number of possible configurations of $\mathbf{Z}$ is $r_{\mathbf{Z}} = \prod_{X_i \in \mathbf{Z}} r_i$.

If $\mathbf{x} \in U_{\mathbf{X}}$ and $\mathbf{Z} \subseteq \mathbf{X}$, then the configuration obtained by considering only the values of the variables in $\mathbf{Z}$, will be denoted as $\mathbf{x}_{\mathbf{Z}}$.

The estimation of maximum likelihood of a conditional probability $P(Y|\mathbf{Z})$ is computed as,

$$P(y|\mathbf{z}) = \frac{N_{y,\mathbf{z}}}{N_{\mathbf{z}}}, \tag{1}$$

where $N_{\mathbf{z}}$ is the number of occurrences of $\mathbf{Z} = \mathbf{z}$ in data $\mathcal{D}$ and $N_{y,\mathbf{z}}$ is the number of occurrences of $Y = y, \mathbf{Z} = \mathbf{z}$. This estimation is good when $N_{\mathbf{z}}$ is large, but its behaviour is not so good when $N_{\mathbf{z}}$ is small. In that case, it is better to use the Laplace correction:

$$P(y|\mathbf{z}) = \frac{N_{y,\mathbf{z}} + 1}{N_{\mathbf{z}} + r}. \tag{2}$$

The Bayesian BDEu score of the variable Y given the set of candidate parents $\mathbf{Z}$ is given by [3]:

$$Score(Y, \mathbf{Z}|\mathcal{D}) = \prod_{\mathbf{z} \in U_{\mathbf{Z}}} \frac{\Gamma(\alpha)}{\Gamma(N_{\mathbf{z}} + \alpha)} \prod_{y \in U_Y} \frac{\Gamma(\alpha/r + N_{y,\mathbf{z}})}{\Gamma(\alpha/r)}, \tag{3}$$

where Γ is the gamma function, $\alpha = s/r_{\mathbf{Z}}$ and s is a parameter (the equivalent sample size). Usually, the logarithm of this score (denoted as $LScore(Y, \mathbf{Z}|\mathcal{D})$) is used. We can use this score to compute a degree of association between Y and a variable X_i given the set of variables $\mathbf{Z}$ (see [1]) in the following way:

$$Dep(Y, X_i, \mathbf{Z}|\mathcal{D}) = LScore(Y, \mathbf{Z} \cup \{X_i\}|\mathcal{D}) - LScore(Y, \mathbf{Z}|\mathcal{D}). \tag{4}$$

The degree of association of Y and X_i conditioned to the presence of variables $\mathbf{Z}$ is the increment in the logarithm of the score of Y when X_i is added to $\mathbf{Z}$.

3   The Mixture of Conditional Distributions Model

When computing the conditional distribution $P(Y|\mathbf{X})$, if the number of variables in $\mathbf{X}$ is large, we have that in general $N_{\mathbf{x}}$ is small for most of $\mathbf{X}$'s configurations, so the maximum likelihood estimation of this probability distribution would be in general poor and even the Laplace correction would not be very accurate as for each configuration $\mathbf{x}$, the estimation of $P(y|\mathbf{x})$ will be based on a short sample (even null in many cases). For these situations we propose the following procedure:

1. Determine a family $\mathbf{Z}_1, \ldots, \mathbf{Z}_k$ of subsets of $\mathbf{X}$, where each $\mathbf{Z}_i$ is of a moderate size.
2. Compute a conditional probability distribution $P_i(Y|\mathbf{Z}_i)$ for each $i = 1, \ldots, k$, by using maximum likelihood or Laplace correction.
3. Estimate $P(Y|\mathbf{X})$ as a convex combination:

$$P(Y|\mathbf{x}) = \sum_{i=1}^{k} \alpha_i P_i(Y|\mathbf{x}_{\mathbf{Z_i}}) \tag{5}$$

where $\alpha_i \in [0,1]$ and $\sum_{i=1}^{k} \alpha_i = 1$. If each $P(Y|\mathbf{Z}_i)$ is a true conditional probability distribution, it is clear that for each configuration $\mathbf{x}$, $P(Y|\mathbf{x})$ is a probability distribution about Y as $P_i(Y|\mathbf{x}_{\mathbf{Z}_i})$ is a probability distribution and we have a convex combination of probability distributions.

Different variants can be obtained depending on the selection of subsets $\mathbf{Z}_1, \ldots, \mathbf{Z}_k$, the estimation procedure, and the coefficients α_i in the convex combination.

Subset Selection

In this paper we have chosen a fixed subset size l, so all the subsets $\mathbf{Z}_i$ are of the same size. We have also chosen k equal to n, the number of variables, so that each variable X_i appears at least in one subset and then its influence is quantified. On average, each variable will appear l times. Subsets are built in a sequential way and two basic procedures have been followed:

1. **The random selection.** We keep a vector counting the number of times that each variable X_i has been selected for a subset $\mathbf{Z}_j$. Each time we are going to select a variable for subset $\mathbf{Z}_j$, we randomly chose a variable among those with a minimum number of previous selections. In this way, we guarantee that each variable is selected exactly the same number of times for subsets l, but apart from this restriction, subsets $\mathbf{Z}_j$ are random.

2. **The association criterion.** Each subset $\mathbf{Z}_i$ is initialized to $\{X_i\}$, and then an iterative procedure is followed for the rest of the $l-1$ variables: in each iteration the variable with maximum value of $Dep(Y, X_j, \mathbf{Z}_i|\mathcal{D})$ is selected among those variables $X_j \in \mathbf{X} \setminus \mathbf{Z}_i$. That is, we start each subset with a different variable, but then the variables adding more information about Y are sequentially added. It is important to remark that it might happen that two sets $\mathbf{Z}_i$ and $\mathbf{Z}_j$ are equal for $i \neq j$, if for example $l = 2$ and the variable with maximum degree of dependence with X_i is X_j and the variable with maximum degree with X_j is X_i.

Small Conditional Probabilities Estimation

Even if the number of variables in $\mathbf{Z}_i$ is small, it is advisable to use Laplace correction, as we can not guarantee that the numbers $N_{\mathbf{Z}_i}$ are large for all the configurations. For example, if $N_{y,\mathbf{z}}$ is 0 and the maximum likelihood estimator is used, then $P(y|\mathbf{z})$ will be estimated as 0.0, which is a too strong assumption. So, some smoothing of the probabilities should be applied [7].

Small Conditional Probabilities Combination

In this step, formula (5) is applied to compute the global conditional probability $P(Y|\mathbf{X})$. Two main approaches are considered:

- **Uniform Combination.** In this case, $\alpha_i = \frac{1}{k}$, i.e. all the conditional probabilities have the same weights.
- **Parameter Optimization.** It is assumed the existence of a hidden variable, H, taking k values, $\{h_1, \ldots, h_k\}$. Y depends also on this hidden variable,

but H is independent of $\mathbf{X}$, and with the hypothesis that if $H = h_i$, then $P(Y|\mathbf{X}, H = h_i) = P(Y|\mathbf{Z}_i, H = h_i)$. In this way,

$$P(Y|\mathbf{X}) = \sum_i P(Y|\mathbf{X}, H = h_i)P(h_i) = \sum_i P(Y|\mathbf{Z}_i, H = h_i)P(h_i) \quad (6)$$

In this way, weights are the probabilities of this hidden variable. Conditional probabilities $P(Y|\mathbf{Z}_i, H = h_i)$ and weights $P(h_i)$ are computed using the **EM** algorithm to obtain a local maximum of the posterior probability (the estimation of the probabilities is done with Laplace correction) [5]. This algorithm starts with an initial estimation of the probabilities $P^0(Y|\mathbf{Z}_i, H = h_i)$, $P^0(h_i)$. Then an iterative algorithm alternating two steps is applied:

- **Expectation.** The value $\hat{N}_{y,\mathbf{z}_i,h_i} = E(N_{y,\mathbf{z}_i,h_i}|P^j(Y|\mathbf{Z}_i, H = h_i), P^j(h_i))$, where $E(N_{y,\mathbf{z},h_i})$ is the expected number of occurrences of $Y = y, \mathbf{Z}_i = \mathbf{z}_i, H = h_i$ computed in a Bayesian network with H and $\mathbf{Z_i}$ as parents of Y and with $P^j(Y|\mathbf{Z}_i, H = h_i), P^j(h_i)$ as parameters. The probabilities $P^0(X_j)$ are not necessary as these variables are always observed in the data $\mathcal{D}$. Analogously, $\hat{N}_{h_i} = E(N_{h_i}|P^j(Y|\mathbf{Z}_i, H = h_i), P^j(h_i))$ is also computed.

- **Maximization.** A new value of the parameters is obtained by considering:

$$P^{j+1}(Y|\mathbf{Z}_i, H = h_i) = \frac{\hat{N}_{y,\mathbf{z}_i,h_i} + 1}{\sum_{y \in U_Y} \hat{N}_{y,\mathbf{z}_i,h_i} + r_Y}, \quad (7)$$

$$P^{j+1}(H = h_i) = \frac{\hat{N}_{h_i} + 1}{\sum_i \hat{N}_{h_i} + k}. \quad (8)$$

The iteration continues till the changes in probability estimations are below a given threshold.

It would be easy to adapt this algorithm to the case in which missing data are also present in the dataset, but this description is given for fully observed data.

4   Data Structures

After probabilities are estimated they should be introduced in a Bayesian network for posterior computations. If we compute $P(Y|\mathbf{X})$ as in Eq. (5) and store it as a probability table, if the number of variables in $\mathbf{X}$ is large, then the necessary space to store this table is huge, and the computations with these tables would be unfeasible. Even, the application of the **EM** algorithm that needs to compute probabilities given a current estimation of the parameters can have efficiency problems. But we can represent these conditional probabilities in an efficient way if we use alternative potential representations, able of taking advantage of the associated asymmetrical independences associated with the model, as we have that given $H = h_i$, then Y will be independent of $\mathbf{X} \setminus \mathbf{Z}_i$ given $\mathbf{Z_i}$. In this paper we have considered the use of probability trees [4], though other structures such as max-product networks could also be employed [13].

A probability tree T for a set of variables $\mathbf{Z}$ is a tree where each internal node. is labeled with a variable $X \in \mathbf{Z}$ and has a child for each one of the values in U_X, and each leaf contains a real number. It represents a potential, i.e. a mapping $t : U_{\mathbf{Z}} \to \mathbb{R}$. The value $t(\mathbf{z})$ is the number in the leaf that is obtained by starting in the root node and following for each internal node $X \in \mathbf{Z}$ the path consistent with the child corresponding to the value of X in configuration $\mathbf{z}$.

Probability trees can represent probability tables and take advantage of asymmetrical independence relationships to obtain more compact encodings of probability potentials. It is also possible to directly compute with these data structures obtaining posterior conditional probabilities [4].

The representation of the conditional probabilities estimated as in Eq. (5) is done by introducing an auxiliary variable H with values $U_H = \{h_1, \ldots, h_k\}$. This auxiliary variable is independent of variables $\mathbf{X}$ and has an associated potential $t_H : U_H \to [0, 1]$ given by $t(h_i) = \alpha_i$. Then, a probability tree is used to represent the conditional probability. The root will be variable H and for each value h_i it will have a child representing the conditional probability $P(Y|\mathbf{Z}_i)$. This child will be a tree that will only be branched by variables in $\mathbf{Z}_i$ and therefore will be at most of size $r \prod_{X_i \in \mathbf{Z_i}} r_i$ and the full potential tree of size upper bounded by $r \sum_{i=1}^{k} \prod_{X_i \in \mathbf{Z_i}} r_i$. This size is not exponential in the number of variables in $\mathbf{X}$ and it is possible to efficiently compute with it, if the size of sets $\mathbf{Z}_i$ is moderated.

The efficiency of the inference algorithm using this representation on a particular case will depend of the network structure and the observation set, but as it has been shown in [4] it is possible to compute in many practical cases in which a computation with probability tables is impossible. In particular, the computations associated with the EM algorithm have been possible because we have used this data structure.

5   Experiments

For the experiments we have used 31 datasets from UCI Machine Learning Repository [10]. Details can be found in Table 1. Data were preprocessed as in [2]: continuous variables are discretized and instances with missing or undefined values are removed. The basic idea of the experiments is to estimate a conditional distribution of the class (variable Y) given the attributes (variables $\mathbf{X}$). We assume that the class is dependent on all the attributes. This is usually true in these datasets as the attributes are considered among variables that are potentially relevant to the class. It is possible that other graph structure can provide a better representation of the data (for example assuming that the arcs are from the class to the attributes as in most of Bayesian classifiers), but this is not what we are investigating in this paper: we are testing parameter estimation procedures for a given graph. Experiments have been carried out in Elvira platform [6].

Table 1. Datasets UCI repository

Database	N	Attributes	Classes
adult-d-nm	45222	14	2
australian-d	690	14	2
breast-no-missing	682	10	2
car	1728	6	4
chess	3196	36	2
cleve-no-missing-d	296	13	2
corral-d	128	6	2
crx-no-missing-d	653	15	2
diabetes-d-nm	768	8	2
DNA-nominal	3186	60	3
flare-d	1066	10	2
german-d	1000	20	2
glass2-d	163	9	2
glass-d	214	9	7
heart-d	270	13	2
hepatitis-no-missing-d	80	19	2
iris-d	150	4	3
letter	20000	16	26
lymphography	148	18	4
mofn-3-7-10-d	1324	10	2
nursery	12960	8	5
mushroom	8124	22	2
pima-d	768	8	2
satimage-d	6435	36	6
segment-d	2310	19	7
shuttle-small-d	5800	9	7
soybean-large-no-missing-d	562	35	19
splice	3190	60	3
vehicle-d-nm	846	18	4
vote	435	16	2
waveform-21-d	5000	21	3

To evaluate the evolution of the quality of the estimation as a function of the number of attributes, we do not use all the attributes in the dataset, using only a maximum of n attributes of the total number of attributes of the dataset, m: the first n attributes are selected and the other $m - n$ attributes are discarded. We have carried out a 10-fold cross validation procedure. To evaluate a method with a new vector of observations $(y, \mathbf{x})$ we compute the logarithm of the conditional probability estimation: $\log(P(y|\mathbf{x})$ (log-likelihood of the conditional probability).

The first experiment evaluates the different procedures against the classical estimation of the probabilities with Laplace correction (2). Table 2 shows the averages of the logarithms of the conditional probabilities when the number of attributes is $n = 7$ and our procedure is used with 7 subsets $\mathbf{Z_i}$ of size 3. **Ramdon-EM** stands for random selection of subsets $\mathbf{Z}_i$ and application of the **EM**-estimation procedure. **Optimal-EM** and **Optimal-Uniform** stand for using the association degree to compute subsets $\mathbf{Z}_i$, with **EM** algorithm and uniform weights, respectively.

Friedman non-parametric test shows a high significance with a p-value $= 3.849e - 06$. A post-hoc analysis using Nemenyi with Holm adjustment provides significant differences of the best procedure, **Optimal-Uniform** with all the other procedures except with **Optimal-EM**. The classical procedure is significantly worse than **Optimal-EM** and **Optimal-Uniform**.

The second experiment evaluates the evolution of the classical and **Optimal-Uniform** with respect to the size of $\mathbf{X}$ by repeating the 10-fold cross validation for the different datasets by including an increasing number of candidate attributes. We keep a size of 3 for the sizes of subsets $\mathbf{Z_i}$. We only report the average log-likelihood of the conditional probability in Fig. 1. The classical method is only evaluated till 7 candidates as the method is very slow and we have memory problems with more parents. It can be seen that our mixture model takes advantage of having more candidates and the average increases with the inclusion of more variables. However, the classical method deteriorates with the number of parents as the probability tables are too large to be estimated with precision from available data.

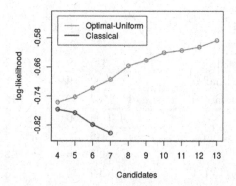

Fig. 1. Evolution as a function of the number of parents

Finally, the third experiment compares the **Optimal-Uniform** and the **Optimal-EM** procedures when the size of sets $\mathbf{Z}_i$ is 1. The reason is that the performance of the uniform combination was better when the size of $\mathbf{Z}_i$ was 3.

Table 2. Results experiment 1

Database	Random-EM	Optimal-EM	Optimal-uniform	Classical
adult-d-nm	−0.3774	−0.3701	−0.3795	−0.4212
australian-d	−0.5853	−0.5753	−0.5763	−0.6300
breast-no-missing	−0.3609	−0.3384	−0.2271	−0.4577
car	−0.7217	−0.5963	−0.4343	−1.3863
chess	−0.6562	−0.6539	−0.6552	−0.6371
cleve-no-missing-d	−0.5596	−0.5510	−0.5420	−0.5403
corral-d	−0.4890	−0.4556	−0.3747	−0.4131
crx-no-missing-d	−0.6121	−0.6299	−0.6285	−0.6473
diabetes-d-nm	−0.4985	−0.4842	−0.4829	−0.4920
DNA	−1.0212	−1.0228	−1.0230	−1.0215
DNA-nominal	−0.9895	−0.9906	−0.9824	−1.0147
flare-d	−0.4008	−0.3944	−0.3948	−0.4163
german-d	−0.5620	−0.5284	−0.5268	−0.6396
glass2-d	−0.5136	−0.4978	−0.4495	−0.4446
glass-d	−1.2742	−1.2706	−0.9690	−1.0738
heart-d	−0.5327	−0.5243	−0.5139	−0.4937
hepatitis-no-missing-d	−0.5084	−0.4520	−0.4050	−0.4864
iris-d	−0.4082	−0.3585	−0.1973	−0.2857
letter-d	−2.6178	−2.5526	−2.4708	−2.4193
lymphography	−0.9744	−0.8648	−0.6804	−0.9352
mofn-3-7-10-d	−0.3470	−0.3815	−0.4099	−0.3572
mushroom	−0.1722	−0.0595	−0.0303	−0.0386
nursery	−1.0298	−0.9554	−0.9210	−1.6174
pima-d	−0.4948	−0.4847	−0.4822	−0.4962
segment-d	−1.3345	−1.1455	−1.0307	−1.1838
shuttle-small-d	−0.3509	−0.2744	−0.1300	−0.3567
soybean-large-no-missing-d	−2.3072	−2.1585	−1.8183	−2.4768
splice	−0.9901	−0.9891	−0.9812	−1.0161
vehicle-d-nm	−0.9402	−0.8775	−0.7426	−0.8249
vote	−0.2610	−0.1963	−0.1434	−0.2676
AVERAGE	−0.7630	−0.7211	−0.6534	−0.7830

Our initial expectations were the reverse. We think that this can be due to over-fitting of **Optimal-EM** when the model is more complex. So we have carried out an experiment comparing these two procedures in a simpler setting. Results can be seen in Table 3. We do not differentiate between optimal and uniform selection, as with size 1, the two criteria are identical. In this case, the use of

Table 3. Results experiment 3

Database	EM estimation	Uniform combination
adult-d-nm.dbc	−0.4085	−0.4773
australian-d.dbc	−0.5850	−0.6332
breast-no-missing.dbc	−0.2092	−0.1967
car.dbc	−0.6719	−0.7412
chess.dbc	−0.6686	−0.6853
cleve-no-missing-d.dbc	−0.5889	−0.6365
corral-d.dbc	−0.5027	−0.5793
crx-no-missing-d.dbc	−0.6214	−0.6515
diabetes-d-nm.dbc	−0.5322	−0.5850
DNA.dbc	−1.0216	−1.0239
DNA-nominal.dbc	−1.0032	−1.0168
flare-d.dbc	−0.3969	−0.4161
german-d.dbc	−0.5365	−0.5731
glass2-d.dbc	−0.5572	−0.5785
glass-d.dbc	−1.3131	−1.2299
heart-d.dbc	−0.5880	−0.6358
hepatitis-no-missing-d.dbc	−0.4675	−0.4405
iris-d.dbc	−0.4067	−0.3765
letter-d.dbc	−2.8527	−2.9689
lymphography.dbc	−0.8610	−0.7895
mofn-3-7-10-d.dbc	−0.4196	−0.4795
mushroom.dbc	−0.4448	−0.5046
nursery.dbc	−1.0642	−1.1369
pima-d.dbc	−0.5349	−0.5858
segment-d.dbc	−1.5168	−1.6195
shuttle-small-d.dbc	−0.3074	−0.3698
soybean-large-no-missing-d.dbc	−2.2550	−2.2249
soybean-large-no-missing-d-2.dbc	−2.3542	−2.2505
splice.dbc	−1.0043	−1.0171
vehicle-d-nm.dbc	−1.0772	−1.1621
vote.dbc	−0.3450	−0.3938
AVERAGE	−0.8425	−0.8703

EM is better than the uniform combination. The differences are significant using a Wilcoxon signed rank test with p-value equal to 0.006217.

Even if we do not provide associated times, it is clear that the uniform combination is always faster than the EM algorithm. However, EM algorithm finishes in a reasonable time each step of the EM-algorithm is linear in the total size of the small conditional potentials and in the dataset size (considering a constant number of small conditioning sets), and the number of iterations is usually small.

6   Conclusions

We have presented a new procedure for estimating a conditional probability of a variable given a large number of variables which is based on a mixture of smaller size conditional probabilities. The procedure outperforms the classical procedure providing better results in the experiments. Another important advantage is that it scales well when increasing the number of conditional variables improving the quality of the predictions when considering more variables. The learned conditional probabilities are of smaller size than the classical probability tables and can be represented by probability trees. It is also possible to compute conditional probabilities in an efficient way using these presentations in a generic Bayesian network.

Future work will include the study of alternative methods to compute the weights and small conditioning sets $\mathbf{Z}_i$ and automatic determination of the optimal size of these subsets, considering also the possibility of mixing subsets of different size. We will try to analyze more in deep the reasons of the differences between the use of EM algorithm and uniform combination in order to determine rules for determining the best procedure in each case. We will investigate whether the use of sets with different sizes, the elimination of duplicate sets, or an optimal selection of a representative family of sets $\mathbf{Z}_i$ according to some global criterion, might have some impact on the performance associated to the use of EM estimation procedure. We also plan to integrate our estimation procedure with others requiring a small number of parameters as noisy-or gates [8], for example by carrying out a preliminary test to decide which method is more appropriate in each concrete situation. This also could be done by means of some scoring procedure as the BIC criterion which can balance the likelihood of observations and the number of parameters of the model.

Acknowledgments. This research was supported by the Spanish Ministry of Economy and Competitiveness under projects TIN2013-46638-C3-2-P and TIN2016-77902-C3-2-P, and the European Regional Development Fund (FEDER).

References

1. Abellán, J., Gómez-Olmedo, M., Moral, S.: Some variations on the PC algorithm. In: Proceedings of the Third European Workshop on Probabilistic Graphical Models (PGM 2006), pp. 1–8 (2006)
2. Acid, S., de Campos, L.M., Castellano, J.G.: Learning bayesian network classifiers: searching in a space of partially directed acyclic graphs. Mach. Learn. **59**, 213–235 (2005)

3. Buntine, W.: Theory refinement in bayesian networks. In: Proceedings of the Seventh Conference on Uncertainty in Artificial Intelligence, pp. 52–60. Morgan Kaufmann, San Francisco (1991)
4. Cano, A., Moral, S., Salmerón, A.: Penniless propagation in join trees. Int. J. Intell. Syst. **15**, 1027–1059 (2000)
5. Dempster, A.P., Laird, N.M., Rubin, D.B.: Maximum likelihood from incomplete data via the EM algorithm. J. R. Stat. Soc. Ser. B **36**, 1–38 (1977)
6. Elvira Consortium: Elvira: an environment for probabilistic graphical models. In: Gmez, J.A., Salmern, A. (eds.) Proceedings of the 1st European Workshop on Probabilistic Graphical Models, pp. 222–230 (2002)
7. Friedman, N., Geiger, D., Goldszmidt, M.: Bayesian network classifiers. Mach. Learn. **29**, 131–163 (1997)
8. Henrion, M.: Some practical issues in constructing belief networks. In: Kanal, L.N., Levitt, T.S., Lemmer, J.F. (eds.) Uncertainty in Artifjcial Intelligence, vol. 3, pp. 161–173. Elsevier, Amsterdam (1989)
9. Nielsen, T.D., Jensen, F.V.: Bayesian Networks and Decision Graphs. Information Science and Statistics. Springer, New York (2007). doi:10.1007/978-0-387-68282-2
10. Lichman, M.: UCI Machine Learning Repository. University of California, School of Information and Computer Science, Irvine, CA (2013). http://archive.ics.uci.edu/ml
11. Liu, Z., Malone, B., Yuan, C.: Empirical evaluation of scoring functions for Bayesian network model selection. BMC Bioinform. **13**(Suppl 15), S14 (2012)
12. Pearl, J.: Probabilistic Reasoning in Intelligent Systems: Networks of Plausible Inference. Morgan Kaufmann, San Mateo (1988)
13. Poon, H., Domingos, P.: Sum-product networks: a new deep architecture. In: IEEE International Conference on Computer Vision Workshops (ICCV Workshops), pp. 689–690 (2011)
14. Spirtes, P., Glymour, C., Scheines, R.: Causation, Prediction and Search. Lecture Notes in Statistics. Springer, New York (1993). doi:10.1007/978-1-4612-2748-9
15. Zagoreski, A., Druzdzel, M.J.: Knowledge engineering for bayesian networks: how common are noisy-MAX distributions in practice. IEEE Trans. Syst. Man Cybern. Syst. **43**, 186–195 (2013)

Closed-Form Solutions in Learning Probabilistic Logic Programs by Exact Score Maximization

Francisco Henrique Otte Vieira de Faria[1(✉)], Fabio Gagliardi Cozman[1], and Denis Deratani Mauá[2]

[1] Escola Politécnica, Universidade de São Paulo, São Paulo, Brazil
franciscohovfaria@gmail.com, fgcozman@usp.br
[2] Instituto de Matemática e Estatística, Universidade de São Paulo, São Paulo, Brazil
denis.maua@gmail.com

Abstract. We present an algorithm that learns acyclic propositional probabilistic logic programs from complete data, by adapting techniques from Bayesian network learning. Specifically, we focus on score-based learning and on exact maximum likelihood computations. Our main contribution is to show that by restricting any rule body to contain at most two literals, most needed optimization steps can be solved exactly. We describe experiments indicating that our techniques do produce accurate models from data with reduced numbers of parameters.

Keywords: Probabilistic logic programming · Score-based structure learning of bayesian networks

1 Introduction

The goal of this paper is to present techniques that learn probabilistic logic programs (PLPs) from complete data. Probabilistic logic programs have been explored for some time [10,13,14,16,20], and are now the object of significant literature [18,19]; yet there is much to be developed when it comes to rule learning. In this paper we wish to examine the extent to which, for some classes of PLPs, we can find the exact optimal PLP with respect to some score. That is, we look for classes of PLPs have some guarantees concerning the optimization of scores given data.

With this broad goal in mind, here we focus on algorithms that learn, in the sense of maximization of minimum description length, acyclic sets of propositional rules with at most two literals in their bodies. This is admitedly a restricted class of PLPs, but note that these PLPs can encode all "noisy" Boolean circuits and can serve as starting point for more ambitious investigations.

Our main contribution is to show that, by focusing on this restricted class of PLPs, we can *exactly*, in *constant time*, solve most optimizations involved in description length minimization.

Because acyclic propositional PLPs are intimately related to Bayesian networks [17], we employ insights from Bayesian network learning, in particular

S. Moral et al. (Eds.): SUM 2017, LNAI 10564, pp. 119–133, 2017.
DOI: 10.1007/978-3-319-67582-4_9

resorting to score-based learning where the score is based on minimum description length [11]. However most of our arguments apply to any decomposable score that depends on likelihood; for example, analogues of K2 and BDeu scores [11] could be easily adopted.

We briefly review the main features of relevant PLPs in Sect. 2. We then introduce the score maximization problem we face (Sect. 3); this follows by direct adaptation of methods from Bayesian network learning. In Sects. 4 and 5 we derive an algorithm for score maximization that relies both on exact maximization of polynomial equations for local optimization and on constraint programming for global optimization. Experiments and results indicating that our method is successful in recovering accurate models are reported in Sect. 6.

2   Probabilistic Logic Programs: A (very) Brief Review

Take a fixed vocabulary consisting of logical variables $X, X_1, \ldots$, predicates $r, r_r, \ldots$, and constants $a, b, \ldots$. A *term* is a constant or a logical variable; an *atom* is written as $r(t_1, \ldots, t_n)$, where r is a predicate of arity n and each t_i is a term (a 0-arity atom is written as r). An atom is *ground* if it does not contain logical variables. A *normal logic program* consists of rules

$$A_0 :- A_1, \ldots, A_m, \mathbf{not}\ A_{m+1}, \ldots, \mathbf{not}\ A_n \cdot,$$

where the A_i are atoms. The *head* of this rule is A_0; the right-hand side is the *body*. A rule without a body, written A_0., is a *fact*. A *literal* is either A (positive) or **not** A (negative), where A is an atom.

In this paper we only consider propositional programs; that is, programs without logical variables.

The *Herbrand base* is the set of all ground atoms built from constants and predicates in a program. We do not consider functions in this paper, hence every Herbrand base is finite.

The *dependency graph* of a program is a directed graph where each predicate is a node, and where there is an edge from a node B to a node A if there is a rule where A appears in the head and B appears in the body (if B appears right after **not**, the edge is *negative*; otherwise, it is *positive*). The *grounded* dependency graph is the dependency graph of the propositional program obtained by grounding.

An acyclic program is one with an acyclic grounded dependence graph. In this paper we only consider acyclic programs.

There are several ways to combine logic programming and probabilities; in this work we adopt a popular combination associated with Sato's distribution semantics [16,20]. A *probabilistic logic program*, abbreviated PLP, is a pair $\langle \mathbf{P}, \mathbf{PF} \rangle$ consisting of a normal logic program $\mathbf{P}$ and a set of *probabilistic facts* $\mathbf{PF}$. A probabilistic fact is a pair consisting of an atom A and a probability value α, written as $\alpha :: A$. (here we adopt the syntax of the ProbLog package [8]). Note that we allow a probabilistic fact to contain logical variables.

To build the semantics of a PLP, we first take its grounding. Suppose we have a PLP with n grounded probabilistic facts (some of them may be given as propositional probabilistic facts, others are obtained by grounding non-propositional probabilistic facts). There are then 2^n ways to select subsets of these propositional probabilistic facts. For each such subset, we can construct the normal logic program consisting of the non-probabilistic part of the PLP plus the atoms in the selected probabilistic facts. That is, for each probabilistic fact $\alpha :: A.$, either keep fact $A.$ with probability α, or erase $A.$ with probability $1 - \alpha$. A *total choice* is simply a subset of the set of ground probabilistic facts that is selected to be kept (other grounded probabilistic facts are discarded). So, for any total choice θ we obtain a normal logic program, denoted by $\mathbf{P} \cup \mathbf{PF}^{\downarrow\theta}$, with probability $\prod_{A_i \in \theta} \alpha_i \prod_{A_i \notin \theta} (1 - \alpha_i)$. Hence the distribution over total choices induces a distribution over logic programs.

Note that if $\mathbf{P}$ is acyclic, then for any total choice we have that $\mathbf{P} \cup \mathbf{PF}^{\downarrow\theta}$ is acyclic. So the semantics of the whole PLP is relatively simple to describe: the probability distribution over total choices induces a probability distribution over interpretations, such that for each fixed total choice we obtain the truth assignment of all atoms by applying the rules in appropriate order (that is, if all atoms in the body of a rule are true, the head is true, and this is the only way to render a head true).

A common pattern in PLPs is a pair rule/probabilistic fact such as

$$A_0 :- A_1, \ldots, A_m, \mathbf{not}\ A_{m+1}, \ldots, \mathbf{not}\ A_n, F., \quad \alpha :: F.,$$

meaning that with probability $1 - \alpha$ the whole rule is not "activated". We write such a construct as

$$\alpha :: A_0 :- A_1, \ldots, A_m, \mathbf{not}\ A_{m+1}, \ldots, \mathbf{not}\ A_n.,$$

again adopting the syntax of ProbLog [8].

Example 1. *Here is an acyclic propositional PLP:*

$0.1 :: \mathsf{burglary}.$ $0.2 :: \mathsf{earthquake}.$
$0.9 :: \mathsf{alarm} :- \mathsf{burglary}, \mathsf{earthquake}.$
$0.8 :: \mathsf{alarm} :- \mathsf{burglary}, \mathbf{not}\ \mathsf{earthquake}.$
$0.7 :: \mathsf{alarm} :- \mathbf{not}\ \mathsf{burglary}, \mathsf{earthquake}.$
$0.1 :: \mathsf{alarm} :- \mathbf{not}\ \mathsf{burglary}, \mathbf{not}\ \mathsf{earthquake}.$

The dependency graph is presented at the right.   □

The class of logic programs where each rule has at most k atoms in the body is denoted LP(k) [5]; we analogously write PLP(k) to denote the class of PLPs where each rule has at most k literals in the body. And we use AP-PLP to refer to acyclic propositional PLPs. In this paper we focus on the class AP-PLP(2).

3   Learning by Score Maximization

One may wish to learn, from data, both the rules of a PLP and the associated probabilities; that is, both the *structure* and the *parameters* of the PLP. A general strategy in structure learning is to add probabilistic estimation to Inductive Logic Programming; usually such a strategy is referred to as Probabilistic Inductive Logic Programming [18,19]. Typically such mix of probabilities and logic requires a search over the space of rules, under the assumption that some examples are "positive" and must receive high probability, while other examples are "negative" and must receive low probability. Search schemes vary and are almost universally based on heuristic measures, to guarantee that large datasets can be processed [1,6,7,25].

Another general strategy, when learning a probabilistic model, is to maximize a score that quantifies the fit between model and data. This is the strategy most often employed to learn the the structure of Bayesian networks, and the strategy adopted in this paper. To grasp the main ideas behind score-based structure learning, we first review the interplay between Bayesian networks and probabilistic logic programs.

3.1   Bayesian Networks and Probabilistic Logic Programs

As noted in the Introduction, acyclic propositional PLPs are closely related to Bayesian networks [17]. Recall that a Bayesian network consists of a directed acyclic graph where each node is a random variable, and a probability distribution over the same random variables, such that the distribution satisfies the following Markov condition: a variable X is independent of its nondescendants nonparents given its parents [15]. For any Bayesian network, its directed acyclic graph is referred to as its "structure". The parents of a variable X, denoted by PA[X], are the nodes/variables that point to X. In this paper every random variable has finitely many values (indeed all of them are binary). When a conditional probability distribution over random variables with finitely many values is encoded using a table, the latter table is often referred to as a *CPT*.

Any Bayesian network over binary variables can be encoded by an acyclic propositional PLP; conversely, any acyclic propositional PLP can be viewed as a Bayesian network. The last statement should be clear from Example 1: the Bayesian network described by the PLP has the structure given by the dependency graph, and the parameters of the network are just the probabilities associated with probabilistic facts and rules. The converse is equally simple to show, and consists of easily translating the probability assignments into rules and probabilistic facts, as argued by Poole [16,17].

The "structure" of an acyclic PLP is related to the "structure" of the Bayesian network associated to the PLP. In fact, the dependency graph of the grounded PLP is the structure of the corresponding Bayesian network. However, a PLP can specify significantly more detail about the underlying probability distributions. Suppose, for instance, that the distribution of a binary variable X, with parents Y and Z, is given by a NoisyOr gate [15]; that is, $X = YY' + ZZ' - YY'ZZ'$, with $\mathbb{P}(Y' = 1) = \alpha$ and $\mathbb{P}(Z' = 1) = \beta$. In this case the conditional probability distribution of X given (Y, Z) is fully specified by two numbers (α and β), instead of the four numbers that a complete specification requires. Note that a small set of probabilistic facts and rules would have no trouble in encoding exactly this NoisyOr gate with the needed two parameters. This is attractive: if a distribution can be at all captured by a small number of parameters, a PLP may be able to do so.

Of course, there are other ways to capture conditional distributions with "local" structure; that is, distributions that require few parameters to yield the probabilities for some variable given its parents. One notable example in the literature is the use of trees to represent conditional distributions [9]. The representation of a conditional probability distribution using trees is sometimes referred to as a CPT-tree [2]. Now it should be clear that CPT-trees and probabilistic rules do not have the same expressivity; for instance a CPT-tree requires three parameters to specify the NoisyOr gate in the previous paragraph (assuming a convention that leaves unspecified branches as zero), while rules can specify the NoisyOr gate with two parameters. So the question as to whether representations based on probabilistic rules are more compact than other representation is meaningful, and this is the sort of abstract question we wish to address with the current paper.

3.2 Score-Based Structure Learning of Bayesian Networks

Several successful structure learning methods are based on score maximization, where a score $s(B, D)$ gets a Bayesian network structure B and a dataset D, and yields a number that indicates the fit between both. We assume that D is complete (that is, there is no missing data) and consists of N observations of all random variables of interest. Sensible scores balance the desire to maximize likelihood against the need to constrain the number of learned parameters. It is well-known that if the one maximizes only the likelihood, then the densest networks are always obtained [11]. One particularly popular score is based on minimum description length guidelines; the score is:

$$s_{\mathsf{MDL}}(B, D) = \mathsf{LL}_D(B) - \frac{|B| \log N}{2}, \tag{1}$$

where: $|B|$ is the number of parameters needed to specify the network, and $\mathsf{LL}_D(B)$ is the log-likelihood at the maximum likelihood estimates (that is, the logartithm of $p(D, \Theta^{B,D}|B)$ with p denoting the probability density of observations given a Bayesian network structure B for probability values $\Theta^{B,D}$. The latter values are obtained again by likelihood maximization; that is, $\Theta^{B,D} = \arg\max_\Theta p(\Theta, D|B)$. We adopt the s_{MDL} score throughout this paper.

The MDL score, as other popular scores such as the K2 and BDeu scores, is *decomposable*; that is, the score is a sum of *local scores*, each one a function of a variable and its parents. We call *family* a set consisting of a variable and its parents.

The current technology on structure learning of Bayesian networks can handle relatively large sets of random variables [3,4,22]. Most existing methods proceed in two steps: first calculate the local score for every possible family; then maximize the global score, usually either by integer programming [4] or by constraint programming [22]. When one deals with structure learning of Bayesian networks where conditional probability distributions are encoded by CPTs, then maximum likelihood estimates $\Theta^{B,D}$ are obtained in closed form: they are, in fact, simply relative frequencies. If we denote by θ_{ijk} the probability $\mathbb{P}(X_i = x_{ij}|\text{PA}[X_i] = \pi_k)$, where π_k is the kth configuration of the parents of X_i, then the maximum likelihood estimate is N_{ijk}/N_{ij}, where N_{ijk} is the number of times the configuration $\{X_i = x_{ij}, \text{PA}[X_i] = \pi_k\}$ occurs in D, and N_{ik} is the number of times the configuration $\{\text{PA}[X_i] = \pi_k\}$ occurs in D. Thus the calculation of the scores is not really taxing; the real computational effort is spent running the global optimization step.

4   A Score-Based Learning Algorithm for the Class AP-PLP(2)

Our goal is to learn a PLP that maximizes the MDL score with respect to the complete dataset D, with the restriction that resulting PLPs must belong to the class AP-PLP(2). We do so by following the two-step scheme discussed in the previous section: first, we must compute the local score for each possible family, and then we must run a global maximization step to obtain the whole PLP.

We must of course translate the language of acyclic propositional PLPs into the language of Bayesian networks. Each proposition in our vocabulary is viewed as a binary random variable (0 is false and 1 is true), and the propositions that appear in the body of a rule are the parents of the proposition in the head of the rule. This is clearly consistent with the correspondence between PLPs and Bayesian networks. Our dataset D is therefore a collection of N observations of the propositions/variables of interest.

Because the MDL score is decomposable, we can globally maximize it by first maximizing each local score separately, and then running a global maximization step that selects a family for each variable. But there is a difference between usual Bayesian network learning and PLP learning: even within a family, we must choose the rule set that relates the head of the family with its parents.

Example 2. *Suppose we have propositions $A_1, A_2, \ldots, A_n$. We must contemplate every possible family; for instance, A_1 may be a root (no parents), or A_1 may be the child of A_2, or the child of A_3, or the child of A_2 and A_3, and so on (we are restricted to at most two parents). Suppose we focus on the family $\{A_1, A_2, A_3\}$ for proposition/variable A_1; that is, A_1 is the head and $\{A_2, A_3\}$ appear in the bodies. How many possible rule sets can we build? Indeed, many. For instance, we may have a simple rule such as*

$$\theta :: A_1 :- A_2, A_3.$$

A_2	A_3	$\mathbb{P}(A_1 = \text{true}|A_2, A_3)$
false	false	0
false	true	0
true	false	0
true	true	θ

This single rule is equivalent to the following CPT:
   Or we may have three rules such as

$$\theta_1 :: A_1. \qquad \theta_2 :: A_1 :-A_2, \textbf{not } A_3. \qquad \theta_3 :: A_1 :-\textbf{not } A_2, \textbf{not } A_3. \qquad (2)$$

These three rules are equivalent to the following CPT:

A_2	A_3	$\mathbb{P}(A_1 = \text{true}|A_2, A_3)$
false	false	$\theta_1 + \theta_3 - \theta_1\theta_3$
false	true	θ_1
true	false	$\theta_1 + \theta_2 - \theta_1\theta_2$
true	true	θ_1

Two lines of this table contain nonlinear functions of the parameters, a fact that complicates the likelihood maximization step.   □

How many different rule sets we must consider? Suppose first that we have a family containing only the head A. Then there is a single rule, the probabilistic fact $\theta::A.$. If we instead have a family containing the head A and the body proposition B, there are six other options to consider:

		$\theta_1 :: A :-B.$
$\theta :: A :-B.$	$\theta :: A :-\textbf{not } B.$	$\theta_2 :: A.$
$\theta_1 :: A :-\textbf{not } B.$	$\theta_1 :: A :-B.$	$\theta_1 :: A :-B.$
$\theta_2 :: A.$	$\theta_2 :: A :-\textbf{not } B.$	$\theta_2 :: A :-\textbf{not } B.$
		$\theta_3 :: A.$

Now suppose we have a head A with parents B and C. First, there are 9 possible rules where A is the head and no proposition other than B or C appears in the body.[1] Each one of the 2^9 subsets of these rules is a possible rule set for this family; however, 13 of these subsets do not mention either B or C. Thus there are $2^9 - 13 = 499$ new rule sets to evaluate.

In any case, for each rule set we consider, we must maximize a likelihood function that may be nonlinear on the parameters. For instance, take the set of

---

[1] These 9 rules are: $\theta :: A.$, $\theta :: A :-B.$, $\theta :: A :-\textbf{not } B.$, $\theta :: A :-C.$, $\theta :: A :-\textbf{not } C.$, $\theta::A :-B, C.$, $\theta :: A :-B, \textbf{not } C.$, $\theta :: A :-\textbf{not } B, C.$, $\theta :: A :-\textbf{not } B, \textbf{not } C.$.

three rules in Expression (2). Denote by N_{000} the number of times that $\{A_1 =$ false$, A_2 =$ false$, A_3 =$ false$\}$ appear in the dataset; by N_{001} the number of times that $\{A_1 =$ false$, A_2 =$ false$, A_3 =$ true$\}$ appear in the dataset, and likewise each N_{ijk} stands for the number of times a particular configuration of A_1, A_2 and A_3 appear in the dataset. Then the local likelihood that must be maximized for this candidate family is

$$(\theta_{1;3})^{N_{100}}(1 - \theta_{1;3})^{N_{000}}(\theta_{1;2})^{N_{110}}(1 - \theta_{1;2})^{N_{010}}\theta_1^{N_{101}+N_{111}}(1 - \theta_1)^{N_{001}+N_{011}},$$

where we use, here and in the remainder of the paper, $\theta_{i;j}$ to denote $\theta_i + \theta_j - \theta_i\theta_j$.

Hence, even at the local level, we face a non-trivial optimization problem. We discuss the solution of this problem in the next section. For now we assume that the problem has been solved; that is, each family is associated with a local score (the log-likelihood of that family, with parameters that maximize likelihood, minus a penalty on the number of parameters). Once the local score are ready, we resort to the constraint-programming algorithm (CPBayes) by Van Beek and Hoffmann [22] to run the global optimization step, thus selecting families so as to have an acyclic PLP. Clearly a selection of families leads to a PLP, as each family is associated with the rule set that maximizes the local score.

The CPBayes algorithm defines a set of constraints that must be satisfied in the Bayesian network learning problem, and seeks for an optimal solution based on a depth-first BnB search. When trying to expand a node in the search tree, two conditions are verified: (1) whether constraints are satisfied, and (2) whether a lower bound estimate of the cost does not exceed the current upper bound. The constraint model includes dominance constraints, symmetry-breaking constraints, cost-based pruning rules, and a global acyclicity constraint. We remark that other approaches for the global optimization can be used, and our contribution is certainly not due to our use of CPBayes in the global optimization step. Thus we do not dwell on this second step.

5   Computing the Local Score

In this section we address the main novel challenge posed by score-based learning of PLPs; namely, the computation of local scores. As discussed in the previous section, this is not a trivial problem for two reasons. First, there are too many sets of rules to consider. Second, likelihood maximization for each rule set may have to deal with nonlinear expressions for probability values.

We deal with the first problem by pruning rule sets; that is, by developing techniques that allow us to quickly eliminate many candidate rule sets.

First of all, we can easily discard rule sets that assign zero probability to some configuration observed in the dataset (for instance, the first rule in Example 2 can be discarded if we observe $\{A_2 =$ false$, A_3 =$ false$\}$).

Second, and more importantly, suppose that we are learning rules with at most k literals in the body. With 2^k rules we can have a rule for each configuration of the parents: Example 1 illustrates this scenario. Note that for such "disjoint" rules, likelihood maximization is simple as it is the same as for usual CPT.

And because *any* CPT can be exactly built with such 2^k rules, any set of rules with more than 2^k rules cannot have higher likelihood, and thus cannot be optimal (as the penalty for the number of parameters increases). In fact, any other rule set with 2^k rules that are not disjoint can be also discarded; these sets can only produce the same likelihood, and will pay the same penalty on parameters, but they will be more complex to handle. Thus we must only deal with rule sets with at most $2^k - 1$ rules, plus the one set of 2^k "disjoint" rules. Hence:

- If we have a family with $k = 1$, we only need to look at sets of one rule plus one set containing two disjoint rules; that is, we only have to consider:

$$\theta :: A :-B. \qquad \theta :: A :-\textbf{not } B. \qquad \begin{array}{l} \theta_1 :: A :-B. \\ \theta_2 :: A :-\textbf{not } B. \end{array}$$

Note that the last of these three rule sets corresponds to a typical CPT, while the first two rule sets genuinely reduce the number of parameters in the model. Estimates that maximize likelihood are easily computed in all three cases.

- Now if we have a family with $k = 2$, we only need to look at sets of up to three rules, plus one set containing four disjoint rules. There are 4 sets consisting each of one rule, 30 sets consisting of two rules each, and 82 sets consisting of three rules each (we cannot list them all here due to space constraints). In this case probability values may have nonlinear expressions as discussed in connection with Expression (2). Thus we still have a challenging optimization problem, where we must find a rule set out of many.

To address the difficulty mentioned in the previous sentence, we resort to a third insight: many of the rule sets obtained for $k = 2$ are actually restricted versions of a few patterns. As an example, consider Table 1. There we find four different rule sets, some with two rules, and one with three rules. The form of their likelihoods is the same, sans some renaming of parameters. Note that the maximum likelihood of the first three rule sets can always be attained by the likelihood of the last rule set; consequently, it makes sense only to retain the last pattern, which consists of disjoint rules.

By doing this additional pruning, we reach 14 distinct rule sets; amongst them we must find a rule set that maximizes likelihood. The remaining difficulty is that probability values are nonlinear functions of parameters, as we have already indicated. However, most of them lead to likelihood expressions that can be *exactly* maximized. For instance, consider the following expression, a likelihood pattern that several rule sets (consisting of two rules) produce:

$$\theta_1^{N_0}(1 - \theta_1)^{N_1}(\theta_1 + \theta_2 - \theta_1\theta_2)^{M_0}(1 - (\theta_1 + \theta_2 - \theta_1\theta_2))^{M_1};$$

this function is maximized by:

$$\theta_1 = \frac{N_0}{N_0 + N_1}, \qquad \theta_2 = \frac{N_1 M_0 - N_0 M_1}{N_1 M_0 + N_1 M_1}.$$

Table 1. A probability pattern shared by several rule sets; first column presents the rule sets, and following columns display the probability values for the configurations of A_1, A_2, A_3

Rule sets	0,0,0	0,0,1	0,1,0	0,1,1	1,0,0	1,0,1	1,1,0	1,1,1
$\theta_1 :: A_1 :-A_2.$ $\theta_2 :: A_1 :-A_2, A_3.$ $\theta_3 :: A_1 :-A_2, \mathbf{not}\ A_3.$	1	1	$1-\theta_{1;3}$	$1-\theta_{1;2}$	0	0	$\theta_{1;3}$	$\theta_{1;2}$
$\theta_1 :: A_1 :-A_2.$ $\theta_2 :: A_1 :-A_2, A_3.$	1	1	$1-\theta_1$	$1-\theta_{1;2}$	0	0	θ_1	$\theta_{1;2}$
$\theta_1 :: A_1 :-A_2.$ $\theta_3 :: A_1 :-A_2, \mathbf{not}\ A_3.$	1	1	$1-\theta_{1;2}$	$1-\theta_1$	0	0	$\theta_{1;2}$	θ_1
$\theta_2 :: A_1 :-A_2, A_3.$ $\theta_3 :: A_1 :-A_2, \mathbf{not}\ A_3.$	1	1	$1-\theta_2$	$1-\theta_1$	0	0	θ_2	θ_1

Similarly, consider the following likelihood pattern (produced by rules sets consisting of three rules):

$$\theta_1^{N_0}(1-\theta_1)^{N_1}(\theta_1+\theta_2-\theta_1\theta_2)^{M_0}(1-(\theta_1+\theta_2-\theta_1\theta_2))^{M_1}$$

$$(\theta_1+\theta_3-\theta_1\theta_2)^{Q_0}(1-(\theta_1+\theta_3-\theta_1\theta_3))^{Q_1};$$

this function is maximized by:

$$\theta_1 = \frac{N_0}{N_0+N_1}, \qquad \theta_2 = \frac{N_1M_0-N_0M_1}{N_1M_0+N_1M_1}, \qquad \theta_3 = \frac{N_1Q_0-N_0Q_1}{N_1Q_0+N_1Q_1}.$$

Due to space constraints we omit the complete list of likelihood patterns and their maximizing parameters. There are only four patterns that do not seem to admit closed form solutions. Here is one example:

$$\theta_1^{N_0}(1-\theta_1)^{N_1}(\theta_1+\theta_2-\theta_1\theta_2)^{M_0}(1-(\theta_1+\theta_2-\theta_1\theta_2))^{M_1}$$
$$(\theta_1+\theta_3-\theta_1\theta_3)^{Q_0}(1-(\theta_1+\theta_3-\theta_1\theta_3))^{Q_1}(\theta_1+\theta_2+\theta_3-\theta_1\theta_2-\theta_1\theta_3-\theta_2\theta_3+\theta_1\theta_2\theta_3)^{R_0}$$
$$(1-(\theta_1+\theta_2+\theta_3-\theta_1\theta_2-\theta_1\theta_3-\theta_2\theta_3+\theta_1\theta_2\theta_3))^{R_1}.$$

By taking logarithms and derivatives, these maximization problems can then turned into the solution of systems of polynomial equations. Such systems can be solved exactly but slowly, or approximately by very fast algorithms (as we comment in the next section).

The procedure we have developed is summarized by Algorithm 1. We firstly generate all possible combinations of rules for all possible families, a possible family consisting of a variable and its parent candidates. Combinations with ensured lower score or zero likelihood are then pruned and parameters are locally optimized for each of the combinations left. Each family is then associated with the combination of rules that gives it the highest score. Finally, a global score maximization algorithm is used to select the best family candidates.

Algorithm 1. Learning algorithm for class AP-PLP(2).

```
1: collect variables from dataset
2: for each family of variables in dataset do
3:    build all possible rules
4:    build all possible combinations of rules
5:    gather rule sets into patterns
6:    for each pattern do
7:       prune combinations with ensured lower score
8:       prune combinations with zero likelihood
9:       for each combination left do
10:         if there is an exact solution to the likelihood maximization problem then
11:            calculate parameters
12:         else
13:            run numeric (exact or approximate) likelihood maximization
14:       calculate local scores
15:    for each family do
16:       associate best rule set with family
17: call CPBayes algorithm to maximize global score
```

6 Experiments

To validate our methods, we have implemented the learning algorithm described previously, and tested it with a number of datasets. Our goal was to examine whether the algorithm actually produces sensible PLPs with less parameters than corresponding Bayesian networks based on explicit CPTs.

The algorithm was implemented in Python and experiments were performed on a Unix Machine with Intel core i5 (2.7 GHz) processor and 8 GB 1867 MHz DDR3 SDRAM. For local optimization of the likelihood scores, in the few cases where that was needed, we used, and compared, two different algorithms: (1) Limited-memory BFGS (L-BFGS) and (2) the Basin-hopping algorithm [23]. Both methods are implemented in the Python library scipy.optimize. The L-BFGS algorithm approximates the BroydenFletcherGoldfarbShanno (BFGS) algorithm [24], which is an iterative method for solving unconstrained nonlinear optimization problems. The L-BFGS algorithm represents with a few vectors an approximation to the inverse Hessian matrix; this approach leads to a significant reduction on memory use. Nevertheless, it has a quite strong dependence on the initial guess. The Basin-hopping is a stochastic algorithm that usually provides a better approximation of the global maximum. The algorithm iteratively chooses an initial guess at random, proceeds to the local minimization and finally compares the new coordinates with the best ones found so far. This algorithm is however much more time-consuming.

To begin, consider a fairly standard dataset that describes diagnoses of cardiac Single Proton Emission Computed Tomography (SPECT) images [12]. The training dataset contains 80 instances, while the testing dataset contains 187 instances. Examples have 23 binary attributes and there is no missing data. The learning algorithm was tested with the same optimization methods and local structure learning approaches. We compare results obtained for two different local structure learning approaches: (1) accepting only combinations of rules that encode complete probability tables and (2) or any combination of rules. Results obtained are listed in Table 2.

Table 2. Heart Diagnosis experiments

L-BFGS							
Training set	Testing set	Complete CPT			Any combination of rules		
# Instances	# Instances	MDL	Log-Likelihood	Parameters	MDL	Log-Likelihood	Parameters
80	187	−1341.73	−1281.78	63	−1316.18	−1263.85	55
Basin-hopping							
Training set	Testing set	Complete CPT			Any combination of rules		
# Instances	# Instances	MDL	Log-Likelihood	Parameters	MDL	Log-Likelihood	Parameters
80	187	−1341.73	−1281.78	63	−1316.18	−1263.85	55

We observe the significant reduction of the number of parameters needed for representation. In addition, results obtained with both optimization algorithms are the exactly same.

We then present results with data generated from a (simulated) faulty Boolean circuit. The purpose of this experiment is to investigate whether our methods can capture nearly-deterministic systems with less parameters than a typical Bayesian network. We should expect so: rules can encode deterministic relationships compactly, so they should lead to a reduction in the number of necessary parameters.

We simulated a digital circuit for addition in the binary numeral system. We consider the addition of two 4-bit numbers and, therefore, 24 logic gates (XOR, OR, AND). The circuit is used to generate binary datasets, where attributes

Table 3. Binary Adder experiments

L-BFGS							
Training set	Testing set	Complete CPT			Any combination of rules		
# Instances	# Instances	MDL	Log-Likelihood	Parameters	MDL	Log-Likelihood	Parameters
30	10000	−317635.87	−317590.82	61	−190628.02	−190592.57	48
60	10000	−282916.56	−282860.54	63	−211580.75	−211535.40	51
90	10000	−231156.16	−231096.56	61	−200133.73	−200086.83	48
120	10000	−281634.57	−281571.16	61	−197082.89	−197029.87	51
250	10000	−244964.95517	−244887.02	65	−251550.34	−251489.19	51
500	10000	−228706.11	−228617.04	66	−217679.01	−217608.84	52
1000	10000	−188356.10	−188236.10	80	−177142.65	−177049.65	62
Basin-hopping							
Training set	Testing set	Complete CPT			Any combination of rules		
# Instances	# Instances	MDL	Log-Likelihood	Parameters	MDL	Log-Likelihood	Parameters
30	10000	−344625.96	−344580.91	61	−190687.47	−190652.02	48
1000	10000	−188356.10	−188236.10	80	−177142.52	−177049.52	62

correspond to the gates outputs. The adder input values are randomly chosen and each gate is associated with a 1% probability of failure, i.e. the likelihood that a gate outputs 0 or 1 at random. Training datasets contain 30, 60, 90, 120, 250, 500 and 1000 instances, while testing datasets contain 10000 instances in all cases. The learning algorithm is tested with both L-BFGS and Basin-hopping optimization methods. Results obtained are listed in Table 3.

We note L-BFGS and Basin-hopping optimization methods perform fairly similar. However, as Basin-hopping is much more time consuming, most tests are run with L-BFGS. For smaller datasets, the algorithm proposed in this paper scores better and requires fewer parameters. For larger datasets, both approaches tend to converge in terms of score, but there is still a significant reduction on the number of parameters.

7 Conclusion

We have described techniques that can learn a PLP from a complete dataset by exact score maximization. Despite the attention that has been paid to PLPs in recent years, it does not seem that exact score-based learning has been attempted so far. This paper offers initial results on such an enterprise; the main contribution is to present cases where closed-form solutions are viable.

The techniques proposed in this paper apply to a restricted albeit powerful class of PLPs. In essence, we have shown that for this class it is possible to maximize the MDL score using a host of insights that simplify computation.

The class we have focused on is the class of acyclic propositional PLPs where rules have at most two literals in their bodies. As acyclic propositional PLPs are closely related to Bayesian networks, we were able to bring results produced for Bayesian network learning into the challenging task of learning probabilistic programs. However, PLPs have features of their own; an advantage is that they can capture conditional distributions with less parameters than usual CPTs; a disadvantage is that learning requires more complex optimization. Our results identify a powerful class of PLPs for which the complexity of optimization can be kept under control.

We have also implemented and tested our methods. We have shown that learned PLPs contain less parameters than the correponding CPT-based Bayesian networks, as intuitively expected. Whenever the model is nearly deterministic, the expressive power of rules leads to improved accuracy.

In future work we intend to extend our techniques to relational but still acyclic programs, and finally to relational and even cyclic programs. For those cases non-trivial extensions will have to be developed as the direct relationship with Bayesian network learning will be lost.

Acknowledgement. The first author is supported by a scholarship from Toshiba Corporation. The second and third authors are partially supported by CNPq. This work was partly supported by the São Paulo Research Foundation (FAPESP) grant 2016/01055-1 and the CNPq grants 303920/2016-5 and 420669/2016-7; also by

São Paulo Research Foundation (FAPESP) grant 2016/18841-0 and CNPq grant 308433/2014-9; finally by FAPESP 2015/21880-4.

References

1. Bellodi, E., Riguzzi, F.: Structure learning of probabilistic logic programs by searching the clause space. Theory Pract. Logic Program. **15**(2), 169–212 (2015)
2. Boutilier, C., Friedman, N., Goldszmidt, M., Koller, D.: Context-specific independence in Bayesian networks. In Conference on Uncertainty in Artificial Intelligence, pp. 115–123 (1996)
3. De Campos, C., Ji, Q.: Efficient learning of Bayesian networks using constraints. J. Mach. Learn. Res. **12**, 663–689 (2011)
4. Cussens, J.: Bayesian network learning with cutting planes. In Conference on Uncertainty in Artificial Intelligence, pp. 153–160 (2011)
5. Dantsin, E., Eiter, T., Voronkov, A.: Complexity and expressive power of logic programming. ACM Comput. Surv. **33**(3), 374–425 (2001)
6. De Raedt, L., Thon, I.: Probabilistic rule learning. In: Frasconi, P., Lisi, F.A. (eds.) ILP 2010. LNCS, vol. 6489, pp. 47–58. Springer, Heidelberg (2011). doi:10.1007/978-3-642-21295-6_9
7. De Raedt, L., Dries, A., Thon, I., Van den Broeck, G., Verbeke, M.: Inducing probabilistic relational rules from probabilistic examples. In: International Joint Conference on Artificial Intelligence (2015)
8. Fierens, D., Van den Broeck, G., Renkens, J., Shrerionov, D., Gutmann, B., Janssens, G., de Raedt, L.: Inference and learning in probabilistic logic programs using weighted Boolean formulas. Theory Pract. Logic Program. **15**(3), 358–401 (2014)
9. Friedman, N., Goldszmidt, M.: Learning Bayesian networks with local structure. In: Conference on Uncertainty in Artificial Intelligence, pp. 252–262 (1998)
10. Fuhr, N.: Probabilistic datalog – a logic for powerful retrieval methods. In: Conference on Research and Development in Information Retrieval, Seattle, Washington, pp. 282–290 (1995)
11. Dean, A., Voss, D., Draguljić, D.: Principles and techniques. Design and Analysis of Experiments. STS, pp. 1–5. Springer, Cham (2017). doi:10.1007/978-3-319-52250-0_1
12. Lichman, M.: UCI Machine Learning Repository. University of California, Irvine, School of Information and Computer Sciences (2013)
13. Lukasiewicz, T.: Probabilistic logic programming. In: European Conference on Artificial Intelligence, pp. 388–392 (1998)
14. Ng, R., Subrahmanian, V.S.: Probabilistic logic programming. Inf. Comput. **101**(2), 150–201 (1992)
15. Pearl, J.: Probabilistic Reasoning in Intelligent Systems: Networks of Plausible Inference. Morgan Kaufmann, San Mateo (1988)
16. Poole, D.: Probabilistic Horn abduction and Bayesian networks. Artif. Intell. **64**, 81–129 (1993)
17. Poole, D.: Probabilistic relational learning and inductive logic programming at a global scale. In: Frasconi, P., Lisi, F.A. (eds.) ILP 2010. LNCS, vol. 6489, pp. 4–5. Springer, Heidelberg (2011). doi:10.1007/978-3-642-21295-6_3
18. De Raedt, L. (ed.): Logical and Relational Learning. Springer, Heidelberg (2008)
19. Riguzzi, F., Bellodi, E., Zese, R.: A history of probabilistic inductive logic programming. Front. Robot. AI **1**, 1–5 (2014)

20. Sato, T.: A statistical learning method for logic programs with distribution semantics. In: International Conference on Logic Programming, pp. 715–729 (1995)
21. Sato, T., Kameya, Y.: Parameter learning of logic programs for symbolic-statistical modeling. J. Artif. Intell. Res. **15**, 391–454 (2001)
22. van Beek, P., Hoffmann, H.-F.: Machine learning of Bayesian networks using constraint programming. In: Pesant, G. (ed.) CP 2015. LNCS, vol. 9255, pp. 429–445. Springer, Cham (2015). doi:10.1007/978-3-319-23219-5_31
23. Wales, D.J., Doye, J.P.: Global optimization by Basin-Hopping and the lowest energy structures of Lennard-Jones clusters containing up to 110 atoms. J. Phys. Chem. A **101**, 5111–5116 (1997)
24. Wales, D.J.: Energy Landscapes. Cambridge University Press, Cambridge (2003)
25. Yang, F., Yang, Z., Cohen, W.W.: Differentiable learning of logical rules for knowledge base completion. Technical report arxiv.org/abs/1702.08367, Carnegie Mellon University (2017)

Fault Tolerant Direct NAT Structure Extraction from Pairwise Causal Interaction Patterns

Yang Xiang[✉]

University of Guelph, Guelph, Canada
yxiang@uoguelph.ca

Abstract. Non-impeding noisy-And Trees (NATs) provide a general, expressive, and efficient causal model for conditional probability tables (CPTs) in discrete Bayesian networks (BNs). A CPT may be directly expressed as a NAT model or compressed into a NAT model. Once CPTs are NAT-modeled, efficiency of BN inference (both space and time) can be significantly improved. The most important operation in NAT modeling CPTs is extracting NAT structures from interaction patterns between causes. Early method does so through a search tree coupled with a NAT database. A recent advance allows extraction of NAT structures from full, valid causal interaction patterns based on bipartition of causes, without requiring the search tree and the NAT database. In this work, we extend the method to direct NAT structure extraction from partial and invalid causal interaction patterns. This contribution enables direct NAT extraction from all conceivable application scenarios.

Keywords: Graphical models · Probabilistic inference · Machine learning · Bayesian networks · Causal models · Non-impeding noisy-AND trees

1 Introduction

Conditional independence encoded in BNs avoids combinatorial explosion in the number of variables. However, BNs are still subject to exponential growth of space and inference time in the number of causes per effect variable in each CPT. A number of space-efficient local models exist, that allow efficient encoding of dependency between an effect and its causes. They include noisy-OR [Pea88], noisy-MAX [Hen89, Die93], context-specific independence (CSI) [BFGK96], recursive noisy-OR [LG04], Non-Impeding Noisy-AND Tree (NIN-AND Tree or NAT) [XJ06], DeMorgan [MD08], tensor-decomposition [VT12], and cancellation model [WvdGR15]. These local models not only reduce the space and time needed to acquire numerical parameters in CPTs, they can also be exploited to significantly reduce inference time, e.g., by exploiting CSI in arithmetic circuits (ACs) and sum-product networks (SPNs) [Dar03, PD11, ZMP15], or by exploiting causal independence in NAT models [XJ16b].

We consider expressing BN CPTs as or compressing them into multi-valued NAT models [Xia12]. Merits of NAT models include being based on simple causal

© Springer International Publishing AG 2017
S. Moral et al. (Eds.): SUM 2017, LNAI 10564, pp. 134–148, 2017.
DOI: 10.1007/978-3-319-67582-4_10

interactions (reinforcement and undermining), expressiveness (recursive mixture, multi-valued), and generality (generalizing noisy-OR, noisy-MAX [XJ16b] and DeMorgan [Xia12]). In addition, they support much more efficient inference. As shown in [XJ16b], two orders of magnitude speedup in lazy propagation is achieved in NAT-modeled BNs. Since causal independence encoded in a NAT model is orthogonal to CSI, NAT models provide an alternative to CSI for efficient probabilistic inference in BNs.

A NAT model over an effect and n causes consists of a NAT topology and a set of numerical parameters (whose cardinality is linear in n). It compactly represents a BN CPT. In a NAT model, each pair of causes either undermines each other in causing the effect, or reinforcing each other. Hence, the interaction can be specified by one bit with values u (undermining) or r (reinforcing). The collection of such bits defines a pairwise causal interaction (PCI) pattern [XLZ09]. A PCI pattern may be *full* (with one bit for each cause pair) or *partial* (with some missing bits). It has been shown that a full PCI pattern uniquely identifies a NAT [XLZ09,XT14]. This property enables PCI patterns to play an important role for acquisition of NAT topology both in compressing a BN CPT into a NAT model and in learning a BN CPT as a NAT model from data. The corresponding computation takes as input a PCI pattern and returns a compatible (defined below) NAT topology. We term this operation as *NAT structure extraction* from PCI.

For instance, in compressing a target BN CPT into a NAT model, the following method has been applied [XL14,XJ16a]. A partial PCI pattern is first obtained from the target CPT. From the pattern, compatible candidate NATs are extracted through a search tree coupled with a NAT database. Which candidate NAT becomes the final choice is determined by parameterization. For each n value, a NAT database is needed that stores all alternative NATs for n causes. Its size grows super-exponentially in n (see below), and hence it is the source of a computational burden, both offline and online. For instance, it takes 40 h to generate (offline) the NAT database for $n = 9$ and its search tree [XL14].

An arbitrary bit pattern (either partial or full) may not have a corresponding NAT. Such a pattern is *invalid* (defined below). A recent advance [Xia17] proposed a method for NAT structure extraction from full and valid PCI patterns without needing a search tree and the NAT database. In this paper, we extend the method along two directions. First, we relax the requirement of full PCI patterns so that NAT structures can be extracted from partial PCI patterns. Second, we relax the requirement of valid PCI patterns so that the input can be an invalid pattern and a NAT is extracted whose PCI pattern is closest to the input pattern. These advancements enable NAT structure extraction in all conceivable application scenarios: valid full PCI patterns, valid partial patterns, invalid full patterns, and invalid partial patterns. All of them are through direct extraction, i.e., without need of the search tree and the NAT database.

Section 2 reviews background on NAT models. The task of fault tolerant, direct NAT structure extraction is specified in Sect. 3. Sections 4 and 5 present theoretical results that the rest of the paper depends on. Direct extraction from full, possibly invalid PCI patterns is covered in Sect. 6 and extraction from partial, possibly invalid patterns is presented in Sect. 7. The experimental results are reported in Sect. 8.

2   Background

This section briefly reviews background on NAT models. More details can be found in [Xia12]. A NAT model is defined over an effect e and a set of n causes $C = \{c_1, ..., c_n\}$ that are multi-valued and graded, where $e \in \{e^0, ..., e^\eta\}$ $(\eta \geq 1)$ and $c_i \in \{c_i^0, ..., c_i^{m_i}\}$ $(m_i \geq 1)$. C and e form a single family in a BN. Values e^0 and c_i^0 are *inactive*. Other values (may be written as e^+ or c_i^+) are *active* and a higher index means higher intensity (graded).

A causal event is a *success* or *failure* depending on if e is active at a given intensity, is *single-* or *multi-causal* depending on the number of active causes, and is *simple* or *congregate* depending on the value range of e. More specifically,

$$P(e^k \leftarrow c_i^j) = P(e^k | c_i^j, c_z^0 : \forall z \neq i) \quad (j > 0)$$

is the probability of a *simple single-causal success*.

$$P(e \geq e^k \leftarrow c_1^{j_1}, ..., c_q^{j_q}) = P(e \geq e^k | c_1^{j_1}, ..., c_q^{j_q}, c_z^0 : c_z \in C \setminus X),$$

is the probability of a *congregate multi-causal success*, where $j_1, ..., j_q > 0$, $X = \{c_1, ..., c_q\}$ $(q > 1)$, and it may be denoted as $P(e \geq e^k \leftarrow \underline{x}^+)$. Interactions among causes may be reinforcing or undermining as defined below.

Definition 1. *Let e^k be an active effect value, $R = \{W_1, ..., W_m\}$ $(m \geq 2)$ be a partition of a set $X \subseteq C$ of causes, $S \subset R$, and $Y = \cup_{W_i \in S} W_i$. Sets of causes in R reinforce each other relative to e^k, iff $\forall S$ $P(e \geq e^k \leftarrow \underline{y}^+) \leq P(e \geq e^k \leftarrow \underline{x}^+)$. They undermine each other iff $\forall S$ $P(e \geq e^k \leftarrow \underline{y}^+) > P(e \geq e^k \leftarrow \underline{x}^+)$.*

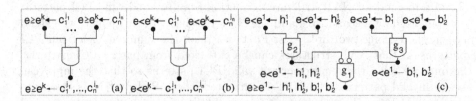

Fig. 1. A direct NIN-AND gate (a), a dual NIN-AND gate (b), and a NAT (c)

A NAT has multiple NIN-AND gates. A *direct* gate involves disjoint sets of causes $W_1, ..., W_m$. Each input event is a success $e \geq e^k \leftarrow \underline{w}_i^+$ $(i = 1, ..., m)$, e.g., Fig. 1 (a) where each W_i is a singleton. The output event is $e \geq e^k \leftarrow \underline{w}_1^+, ..., \underline{w}_m^+$. Its probability is

$$P(e \geq e^k \leftarrow \underline{w}_1^+, ..., \underline{w}_m^+) = \prod_{i=1}^{m} P(e \geq e^k \leftarrow \underline{w}_i^+),$$

which encodes undermining causal interaction. Each input event of a *dual* gate is a failure $e < e^k \leftarrow \underline{w}_i^+$, e.g., Fig. 1 (b). The output event is $e < e^k \leftarrow \underline{w}_1^+, ..., \underline{w}_m^+$. Its probability is

$$P(e < e^k \leftarrow \underline{w}_1^+, ..., \underline{w}_m^+) = \prod_{i=1}^{m} P(e < e^k \leftarrow \underline{w}_i^+),$$

which encodes reinforcement. Figure 1(c) shows a NAT, where causes h_1 and h_2 reinforce each other, so do b_1 and b_2, but the two groups undermine each other.

A NAT can be depicted simply by a Root-Labeled-Tree (RLT).

Definition 2. *Let T be a NAT. The RLT of T is a directed graph obtained from T as follows. (1) Delete each gate and direct its inputs to output. (2) Delete each non-root label. (3) Replace each root label by the corresponding cause.*

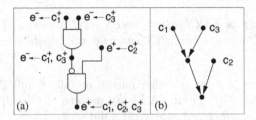

Fig. 2. A NAT (a) and its RLT (b)

Figure 2 shows a NAT and its RLT. The leaf of RLT corresponds to leaf gate of the NAT. When the leaf gate is dual (or direct), the leaf of RLT is said to be dual (or direct). The leaf gate of a NAT is at level-one. A gate that feeds into the leaf gate is at level-two, and so on. We refer to levels of nodes of a RLT similarly. All gates in the same level have the same type (dual or direct) and gates in adjacent levels differ. An RLT and a leaf type uniquely specifies a NAT.

A NAT T has a single leaf z. For $n \geq 2$, leaf z has at least two parents. Each parent v of z is the leaf of a subtree *induced* by z. If v is a root, then v is a *root parent* of z, and the induced subtree is trivial. In Fig. 2(b), there are two subtrees induced by the leaf. One subtree is trivial, where c_2 is a root parent of the leaf, and the *root set* of the subtree is $\{c_2\}$. The root set of the other subtree is $\{c_1, c_3\}$.

Each NAT uniquely defines pairwise causal interaction between each pair of causes c_i and c_j $(i \neq j)$, denoted by a PCI bit $\pi(c_i, c_j) \in \{u, r\}$. The value $\pi(c_i, c_j)$ is defined by the common gate of c_i and c_j at the highest level [XLZ09]. The NAT in Fig. 1(c) has $\pi(h_1, h_2) = r$ since g_2 is dual and $\pi(h_1, b_2) = u$ since g_1 is direct. A collection of PCI bits is a PCI pattern π. If π includes one bit for each cause pair, it is a *full* pattern. Otherwise, it is *partial*.

3   Fault Tolerant Direct NAT Structure Extraction

It has been shown that a full PCI pattern uniquely identifies a NAT [XLZ09, XT14]. This enables PCI patterns to play an important role for acquisition of NAT topology in compressing a BN CPT into a NAT model and in learning a BN CPT from data as a NAT model. In either case, the input is a PCI pattern and the output is a NAT. We refer to the task as *NAT structure extraction*. Input patterns to the task can be classified as follows. First, we relate two PCI patterns over the same set of causes.

Definition 3. *Let π and ψ be PCI patterns over a set C of causes. If for each pair of causes c_i and c_j $(i \neq j)$ such that both $\pi(c_i, c_j)$ and $\psi(c_i, c_j)$ are defined, $\pi(c_i, c_j) = \psi(c_i, c_j)$ holds, then π and ψ are* **compatible**. *Otherwise, they are* **incompatible**.

Either π or ψ may be partial or full. Compatibility is determined by PCI bits that are defined under both π and ψ. Next, we relate a PCI pattern and a NAT over the same set of causes.

Definition 4. *Let π be a PCI pattern over a set C of causes. Then π is* **valid** *if there exists a NAT over C whose PCI pattern ψ is compatible with π. Otherwise, π is* **invalid**.

In the definition, π may be either partial or full. A full PCI pattern over a set of n causes has $C(n, 2)$ bits. A binary pattern of $C(n, 2)$ bits has $2^{C(n,2)}$ variations, not all of which are necessarily valid.

For $n = 2$, there are 2 NATs. A PCI pattern has $C(2, 2) = 1$ bit. Hence, every PCI pattern is valid. For $n = 3$, there are 8 NATs. A PCI pattern has $C(3, 2) = 3$ bits. Hence, every PCI pattern is valid. For $n = 4$, there are 52 NATs. There are $2^{C(4,2)} = 2^6 = 64$ binary patterns, of which $64 - 52 = 12$ patterns are invalid. For $n = 7$, there are $2^{21} = 2,097,152$ binary patterns and 78,416 NATs [XZL09]. The number of invalid full PCI patterns is 2,018,736.

The extraction task has been performed in the context of compressing BN CPTs into NAT models, where PCI patterns are obtained from CPTs and then NATs are extracted [XL14, XJ16a]. The extraction [XL14, XJ16a] relies on a search tree coupled with a NAT database. For each n value, a NAT database stores all alternative NATs over n causes, and the search tree retrieves one or more NATs given a valid PCI pattern [XL14]. We refer to the method as *search tree based* extraction.

The size of the database and the search tree grow super-exponentially in n. Although constructed offline, they are the source of a computational burden. For $n = 9$, there are 25,637,824 NATs and it takes 40 h to generate the database and the search tree [XL14]. Although NAT models are local models (one BN family per model), and hence n does not grow unbounded due to conditional independence encoded in BNs, it is costly and difficult to generate databases and search trees when n grows beyond 9.

To alleviate these costs, a method is proposed recently [Xia17] for NAT extraction without need of the NAT databases and the search tree. We refer to the method as *direct* extraction. The method requires a full, valid PCI pattern as the input. When a PCI pattern is obtained from a CPT, the pattern is full if the CPT is a NAT model, and is partial otherwise. In this work, we extend the direct method to allow partial input patterns.

When a PCI pattern is obtained from a CPT, there is no guarantee that it is valid. Therefore, the full spectrum of input for NAT extraction includes full and partial, as well as valid and invalid PCI patterns. We refer to NAT structure extraction from invalid PCI patterns as being *fault tolerant*. Existing NAT extraction [XL14,XJ16a] does not explicitly consider the case when input PCI patterns are invalid. Fault tolerant NAT elicitation was considered in [Xia10]. However, it does not provide algorithms for detecting invalid PCI patterns and generating NATs accordingly. In this work, we develop such an algorithm for fault tolerant and direct NAT extraction.

4   Bipartitions of Causes in NAT Models

The direct method for NAT structure extraction initiated in [Xia17] is based on bipartitions of causes. Below, we reformulate some relevant concepts and results from [Xia17] for better clarity and extend them for the purpose of this work.

Definition 5. *Let C ($|C| \geq 2$) be a set of causes, X and Y be non-empty subsets of C where $X \cap Y = \emptyset$ and $X \cup Y = C$, and π be a full PCI pattern over C. Then $\{X, Y\}$ is a **uniform causal bipartition** of C under π if one of the following holds.*

1. $\forall x \in X, \forall y \in Y, \pi(x, y) = r$
2. $\forall x \in X, \forall y \in Y, \pi(x, y) = u$

For $C = \{x, y\}$, π has a single bit. Hence, it is trivially true that $\{\{x\}, \{y\}\}$ forms a uniform causal bipartition. For $C = \{x, y, z\}$, every PCI pattern is valid (see Sect. 3), and has a NAT T. At least one cause, say x, is the parent of the leaf in T, and $\{\{x\}, \{y, z\}\}$ is a uniform causal bipartition.

Bipartitions in Def. 5 are based on causal interactions. Bipartitions in Def. 6 below are based on NAT topology.

Definition 6. *Let T be a NAT over C. Let $\{X, Y\}$ be a bipartition of C, where $X \neq \emptyset$, $Y \neq \emptyset$, $X \cap Y = \emptyset$, and $X \cup Y = C$. If for each leaf-induced subtree of T and its root set R, either $R \subseteq X$ or $R \subseteq Y$ holds, then $\{X, Y\}$ is a **subtree-consistent bipartition** of C with respect to T.*

In Fig. 2, $\{\{c_2\}, \{c_1, c_3\}\}$ is a subtree-consistent bipartition of $C = \{c_1, c_2, c_3\}$, but $\{\{c_1\}, \{c_2, c_3\}\}$ is not.

Theorem 1 below relates the two types of bipartitions defined. It is equivalent to Theorem 1 in [Xia17] but with better clarity.

Theorem 1. *Let T be a NAT over C and π be the PCI pattern of T. Every subtree-consistent bipartition of C is a uniform causal bipartition.*

Theorem 2 below strengthens Theorem 1 with the existence of subtree-consistent bipartitions. It will be used later to justify a main result of this work.

Theorem 2. *Every NAT over a set C ($|C| \geq 2$) of causes has at least one subtree-consistent bipartition of C.*

Proof: Since $|C| \geq 2$, the leaf of T has at least two parents. Let x be such a leaf parent. If x is a root, then $\{\{x\}, C \setminus \{x\}\}$ is a subtree-consistent bipartition. Otherwise, x is the leaf of a subtree. Let X be the root set of the subtree. Then $\{X, C \setminus X\}$ is a subtree-consistent bipartition.   □

5 PCI Core and Invalid PCI Patterns

NAT extraction from invalid PCI patterns necessitates operations different from extraction from valid patterns. Activation of such operations in turn necessitates detection of invalid patterns. Below we analyze conditions for such detection. First, we formalize necessary concepts.

Definition 7. *Let π be a full PCI pattern over a set C ($|C| \geq 2$) of causes. If there exists no uniform causal bipartition under π, then π is a **PCI core** and C is the **domain** of the PCI core.*

From Sect. 3, there exists no PCI core when $n = 2$ and 3. Following [Xia17], we analyze a PCI pattern equivalently through its PCI matrix, and denote both by π interchangeably. Consider the PCI matrix π in Fig. 3 (left).

	a	b	c	d
a		r	u	r
b	r		r	u
c	u	r		u
d	r	u	u	

	e	a	b	c	d
e		u	u	u	u
a	u		r	u	r
b	u	r		r	u
c	u	u	r		u
d	u	r	u	u	

Fig. 3. PCI matrix over $C = \{a, b, c, d\}$ (left) and one over $C = \{e, a, b, c, d\}$ (right)

Since the row indexed by a is not uniform, $\{\{a\}, \{b, c, d\}\}$ is not a uniform causal partition. In fact, none of the bipartitions $\{X, Y\}$ is when $|X| = 1$. For $X = \{a, d\}$ and $Y = \{b, c\}$, consider cells at the intersection of rows indexed by X and columns indexed by Y. The first row (r, u) in the intersection is non-uniform. Hence, $\{\{a, d\}, \{b, c\}\}$ is not a uniform causal partition. In fact, none of the bipartitions $\{X, Y\}$ is when $|X| = 2$. Hence, π is a PCI core. This shows that the smallest PCI core (over the least number of causes) occurs when $n = 4$.

Definition 8. *Let π be a PCI pattern over C. A PCI pattern ψ over $X \subseteq C$ ($|X| \geq 2$) is a sub-pattern of π if, for every $x, y \in X$, $\psi(x, y) = \pi(x, y)$.*

Note that π is a trivial sub-pattern of itself. Consider the PCI matrix π in Fig. 3 (right). The partition $\{\{e\}, \{a, b, c, d\}\}$ is causally uniform. Hence, π is not a PCI core. If we delete the row and the column indexed by e, the remainder is identical to the matrix in the left. Since the sub-pattern over $\{a, b, c, d\}$ is a PCI core, no other uniform causal partition of C exists under π.

Theorem 3 below reveals a fundamental condition of invalid PCI patterns.

Theorem 3. *A PCI pattern π over C ($|C| \geq 2$) is invalid iff π contains a sub-pattern ψ that is a PCI core.*

Proof: [Sufficiency] Assume that π contains a sub-pattern ψ over $S \subseteq C$ that is a PCI core. Since the smallest PCI core has 4 causes, $|S| \geq 4$. We show that there exists no NAT whose PCI pattern equals to π.

We prove by contradiction. Suppose that a NAT T over C exists with PCI pattern π. By Theorem 2, T has a subtree-consistent bipartition $\{X, Y\}$ of C. Either X and Y split S (possible since $|S| \geq 4$) or they don't. We consider each case below.

(Case 1). If X and Y split S, denote $S_X = X \cap S \neq \emptyset$ and $S_Y = Y \cap S \neq \emptyset$, where $S_X \cup S_Y = S$. Let ψ be the sub-pattern of π over S. By Theorem 1, $\{X, Y\}$ is a uniform casual bipartition of C under π. Hence, $\{S_X, S_Y\}$ is also a uniform casual bipartition of S under ψ: a contradiction to the assumption that ψ is a PCI core.

(Case 2). If X and Y do not split S, then S is contained in one of them, say X. From $|S| \geq 4$, we have $|X| \geq 4$. Since $\{X, Y\}$ is a subtree-consistent bipartition of C, either Y is the root set of a subtree T' induced by the leaf of T, or Y is made of root sets of multiple such subtrees. In either case, we remove each subtree induced by the leaf of T whose root set is contained in Y, and refer to the reduced tree by T'. If the leaf z of T is left with a single parent z' due to the removal, we remove z from T so that z' becomes the leaf of T'. The resultant T' is a well-defined NAT over $X \subset C$ and $|X| \geq 4$.

Since C is finite and the reduction produces a NAT over a proper subset of causes, by processing a subtree-consistent bipartition in T' recursively, Case 1 must be true eventually.

[Necessity] Suppose a PCI pattern π over C does not correspond to any NAT. We show that π contains a sub-pattern ψ that is a PCI core. We prove by contraposition. Assume that π does not contain any PCI core. We show by induction on $|C|$ that a NAT can be constructed with PCI pattern π.

For $|C| = 2$, say, $C = \{x, y\}$, since π is not a PCI core, the only bipartition $\{\{x\}, \{y\}\}$ is a uniform causal bipartition of C. Hence, a tree T with the leaf z and root parents x and y is a NAT over C.

Assume that for $|C| = k \geq 2$, if PCI pattern π over C does not contain a PCI core, then a NAT can be constructed with pattern π.

Consider $|C| = k + 1$ where PCI pattern π over C does not contain a PCI core. Since π is not a PCI core, there exists a uniform causal bipartition $\{X, Y\}$

of C, where $|X| \leq k$ and $|Y| \leq k$. Since $k + 1 \geq 3$, X and Y cannot both be singletons. Either exactly one of them is a singleton (Case a) or none of them is a singleton (Case b). We construct a NAT with pattern π in each case below.

(Case a). Suppose that X is a singleton $\{x\}$. Since π does not contain a PCI core, neither the sub-pattern ψ of π over Y does. Since $|Y| = k$, by inductive assumption, a NAT T_Y can be constructed with PCI pattern ψ. Denote the leaf of T_Y by z.

If z is direct and the uniform causal interaction relative to bipartition $\{\{x\}, Y\}$ is u, add the root parent x to z in T_Y. The resultant tree T is a NAT with PCI pattern π. The processing is similar if z is dual and the interaction relative to $\{\{x\}, Y\}$ is r.

If z is direct and the causal interaction relative to bipartition $\{\{x\}, Y\}$ is r, create a tree T with leaf v whose two parents are x and z. The resultant tree T is a NAT with PCI pattern π. The processing is similar if z is dual and the interaction relative to $\{\{x\}, Y\}$ is u.

(Case b). Suppose that none of X and Y is singleton. Let π_X (π_Y) be the sub-pattern of π over X (Y). Since π does not contain a PCI core, neither π_X nor π_Y does. Since $|X| \leq k$ ($|Y| \leq k$), by inductive assumption, a NAT T_X (T_Y) can be constructed with PCI pattern π_X (π_Y). Denote the leaf of T_X (T_Y) by z_X (z_Y).

If z_X and z_Y are both direct and the uniform causal interaction relative to bipartition $\{X, Y\}$ is u, merge T_X and T_Y by adding all parents of z_Y as parents of z_X and deleting z_Y. The resultant tree T is a NAT with PCI pattern π. The processing is similar if z_X and z_Y are both dual and the uniform causal interaction relative to bipartition $\{X, Y\}$ is r.

If z_X is direct, z_Y is dual, and the uniform causal interaction relative to bipartition $\{X, Y\}$ is u, merge T_X and T_Y by making z_Y a parent of z_X. The resultant tree T is a NAT with PCI pattern π. The processing is similar for other cases where the types of z_X and z_Y differ. $\square$

Theorem 3 establishes that the necessary and sufficient condition of an invalid PCI pattern π is that either π is a PCI core or a sub-pattern of π is.

6   NAT Extraction with Invalid PCI Pattern Detection

We apply formal results from previous sections to algorithms in [Xia17] to extend their functionality as well as to improve their semantic clarity.

Algorithm 1 below extends InteractBtwSets [Xia17] by improving its semantic clarity. As input, it takes a set C of causes, a PCI matrix π over C, and a proper subset $X \subset C$ from which a bipartition $\{X, Y\}$ (line 1) is defined. It determines if $\{X, Y\}$ is a uniform causal bipartition. If so, it returns the NIN-AND gate type that implements the causal interaction. Otherwise, it returns null.

Algorithm 1. $IsUniformCausalBipartition(C, \pi, X)$

1 $Y = C \setminus X$;
2 if $\forall x \in X, \forall y \in Y$, $\pi(x, y) = r$ holds, $gatetype = dual$;

3   else if $\forall x \in X, \forall y \in Y, \pi(x,y) = u$ *holds, gatetype = direct;*
4   else gatetype = null;
5   return gatetype;

TestPciSetNat below extends SetNatByPci [Xia17] on both functionality and semantic clarity. As input, it takes a set C of causes and a full PCI matrix π over C. Unlike [Xia17] where π is assumed valid, π can be either valid or invalid. TestPciSetNat calls IsUniformCausalBipartition to evaluate alternative bipartitions. If π is valid, it returns the respective NAT. Otherwise, invalidity of π is detected and the domain of a PCI core is returned instead (lines 21, 25, and 27). $InNat1$ and $InNat2$ are sets of causes added to the current NAT T. The *Subsets* collects subsets X and Y for each uniform causal bipartition $\{X, Y\}$. An example of matrix reduction (lines 8 and 20) is in Fig. 3, where the matrix in the right over $\{e, a, b, c, d\}$ is *reduced* to the matrix in the left over $\{a, b, c, d\}$. Although NAT is used to refer to T, the actual data structure of T is an RLT (hence, the reference to leaf z).

Algorithm 2. *TestPciSetNat*(C, π)

1   init NAT T with leaf z only; type(z) = null; init set $InNat1 = \emptyset$;
2   for each $x \in C$, do
3   if $\forall y \in C \setminus \{x\}, \pi(x,y) = r$ holds,
4   type(z) = dual; add x to T as a parent of z; $InNat1 = InNat1 \cup \{x\}$;
5   else if $\forall y \in C \setminus \{x\}, \pi(x,y) = u$ holds,
6   type(z) = direct; add x to T as a parent of z; $InNat1 = InNat1 \cup \{x\}$;
7   if $InNat1 = C$, return T;

8   reduce (C, π) to (C', ψ) relative to $InNat1$;
9 $InNat2 = \emptyset$; Subsets $= \emptyset$;
10 for $i = 2$ to $|C'|/2$, do
11   for each $X \subset C'$ where $|X| = i$, do
12   gatetype = IsUniformCausalBipartition(C', ψ, X);
13   if gatetype $\neq$ null,
14   if type(z) = null, assign type(z) = gatetype;
15   if gatetype = type(z), Subsets = Subsets $\cup \{X, C' \setminus X\}$;

16 if Subsets $\neq \emptyset$,
17   for each $X \in$ Subsets,
18   if $\exists V \in$ Subsets such that $X \supseteq V$, remove X from Subsets;
19   for each $X \in$ Subsets, do
20   reduce π to ψ over X; $R = TestPciSetNat(X, \psi)$;
21   if $R = X$, return X;
22   add R to T as a subtree induced by z;
23 $InNat2 =$ union of subsets in Subsets;
24   if $InNat2 = C'$, return T;
25 if $InNat1 \cup InNat2 = \emptyset$, return C;

26 R = TestPciSetNat(C', ψ);
27 if R = C', return C';
28 add R to T as a subtree induced by z;
29 return T;

TestPciSetNat is sound because whenever π is invalid, TestPciSetNat returns the domain of a PCI core. By Theorem 3, either π is a PCI core, which is detected in line 25 with domain C returned, or π contains a PCI core, which is detected in lines 21 and 27 with the corresponding domains X and C' returned. TestPciSetNat is complete because whenever π is valid, TestPciSetNat returns a respective NAT. This can be established similarly as Theorem 3 in [Xia17]. We omit detailed analysis on soundness and completeness due to space.

When π is valid, the time complexity of TestPciSetNat is a function of the respective NAT T. Let z be the leaf of T. If every cause in C is a parent of z, only lines 1 to 7 is run, and the complexity is $O(n^2)$. If no cause is a parent of z, lines 1 to 7 are followed by lines 8 to 15. The number of alternative X (line 11) is $2^{n-1} - n - 1$ and evaluation of each X takes $O(n^2/4)$ time. The complexity is $O(n^2\ 2^n)$. This is also the complexity when π is a PCI core.

If π is valid, some causes are the parents of z, the computation time is between $O(n^2)$ and $O(n^2\ 2^n)$. The same holds if π is invalid and contains a PCI core over a proper subset of C. In summary, the time complexity of TestPciSetNat is a function of π and is between $O(n^2)$ and $O(n^2\ 2^n)$. Note that since a NAT model is over a single BN family, n is not unbounded.

7   NAT Extraction from Partial PCI Patterns

The input to TestPciSetNat is a full PCI pattern. The following algorithm allows the input pattern to be full or partial, and valid or invalid. In particular, input of SetNat includes a set C of causes, a PCI pattern π over C, and a set B (possibly empty) of missing PCI bits. Set B is such that if $\pi(x, y)$ is a missing bit, then $(x, y) \in B$.

Set Π collects full PCI patterns that are compatible with π. In line 5, the full PCI pattern ψ is obtained from the partial pattern π by adding the missing bits according to θ. Variable bsc is a PCI bit switching counter. When π is invalid, it controls the number of bits in π that will be switched.

Algorithm 3. *SetNat(C, π, B)*

1 def = set of defined bits in π;
2 Π = ∅;
3 if B = ∅, Π = {π};
4 else for each instantiation θ of missing bits in B,
5   complete π by θ into ψ; Π = Π ∪ {ψ};
6 for each ψ ∈ Π, do
7   R = TestPciSetNat(C, ψ);

```
8    if R is a NAT, return R;
9   bsc = 1;
10 do
11   for each ψ ∈ Π, do
12     for each combination of bsc bits in def, do
13       get τ from ψ by switching these bits;
14       R = TestPciSetNat(C, τ);
15       if R is a NAT, return R;
16   bsc++;
17 end do
```

When π is full, $B = \emptyset$ and line 3 is run. Otherwise, lines 4 and 5 are run. Each full pattern in Π is processed in lines 6 to 8. If π is valid, one ψ will succeed and the respective NAT will be returned.

Otherwise, π is invalid. Lines 9 to 17 switch some bits in π for each ψ. The number of bits to be switched starts from 1 and increases as needed. Hence, a NAT with the minimum number of PCI bits that differ from π (least incompatible) will be returned.

8   Experiment

We evaluated the algorithms by 16 batches of experiments (see Table 1), running in a ThinkPad X230. Each batch extracts NATs from 100 PCI patterns. In batches 1 to 4, each input PCI pattern is derived from a randomly generated NAT with $n = 8, 12, 16, 20$, respectively. Hence, each pattern is full and valid.

In the remaining batches, each input pattern is derived from a random NAT and is modified in addition. In batches 5 to 8, the pattern is modified by randomly selecting a bit and switching its value. Hence, the pattern is full, but may be invalid. In batches 9 to 12, a randomly selected bit is dropped from each pattern. Hence, the pattern is partial but valid. In batches 13 to 16, for each pattern, one bit is dropped and the value of another bit is switched. Hence, the pattern is partial and may be invalid. The experiment setup includes all combinations of fullness and validity of input patterns, and spans a wide range of n values.

Being able to conduct the experiment at $n = 12, 16, 20$ is itself a demonstration of the advantage of the proposed algorithms. The number of NATs for $n = 9$ is 25,637,824 [XZL09]. Generation of the NAT database and search tree take about 40 h [XL14]. The number of NATs for $n = 10$ is 564,275,648. It would take at least 880 h to generate the NAT database and search tree.

After each NAT is extracted, its PCI pattern is compared with the input pattern. For batches 1 to 4 and 9 to 12, each NAT pattern is compatible with the input pattern. For batches 5 to 8 and 13 to 16, the extracted NAT pattern differs from the input by no more than 1 bit. Hence, our algorithms successfully extract NATs in all possible types of scenarios. Due to space, we skip a more elaborative report and analysis of the experimental results.

Table 1. Summary of experimental batches

Index	n	Valid	Full	#Missing switched	bits	Runtime $\hat{\mu}$ ms	$\hat{\sigma}$ ms
1	8	Yes	Yes	0	0	0.16	1.59
2	12	Yes	Yes	0	0	0.82	3.42
3	16	Yes	Yes	0	0	21.83	21.93
4	20	Yes	Yes	0	0	520.59	437.37
5	8	May not	Yes	1	0	0.36	2.27
6	12	May not	Yes	1	0	22.19	30.24
7	16	May not	Yes	1	0	1117.95	1498.54
8	20	May not	Yes	1	0	63439.55	70736.72
9	8	Yes	No	0	1	0.19	1.60
10	12	Yes	No	0	1	1.32	4.27
11	16	Yes	No	0	1	31.60	30.46
12	20	Yes	No	0	1	855.87	821.17
13	8	May not	No	1	1	0.64	3.09
14	12	May not	No	1	1	34.39	47.41
15	16	May not	No	1	1	2187.95	2999.55
16	20	May not	No	1	1	96184.07	122835.54

9   Conclusion

The main contribution of this paper is a collection of algorithms for direct NAT extraction from partial or invalid PCI patterns, founded on formal analysis. They allow NAT structure extraction in all conceivable scenarios, and enable NAT modeling to be applied more effectively in compressing BN CPTs and in learning compact BN CPTs from data. Integrating these algorithms with the existing CPT compression algorithms is an immediate future work.

Our experiments showed that an incorrect PCI bit in the input pattern is much more costly than a missing PCI bit in NAT extraction. Further research will be devoted to improve efficiency of extraction from invalid PCI patterns.

Acknowledgement. Financial support from the NSERC Discovery Grant is acknowledged.

References

[BFGK96] Boutilier, C., Friedman, N., Goldszmidt, M., Koller, D.: Context-specific independence in Bayesian networks. In: Proceeding of 12th Conference on Uncertainty in Artificial Intelligence, pp. 115–123 (1996)

[Dar03] Darwiche, A.: A differential approach to inference in Bayesian networks. J. ACM **50**(3), 280–305 (2003)

[Die93] Diez, F.J.: Parameter adjustment in Bayes networks: the generalized noisy OR-gate. In: Heckerman, D., Mamdani, A. (eds.) Proceeding of 9th Conference on Uncertainty in Artificial Intelligence, pp. 99–105. Morgan Kaufmann (1993)

[Hen89] Henrion, M.: Some practical issues in constructing belief networks. In: Kanal, L.N., Levitt, T.S., Lemmer, J.F. (eds.) Uncertainty in Artificial Intelligence 3, pp. 161–173. Elsevier Science Publishers (1989)

[LG04] Lemmer, J.F., Gossink, D.E.: Recursive noisy OR - a rule for estimating complex probabilistic interactions. IEEE Trans. Syst. Man Cybern. Part B **34**(6), 2252–2261 (2004)

[MD08] Maaskant, P.P., Druzdzel, M.J.: An independence of causal interactions model for opposing influences. In: Jaeger, M., Nielsen, T.D. (eds.) Proceeding 4th European Workshop on Probabilistic Graphical Models, pp. 185–192. Hirtshals, Denmark (2008)

[PD11] Poon, H., Domingos, P.: Sum-product networks: a new deep architecture. In: Proceeding of 12th Conference on Uncertainty in Artificial Intelligence, pp. 2551–2558 (2011)

[Pea88] Pearl, J.: Probabilistic Reasoning in Intelligent Systems: Networks of Plausible Inference. Morgan Kaufmann, San Francisco (1988)

[VT12] Vomlel, J., Tichavský, P.: An approximate tensor-based inference method applied to the game of minesweeper. In: van der Gaag, L.C., Feelders, A.J. (eds.) PGM 2014. LNCS, vol. 8754, pp. 535–550. Springer, Cham (2014). doi:10.1007/978-3-319-11433-0_35

[WvdGR15] Woudenberg, S., van der Gaag, L.C., Rademaker, C.: An intercausal cancellation model for Bayesian-network engineering. Inter. J. Approximate Reasoning **63**, 3247 (2015)

[Xia10] Xiang, Y.: Acquisition and computation issues with NIN-AND tree models. In: Myllymaki, P., Roos, T., Jaakkola, T. (eds.) Proceeding of 5th European Workshop on Probabilistic Graphical Models, Finland, pp. 281–289 (2010)

[Xia12] Xiang, Y.: Non-impeding noisy-AND tree causal models over multi-valued variables. Int. J. Approximate Reasoning **53**(7), 988–1002 (2012). Oct

[Xia17] Xiang, Y.: Extraction of NAT causal structures based on bipartition. In: Proceeding of 30th International Florida Artificial Intelligence Research Society Conference (2017, in press)

[XJ06] Xiang, Y., Jia, N.: Modeling causal reinforcement and undermining with Noisy-AND trees. In: Lamontagne, L., Marchand, M. (eds.) AI 2006. LNCS, vol. 4013, pp. 171–182. Springer, Heidelberg (2006). doi:10.1007/11766247_15

[XJ16a] Xiang, Y., Jiang, Q.: Compression of general Bayesian Net CPTs. In: Khoury, R., Drummond, C. (eds.) AI 2016. LNCS, vol. 9673, pp. 285–297. Springer, Cham (2016). doi:10.1007/978-3-319-34111-8_35

[XJ16b] Xiang, Y., Jin, Y.: Multiplicative factorization of multi-valued NIN-AND tree models. In: Markov, Z., Russell, I. (eds.) Proceeding of 29th International Florida Artificial Intelligence Research Society Conference, pp. 680–685. AAAI Press (2016)

[XL14] Xiang, Y., Liu, Q.: Compression of Bayesian networks with NIN-AND tree modeling. In: van der Gaag, L.C., Feelders, A.J. (eds.) PGM 2014. LNCS, vol. 8754, pp. 551–566. Springer, Cham (2014). doi:10.1007/978-3-319-11433-0_36

[XLZ09] Xiang, Y., Li, Y., Zhu, Z.J.: Towards effective elicitation of NIN-AND tree causal models. In: Godo, L., Pugliese, A. (eds.) SUM 2009. LNCS, vol. 5785, pp. 282–296. Springer, Heidelberg (2009). doi:10.1007/978-3-642-04388-8_22

[XT14] Xiang, Y., Truong, M.: Acquisition of causal models for local distributions in Bayesian networks. IEEE Trans. Cybern. **44**(9), 1591–1604 (2014)

[XZL09] Xiang, Y., Zhu, Z.J., Li, Y.: Enumerating unlabeled and root labeled trees for causal model acquisition. In: Gao, Y., Japkowicz, N. (eds.) AI 2009. LNCS, vol. 5549, pp. 158–170. Springer, Heidelberg (2009). doi:10.1007/978-3-642-01818-3_17

[ZMP15] Zhao, H., Melibari, M., Poupart, P.: On the relationship between sum-product networks and Bayesian networks. In: Proceeding of 32nd International Conference Machine Learning (2015)

The Altruistic Robot: Do What I Want, Not Just What I Say

Richard Billingsley[✉], John Billingsley, Peter Gärdenfors, Pavlos Peppas,
Henri Prade, David Skillicorn, and Mary-Anne Williams

University of Technology Sydney, Ultimo, Australia
`richard.billingsley@uts.edu.au, john.billingsley@usq.edu.au,`
`peter.gardenfors@lucs.lu.se, pavlos.peppas@gmail.com, prade@irit.fr,`
`skill@cs.queensu.ca, mary-anne@TheMagicLab.org`

Abstract. As autonomous robots expand their application beyond research labs and production lines, they must work in more flexible and less well defined environments. To escape the requirement for exhaustive instruction and stipulated preference ordering, a robot's operation must involve choices between alternative actions, guided by goals. We describe a robot that learns these goals from humans by considering the timeliness and context of instructions and rewards as evidence of the contours and gradients of an unknown human utility function. In turn, this underlies a choice-theory based rational preference relationship. We examine how the timing of requests, and contexts in which they arise, can lead to actions that pre-empt requests using methods we term contemporaneous entropy learning and context sensitive learning. We provide experiments on these two methods to demonstrate their usefulness in guiding a robot's actions.

1 Introduction

A robot must have a mechanism for choosing which actions to carry out, and this mechanism must learn from experience to avoid needing every action to be specified from the outset. In a competitive market for robots, we might assume that market forces would favor the most useful robot, where useful is some blend of being safe, economic and helpful. This motivates our developing a robot model, which we term the Altruistic Robot (AR), that learns through optimization how to choose actions that are useful to humans. The concurrent two processes of the AR are: first, to learn what is most useful, and second, to optimize the implementation of activities it determines make it most useful.

This paper largely focuses on the first process, that of learning to be useful, and we examine mathematically some approaches for accomplishing this using the frameworks of information theory, machine learning and utility theory.

The basis of our approach is that humans have context dependent desires [21] that are largely unknown to both humans and robots. If the human desires are rational, they form a latent context-sensitive human utility function which indicates the level of satisfaction of the human in any given state.

© Springer International Publishing AG 2017
S. Moral et al. (Eds.): SUM 2017, LNAI 10564, pp. 149–162, 2017.
DOI: 10.1007/978-3-319-67582-4_11

Requests, rewards, criticisms and other utterances provide hints of the human's desire to change this state within the prevailing contexts, as each desire occurs to the human. If the robot can use these utterances to learn the human utility function, it can then engage in activities that make it most useful. An AR should therefore not simply do what it is told, but should learn from these utterances what states the humans likes in each time and context, and should learn to fulfill the desires before being asked. So the robot must optimize its ability to learning the human satisfaction function from its gradients, which are indicated by the human's utterances.

Our first approach looks at the timing of requests to learn how to be most useful.

Suppose, for example, every afternoon when Harry comes home, he asks a robot to warm the room. After a few occasions, the robot's model of Harry's satisfaction should include having a warm room when he comes home. After developing the habit of warming the room, the lack of ongoing rewards (no more thank yous) should not deter the robot from assuming that a warm room for home-coming is an expected part of the routine.

By examining the interval between requests and prior events, we develop a model we term Contemporaneous Entropy Learning (CEL) that predicts how the human satisfaction function depends on these prior events. We mathematically compare CEL to classical machine learning (CL) and show a 33% reduction in total number of errors needed to learn the events that lead to the desires.

Our second approach examines the context of requests.

Suppose when Harry comes home, he often (but not always) says I am cold and then requests close the window or turn up the heat. Clearly there is some semantic relationship between these utterances relating to the unspoken desire to be warmer. We examine how using distributed representations [12], like those used in natural language processing, the robot can learn representations for utterances that capture their semantic content. In this way the robot can best predict Harry's latent desire of being warmer, and take the appropriate actions without further requests.

The main contribution of this paper is the AR model for robot behavior, which provides a framework that allows the robot to learn to predict latent human desires, and act accordingly. As the robot creates a better model of each human, it becomes a more social robot in that it aims to learn from social cues the activities it should perform. This framework includes the timing component CEL and the context component we term semantic context learning (SCL).

In this way we provide a pathway for developing a social robot that learns to optimize its usefulness to its master.

2   Altruistic Robot

The activities of a robot are made up of the accomplishments it completes, which are in turn each made up of atomic actions. To provide for flexibility, the desired accomplishments may not be stipulated exactly from the outset, but a choice may be present in the selection of the next one. This choice must be made on the basis of the robot's internal and external state and memory, using a calculation of preference between the available accomplishments that can be made.

We define the Altruistic Robot (AR) as a robot that is learning to be most useful. Its accomplishment selection maximizes an internal utility function that is continually learning to predict the utility functions it perceives humans to possess.

2.1   Background

If a robot acts rationally under Choice Theory [10], there are a set of utility values [23] that can be assigned to each available choice. The robot can then act in a way to maximize its utility by assigning higher probabilities or certainties to performing the actions with higher utilities.

Learning to order choices using preference information is NP-complete [20]. However, there are straightforward methods for learning to compute utilities from pairwise preferences using linear and logistic regression [9].

Neural networks have also been used to fine tune top-k ranking of utilities [3]. Neural networks (NN) can be trained with back-propagation and provide a flexible approach to learning complex functions and have been shown to be able to approximate any function with reasonable accuracy [8].

The main focus of this paper is on providing the robot motivation for completing accomplishments based on evidence of human satisfaction. We use the words accomplishment and action interchangeably.

Attention module. In the ASMO Attention architecture [19], each of n potential activities are assigned a real valued variable v_i representing the utility of performing the ith activity. The robot's processes use these variables to vote for the best set of compatible actions to be performed at the next time interval. Consequently, activities not performed in a while can be boosted [14], while activities with urgent need can be prioritized. This allows greater prioritization and attention to be given to activities that are seen as most beneficial or urgent in any given circumstance.

2.2   Introducing the Altruistic Robot

When social robots are better able to understand humans, they can tailor their actions to be more compatible with their master's wishes. The Altruistic Robot (AR) follows the two concurrent processes of learning the human's desires from utterances and learning how best to satisfy the demands. We will later contrast this with simply following instructions.

Utility theory provides a framework for representing human preferences through scalar quantities. The AR assumes that each human possesses their own utility function $v(x, c, t)$ representing their satisfaction in environment x with state c at time t. While the environment x is known by the robot, the state vector c is latent and contains important but not directly measurable information of desires of the human. Consequently, the robot can only predict and not measure values for c when calculating $v(x, c, t)$.

This function v is independent of the actions a of the robot, but the robot's actions affect future values of x and c. Even though the robot (and possibly the human) may not be aware of the exact function v, the human can instruct the robot by requesting desired actions r that if performed will lead to future values of x and c in which v is increased. From these instructions, the robot must continually learn the shape of $v(x, c, t)$ and which actions will maximize it and when.

The difference between the AR approach and simply following instructions can be seen when the robot does something in addition to the instruction, like also fetching warm clothing when asked to turn up the temperature, or making coffee before being asked to do so. In such cases, the robot has predicted the humans desires for warmth and drink and works to satisfy these desires.

As the robot learns $v(x, t)$ it must also learn how x is affected by its own actions $x' = w(a, x, t)$. For example, if the robot knows the human wants a tidy house, it must also realize the need to wait until it returns home before it can accomplish this. In this way it balances its ability to accomplish a with the resulting increase in v. This may even lead the robot towards preparation actions that improve w, like recharging its batteries, that have no direct bearing on v. Likewise, steep priors to the utility function could be established representing moral values to keep the robot from accidentally learning unacceptable actions.

In the following sections we outline two techniques to begin implementing the AR model, by considering the timing and environment of requests.

3   Learning from Timing

Under the AR model for a robot, all rewards comes through satisfying human desires. The ability to learn how to anticipate such desires allows the AR to be helpful without continual direct instruction.

The usefulness of each activity can vary with time and situation. If you have just made coffee, being given some sugar might be useful, while if you have just finished drinking coffee, having your cup cleaned would be nice. The AR must learn the changing circumstances that make each activity best suited to each time.

When Harry is given coffee and immediately says 'bring me sugar', this is different from Harry getting coffee, waiting an hour or so and then making the request. In the former case it is far more likely that being given coffee was a trigger for Harry's desire for sugar. To accommodate this, we develop a model using exponentially decaying features that learns preferentially from recent events.

3.1   Timing Background

Novianto et al. (2014) notes that it is helpful for a robot to learn to predict stimuli. For example, person h entering a room may be considered as stimulus

X, and if after arriving, they immediately ask for an action r to take place, this may be called stimulus Y. Predicting Y from stimulus X is beneficial because it enables the robot to begin action r in advance of it actually being requested.

Classical learning. Under the classical learning (CL) model [22], weights V are used for predicting unconditioned stimulus Y from the conditioned stimulus X, where each stimulus has the values 1 when present and 0 when not. V is updated by the difference equations:

$$\triangle V_t = \beta R \times \alpha \bar{X} \text{ where } R = \dot{Y}_t = Y_t - Y_{t-1} \tag{1}$$

which updates V proportional to R while making R like a derivative of Y, and:

$$\bar{X}_{t+1} = \delta \bar{X}_t + (1 - \delta)X_t \tag{2}$$

which makes $\bar{X}$ a decaying average of X.

Maximum Entropy Learning. In other machine learning tasks [16] a maximum entropy classifier (MEC) is used for learning parameters to make predictions. When training such classifiers to predict categorical information p with empirical distribution y, the Kullback-Leibler divergence $E = \sum y(x) \log(y(x)/p(x))$ [11] is minimized. For binary features x, the classifier makes predictions $p = \sigma(Wx + b)$ where W is the weight and b the bias and $\sigma(x) = (1 + \exp(-x))^{-1}$.

The gradient between the predictions and actual data y with respect to the weights is then $\frac{dE}{dW} = (y - p) \times x$ where $\times$ is the outer product operator. The expectation maximization algorithm [4] allows stochastic gradient descent [2] with the weight parameters being updated by:

$$\triangle W_t = \beta(y - p) \times x \tag{3}$$

An intuitive view of the effect of the $-p$ term on the gradient $\triangle W_t$ is that it prevents W from growing without bound whenever y and x remained largely unchanged. Initially, when $w \approx 0$ the prediction $p \approx 0$ also, but as the prediction p gets larger, the update $\triangle W_t$ gets smaller as p approaches y

Noting the similarity of Eq. 3 to that of classical learning Eq. 1, we explore the way that the benefits of classical learning can be brought to a maximum entropy framework.

3.2   Introducing Contemporaneous Entropy Learning

Harry's request is evidence of his desire. The timing of Harry's request provide evidence that his desire was triggered by a recent event, since otherwise Harry would have made the request sooner. Here, we examine the relationship between a constant probability of delaying a request and using exponential decaying features.

Suppose the binary valued variable s represents whether a specific event x which occurred t minutes ago was the stimulus for Harry's desire c, which later led him to make request r for an action. With probability $(1 - k)$, Harry was preoccupied during each of the t intervening minutes between desire and request, delaying him from requesting r sooner. So if event x was the trigger, then probability that Harry will first request r exactly t minutes after the event x is: $P(r|s,t) = (1 - k)^{t-1}kP(r|s)$

Likewise, regardless of event x, there is a background probability h that Harry will request r anyway. The probability that Harry will first request r after exactly t minutes in any case is given by:

$$P(r|t) = (1 - h)^{t-1}hP(r)$$

Using Bayes theorem, the robot can predict $P(s)$ from r and t to see how likely it was that x was the stimulus for r:

$$P(s|r,t) = P(r,t|s)\frac{P(s)}{P(r,t)} = k(1 - k)^{t-1}P(r|s)\frac{P(s)}{h(1 - h)^{t-1}P(r)}$$

$$\approx (k - h)e^{-(k-h)t}P(r|s)\frac{P(s)}{P(r)} \tag{4}$$

The conditional probability of a request at time t given there have been none earlier is given by $kP(r|s)$. So if a classifier learns to predict $P(r|s)$, we can use Eq. 4 to predict $P(s|r,t)$.

Gradient weighting. We notice that an error ε in predicting $P(r|s)$ will result in a proportional error $(k - h)e^{-(k-h)t}\varepsilon$ when predicting $P(s|r,t)$. This suggests we should scale the training examples according to their impact in predicting $P(s|r,t)$.

Consequently, if using SGD to learn the classifier's weights, to minimize the loss when predicting s we should multiply each gradient by $f_T(t)$ defined by:

$$f_T(t) = \begin{cases} t < T: & 0 \\ t \geq T: & (k - h)e^{-(k-h)(t-T)} \end{cases} \tag{5}$$

where T is the time the event occurred. For neatness, adjust k to remove the h term.

Multiple events. We now consider learning from repeated experiences. Suppose Harry's desire is a binary function $c(t) \in \{0, 1\}$.

We construct a single time-decaying feature function $f(t)$ of the above form by summing together the individual time-decaying functions $f_T(t)$ from Eq. 5 for each transition of $c(t)$. So:

$$f(t) = \sum_i f_{T_i}(t) \text{ where } c(T_i^-) = 0 \wedge c(T_i^+) = 1.$$

Appendix A shows: $f(t) = \tilde{c}(t) = c(t) - \bar{c}(t)$ where $\bar{c}(t) = \int_0^t c(s)ke^{-k(t-s)}ds$.
A plot for $\bar{c}(t)$ and $\tilde{c}(t)$ is shown in Fig. 1 for a square-wave desire $c(t)$.

Fig. 1. Left: A square input pulse x, Middle: an exponential moving average $\bar{x}$, Right: a contemporaneous feature $\widetilde{x} = x - \bar{x}$ showing recent changes in x

Multiple features. We can extend c to an environment with multiple changing features, each of which could be the cause of the request. If each feature is represented by the elements of a vector $\mathbf{x}$, we can build a time-decaying feature function vector

$$\widetilde{\mathbf{x}}(t) = \mathbf{x}(t) - \bar{\mathbf{x}}(t) \text{ where } \bar{\mathbf{x}} = \int_0^t \mathbf{x}(s)ke^{-k(t-s)}ds$$

When using a maximum entropy classifier with weights $\mathbf{W}$ to predict the desire c, we use the original features $\mathbf{x}$ as follows: $P(c|\mathbf{x}) = \sigma(\mathbf{W}_i(\mathbf{x}))$.

Update weighting. Since c is latent and only indicated by the requests, the weights $\mathbf{W}$ of the classifier should be updated using the decaying features through the equation

$$\triangle\mathbf{W} = \lambda(r - p) \times \widetilde{\mathbf{x}} \tag{6}$$

where r is the empirical requests, and p are the predicted requests. This scales the gradients according to their timeliness of the features.

Once fully trained, the elements of $\sigma(\mathbf{W}_i)$ will represent the probability $\mathbf{x}_i$ will cause a later request r, so $P(s_i|r) = \sigma(\mathbf{W}_i)$. We refer to this logistic model using time-decaying weight updates as a Contemporaneous Entropy Learning (CEL)

3.3   Comparing CL with CEL

We now compare the CEL method with classical learning to show that they initially perform identical learning.

If we examine a continuous version of the classical learning Eq. 2, we have: $\frac{d}{dt}\bar{x}(t) = k(x(t) - \bar{x}(t))$ where $k = 1 - \delta$. and $x(t)$ is the unconditioned stimulus at time t. Appendix A shows: $\bar{x}(t) = r\int_0^t e^{-ks}x(t-s)ds$

The continuous version of the other classical learning Eq. 1 gives: $\frac{dV(t)}{dt} = \beta\frac{dy(t)}{dt} \times \alpha\bar{x}(t)$ where V is the parameter being learned and $y(t)$ is the conditioned stimulus at time t.

Integrating both sides allows us to calculate V at time t following training with inputs of $x(t)$ and $y(t)$:

$$V(t) = \int_0^t \beta\frac{dy(s)}{dt} \times \alpha\bar{x}(s)ds \tag{7}$$

Rearranging Eq. 7 by integrating by parts, where $y(0) = 0$, we get:

$$V = [\beta y \times \alpha \bar{x}(s)]_0^t - \int_0^t \beta y \times \alpha \frac{d\bar{x}}{dt} ds \tag{8}$$

The two components of the reformulated classically learned parameters can be examined separately by setting: $V = U - W$ where U is the first term of Eq. 8 and W the second.

If $y(0) = 0$, then U can be computed from the instantaneous values of $y(t)$ and $\bar{x}(t)$ using: $U = \beta y(t) \times \alpha \bar{x}(t)$

Since U only needs the current values of y and $\bar{x}$, and is independent of their histories, it has no parameters to learn. Meanwhile W can be rearranged as: $W = \int_0^t \beta y(s) \times \alpha k \left(x(s) - \bar{x}(s)\right) ds$

The change in W at time t is given by $\frac{dW}{dt} = \beta y(s) \times \alpha k \left(x(s) - \bar{x}(s)\right)$

So the update used to learn the parameter W with classical learning is given at time t by:

$$\triangle W = \beta y \times \alpha k \widetilde{x} \text{ where } \widetilde{x} = x(t) - \bar{x}(t) \tag{9}$$

If we compare the reformulated CL update Eq. 9 with the CEL update Eq. 6 we see that with the right learning rates $\lambda = \alpha \beta k$, the two equations would be the same by setting $y = (r - p)$.

This means the CEL approach of using a MEC with updates augmented by $\widetilde{x} = x - \bar{x}$ initially performs classical learning when $p \approx 0$ to predict the most useful actions.

The assignment $y = (r - p)$ occurs naturally because as the robot pre-empts the request with probability p, the human's opportunity to make the request declines to $r - p$. This term allows the robot retains habits without ongoing rewards.

3.4   Combination Learning

Because CEL uses a neural network approach, the model can be extended to deep learning [1], auto encoders [17] and other unsupervised machine learning methods such as restricted Boltzmann machines [7]. Such deeper models will allow the robot to predict requests where a combination of triggers is required to create the desire. For example, Harry may only ask for sugar with his coffee when he has not been given sweetener or a cookie. These combinatoric time dependent features will automatically be learned through the CEL learning method.

3.5   CEL Experiments and Results

We compared CEL with CL directly, and compared CL with MEC using an environmental variable detection task.

Following [22] we used rectangular pulses for X the conditioned stimulus and Y the unconditioned stimulus.

In our experiments, the pulses for X and Y repeated 5 times at 100 unit intervals, were 20 units wide. Their onsets were delayed from one another by τ units which was initially set to 30 units.

Figure 2 shows on the left traces of X and Y, together with the learned value of V using CL. On the right it shows traces of X and Y together with $\widetilde{X}$ and V using CEL.

As can be seen the final values of V are almost identical, but the path is discontinuous in the case of CL and continuous for CEL. The advantage of CEL's monotonic V in this example means that CEL is an any-time function in that its current value remains close to its most optimal value so far.

The continuous nature of CEL is because the calculation involves a smooth integral of the product of continuous functions. The discontinuous nature of CL is because its integral includes the product of $\dot{Y}$ which approximates the unbounded Dirac delta function when Y makes step changes. This discontinuity can also be seen because CL can be calculated from $V = W - U$ where U is the continuous CEL function, and $W = \beta Y(t) \times \alpha(i)\bar{X}(t)$ which is discontinuous whenever Y steps.

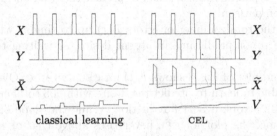

Fig. 2. On the left, from top to bottom, traces of $X, Y, \bar{X}$ and V using classical learning. Note V steps up and down. On the right from top to bottom, traces of $X, Y, \widetilde{X}$ and parameters V using CEL. Note with CEL, V increases monotonically making the robots learning continuous at all times.

Figure 3 shows on the left and middle the final values for V for different values of τ (the delay between X and Y) varying from 5 to 95 for CEL and CL.

As can be seen the outputs are almost identical with only minor variation due to numerical approximations of computing $\dot{Y}$ from the difference of two values of Y. This further shows the relationship between CL and CEL.

Environment variable identification task. We compared CEL with MEC to identify how CEL can accelerate learning above regular logistic regression.

We generated a dataset consisting of 1000 binary Markov random variables $x_i(t)$ to represent the environment. At each time-step, each variable with value 0 had an independent probability of 0.1 to switch to a 1, and would otherwise remain unchanged. Likewise, each variable with value 1 had an independent probability 0.2 to switch to a 0, and would otherwise remain unchanged.

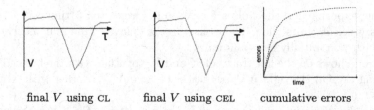

final V using CL   final V using CEL   cumulative errors

Fig. 3. Left and middle, final values of V (y-axis) for different values of τ (x-axis) ranging from 0 to 100, using CL and CEL. Right, Total errors during learning for environment variable identification task for MEC (dotted) and CEL (solid). The MEC made 1342 total errors, while the lower CEL trace indicates 908 total errors. Initially both MEC and CEL yielded the same error rate but CEL learned to stop making errors sooner

The desire c was assigned to one variable as $c = x_0 - z$ where the value z indicated whether the robot had already satisfied the desire. Whenever a desire was present, unsatisfied and unrequested, a request was made with probability 0.1 in each time-step. The robot learned to predict the desire using both an MEC and a CEL for comparison.

We counted the total number of errors $E(t)$ up to time t, where each error was either a the robot predicting an absent desire, or waiting too long that a request was made.

The right hand plot of Fig. 3 shows $E(t)$ for CEL and MEC. Over 30,000 time-steps, CEL was able to learn to predict the desire with 908 total errors compared to MEC with 1342 errors, a 33.4% reduction. While the slope of the two plots start out the same, the plot for CEL never rises above that of MEC and levels out sooner. This shows that CEL was faster at learning to identify the triggers for the desire with fewer errors.

4 Learning from Context

Just as the timing of requests conveys meaning about desires, so too does the context.

To illustrate this, we give an example: At the train station, it is likely that you will do travel related actions, perhaps wait for a train, or leave to go somewhere else. The semantics of the train station make this likely because this it is where people board and depart from trains. Because the train station semantics are shared by most people, most robots accompanying most people would likely experience the similar waiting or leaving related desires.

Likewise, asking for sugar with coffee may indicate a preference for sweetness in other hot drinks. As examples may be sparse, the AR must learn to generalize and identify similarities in situations where possible to help predict the likely desires that each situation entails. The generalization reduces the dimensionality of the feature space formed the complexity in the identification task.

4.1   Context Background

Case based decision making [6] suggests that the best action in any situation can be learned from other similar situations [5]. If an action had a positive outcome in a similar situation before, then there is a good chance that action will do well again. This approach suggests preferences can be identified by first finding the closest prior experience. This can be compared with the back-propagation [18] approach where the neural-network's output for any given input vector will be guided by training examples with similar features. With this in mind we examine ways to measure the similarity of situations.

Word vectors. We compare the process a robot can learn from the patterns of events, situations, requests and rewards with a task in natural language processing.

Semantic vectors [12] can be learned for nouns, verbs, adjectives and such from their co-occurrence distribution within small text windows using unsupervised learning. The representations encapsulate semantics that allows algebra to be performed on the vectors, so Queen = King - Man + Woman.

Each dictionary word is assigned two initial small random vectors $\mathbf{v}$ and $\mathbf{u}$. These are trained with stochastic gradient descent so that the target word v can be predicted from its context with $P(\mathbf{v}|\mathbf{u}) \propto \prod_{u \in \text{window} \backslash v} \exp(\sum_i \mathbf{v}_i \mathbf{u}_i)$.

Using the gradient of $P(v|u)$ between the empirical and predicted windows, u and v are trained to learn the representations. Typically $\mathbf{u}$ takes on a hierarchical representation and is discarded, while semantic similarities are only drawn from $\mathbf{v}$. We propose to build formulaic sentences that represent the situations that each request occurs within. Then using distributed learning we can impute into each of the resulting words a semantic representation that learns the underlying meaning to the words within each request.

4.2   Introducing Semantic Context Learning

The purpose of Semantic Context Learning (SCL) is to develop representations that contain the semantics of situations of places, people, events and requests that enables desires to be predicted from similar situations.

We can build sentences like 'Sam caught train at station' or 'Sam was waiting at station' and use the distributed learning process to learn 'station' often leads to 'waiting' and 'caught train' actions. If Sam repeatedly asks to buy newspapers while waiting, the robot can learn a semantic representations of 'Sam wants newspaper at station' that may generalize this to include 'Sam wants book at airport' due to the words semantic similarities.

While the word2vec [12] vectors $\mathbf{v}$ belong to one space, the vectors $\mathbf{u}$ belong to their dual space and allow the conditional probability $p(\mathbf{v}|\mathbf{u})$ to be computed directly. To identify activities within locations (rather than similar locations to nouns), requires identifying verbs with high affinity to the location nouns. If we simply searched for verbs closest to the location, we would identify locations and actions sharing similar contexts. However, in this case we wanted actions directly belonging to the locations's context, so we compare the original space vectors **u** with dual space vectors **v**.

In the above example, direct labels were provided for Sam, Bob, at station etc. The robots can also learn vectors directly from their environment. Camera image pixel intensities can be passed through an auto encoder, or image recognition software to create compressed features which would then be used when making prediction for the next action.

This would allow the robot to expect different desires and act differently indoors from outdoors, at day or night, in crowds or all alone. If the robot was always asked to keep quiet indoors when surrounded by people, but be louder when outdoors, alone with Bob, the robot can learn to predict latent desires given the environment $P(s|v)$ just as with timing.

4.3   Context Experiments and Results

As a labeled dataset of robot experiences was unavailable, we employed a tagged corpus of predicates [13] consisting of things, events, actions, spacial, goals, functions and generic types.

We removed the triple structure and pre-processed the corpus by removing its source labels, e.g. [OMCS]:, internal quote marks, brackets and numeric identifiers, e.g. #2, while preserving the part-of-speech tags. This made sentences like train.n pass.v car.n.

After generating vectors using word2vec on this corpus, we identified nouns with the highest cosine similarity thereby identifying similarities with places, as shown in Table 1

Table 1. Left - Most similar places by cosine similarity, Right - Similar activities to places by cosine similarity between space and dual space

bus	train	boat	shop	bus	train	boat	shop
school bus	locomotive	sailboat	store	exit	park	dock	vend
bus driver	passenger car	sailboats	grocery	park	drive	board	sell
bus station	track	sailer	grocery store	board	transport	sail	shop
bus stop	engine	skiff	walmart	buckle	ride	motor	stock

To obtain both **u** and **v** we adjusted the word2vec to output both tables generated using 10 draws of negative sampling, training on 10 passes through the corpus. Table 1 - Right shows the verbs with dual-space vectors **u** that are most similar to nouns with vectors **v**.

Although generated from predicates not robotic experiences, this helps demonstrate how likely activities can be predicted from locations by learning through the SCL approach.

5   Conclusion

We have introduced the concept of an Altruistic Robot (AR) that learns to optimize how to be useful. This is achieved by the ability of the robot to learn how to represent an external utility function of its human master that is unknown, but whose gradient and contours are hinted through the timing and context of requests and rewards. The robot then schedules actions to implement requests to the greatest satisfaction of its master.

We have developed the CEL model that uses timing to help learn to model a human utility function from requests, and accelerates learning 33% over maximum entropy classifiers on this type of data. We have compared this model with classical learning (CL), and shown that initially CEL performs a CL in a continuously improving fashion, and demonstrated how CL can be extended into a maximum entropy framework providing a bridge to modern deep learning techniques [1].

We have also presented the SCL model that learns a semantic representation of locations, things and actions, and performed qualitative experiments to predict actions in locations from a predicate corpus.

A   Appendix

$f(t) = \dot{c}(t) \star ke^{-ks}$ where $\star$ denotes convolution. Taking Laplace functions: $\mathcal{L}f = (s\mathcal{L}c)\left(\frac{1}{s+k}\right) \Rightarrow \mathcal{L}f = \left(1 - \frac{k}{s+k}\right)\mathcal{L}x \Rightarrow f(t) = c(t) - k\int_0^t c(t)e^{-k(t-s)}ds$. For $\frac{d\bar{x}}{dt} = k(x - \bar{x})$ we see: $s\mathcal{L}\bar{x} = k(\mathcal{L}x - \mathcal{L}\bar{x}) \Rightarrow \mathcal{L}\bar{x} = \mathcal{L}x\left(\frac{k}{s+k}\right) \Rightarrow \bar{x} = x \star re^{-kt} = r\int_0^t e^{-ks}x(t-s)ds$

References

1. Bengio, Y.: Deep learning of representations for unsupervised and transfer learning. In: Workshop on Unsupervised and Transfer Learning, ICML (2011)
2. Bottou, L.: Large-scale machine learning with stochastic gradient descent. Compstat **2010**, 177–186 (2010)
3. Cao, Z., Qin, T., Liu, T. Y., Tsai, M.F., Li, H.: Learning to rank: from pairwise approach to listwise approach. In Proceedings of the 24th International Conference on Machine Learning, pp. 129–136. ACM (2007)
4. Dempster, A.P., Laird, N.M., Rubin, D.B.: Maximum likelihood from incomplete data via the EM algorithm. J. Roy. Stat. Soc.: Ser. B (Methodol.) 1–38 (1977)
5. Dubois, D., Godo, L., Prade, H., Zapico, A.: On the possibilistic decision model: from decision under uncertainty to case-based decision. Int. J. Uncertainty Fuzziness Knowl. Based Syst. **7**(06), 631–670 (1999)
6. Gilboa, I., Schmeidler, D.: Case-based decision theory. Q. J. Econ. **110**(3), 605–639 (1995)
7. Hinton, G.E., Osindero, S., Teh, Y.W.: A fast learning algorithm for deep belief nets. Neural Comput. **18**(7), 1527–1554 (2006)

8. Hornik, K., Stinchcombe, M., White, H.: Multilayer feedforward networks are uni-versal approximators. Neural Networks **2**(5), 359–366 (1989)
9. Hüllermeier, E., Fürnkranz, J., Cheng, W., Brinker, K.: Label ranking by learning pairwise preferences. Artif. Intell. **172**(16), 1897–1916 (2008)
10. Kreps, D.: Boulder Notes on the Theory of Choice. Westview press, Boulder (1988)
11. Kullback, S., Leibler, R.A.: On information and sufficiency. Ann. Math. Stat. **22**(1), 79–86 (1951)
12. Le, Q.V., Mikolov, T.: Distributed representations of sentences and documents. arXiv preprint arXiv:1405.4053 (2014)
13. Liu, H., Singh, P.: OMCSNet: a commonsense inference toolkit (2003). http://web.media.mit.edu/~hugo/publications
14. Novianto, R., Johnston, B., Williams, M.-A.: Habituation and sensitisation learn-ing in ASMO cognitive architecture. In: Herrmann, G., Pearson, M.J., Lenz, A., Bremner, P., Spiers, A., Leonards, U. (eds.) ICSR 2013. LNCS, vol. 8239, pp. 249–259. Springer, Cham (2013). doi:10.1007/978-3-319-02675-6_25
15. Novianto, R., Williams, M.-A., Gärdenfors, P., Wightwick, G.: Classical con-ditioning in social robots. In: Beetz, M., Johnston, B., Williams, M.-A. (eds.) ICSR 2014. LNCS, vol. 8755, pp. 279–289. Springer, Cham (2014). doi:10.1007/978-3-319-11973-1_29
16. Ratnaparkhi, A.: Learning to parse natural language with maximum entropy mod-els. Mach. Learn. **34**(1–3), 151–175 (1999)
17. Rifai, S., Vincent, P., Muller, X., Glorot, X., Bengio, Y.: Contracting auto-encoders: explicit invariance during feature extraction. In: Proceedings of the Twenty-eight International Conference on Machine Learning (ICML 2011) (2011)
18. Rumelhart, D.E., Hinton, G.E., Williams, R.J.: Learning representations by back-propagating errors. Nature **323**(6088), 533–536 (1986)
19. Samsonovich, A.V., et al.: Attention in the ASMO cognitive architecture. In: Bio-logically Inspired Cognitive Architectures 2010: Proceedings of the First Annual Meeting of the BICA Society, vol. 221, p. 98. IOS Press (2010)
20. Cohen, W.W., Schapire, R.E., Singer, Y.: Learning to order things. Adv. Neural Inf. Process. Syst. **10**, 451 (1998)
21. Scitovsky, T.: Human desire and economic satisfaction: essays on the frontiers of economics. Wheatsheaf Books (1986)
22. Sutton, R.S., Barto, A.G.: Time-derivative models of pavlovian reinforcement (1990)
23. von Neumann, J., Morgenstern, O.: Theory of Games and Economic Behavior, vol. 60. Princeton University Press Princeton, Princeton (1944)

Expressivity of Possibilistic Preference Networks with Constraints

Nahla Ben Amor[1], Didier Dubois[2], Héla Gouider[1(✉)], and Henri Prade[2]

[1] LARODEC Laboratory, Université de Tunis, 41 rue de la Liberté,
2000 Le Bardo, Tunisia
nahla.benamor@gmx.fr, gouider.hela@gmail.com
[2] IRIT – CNRS, 118, route de Narbonne, 31062 Toulouse Cedex 09, France
{dubois,prade}@irit.fr

Abstract. Among several graphical models for preferences, CP-nets are often used for learning and representation purposes. They rely on a simple preference independence property known as the *ceteris paribus independence*. Our paper uses a recent symbolic graphical model, based on possibilistic networks, that induces a preference ordering on configurations consistent with the ordering induced by CP-nets. Ceteris paribus preferences in the latter can be retrieved by adding suitable constraints between products of symbolic weights. This connection between possibilistic networks and CP-nets allows for an extension of the expressive power of the latter while maintaining its qualitative nature. Elicitation complexity is thus kept stable, while the complexity of dominance and optimization queries is cut down.

1 Introduction

Various graphical models have been proposed in the literature in order to represent preferences in an intuitive manner. A survey of such approaches is in [3]. We may roughly distinguish between (i) quantitative models such as GAI networks [15] that use numerical utility functions (ii) qualitative models where preferences are contextually expressed by local comparisons between attribute values. The latter request less assessment effort from the user.

Among qualitative models, CP-nets [7] are the most popular. They provide a well-developed compact representation setting for preference modeling. The CP-net representation consists in a directed graph expressing conditional preference statements, interpreted under the *ceteris paribus* assumption. As an effect of the systematic application of this assumption, it has been observed that priority in the network is given to parent decision variables over children ones, a feature not deliberately required.

The more recently introduced π-pref nets [2,4] may be also classified as qualitative models. Indeed, similarly to CP-nets, this model represents local preferences in terms of conditional comparisons between variable assignments. π-pref nets are inspired by product-based numerical possibilistic networks [14] but they

© Springer International Publishing AG 2017
S. Moral et al. (Eds.): SUM 2017, LNAI 10564, pp. 163–177, 2017.
DOI: 10.1007/978-3-319-67582-4_12

use symbolic (non-instantiated) possibility weights to model conditional preference tables. Additional information about the relative strength of preferences can be taken into account by adding constraints between these weights.

The paper proves that a π-pref net is able to capture ceteris paribus preferences between solutions induced by a CP-net, if suitable constraints between products of symbolic weights are added. These constraints explicitly express the higher importance of parent decision variables over their children nodes in the π-pref net. In [4], it was proved that π-pref nets orderings exactly correspond to a Pareto ordering over vectors expressing levels of satisfaction for each variable. We show that this ordering of configurations is refined by the ordering obtained by comparing sets of satisfied preference tables. These results show that the setting of π-pref nets is closely related to CP-nets since *ceteris paribus* constraints can be expressed by specific inequality constraints between products of symbolic weights.

The paper is organized as follows. Section 2 provides a brief background on CP-nets, while Sect. 3 introduces π-pref nets based on possibilistic networks with symbolic weights. Section 4 investigates conditions that enable preferences expressed by π-pref nets to get closer to CP-nets orderings. Section 5 presents related work, especially CP-theories [16], and the conclusion briefly compares the formalisms in terms of expressive power and query complexity.

2 CP-nets

Let $\mathcal{V} = \{A_1, \ldots, A_n\}$ be a set of Boolean variables, each taking values denoted, e.g., by a_i or $\neg a_i$. Each variable A_i has a value domain D_{A_i}. Ω denotes the universe of discourse, which is the Cartesian product of all variable domains in $\mathcal{V}$. Each element ω_i of Ω is called a *configuration*.

The user is assumed to express preferences under the form of comparisons between values of each variable, conditioned on some other instantiated variables. CP-nets deal with strict *preference statements*. Unconditional statements are of the form: "I prefer a^+ to a^-", where $a^+, a^- \in \{a, \neg a\}$ and $a^- = \neg a^+$, and we denote them by $a^+ \succ a^-$. When $A = a^+$, we say that the *quality* of the choice for A is good, and is bad otherwise. If the preference on A depends on other variables $\mathcal{P}(A)$ called the *parents* of A, and $p(A)$ is an instantiation of $\mathcal{P}(A)$, conditional preference statements are of the form "in the context $p(A)$, I prefer a^+ to a^-", denoted by $p(A) : a^+ \succ a^-$. To each variable we associate a table representing the local preferences on its domain values in each parent context (the value of a^+, respectively a^-, depends on the parents context).

Example 1. *Consider a preference specification about a holiday house in terms of 4 decision variables $\mathcal{V} = \{T, S, P, C\}$ standing for type, size, place and car park respectively, with values $T \in \{flat\ (t_1),\ house\ (t_2)\}$, $S \in \{big\ (s_1),\ small\ (s_2)\}$, $P \in \{downtown\ (p_1),\ outskirt\ (p_2)\}$ and $C \in \{car\ (c_1),\ nocar\ (c_2)\}$. Preference on T is unconditional, while all the other preferences are conditional as follows: $t_1 \succ t_2,\ t_1 : p_1 \succ p_2,\ t_2 : p_2 \succ p_1,\ p_1 : c_1 \succ c_2,\ p_2 : c_2 \succ c_1,\ t_1 : s_2 \succ s_1,\ t_2 : s_1 \succ s_2.$*

Definition 1. *A (conditional) preference network is a directed acyclic graph with nodes $A_i, A_j \in \mathcal{V}$, s.t. each arc from A_j to A_i expresses that the preference about A_i depends on A_j. Each node A_i is associated with a preference table CPT_i that associates strict preference statements $p(A_i) : a_i^+ \succ a_i^-$ between the two values of A_i conditional to each possible instantiation $p(A_i)$ of the parents $\mathcal{P}(A_i)$ of A_i.*

The preference statements of Example 1 correspond to the CP-net of Fig. 1.

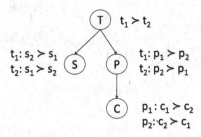

Fig. 1. Preference network for Example 1

Preference networks can be viewed as a qualitative counterpart of Bayesian nets. CP-nets [7,8] are preference networks relying on the *ceteris paribus* preferential independence assumption. Namely, a CP-net induces a partial order $\succ_{CP}$ between configurations, based on this preferential independence assumption: a value is preferred to another in a given context, everything else being equal. Given $\mathcal{U} \subseteq \mathcal{V}$ and $\omega \in \Omega$, $\omega_{\mathcal{U}}$ denotes the restriction of ω to variables in $\mathcal{U}$.

Definition 2 (*Ceteris Paribus*). *Each strict preference statement $p(A_i) : a_i^+ \succ a_i^-$, is translated into $\omega \succ_{CP} \omega'$, whenever $\omega_{\{A_i\}} = a_i^+, \omega'_{\{A_i\}} = a_i^-$, and $\omega_{\mathcal{V} \setminus \{A_i\}} = \omega'_{\mathcal{V} \setminus \{A_i\}}$, and $\omega_{\mathcal{P}(A_i)} = \omega'_{\mathcal{P}(A_i)} = p(A_i)$.*

Due to the ceteris paribus assumption, configurations compared in the preference statements differ by a single flip, and switching A_i from a_i^+ to a_i^- is called a *worsening flip*. We get a directed acyclic graph of configurations (the *configuration graph*) with a unique top corresponding to the best configuration $(A_i = a_i^+, \forall i)$ and a unique bottom corresponding to the worst one $(A_i = a_i^-, \forall i)$. The worsening flip graph for Example 1 is represented in Fig. 2.

The configuration graphs induced by CP-nets are partial in general, and many configurations remain incomparable, for instance $t_1 p_1 c_1 s_1$ and $t_1 p_1 c_2 s_2$ are not comparable in the worsening graph of Fig. 2. Moreover, in the CP-nets semantics, parent preferences look more important than children ones, for example, the preferences of the node P are more important than C and the preferences of the root T are more important than all the other nodes.

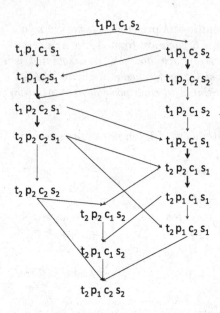

Fig. 2. CP-net preferences for Example 1 up to transitive closure (5 bold arrows represent ceteris paribus preference relations that are not recovered by π-pref net, 8 one-flip comparisons over 32 can be recovered by transitivity, e.g. from $t_1p_1c_1s_2$ to $t_2p_1c_1s_2$).

3 π-Pref nets

Possibility theory [11] can be used for representing preferences. It relies on the idea of a possibility distribution π, i.e., a mapping from a universe of discourse Ω to the unit interval $[0, 1]$. Possibility degrees $\pi(\omega)$ estimate to what extent the configuration ω is not unsatisfactory. π-pref nets are based on possibilistic networks [5], using conditional possibilities of the form $\pi(a_i|p(A_i)) = \frac{\Pi(a_i \wedge p(A_i))}{\Pi(p(A_i))}$, where $\Pi(\varphi) = \max_{\omega \models \varphi} \pi(\omega)$.

Definition 3 ([2,4]). *A possibilistic preference network (π-pref net) is a preference network in the sense of Definition 1, where each preference statement $p(A_i) : a_i^+ \succ a_i^-$ is associated to a conditional possibility distribution such that $\pi(a_i^+|p(A_i)) = 1 > \pi(a_i^-|p(A_i)) = \alpha_{A_i|p(A_i)}$, and $\alpha_{A_i|p(A_i)}$ is a non-instantiated variable on $[0, 1)$ we call symbolic weight.*

One may also have indifference statements $p(A_i) : a_i \sim \neg a_i$, expressed by $\pi(a_i|p(A_i)) = \pi(\neg a_i|p(A_i)) = 1$.

On top of the preferences encoded by a π-pref net, a set $\mathcal{C}$ of additional equality or inequality constraints between symbolic weights or products of symbolic weights can be provided by the user. Such constraints may represent, for instance, the relative strength of preferences associated to different instantiations of parent variables of the same variable.

π-pref nets induce a partial ordering between configurations based on the comparison of their degrees of possibility in the sense of a joint possibility distribution computed using the product-based chain rule, expressing a satisfaction erosion effect:

$$\pi(A_i, \ldots, A_n) = \prod_{i=1,\ldots,n} \pi(A_i | p(A_i)) \tag{1}$$

The preferences in the obtained configuration graph are of the form $\omega \succ_\pi \omega'$ if and only if $\pi(\omega) > \pi(\omega')$ *for all instantiations of the symbolic weights.*

Example 2. *Consider preference statements in Example 1. Conditional possibility distributions are as follows:* $\pi(t_1) = 1$, $\pi(t_2) = \alpha$, $\pi(p_1|t_1) = \pi(p_2|t_2) = 1$, $\pi(p_2|t_1) = \beta_1$, $\pi(p_1|t_2) = \beta_2$, $\pi(s_1|t_1) = \gamma_1$, $\pi(s_2|t_2) = \gamma_2$, $\pi(s_2|t_1) = \pi(s_1|t_2) = 1$, $\pi(c_1|p_1) = \pi(c_2|p_2) = 1$, $\pi(c_2|p_1) = \delta_1$ *and* $\pi(c_1|p_2) = \delta_2$. *Applying the product-based chain rule, we can compute the joint possibility distribution relative to T, P, C and S. Figure 3 represents with thin arrows the configuration graph induced from the joint possibility distribution. Clearly, the configuration* $t_1 p_1 c_1 s_2$ *is the root (since it is the unique one with degree* $\pi(t_1 p_1 c_1 s_2) = 1$).

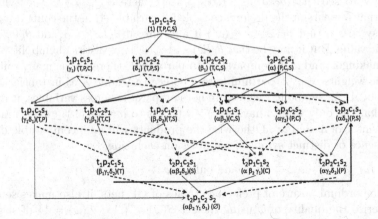

Fig. 3. Configuration graph of Example 1. Thin arrows reflect $\succ_\pi$, dotted arrows compare sets $\mathcal{S}(\omega_i)$ of Sect. 4, and bold arrows reflect additional ceteris paribus comparisons recovered by the constraints, also in bold on Fig. 2. Values under 1st (resp. 2nd) brackets correspond to joint possibility degrees (resp. sets $\mathcal{S}(\omega_i)$)

In the following, we compare configuration graphs induced by both CP-nets and π-pref nets. Clearly, they are different when no additional constraints are assumed between symbolic weights. For instance, if we consider the previous example we can check that in contrast with the CP-net of Fig. 2 where $t_1 p_1 c_2 s_1 \succ_{CP} t_1 p_2 c_2 s_1$, the π-pref net fails to compare them since there is no inequality constraint between $\pi(t_1 p_1 c_2 s_1) = \gamma_1 \delta_1$ and $\pi(t_1 p_2 c_2 s_1) = \gamma_1 \beta_1$.

4   Main Results

In this section, we show that the configuration graph of any CP-net is consistent with the configuration graph of the π-pref net without local indifference, based on the same preference network, provided that some constraints on products of symbolic weights are added to the π-pref net, in order to restore the ceteris paribus priorities. Precisely, the added constraints reflect the higher importance of parent nodes with respect to their children. Under an additional property whose validity can only be conjectured at this point, π-pref net would capture CP-nets exactly.

4.1   Consistency Between CP-nets and π-pref nets

In the following, we first recall that the ordering between configurations induced by a π-pref net corresponds to the Pareto ordering between the vectors $\omega = (\theta_1(\omega), \ldots, \theta_n(\omega))$ where $\theta_i(\omega) = \pi(\omega_{A_i} | \omega_{\mathcal{P}(A_i)})$, $i = 1, \ldots, n$. The Pareto ordering is defined by

$$\omega \succ_{Pareto} \omega' \text{ iff } \forall\, k,\ \theta_k(\omega) \geq \theta_k(\omega') \text{ and } \theta_i(\omega) > \theta_i(\omega') \text{ for some } i.$$

It is easy to see that $\theta_i(\omega) \in \{1, \alpha_{A_i | p(A_i)}\}$ where $\alpha_{A_i | p(A_i)}$ is the symbolic weight that appears in the preference table for variable A_i in the context $\omega_{\mathcal{P}(A_i)}$. It is easy to see that $\theta_k(\omega) > \theta_k(\omega')$ if and only if $\theta_k(\omega) = 1$ and $\theta_k(\omega')$ is a symbolic value. But it may be that $\theta_k(\omega)$ and $\theta_k(\omega')$ are distinct symbolic values, hence making ω and ω' incomparable. In particular, there are as many different symbolic weights $\alpha_{A | p(A)}$ pertaining to a boolean variable A as instantiations of parents of A. As symbolic weights are not comparable across variables, it is easy to see that the only way to have $\pi(\omega) \geq \pi(\omega')$ is to have $\theta_k(\omega) \geq \theta_k(\omega')$ in each component k of ω and ω'. Otherwise the products will be incomparable due to the presence of distinct symbolic variables on each side. So,

$$\omega \succ_{\pi} \omega' \text{ if and only if } \omega \succ_{Pareto} \omega'$$

Given the ordinal nature of preference tables of CP-nets, it also makes sense to characterize the quality of ω using the set $\mathcal{S}(\omega) = \{A_i : \theta_i(\omega) = 1\}$ of satisfied preference statements (one per variable). It is then clear that the Pareto ordering between configurations induced by the preference tables is refined by comparing these satisfaction sets:

$$\omega \succ_{Pareto} \omega' \Rightarrow \mathcal{S}(\omega') \subset \mathcal{S}(\omega) \tag{2}$$

since if two configurations contain variables having bad assignments in the sense of the preference tables, the corresponding symbolic values may differ if the contexts for assigning a value to this variable differ.

Example 3. *To see that this inclusion-based ordering is stronger than the π-pref net ordering, consider Fig. 3 where $\pi(t_1 p_2 c_1 s_2) = \beta_1 \delta_2$ with $\mathcal{S}(t_1 p_2 c_1 s_2) = \{T, S\}$ and $\pi(t_2 p_1 c_2 s_1) = \alpha \beta_2 \delta_1$ with $\mathcal{S}(t_2 p_1 c_2 s_1) = \{S\}$. We do have that $\mathcal{S}(t_1 p_2 c_1 s_2) \supset \mathcal{S}(t_2 p_1 c_2 s_1)$, but $\beta_1 \delta_2$ is not comparable with $\alpha \beta_2 \delta_1$. Dotted and thin arrows of Fig. 3 represent the configuration graph induced by comparing sets $\mathcal{S}(\omega)$.*

It is noticeable that if the weights $\alpha_{A_i|p(A_i)}$ reflecting the satisfaction level due to assigning the bad value to A_i in the context $p(A_i)$ do not depend on the context, then we have an equivalence in Eq. (2):

Proposition 1. *If* $\forall i = 1, \ldots, n, \alpha_{A_i|p(A_i)} = \alpha_i, \forall p(A_i) \in \mathcal{P}(A_i),$ *then*

$$\omega \succ_{Pareto} \omega' \iff \mathcal{S}(\omega') \subset \mathcal{S}(\omega).$$

Proof: Suppose $\mathcal{S}(\omega') \subset \mathcal{S}(\omega)$ then if $A \in \mathcal{S}(\omega')$ we have $\theta_i(\omega) = \theta_i(\omega') = 1$; if $A \in \mathcal{S}(\omega) \setminus \mathcal{S}(\omega')$, then $\theta_i(\omega') = \alpha_i, \theta_i(\omega) = 1$ and $\theta_i(\omega') = \alpha_i = \theta_i(\omega)$ otherwise. This implies $\omega \succ_{Pareto} \omega'$.

The inclusion-based ordering $\mathcal{S}(\omega') \subset \mathcal{S}(\omega)$ does not depend on the parent variables context but only on the fact that a variable has a good or a bad value. Similarly, when the symbolic weights no longer depend on parents instantiations, there is only one symbolic weight per variable. So, the above result is not surprising.

Example 4. *Using the same nodes as in Example 3, the unique weight assumption enforces* $\beta_1 = \beta_2 = \beta$ *and* $\delta_1 = \delta_2 = \delta$, *which yields* $\pi(t_1p_2c_1s_2) = \beta\delta > \pi(t_2p_1c_2s_1) = \alpha\beta\delta.$

In the following, we assume that the components of vector ω are linearly ordered in agreement with the partial ordering of variables in the symbolic preference network, namely, if $i < j$ then A_i is not a descendant of A_j in the preference net (i.e. topological ordering). For instance in the preference net of Fig. 1, we can use the ordering (T, P, C, S).

Let us first prove that, in the configuration graphs induced by a CP-net and the corresponding π-pref net, there cannot be any preference reversals between configurations. Let $\mathcal{C}h(A)$ denote the children set of $A \in \mathcal{V}$.

Lemma 1. *Let* ω *and* ω' *be two configurations such that* $\omega \succ_{CP} \omega'$ *and* ω *and* ω' *differ by one flip of a variable* A_i *then* $\mathcal{S}(\omega) \subset \mathcal{S}(\omega')$ *is not possible.*

Proof: Compare $\mathcal{S}(\omega)$ and $\mathcal{S}(\omega')$. It is clear that $A_i \notin \mathcal{S}(\omega')$ (otherwise the flip would not be improving) and $\mathcal{S}(\omega) = (\mathcal{S}(\omega') \cup \{A_i\} \cup Ch_-^+(A_i)) \setminus Ch_+^-(A_i)$, where $Ch_-^+(A_i)$ is the set of variables that switch from a bad to a good value when going from ω' to ω, and $Ch_+^-(A_i)$ is the set of variables that switch from a good to a bad value when going from ω' to ω. It is clear that it can never be the case that $\mathcal{S}(\omega) \subset \mathcal{S}(\omega')$, indeed A_i is in $\mathcal{S}(\omega)$ and not in $\mathcal{S}(\omega')$ by construction. But $\mathcal{S}(\omega')$ may contain variables not in $\mathcal{S}(\omega)$ (those in $Ch_+^-(A_i)$ if not empty). So either $\mathcal{S}(\omega') \subset \mathcal{S}(\omega)$ or the two configurations are not Pareto-comparable. □

In the following, given two configurations ω and ω', let $\mathcal{D}^{\omega,\omega'}$ be the set of variables which bear different values in ω and ω'.

Proposition 2. *If* $\omega \succ_{CP} \omega'$ *then* $\mathcal{S}(\omega) \subset \mathcal{S}(\omega')$ *is not possible.*

Proof: If $\omega \succ_{CP} \omega'$, then there is a chain of improving flips $\omega_0 = \omega' \prec_{CP}$ $\omega_1 \prec_{CP} \cdots \prec_{CP} \omega_k = \omega$. Applying the above Lemma, $\mathcal{S}(\omega_i) = (\mathcal{S}(\omega_{i-1}) \cup \{V_{i-1}\} \cup Ch_-^+(V_{i-1})) \setminus (Ch_+^-(V_{i-1})$ for some variable $V_{i-1} = A_j$. By the above Lemma, we cannot have $\mathcal{S}(\omega_{i-1}) \subset \mathcal{S}(\omega_i)$. Suppose we choose the chain of improving flips by flipping at each step a top variable A_j in the preference net, among the ones to be flipped, i.e. $j = \min\{\ell : A_\ell \in \mathcal{D}^{\omega_{i-1},\omega}\}$. It means that when following the chain of improving flips, the status of each flipped variable will not be questioned by later flips, as no flipped variable will be a child of variables flipped later on. So $\mathcal{S}(\omega)$ will contain some variables not in $\mathcal{S}(\omega')$, so $\mathcal{S}(\omega) \subset \mathcal{S}(\omega')$ is not possible.

The previous results show that it is impossible to have a preference reversal between the CP-net ordering and the inclusion ordering, which implies that no preference reversal is possible between CP-net ordering and the π-pref net ordering. It suggests that we can try to add ceteris paribus constraints to a π-pref net and so as to capture the preferences expressed by a CP-net.

As previously noticed, in CP-nets, parent preferences look more important than children ones. This property is not ensured by π-pref nets where all violations are considered having the same importance. Indeed, we can check from Figs. 2 and 3 that the two configuration graphs built from the same preference statements of Example 1 are different. In the following, we lay bare local constraints between each node and its children that enable ceteris paribus to be simulated. Let $D_{\mathcal{P}(A)} = \times_{A_i \in \mathcal{P}(A)} D_{A_i}$ denote the Cartesian product of domains of variables in $\mathcal{P}(A)$, $\alpha_{A|p(A)} = \pi(a^-|p(A))$ and $\gamma_{C|p(C)} = \pi(c^-|p(C))$.

Proposition 3. *Suppose a CP-net and a π-pref net built from the same preference statements. Let us add to the latter all constraints induced by the condition:*
$\forall A \in \mathcal{V}$ *s.t.* $Ch(A) \neq \emptyset$:

$$\max_{p(A) \in D_{\mathcal{P}(A)}} \alpha_{A|p(A)} < \prod_{C \in Ch(A)} \min_{p(C) \in D_{\mathcal{P}(C)}} \gamma_{C|p(C)} \tag{3}$$

Let $\succ_\pi^+$ be the resulting preference ordering built from the preference tables and applying constraints between symbolic weights of the form of Eq. 3, then, $\omega \succ_{CP}$ $\omega' \Rightarrow \omega \succ_\pi^+ \omega'$.

Proof: The relation $\succ_{CP}$ is determined by comparing configurations ω, ω' of the form $\omega = a^+ \wedge p(A) \wedge r$ and $\omega' = a^- \wedge p(A) \wedge r$ (where $R = \mathcal{V} \setminus (A \cup \mathcal{P}(A))$), that differ by one flip of variable A. So the local preference $p(A) : a^+ \succ a^-$ is equivalent to have $\omega \succ_{CP} \omega'$ under ceteris paribus assumption, and also equivalent to have $\pi(a^-|p(A)) < \pi(a^+|p(A)) = 1$ in the corresponding π-pref net based on the same preference tables.

Now let us show that $\forall A \in \mathcal{V}$, $\forall p(A) \in D_{\mathcal{P}(A)}$ and every instantiation r of the variables in $\mathcal{V} \setminus (\{A\} \cup \mathcal{P}(A))$, the local preference $\pi(a^-|p(A)) < \pi(a^+|p(A))$ implies $\pi(a^- \wedge p(A) \wedge r) < \pi(a^+ \wedge p(A) \wedge r)$ under the condition expressed by Eq. (3). Consider the instantiation $ch(A) = \wedge_{C \in Ch(A)} \omega_C$, where $\omega_C \in \{c, \neg c\}$, of the children of A such that $ch(A) \wedge o = r$ (i.e. $O = \mathcal{V} \setminus (A \cup \mathcal{P}(A) \cup Ch(A))$.

The chain rule states (Eq. (1)):
$$\pi(\omega') = \prod_{B \in \mathcal{V}} \pi(\omega'_B | \omega'_{\mathcal{P}(B)}) =$$
$$\pi(a^- | p(A)) \cdot \prod_{C \in Ch(A)} \pi(\omega'_C | p'(C)) \cdot \prod_{B \notin \{A\} \cup Ch(A)} \pi(\omega'_B | \omega'_{\mathcal{P}(B)}).$$

Clearly the last term does not depend on A and is thus a constant β. So
$$\pi(\omega') = \beta \cdot \alpha_{A | p(A)} \cdot \prod_{C \in Ch(A)} \pi(\omega'_C | p'(C)).$$
Likewise, since $\omega = a^+ \wedge p(A) \wedge ch(A) \wedge o$, we have
$$\pi(\omega) = \beta \cdot \prod_{C \in Ch(A)} \pi(\omega_C | p(C)), \text{ since } \pi(a^+ | p(A)) = 1. \text{ Note that while } p'(C)$$
is of the form $a^- \wedge p_{-A}(C)$ where $\mathcal{P}_{-A}(C)$ is the set of parents of C but for A, $p(C)$ is of the form $a^+ \wedge p_{-A}(C)$.

So the inequality $\pi(\omega) > \pi(\omega')$, present in the CP-net, requires:

$$\prod_{C \in Ch(A)} \pi(\omega_C | p(C)) > \alpha_{A | p(A)} \cdot \prod_{C \in Ch(A)} \pi(\omega'_C | p'(C)).$$

Condition (3) implies $\alpha_{A | p(A)} < \prod_{C \in Ch(A)} \pi(\omega_C | p(C))$, which implies the above inequality. It proves that, under Condition (3), $\omega \succ_{CP} \omega'$ implies $\omega \succ_{\pi}^+ \omega'$. $\square$

This proposition ensures that the ordering induced by the joint possibility distribution of a π-pref net enhanced by constraints of the form (3) can refine the CP-net ordering having the same preference tables, provided that suitable constraints are added at each node $A \in \mathcal{V}$ between the local conditional possibility distribution at this node and the product of possibility degrees of the children of A. It comes down to constraints between each symbolic weight and a product of other ones. Indeed the less preferred value, $\min(\pi(a | p(A)), \pi(\neg a | p(A)))$, of A in the context of the parents $p(A)$ of A is a symbolic weight (non instantiated possibility degree). In other words, the inequality ensures that the less preferred value of each A given $p(A)$ is strictly less preferred than the product of the less preferred values of the children of A. This result is the symbolic counterpart of the one in [13], using preference networks with numerical ranking functions.

Example 5. *In the graph of Fig. 3 induced by the π-pref net of Example 2, Proposition 3 leads us to add conditions $\alpha < \min(\gamma_1, \gamma_2) \cdot \min(\beta_1, \beta_2)$ and $\max(\beta_1, \beta_2) < \min(\delta_1, \delta_2)$. Clearly these conditions are too strong here. First some of the products like $\gamma_1 \beta_2$ never appear in Fig. 3. Moreover, the reader can check that adding constraints $\beta_1 \gamma_1 > \alpha$ and $\beta_i < \delta_i, (i = 1, 2)$ turns the configuration graph of Fig. 3 into the CP-net-induced configuration graph of Fig. 2.*

Instead of imposing priority of parents over children, we can also add the ceteris paribus constraints to the π-pref net directly, considering only worsening flips. Let ω, ω' differ by one flip, and such that none of $\omega \succ_{\pi} \omega'$, $\omega' \succ_{\pi} \omega$ holds, and moreover, $\omega \succ_{CP} \omega'$. We must enforce the condition $\pi(\omega) > \pi(\omega')$. Suppose the flipping variable is A. Clearly, $A \in S(\omega)$, but $A \notin S(\omega')$. Let α be the possibility degree of A when it takes the bad value in context $\omega_{p(A)}$ (it is 1 when it takes the good value). When flipping A from a good to a bad value, only the quality of the children variables $Ch(A)$ of A may change. $Ch(A)$ can be partitioned into at most 4 sets, $Ch_-^-(A)$ (resp. $Ch_+^-(A), Ch_-^+(A), Ch_+^+(A)$), which represents the set of children of A whose values remain bad (resp. change from good to bad, from bad to good, and stay good) when flipping A from a^+ to a^-. Strictly speaking these sets depend upon ω. Then it can be easily checked that:

$$\pi(\omega) = 1 \cdot \prod_{C_i \in Ch^+_-(A)} \gamma_i \cdot \prod_{C_j \in Ch^-_-(A)} \gamma_j \cdot \beta$$

$$\pi(\omega') = \alpha \cdot \prod_{C_k \in Ch^-_+(A)} \gamma_k \cdot \prod_{C_j \in Ch^-_-(A)} \gamma_j \cdot \beta$$

where β is a product of symbols, pertaining to nodes other than A and its children, that remain unchanged by the flip of A. Then the constraint $\pi(\omega) > \pi(\omega')$ comes down to the inequality:

$$\prod_{C_i \in Ch^+_-(A)} \gamma_i > \alpha \cdot \prod_{C_k \in Ch^-_+(A)} \gamma_k \qquad (4)$$

where symbols appearing on one side do not appear on the other side. Such constraints are clearly weaker than Condition (3) but are sufficient to retrieve all the preferences of the CP-net. Note that the preferences $\omega \succ_\pi \omega'$ and $\omega \succ_{CP} \omega'$ conjointly hold in both approaches whenever A has no child node, and more generally whenever the worsening flip on A corresponds to no child variable moving from a bad to a good state, i.e. $Ch^+_-(A) = \emptyset$. In fact, condition (4) holds for all preference arcs in the configuration graph of the CP-net, whether this preference appears in the π-pref net or not. We get the following result.

Proposition 4. *Consider a CP-net and the preference relation $\succ^+_\pi$ on configurations built from the same preference tables by adding all constraints of the form (4) between configurations differing by one flip to the preferences of the form $\omega \succ_\pi \omega'$. Then:*

$$\omega \succ_{CP} \omega' \Rightarrow \omega \succ^+_\pi \omega'$$

Proof: Indeed, first the preferences according to $\succ_{CP}$ and $\succ_\pi$ do not contradict each other, per Proposition 2. Then we add constraints to the π-pref net for all CP-net worsening flips that are not captured by $\succ_\pi$, using constraints (4). So we have then captured the whole preference graph of the CP-net, plus possibly other preferences between configurations.

In the transformation of a CP-net into a π-pref net, we keep the same graphical structure and the tables are filled directly from the preference statements of the CP-net. Besides, we must point out that, when mimicking CP-nets, constraints are not elicited from the user but computed directly from the graph structure.

Example 6. *The above constraints (4) that must be added to the π-pref configuration graph of Fig. 3 are precisely those found to be necessary and sufficient in*

Example 5 to recover the CP-net ordering, i.e., $\alpha < \beta_1\gamma_1$, $\beta_1 < \delta_1$ and $\beta_2 < \delta_2$. Note that the number of additional constraints to be added to capture the CP-net comparisons missed by the π-pref net is quite small. For instance, the number of constraints here is 4 against 120 potential comparisons.

So, in the example, we exactly capture the preference graph of a CP-net using additional constraints between products of symbolic weights. The above considerations thus encourage us to study whether π-pref nets without constraints are *refined* by CP-nets, namely if the configuration graph of the former contains less strict preferences between configurations than the one of the latter, so that adding the constraints (4) are enough to simulate a CP-net by a π-pref net with constraints. Note that if it were not the case, it would mean that CP-nets do not respect Pareto-ordering.

4.2   Towards Exact Representations of CP-nets by π-pref nets

In this subsection, we consider the inclusion-ordering. One may wonder if there may exist some configurations that can be compared by the inclusion-based ordering, while they remain incomparable for CP-nets. This is not the case in our running example.

Example 7. *Consider the top configuration $\omega' = t_1p_1c_1s_2$ which inclusion-dominates $\omega = t_2p_2c_2s_1$ in the π-pref net configuration graph in Fig. 3, since the former has good values for all variables and only the value t_2 is bad in the latter, i.e., $\mathcal{S}(\omega') = \{T, P, C, S\}$ and $\mathcal{S}(\omega) = \{P, C, S\}$. But the two configurations are far away in terms of flips since $\mathcal{D}^{\omega,\omega'} = \{T, P, C, S\}$. They can, however, be related by a chain of worsening flips. Namely, as $\mathcal{S}(\omega') \setminus \mathcal{S}(\omega) = \{T\}$, we must flip T first, and $\omega_1 = t_1p_2c_2s_1$, with $\mathcal{S}(\omega_1) = \{T, C\}$ so $\mathcal{S}(\omega') \setminus \mathcal{S}(\omega_1) = \{P, S\}$ and $\mathcal{D}^{\omega_1,\omega'} = \{P, C, S\}$. We now must flip P and get $\omega_2 = t_1p_1c_2s_1$ with $\mathcal{S}(\omega_2) = \{T, P\} = \mathcal{D}^{\omega_2,\omega'}$. As $\mathcal{S}(\omega') \setminus \mathcal{S}(\omega_2) = \{C, S\}$, we must flip C, and $\omega_3 = t_1p_1c_1s_1$, with $\mathcal{S}(\omega_3) = \{T, P, C\}$ so $\mathcal{S}(\omega') \setminus \mathcal{S}(\omega_3) = \{S\} = \mathcal{D}^{\omega_3,\omega'}$. We now must flip S and get $\omega_4 = t_1p_1c_1s_2 = \omega'$.*

The question whether the preference ordering of configurations induced by CP-nets is consistent with the ordering between the sets of variables that take good values in agreement with the preference tables seems to have been overlooked so far in the CP-net literature. The inclusion ordering between sets of variables with satisfactory values is intuitive in the sense that if a configuration ω violates all the preference statements violated by another configuration ω' plus some other(s), then ω' should indeed be strictly preferred to ω. The consistency of CP-nets with inclusion, namely the property

$$\mathcal{S}(\omega_1) \subset \mathcal{S}(\omega_2) \Rightarrow \omega_2 \succ_{CP} \omega_2 \qquad (*)$$

can be naturally conjectured since the opposite case would cast a doubt on the rationality of such networks. Proposition 2 proves a weak consistency between them. However, at this stage providing a formal complete proof looks tricky and besides, is not directly related to the expressivity of π-pref nets, the very topic of this paper. The results in the following are conditioned by the truth of the conjecture, or are restricted to those CP-nets that agree with the inclusion-based orderings. Based on this assumption, Proposition 5 indicates that the CP-net ordering refines, hence is consistent with, the ordering induced by a π-pref net built from the same preference specification. This is because the inclusion-ordering refines the Pareto (or π-pref net) ordering.

Proposition 5. *Consider a CP-net that refines the inclusion-based ordering and a π-pref net built from the same preference statements, we have:*

$$\omega' \succ_\pi \omega \Rightarrow \omega' \succ_{CP} \omega$$

Let us now prove that, if the conjecture (*) is valid, we are able to *exactly* induce the CP-net ordering from the π-pref net ordering by adding suitable constraints between symbolic weights or their products. First, we have seen that we can add to the π-pref net configuration graph all missing preference statements induced by the CP-net and not already present in the π-pref net configuration graph. These statements concern all pairs (ω, ω') that differ by one flip and such that $\pi(\omega)$ and $\pi(\omega')$ are not comparable. Note that adding such preference statements to the Pareto configuration graph in case of Pareto-incomparability yields the CP-net configuration graph (up to transitive closure).

The question remains whether we can express the latter in terms of additional constraints between symbolic weights or products thereof.

Proposition 6. *Consider a CP-net that refines the inclusion-based ordering and the preference relation $\succ_\pi^+$ on configurations built from the same preference tables by enforcing all constraints of the form (4) between configurations differing by one flip. Then:*

$$\omega \succ_{CP} \omega' \Leftrightarrow \omega \succ_\pi^+ \omega'$$

Proof: ($\Rightarrow$) This direction is proved by Proposition 4. ($\Leftarrow$) As $\omega \succ_\pi \omega' \Rightarrow \omega \succ_{CP} \omega'$ by assumption, adding ceteris paribus constraints corresponding to worsening flips to $\succ_\pi$ will not produce by transitivity any preference relation not in $\succ_{CP}$. □

It is clear that, beside ceteris paribus constraints, other constraints could be added to a π-pref net, that cannot be expressed by a CP-net, and that account for different types of preference information. This fact suggests that π-pref nets with constraints have a better expressive power and are more flexible than CP-nets (known as a powerful qualitative model), and provide a general class of qualitative graphical models where the *ceteris paribus* ordering could be further refined without going numerical (i.e. unlike UCP-nets). It is clear therefore that the constraints added to refine this order should, in this case, be consistent with *ceteris paribus*. Finally, π-pref nets are sometimes able to represent preference orderings when CP-nets fail to do it, as shown in the example below [1].

Example 8. *Let us consider two binary variables A and B standing respectively for "vacations" and "good weather". Suppose that we have the following preference ordering: $ab \succ \neg a \neg b \succ a \neg b \succ \neg ab$. We observe that this complete order cannot be represented by a CP-net. In fact, given two variables we can define two possible structures: either A depends on B or conversely. But, none of them are capable to capture this total ordering in the CP-net setting. Indeed, this total order exhibits a violation of the Ceteris Paribus principle. However, such preferences can be represented by a joint possibility distribution such that: $\pi(ab) > \pi(\neg a \neg b) > \pi(a \neg b) > \pi(\neg ab)$. Thus, we have $\top : a \succ \neg a,\ a : b \succ \neg b$ and $\neg a : \neg b \succ b$. It corresponds to a network with two nodes and their corresponding conditional possibility distributions are: $\pi(a) = 1$, $\pi(\neg a) = \alpha$, $\pi(b|a) = 1$, $\pi(b|\neg a) = \gamma$, $\pi(\neg b|a) = \beta$ and $\pi(\neg b|\neg a) = 1$. This yields $\pi(ab) = 1 > \pi(\neg a \neg b) = \alpha > \pi(a \neg b) = \beta > \pi(\neg ab) = \alpha\gamma$ taking $\alpha > \beta$ and $\beta = \gamma$.*

5 Related Works

Despite the existence of various graphical models for preferences [3], only few works have been concerned in comparing their expressive power. We can, in particular, mention two interesting results. The first concerns OCF-nets, which are preference networks where possibility distributions are replaced by ranking (ordinal conditional) functions (OCF) valued in the set of integers and the chain rule is additive. These functions may be transformed into possibility distributions [12]. [13] proved that OCF-nets can refine CP-net orderings. Precisely, OCF-nets will always lead to total orderings that are compatible with CP-nets. To do so they use a set of particular constraints to be imposed on their integer weights, which basically correspond to our constraints (3), albeit between numerical values. In contrast, the use of symbolic weights in our approach preserves the partiality of the ordering, and, if the CP-net order does refine the inclusion ordering, the CP-net configuration graph can be exactly recovered. Moreover the use of symbolic weights does not commit us to the choice of particular numerical values.

There were several attempts to represent CP-net orderings using a possibilistic logic base (a logical counterpart of π-pref nets), where the product is replaced by the minimum. See [10] for a bibliography and a discussion. It was observed that an exact logical representation of CP-nets was not possible when variables in the net have several children variables even though good approximations could be built. This is because additional constraints in this framework compare individual symbolic weights, not product thereof.

Some extensions of CP-nets can be considered as akin to π-pref nets. TCP-nets [9] also add priority constraints between variable nodes, that we can render in π-pref nets by inequalities between symbolic weights pertaining to different CP-tables. Utility-enhanced CP-nets (UCP-nets) [6] add additive utility functions to CP-nets in order to encode total orderings consistent with the ceteris paribus assumption. To do so, linear constraints are added on utility values that are somewhat similar to constraints (4). They express that for any variable, given an instantiation of its parents, the utility gain in choosing the good value rather than the bad one in this context, should be more important than the maximum value of the sum of the possible utility loss for its children over all possible instantiations of the other related variables. Up to a log transformation this is like comparing products.

π-pref nets can also be compared with so-called CP-theories [16]. The latter interpret conditional preference statements assuming they hold *irrespectively* of the values of other variables. It means that any configuration ω such that $\omega_A = a^+$ and $\omega_{\mathcal{P}(A)} = p(A)$ is preferred to any configuration ω such that $\omega_A = a^-$ and $\omega_{\mathcal{P}(A)} = p(A)$. In terms of possibility functions, it reads $\Delta(p(A) \wedge a^+) > \Pi(p(A) \wedge a^-)$, where $\Delta(\varphi) = \min_{\omega \models \varphi} \pi(\omega)$. In [16] are studied hybrid nets where some variables are handled ceteris paribus, while the preference holds irrespectively of other variables. In π-pref nets preference statements are interpreted by $\pi(a^+|p(A)) > \pi(a^-|p(A))$ which is provably equivalent to $\Pi(p(A) \wedge a^+) > \Pi(p(A) \wedge a^-)$, i.e. comparing best configurations. It is clear that if $\omega \succ_{CP} \omega'$ holds, then $\omega \succ \omega'$ holds in a CP-theory, where conditional preference holds irrespectively of other variables, because the CP-theory generates more preference constraints between configurations, including the ones induced by the *ceteris paribus* assumption. Constraints induced by CP theories can thus be captured in π-pref nets by adding more constraints between products of symbolic weights.

6 Conclusion

In this paper, we have explored the expressive power of π-pref nets. First, we have proved that the CP-net configuration orderings cannot contradict those of the π-pref nets and we found suitable additional constraints to refine π-pref net orderings in order to encompass ceteris paribus constraints of CP-nets. CP-nets would then be exactly captured by π-pref nets with constraints if their configuration graph did refine the inclusion-based ordering. This indicates that CP-nets potentially represent a subclass of π-pref nets with constraints. One may further refine CP-net preferences by adding more constraints between symbolic weights appearing in π-pref nets. For instance, one may introduce priorities between two parents nodes or between two child nodes.

Regarding query processing, finding an optimal configuration is straightforward, for both CP-nets and π-pref nets. In fact, it consists in traversing the network from root to leaves and choose the best value for each variable depending on its parents configuration. The complexity is linear with the size of the network. As to comparing two configurations, the dominance query for CP-nets consists in finding a chain of worsening flips from one configuration to the other. It is NP-complete to PSPACE-complete depending on the graph structure [8]. For π-pref nets, without constraints this query comes down to a Pareto-comparison of vectors symbolic weights. If there are constraints, the approach requires a reordering of coefficients and the complexity is at most equal to $O(n!)$ [2]. Dominance and optimization queries on instantiated π-pref nets respecting constraints will have the same complexities as for UCP-nets.

References

1. Amor, N.B., Dubois, D., Gouider, H., Prade, H.: Possibilistic networks: a new setting for modeling preferences. In: Straccia, U., Calì, A. (eds.) SUM 2014. LNCS, vol. 8720, pp. 1–7. Springer, Cham (2014). doi:10.1007/978-3-319-11508-5_1
2. Amor, N.B., Dubois, D., Gouider, H., Prade, H.: Possibilistic conditional preference networks. In: Destercke, S., Denoeux, T. (eds.) ECSQARU 2015. LNCS, vol. 9161, pp. 36–46. Springer, Cham (2015). doi:10.1007/978-3-319-20807-7_4
3. Amor, N.B., Dubois, D., Gouider, H., Prade, H.: Graphical models for preference representation: an overview. In: Schockaert, S., Senellart, P. (eds.) SUM 2016. LNCS, vol. 9858, pp. 96–111. Springer, Cham (2016). doi:10.1007/978-3-319-45856-4_7
4. Amor, N.B., Dubois, D., Gouider, H., Prade, H.: Preference modeling with possibilistic networks and symbolic weights: a theoretical study. In: Proceedings of ECAI 2016, pp. 1203–1211 (2016)
5. Benferhat, S., Dubois, D., Garcia, L., Prade, H.: On the transformation between possibilistic logic bases and possibilistic causal networks. Int. J. Approx. Reas. **29**, 135–173 (2002)
6. Boutilier, C., Bacchus, F., Brafman, R.I.: UCP-networks: a directed graphical representation of conditional utilities. In: Proceedings of UAI 2001, pp. 56–64 (2001)
7. Boutilier, C., Brafman, R.I., Hoos, H., Poole, D.: Reasoning with conditional ceteris paribus preference statements. In: Proceedings of UAI 1999, pp. 71–80 (1999)
8. Boutilier, C., Brafman, R., Domshlak, C., Hoos, H., Poole, D.: CP-nets: a tool for representing and reasoning with conditional ceteris paribus preference statements. J. Artif. Intell. Res. (JAIR) **21**, 135–191 (2004)
9. Brafman, R.I., Domshlak, C.: TCP-nets for preference-based product configuration. In: Proceedings of the 4th Workshop on Configuration (in ECAI-2002), pp. 101–106 (2002)
10. Dubois, D., Hadjali, A., Prade, H., Touazi, F.: Erratum to: database preference queries - a possibilistic logic approach with symbolic priorities. Ann. Math. Artif. Intell. **73**(3–4), 359–363 (2015)
11. Dubois, D., Prade, H.: Possibility Theory: An Approach to Computerized Processing of Uncertainty. Plenum Press, New York (1988)
12. Dubois, D., Prade, H.: Qualitative and semi-quanlitative modeling of uncertain knowledge-a discussion. In: Computational Models of Rationality Essays Dedicated to Gabriele Kern-Isberner on the Occasion of Her 60th Birthday, pp. 280–292. College Publications (2016)
13. Eichhorn, C., Fey, M., Kern-Isberner, G.: CP-and OCF-networks-a comparison. Fuzzy Sets Syst. (FSS) **298**, 109–127 (2016)
14. Fonck, P.: Conditional independence in possibility theory. In: Proceedings of UAI 1994, pp. 221–226 (1994)
15. Gonzales, C., Perny, P.: GAI networks for utility elicitation. In: Proceedings of KR 2004, vol. 4, pp. 224–234 (2004)
16. Wilson, N.: Extending CP-nets with stronger conditional preference statements. In: Proceedings of AAAI 2004, vol. 4, pp. 735–741 (2004)

Assumption-Based Argumentation Equipped with Preferences and Constraints

Toshiko Wakaki[✉]

Shibaura Institute of Technology, 307 Fukasaku, Minuma-ku,
Saitama-city, Saitama 337–8570, Japan
twakaki@sic.shibaura-it.ac.jp

Abstract. Čyras and Toni claimed that assumption-based argumentation equipped with preferences (p_ABA) cannot solve two examples presented by them since the given preferences don't work in their p_ABAs whose underlying ABAs have a unique extension, and hence they proposed ABA$^+$. However in p_ABAs encoded by them, we found that they mistook hypotheses contained in their example for assumptions, while Čyras ignored some constrains contained in another example. Hence against their claim, first this paper shows that p_ABAs in which we expressed the respective knowledge correctly give us solutions of them without any difficulties. Second we present the technique to represent hypotheses in ABA as well as a method to incorporate some kind of constraints in p_ABA. Finally we show a famous non-monotonic reasoning example with preferences that ABA$^+$ leads to incorrect results.

1 Introduction

Assumption-Based Argumentation (ABA) [1,9] is a general-purpose argumentation framework whose arguments are structured. It does not have a mechanism to deal with the given explicit preferences though explicit preferences are often required to resolve conflicts between arguments in human argumentation.

Recently to overcome difficulties of the existing approaches that map the explicit preferences into ABA, we proposed an assumption-based argumentation framework equipped with preferences (p_ABA) [22,23], which incorporates *explicit preferences over sentences* into ABA. As discussed in [23], our approach introducing preferences over sentences in the framework is inspired by *prioritized circumscription* [15,16], namely, the most well-established formalization for commonsense reasoning with preferences that enables us to represent various preferences by means of priorities over its minimized predicates. In regard to the semantics of p_ABA, we provide a method to 'lift' a sentence ordering given in p_ABA to the argument ordering. Accordingly we can freely give any semantics to the proposed p_ABA based on either an argument ordering or a sentence ordering. W.r.t. other frameworks (e.g. $ASPIC^+$ [17,18]), the lift from preferences to argument orderings is also performed. However based on their argument orderings, the altered argumentation framework with a modified successful *attacks* (i.e. *defeat*) is constructed, to which Dung's argument-based

© Springer International Publishing AG 2017
S. Moral et al. (Eds.): SUM 2017, LNAI 10564, pp. 178–193, 2017.
DOI: 10.1007/978-3-319-67582-4_13

semantics is applied. This denotes that the extension of the altered argumentation framework is not always an extension of the initial argumentation framework without preferences but a modified one, which is in conflict with our philosophy based on the idea to treat preferences in prioritized circumscription [15,16]. Now recall that, when every model is a Herbrand model, a model of prioritized circumscription expressed by $Circum(T; P^1 > \cdots > P^k; Z)$ is a minimal one among Herbrand models of the first order theory T w.r.t. the model (i.e. structure) ordering $\leq^{P^1 > \cdots > P^k; Z}$ lifted from the given predicate ordering $P^1 > \cdots > P^k$. Thus in a similar way, we presented a method to lift a sentence ordering (resp. an argument ordering) to the extension ordering ($\sqsubseteq_{ex}$). Thanks to the extension ordering, the semantics of p_ABA is given by $\mathcal{P}$-argument extensions along with $\mathcal{P}$-assumption extensions which are maximal ones w.r.t. the extension ordering $\sqsubseteq_{ex}$ among extensions of its underlying ABA. Thus in a special case of $Circum(T; P^1 > \cdots > P^k; Z)$ where T has a unique model, $Circum(T; P^1 > \cdots > P^k; Z)$ has a unique model which coincides with the model of T since the unique model of T is always minimal w.r.t. $\leq^{P^1 > \cdots > P^k; Z}$ regardless of the given $P^1 > \cdots > P^k$. This means that when T has a unique model, any of other interpretations of T which are inevitably inconsistent is never selected as a model of $Circum(T; P^1 > \cdots > P^k; Z)$ by taking account of priorities. Hence inheriting the same property, in a special case of p_ABA such that its underlying ABA (which satisfies rationality postulates [10]) has a unique extension, p_ABA $\langle \mathcal{L}, \mathcal{R}, \mathcal{A}, \mathcal{C}, \preceq \rangle$ has a unique $\mathcal{P}$ extension which coincides with the extension of its underlying ABA since the unique argument extension of such ABA is maximal w.r.t. $\sqsubseteq_{ex}$ regardless of the given $\preceq$.

As for this property, Čyras and Toni claimed that a p_ABA framework cannot solve two examples (i.e. [5, Example 1], [4, Example 1]) since the given preferences don't work in p_ABAs encoded by them whose underlying ABAs have a unique extension, and hence they proposed ABA$^+$ [5]. However in their p_ABAs, we found that they mistook hypotheses contained in [5, Example 1] for assumptions, while Čyras ignored some constrains contained in another one [4, Example 1]. Hence against their claim, first we show that p_ABAs in which we expressed the respective knowledge correctly give us solutions of them without any difficulties, and the given preferences work well in the p_ABAs where the underlying ABAs have the multiple extensions. Second we present the technique to represent hypotheses in ABA as well as a method to incorporate some kind of constraints in ABA and p_ABA. Finally, and perhaps most importantly, we show a famous non-monotonic reasoning example involving the use of preferences that ABA$^+$ leads to incorrect results, while p_ABA avoids this problem.

This paper is organized as follows. Section 2 gives preliminaries. Section 3 presents how to represent hypotheses in ABA. Section 4 presents an ABA equipped with preferences and constrains. Section 5 discusses related work. Section 6 concludes this paper while showing a counterexample to ABA$^+$.

2   Preliminaries

Definition 1 (ABA). *An assumption-based argumentation framework (an ABA framework, or an ABA, for short) [1,9,13] is a tuple $\langle \mathcal{L}, \mathcal{R}, \mathcal{A}, \mathcal{C} \rangle$, where*

- *$(\mathcal{L}, \mathcal{R})$ is a deductive system, with $\mathcal{L}$ a language consisting of countably many sentences and $\mathcal{R}$ a set of inference rules of the form $b_0 \leftarrow b_1, \ldots, b_m (m \geq 0)$, where b_0 (resp. $b_1, \ldots, b_m$) is called the head (resp. the body) of the rule.*
- *$\mathcal{A} \subseteq \mathcal{L}$, is a (non-empty) set, referred to as assumptions.*
- *$\mathcal{C}$ is a total mapping from $\mathcal{A}$ into $2^{\mathcal{L}} \setminus \{\emptyset\}$, where each $c \in \mathcal{C}(\alpha)$ is a contrary of $\alpha \in \mathcal{A}$.*

We enforce that ABA frameworks are flat, namely assumptions do not occur as the heads of rules. For a special case such that each assumption has the unique contrary sentence (i.e. $|\mathcal{C}(\alpha)|=1$ for $\forall \alpha \in \mathcal{A}$), an ABA framework is usually defined as $\langle \mathcal{L}, \mathcal{R}, \mathcal{A}, ^- \rangle$, where a total mapping $^-$ from $\mathcal{A}$ into $\mathcal{L}$ is used.

In ABA, an *argument for (the claim) $c \in \mathcal{L}$ supported by $K \subseteq \mathcal{A}$ ($K \vdash c$ in short) is a (finite) tree with nodes labelled by sentences in $\mathcal{L}$ or by τ, and *attacks* against arguments are directed at the assumptions in their supports as follows.

- An argument $K \vdash c$ *attacks* an assumption α iff $c \in \mathcal{C}(\alpha)$.
- An argument $K_1 \vdash c_1$ *attacks* an argument $K_2 \vdash c_2$ iff $c_1 \in \mathcal{C}(\alpha)$ for $\exists \alpha \in K_2$.

Corresponding to ABA $\mathcal{F} = \langle \mathcal{L}, \mathcal{R}, \mathcal{A}, \mathcal{C} \rangle$, the abstract argumentation framework $AF_{\mathcal{F}} = (AR, attacks)$ is constructed based on *arguments* and *attacks* addressed above, and all argumentation semantics [7] can be applied to $AF_{\mathcal{F}}$. For a set *Args* of arguments, let $Args^+ = \{A|$ there exists an argument in *Args* that *attacks* $A\}$. *Args* is *conflict-free* iff $Args \cap Args^+ = \emptyset$. *Args defends* an argument A iff each argument that attacks A is attacked by an argument in *Args*.

Definition 2 [2,7,9,13]. *Let $\langle \mathcal{L}, \mathcal{R}, \mathcal{A}, \mathcal{C} \rangle$ be an ABA framework, and AR the associated set of arguments. Then $Args \subseteq AR$ is: admissible iff Args is conflict-free and defends all its elements; a complete argument extension iff Args is admissible and contains all arguments it defends; a preferred (resp. grounded) argument extension iff it is a (subset-)maximal (resp. (subset-)minimal) complete argument extension; a stable argument extension iff it is conflict-free and $Args \cup Args^+ = AR$; an ideal argument extension iff it is a (subset-)maximal complete argument extension that is contained in each preferred argument extension.*

The various ABA semantics [1] is also described in terms of sets of assumptions.

- A set of assumptions *Asms attacks* an assumption α iff *Asms* enables the construction of an argument for the claim $\exists c \in \mathcal{C}(\alpha)$.
- A set of assumptions $Asms_1$ *attacks* a set of assumptions $Asms_2$ iff $Asms_1$ *attacks* some assumption $\alpha \in Asms_2$.

For a set of assumptions *Asms*, let $Asms^+ = \{\alpha \in \mathcal{A}|Asms$ attacks $\alpha\}$. *Asms* is *conflict-free* iff $Asms \cap Asms^+ = \emptyset$. *Asms defends* an assumption α iff each set of assumptions that attacks α is attacked by *Asms*. Assumption extensions are defined like argument extensions as follows.

Definition 3 [2,9,13]. *Let $\langle \mathcal{L}, \mathcal{R}, \mathcal{A}, \mathcal{C} \rangle$ be an ABA framework. Then $Asms$ is: admissible iff $Asms$ is conflict-free and defends all its elements; a complete assumption extension iff $Asms$ is admissible and contains all assumptions it defends; a preferred (resp. grounded) assumption extension iff it is a (subset-) maximal (resp. (subset-)minimal) complete assumption extension; a stable assumption extension iff it is is conflict-free and $Asms \cup Asms^+ = \mathcal{A}$; an ideal assumption extension iff it is a (subset-)maximal complete assumption extension that is contained in each preferred assumption extension.*

Let $Sname \in \{complete, preferred, grounded, stable, ideal\}$. It is shown that there is a one-to-one correspondence between assumption extensions and argument extensions of a given ABA $\langle \mathcal{L}, \mathcal{R}, \mathcal{A}, \mathcal{C} \rangle$ under the $Sname$ semantics as follows.

Theorem 1 [2,22,23]. *Let $\langle \mathcal{L}, \mathcal{R}, \mathcal{A}, \mathcal{C} \rangle$ be an ABA framework, AR be the set of all arguments that can be constructed using this ABA framework, and $\mathtt{Asms2Args} : 2^{\mathcal{A}} \to 2^{AR}$ and $\mathtt{Args2Asms} : 2^{AR} \to 2^{\mathcal{A}}$ be functions such that,*

$$\mathtt{Asms2Args}(Asms) = \{K \vdash c \in AR \mid K \subseteq Asms\},$$

$$\mathtt{Args2Asms}(Args) = \{\alpha \in \mathcal{A} \mid \alpha \in K \text{ for an argument } K \vdash c \in Args\}.$$

Then if $Asms \subseteq \mathcal{A}$ is a $Sname$ assumption extension, then $\mathtt{Asms2Args}(Asms)$ is a $Sname$ argument extension, and if $Args \subseteq AR$ is a $Sname$ argument extension, then $\mathtt{Args2Asms}(Args)$ is a $Sname$ assumption extension.

Proof. In [23], proofs for $Sname \in \{complete, preferred, grounded, stable\}$ are given. For $Sname = ideal$, it is also easily proved like [2,8]. □

For notational convenience, let $claim(Ag)$ stand for the claim c of an argument Ag such that $K \vdash c$, and $\mathtt{Concs}(E) = \{c \in \mathcal{L} \mid K \vdash c \in E\}$ for an extension E.

Definition 4 (ABA equipped with preferences [22,23]**).** *An assumption-based argumentation framework equipped with preferences (a p_ABA framework, or p_ABA for short) is a tuple $\langle \mathcal{L}, \mathcal{R}, \mathcal{A}, \mathcal{C}, \preceq \rangle$, where*
- *$\langle \mathcal{L}, \mathcal{R}, \mathcal{A}, \mathcal{C} \rangle$ is an ABA framework,*
- *$\preceq \subseteq \mathcal{L} \times \mathcal{L}$ is a sentence ordering called a priority relation, which is a preorder, that is, reflexive and transitive. As usual, $c' \prec c$ iff $c' \preceq c$ and $c \not\preceq c'$. For any sentences $c, c' \in \mathcal{L}$, $c' \preceq c$ (resp. $c' \prec c$) means that c is at least as preferred as c' (resp. c is strictly preferred to c').*

For a special case such that each assumption has the unique contrary sentence (i.e. $|\mathcal{C}(\alpha)|=1$ for $\forall \alpha \in \mathcal{A}$), a p_ABA framework may be represented as $\langle \mathcal{L}, \mathcal{R}, \mathcal{A}, {}^-, \preceq \rangle$ instead of $\langle \mathcal{L}, \mathcal{R}, \mathcal{A}, \mathcal{C}, \preceq \rangle$, where a total mapping ${}^-$from $\mathcal{A}$ into $\mathcal{L}$ is used instead of a total mapping $\mathcal{C}$ from $\mathcal{A}$ into $2^{\mathcal{L}} \setminus \{\emptyset\}$ like ABA.

For p_ABA $\langle \mathcal{L}, \mathcal{R}, \mathcal{A}, \mathcal{C}, \preceq \rangle$, let $\mathcal{F} = \langle \mathcal{L}, \mathcal{R}, \mathcal{A}, \mathcal{C} \rangle$ be the associated ABA and $AF_{\mathcal{F}} = (AR, attacks)$. Then $\leq \subseteq AR \times AR$ is the argument ordering over AR constructed from $\preceq$ as follows.

For any arguments $Ag_1, Ag_2 \in AR$ such that $K_1 \vdash c_1$ and $K_2 \vdash c_2$,

$$Ag_1 \leq Ag_2 \quad \text{iff} \quad c_1 \preceq c_2 \qquad \text{for } c_i = claim(Ag_i) \ (1 \leq i \leq 2)$$

Definition 5 (Preference relations $\sqsubseteq_{ex}$). *Given p_ABA $\langle \mathcal{L}, \mathcal{R}, \mathcal{A}, \mathcal{C}, \preceq \rangle$, let $\mathcal{E}$ be the set of Sname argument extensions of the AA framework $AF_{\mathcal{F}} = (AR, attacks)$ corresponding to the ABA framework $\mathcal{F} = \langle \mathcal{L}, \mathcal{R}, \mathcal{A}, \mathcal{C} \rangle$ under Sname semantics and $f : 2^{AR} \times 2^{AR} \rightarrow 2^{AR}$ be the function s.t. $f(U,V) = \{X| \; claim(X) = claim(Y) \; \text{for} \; X \in U, \; Y \in V\}$. Then $\sqsubseteq_{ex}$ over $\mathcal{E}$ (i.e. $\sqsubseteq_{ex} \subseteq \mathcal{E} \times \mathcal{E}$) is defined as follows [22,23]. For any Sname argument extensions, E_1, E_2 and E_3 from $\mathcal{E}$,*

1. *$E_1 \sqsubseteq_{ex} E_1$,*
2. *$E_1 \sqsubseteq_{ex} E_2$ if for some argument $Ag_2 \in E_2 \setminus \Delta_2$,*
   (i) there is an argument $Ag_1 \in E_1 \setminus \Delta_1$ s.t. $claim(Ag_1) \preceq claim(Ag_2)$ and,
   (ii) there is no argument $Ag_3 \in E_1 \setminus \Delta_1$ s.t. $claim(Ag_2) \prec claim(Ag_3)$,
       where $\Delta_1 = f(E_1, E_2)$ and $\Delta_2 = f(E_2, E_1)$,
3. *if $E_1 \sqsubseteq_{ex} E_2$ and $E_2 \sqsubseteq_{ex} E_3$, then $E_1 \sqsubseteq_{ex} E_3$;*

$\sqsubseteq_{ex}$ is a preorder. We write $E_1 \sqsubset_{ex} E_2$ if $E_1 \sqsubseteq_{ex} E_2$ and $E_2 \not\sqsubseteq_{ex} E_1$ as usual.

The preference relation $\sqsubseteq_{ex}$ can be also defined by using the argument ordering $\leq$ in a way that $claim(Ag_1) \preceq claim(Ag_2)$ and $claim(Ag_2) \prec claim(Ag_3)$ is replaced with $Ag_1 \leq Ag_2$ and $Ag_2 < Ag_3$ in item no. 2 of Definition 5 [22,23].

Let $Sname \in \{complete, preferred, grounded, stable, ideal\}$. The semantics of p_ABA is given by $Sname$ $\mathcal{P}$ extensions which are the maximal ones w.r.t. $\sqsubseteq_{ex}$ among $Sname$ extensions as follows.

Definition 6 ($\mathcal{P}$-extensions [22,23]). *Given a p_ABA framework $\langle \mathcal{L}, \mathcal{R}, \mathcal{A}, \mathcal{C}, \preceq \rangle$, let $\mathcal{E}$ be the set of Sname argument extensions of $AF_{\mathcal{F}} = (AR, attacks)$ corresponding to the ABA framework $\langle \mathcal{L}, \mathcal{R}, \mathcal{A}, \mathcal{C} \rangle$ under Sname semantics. Then a Sname argument extension $E \in \mathcal{E}$ is called a Sname $\mathcal{P}$-argument extension of the p_ABA framework if $E \sqsubseteq_{ex} E'$ implies $E' \sqsubseteq_{ex} E$ (with respect to $\preceq$) for any $E' \in \mathcal{E}$. In other words, E is a Sname $\mathcal{P}$-argument extension of a p_ABA iff there is no Sname argument extension $E' \in \mathcal{E}$ such that $E \sqsubset_{ex} E'$. For a Sname $\mathcal{P}$-argument extension E, Args2Asms(E) is called a Sname $\mathcal{P}$-assumption extension. Both a Sname $\mathcal{P}$-argument extension and a $\mathcal{P}$-assumption extension may be called a Sname $\mathcal{P}$ extension for short.*

3   Representing Hypotheses in ABA

In logic programming, NAF literals are used to perform non-monotonic and default reasoning, while hypotheses (i.e. abducibles, abducible facts) are used to perform abductive reasoning or hypothetical reasoning. In [14,20], it is shown that an abductive logic program (or an abductive program) can be transformed into a logic program without abducibles, where for each abducible a, a new atom a' is introduced representing the complement of a and a new pair of rules:

$$a \leftarrow not \; a', \qquad a' \leftarrow not \; a$$

is added to the program. Such transformation for hypotheses was also used to compute abductive argumentation [21].

In ABA, hypotheses are different from assumptions as abducible literals are different from NAF literals in logic programming. Hence when hypotheses (or abducible facts) are contained in the knowledge, each one, say a, can be also expressed by a new pair of rules in ABA as follows:

$$a \leftarrow \delta, \qquad a' \leftarrow \delta'$$

where a' is a newly introduced sentence representing the complement of a, while δ, δ' are newly introduced assumptions such that $\overline{\delta} = a'$, $\overline{\delta'} = a$.

Example 1. Consider the example shown in [5, Example 1] as follows:
"Zed wants to go out and two of his friends, Alice and Bob, are available. Best, Zed would take them both, but as far as he knows, Bob does not like Alice, although she does not have anything against Bob. If Zed offers to both of them at the same time, Bob may be in the awkward position to refuse Alice's company. Offering separately, Alice is up for all three going, while Bob insists on cutting Alice out. Zed may opt for the latter option. However, had Zed a preference between the two, - say Alice were a better friend of his - then he would go out with her."   □

In what follows, a (resp. b) denotes that Alice (resp. Bob) might go out with Zed, while $\neg a$ (resp. $\neg b$) denotes the negation of a (resp. b). Since Alice (resp. Bob) might go out with Zed or might not, a and b are not assumptions but hypotheses. Hence the situation about them is expressed by the following rules:

$$a \leftarrow \alpha, \qquad \neg a \leftarrow \alpha', \qquad b \leftarrow \beta, \qquad \neg b \leftarrow \beta',$$

where $\mathcal{A} = \{\alpha, \alpha', \beta, \beta'\}$, $\overline{\alpha} = \neg a$, $\overline{\alpha'} = a$, $\overline{\beta} = \neg b$ and $\overline{\beta'} = b$. The opted option such that Bob insists on cutting Alice out is expressed by $\neg a \leftarrow b$. Preferences such that best, Zed would take them both, but he prefers Alice to Bob are expressed by:

$$\{\neg a, \neg b\} \preceq \{b\} \preceq \{a\} \preceq \{a, b\} \tag{1}$$

According to [23, Definition 25], preferences between conjunctive knowledge shown above can be encoded in p_ABA $\langle \mathcal{L}, \mathcal{R}, \mathcal{A}, \bar{}, \preceq \rangle$ by introducing new rules:

$$c_1 \leftarrow a, b, \qquad c_2 \leftarrow \neg a, \neg b,$$

along with preferences:

$$c_2 \preceq b \preceq a \preceq c_1 \qquad (a, b, c_1, c_2 \in \mathcal{L} \setminus \mathcal{A}) \tag{2}$$

Then based on p_ABA consisting of $\mathcal{R} = \{\neg a \leftarrow b, \; a \leftarrow \alpha, \; \neg a \leftarrow \alpha', \; b \leftarrow \beta, \; \neg b \leftarrow \beta', \; c_1 \leftarrow a, b, \; c_2 \leftarrow \neg a, \neg b\}$, $\mathcal{A} = \{\alpha, \alpha', \beta, \beta'\}$, $\overline{\alpha} = \neg a$, $\overline{\alpha'} = a$, $\overline{\beta} = \neg b$, $\overline{\beta'} = b$ and $c_2 \preceq b \preceq a \preceq c_1$, arguments and *attacks* are constructed as follows:

- $A : \{\alpha\} \vdash a$
- $A' : \{\alpha'\} \vdash \neg a$
- $B : \{\beta\} \vdash b$
- $B' : \{\beta'\} \vdash \neg b$
- $X : \{\beta\} \vdash \neg a$
- $C_1 : \{\alpha, \beta\} \vdash c_1$
- $C_2 : \{\alpha', \beta'\} \vdash c_2$
- $\xi : \{\alpha\} \vdash \alpha$
- $\xi' : \{\alpha'\} \vdash \alpha'$
- $\eta : \{\beta\} \vdash \beta$
- $\eta' : \{\beta'\} \vdash \beta'$

$attacks = \{(A, A'), (A, \xi'), (A, C_2), (A', A), (A', \xi), (A', C_1), (X, A), (X, \xi), (X, C_1),$
$\quad (B, B'), (B, \eta'), (B, C_2), (B', B), (B', \eta), (B', X), (B', C_1)\}.$

The underlying ABA has the preferred (resp. stable) argument extensions E_i $(1 \leq i \leq 3)$ as follows:

$$
\begin{aligned}
E_1 &= \{A', B, X, \xi', \eta\}, && \text{with } \mathrm{Concs}(E_1) = \{\neg a, b, \alpha', \beta\}, \\
E_2 &= \{A, B', \xi, \eta'\}, && \text{with } \mathrm{Concs}(E_2) = \{a, \neg b, \alpha, \beta'\}, \\
E_3 &= \{A', B', C_2, \xi', \eta'\}, && \text{with } \mathrm{Concs}(E_3) = \{\neg a, \neg b, c_2, \alpha', \beta'\}.
\end{aligned}
$$

Due to $E_3 \sqsubseteq E_1 \sqsubseteq E_2$ derived from (2), E_2 is the unique preferred (resp. stable) $\mathcal{P}$-argument extension in the p_ABA. Hence against Čyras and Toni's claim [5], E_2 gives us the solution that Zed would go out with Alice and without Bob.

Remark: In [5], Čyras and Toni expressed this example by p_ABA consisting of $\mathcal{R} = \{\overline{\alpha} \leftarrow \beta\}$, $\mathcal{A} = \{\alpha, \beta\}$, $\beta \preceq \alpha$. Then they claimed that regarding arguments A for Alice and B for Bob, A for Alice cannot be obtained from the extension of p_ABA encoded by them since its underlying ABA has the unique extension $\{B\}$, where $attacks = \{(B, A)\}$ and B < A. Therefore the reason why they could not obtain the solution based on p_ABA is that they mistook the hypothetical knowledge a, b for assumptions α, β; and it is not due to the property of p_ABA.

4   Assumption-Based Argumentation Equipped with Preferences and Constraints

4.1   Problematic Knowledge Representation

As addressed in introduction, Čyras [4] claimed that p_ABA cannot solve the example called "Cakes" presented by him, whose scenario is shown as follows.

Example 2 (Cakes [4, Example 1]). There are three pieces of cakes on a table: a piece of Almond cake, a Brownie, and a piece of Cheesecake. You want to get as many cakes as possible, and the following are the rules of the game.

1. You can take cakes from the table in two 'rounds':
    (a) In the first round you can take at most two cakes;
    (b) In the second round you can take at most one cake.
2. If you take Almond cake and Cheesecake in the first round, Brownie will not be available in the second round. (Nothing is known about other possible combinations.)
3. Finally, very importantly, suppose that you prefer Brownie over Almond cake. (No other preferences.)

Which pair(s) of cakes would you choose in the first round?   □

The solution of *Cakes* is that "either a pair of Brownie cake and Cheesecake or a pair of Almond and Brownie cakes is chosen in the first round".

In [4], Čyras expressed the knowledge of *Cakes* in p_ABA consisting of inference rules $\mathcal{R} = \{\overline{b} \leftarrow a, c\}$, assumptions $\mathcal{A} = \{a, b, c\}$ and preference $a < b$, and claimed that p_ABA cannot obtain its solution since the given preference doesn't work in his p_ABA whose underlying ABA has a unique extension. On the other hand, he expressed the knowledge in $ASPIC^+$ consisting of the strict rules $\mathcal{R}_s = \{a, c \rightarrow \neg b, a, b \rightarrow \neg c, b, c \rightarrow \neg a\}$, premises $\mathcal{K}_p = \{a, b, c\}$ and preference $a < b$, and concluded that the $ASPIC^+$ cannot obtain the solution since three extensions exist under the Elitist comparison (resp. the Democratic comparison). Now recall the Prakken and Modgil's result [17,18] that ABA is a special case of $ASPIC^+$ with only strict inference rules $\mathcal{R}_s$, premises $\mathcal{K}_p$ [17] (or assumptions $\mathcal{K}_a$ [18]) and no preferences. However for *Cakes*, there is no correspondence between his $ASPIC^+$ except preferences (i.e. $\mathcal{R}_s$) and the underlying ABA of his p_ABA (i.e. $\mathcal{R}$). This indicates that Čyras' knowledge representation for *Cakes* is problematic. Thereby for *Cakes*, let us construct the p_ABA from his $ASPIC^+$ according to Prakken and Modgil's result. Then we obtain p_ABA $\langle \mathcal{L}, \mathcal{R}, \mathcal{A}, ^-, \preceq \rangle$, where $\mathcal{R} = \{\neg b \leftarrow a, c, \neg c \leftarrow a, b, \neg a \leftarrow b, c\}$, $\mathcal{A} = \{a, b, c\}$, $\overline{a} = \neg a$, $\overline{b} = \neg b$, $\overline{c} = \neg c$ and $a \preceq b$. We can construct arguments A: $\{a\} \vdash a$, B: $\{b\} \vdash b$, C: $\{c\} \vdash c$, A': $\{b, c\} \vdash \neg a$, B': $\{a, c\} \vdash \neg b$, C': $\{a, b\} \vdash \neg c$, and obtain $attacks = \{(C', C), (C', A'), (C', B'), (B', B), (B', A'), (B', C'), (A', A), (A', B'), (A', C')\}$. Its associated ABA has three extensions: $E_1 = \{C, A, B'\}$, $E_2 = \{B, C, A'\}$, $E_3 = \{A, B, C'\}$. Since $E_1 \sqsubseteq_{ex} E_2$ is derived due to $a \preceq b$, both E_2 and E_3 (resp. Args2Asms(E_2) $= \{b, c\}$, Args2Asms(E_3) $= \{a, b\}$) is obtained as the preferred and stable $\mathcal{P}$-argument extensions (resp. $\mathcal{P}$-assumption extensions). This means that the solution is obtained based on the p_ABA reconstructed from his $ASPIC^+$.

However it should be noted that the constraints no. 2 and no. 1 (b) in *Cakes* are not expressed in $\mathcal{R}_s$ of his $ASPIC^+$, while constraints no. 1 (a) and no. 1 (b) are not expressed in $\mathcal{R}$ of his p_ABA. In the following, we show that the solution of *Cakes* can be obtained from each of three different p_ABAs respectively where the knowledge of *Cakes* is expressed in three different ways.

4.2   Solving Cakes Example Based on the Semantics of P_ABA

p_ABA with the contrary function $\mathcal{C}$ gives us the solution of *Cakes* as follows.

Example 3 (Cont. Example 2). Suppose that a, b, c stand for a piece of Almond cake, a piece of Brownie, and a piece of Cheesecake respectively, while a_i (resp. b_i, c_i) ($1 \leq i \leq 2$) stands for the solution of the problem such that Almond cake (resp. Brownie cake, Cheesecake) is taken at i-th round, where $\mathcal{A} = \{a_1, b_1, c_1, a_2, b_2, c_2\}$. Moreover the symbol t_x1 (resp. t_x2) denotes the operation such that $x \in \{a, b, c\}$ is taken in the first (resp. second) round, and the symbol t_xy1 denotes the operation such that both $x \in \{a, b, c\}$ and $y \in \{a, b, c\}$ where $x \neq y$ are taken in the first round according to the rules of the game. Then the cake example is modeled in p_ABA $\mathcal{F}_{pABA} = \langle \mathcal{L}, \mathcal{R}, \mathcal{A}, \mathcal{C}, \preceq \rangle$, where

$-\ \mathcal{R} = \{t\_ab1 \leftarrow a_1, b_1, \quad t\_bc1 \leftarrow b_1, c_1, \quad t\_ca1 \leftarrow c_1, a_1, \quad t\_a1 \leftarrow a_1,$
$\qquad\qquad t\_b1 \leftarrow b_1, \quad t\_c1 \leftarrow c_1, \quad t\_a2 \leftarrow a_2, \quad t\_b2 \leftarrow b_2, \quad t\_c2 \leftarrow c_2\}$

- $\mathcal{A} = \{a_1, b_1, c_1, a_2, b_2, c_2\}$,
- $\mathcal{C}(a_1) = \{t\_a2, t\_bc1\}$, $\mathcal{C}(b_1) = \{t\_b2, t\_ca1\}$, $\mathcal{C}(c_1) = \{t\_c2, t\_ab1\}$,
  $\mathcal{C}(a_2) = \{t\_a1, t\_b2, t\_c2\}$, $\mathcal{C}(b_2) = \{t\_b1, t\_a2, t\_c2, t\_ca1\}$,
  $\mathcal{C}(c_2) = \{t\_c1, t\_a2, t\_b2\}$ and $a_i \preceq b_j$, $a_i \preceq a_i$, $b_j \preceq b_j$ $(1 \leq i, j \leq 2)$.

15 arguments are constructed in $\mathcal{F}_{\text{pABA}}$ as follows.

- $A_1 : \{a_1\} \vdash t\_a1$ • $A_2 : \{a_2\} \vdash t\_a2$ • $B_1 : \{b_1\} \vdash t\_b1$
- $B_2 : \{b_2\} \vdash t\_b2$ • $C_1 : \{c_1\} \vdash t\_c1$ • $C_2 : \{c_2\} \vdash t\_c2$
- $AB : \{a_1, b_1\} \vdash t\_ab1$ • $BC : \{b_1, c_1\} \vdash t\_bc1$ • $CA : \{c_1, a_1\} \vdash t\_ca1$
- $\alpha_1 : \{a_1\} \vdash a_1$ • $\alpha_2 : \{a_2\} \vdash a_2$ • $\beta_1 : \{b_1\} \vdash b_1$ • $\beta_2 : \{b_2\} \vdash b_2$
- $\gamma_1 : \{c_1\} \vdash c_1$ • $\gamma_2 : \{c_2\} \vdash c_2$

Then the associated ABA of $\mathcal{F}_{\text{pABA}}$ has preferred and stable extensions as follows:

$$E_1 = \{A_1, C_1, \alpha_1, \gamma_1, CA\}, \qquad \text{with } \texttt{Args2Asms}(E_1) = \{a_1, c_1\}$$
$$E_2 = \{A_2, B_1, C_1, \alpha_2, \beta_1, \gamma_1, BC\}, \quad \text{with } \texttt{Args2Asms}(E_2) = \{a_2, b_1, c_1\}$$
$$E_3 = \{A_1, B_1, C_2, \alpha_1, \beta_1, \gamma_2, AB\}, \quad \text{with } \texttt{Args2Asms}(E_3) = \{a_1, b_1, c_2\}$$

Since $E_1 \sqsubseteq_{ex} E_2$ is derived due to $\alpha_i \leq \beta_j$ or $a_i \preceq b_j$ $(1 \leq i, j \leq 2)$, both E_2 and E_3 (resp. $\{a_2, b_1, c_1\}$ and $\{a_1, b_1, c_2\}$) are obtained as preferred and stable $\mathcal{P}$-argument extensions (resp. $\mathcal{P}$-assumption extensions) in $\mathcal{F}_{\text{pABA}}$. Hence E_2 and E_3 give us solution of *Cakes* that either a pair of Brownie cake and Cheesecake or a pair of Almond and Brownie cakes is chosen in the first round.

4.3   Assumption-Based Argumentation Equipped with Preferences and Constraints

In this subsection, we present a general method to express some kind of constraints in ABA $\langle \mathcal{L}, \mathcal{R}, \mathcal{A}, ^- \rangle$ as well as p_ABA $\langle \mathcal{L}, \mathcal{R}, \mathcal{A}, ^-, \preceq \rangle$. Moreover we show that the solution of *Cakes* is also obtained by applying the method to p_ABA.

Definition 7 (Constraints). *Given an ABA framework $\langle \mathcal{L}, \mathcal{R}, \mathcal{A}, ^- \rangle$, a rule without head of the form:*

$$\leftarrow a_1, \ldots, a_m \qquad (\text{or equivalently } \leftarrow \{a_1, \ldots, a_m\})$$

is called a constraint, where $a_i \in \mathcal{A}$ $(1 \leq i \leq m)$.

In general, let $b_i \in \mathcal{L}$ such that there exists an argument $B_i \vdash b_i$ where $B_i \neq \emptyset$. Then $\leftarrow b_1, \ldots, b_m$, or equivalently $\leftarrow \bigcup_{i=1}^{m} \{b_i\}$ stands for a set of the constraints $\leftarrow \bigcup_{i=1}^{m} B_i$ obtained by replacing $\{b_i\}$ with $B_i \subseteq \mathcal{A}$ in every possible way.

Satisfaction of constraints is defined as follows.

Definition 8 (Satisfaction).

- *A set of assumptions $Asms \subseteq \mathcal{A}$ satisfies a constraint $\leftarrow a_1, \ldots, a_m$ iff $\{a_1, \ldots, a_m\} \not\subseteq Asms$ holds.*
- *A set of assumptions $Asms \subseteq \mathcal{A}$ satisfies a set of constraints C iff $\{a_1, \ldots, a_m\} \not\subseteq Asms$ holds for $\forall \leftarrow a_1, \ldots, a_m \in \mathsf{C}$.*

Definition 9 (ABA equipped with constraints). *Given an ABA framework* $\mathcal{F} = \langle \mathcal{L}, \mathcal{R}, \mathcal{A}, ^- \rangle$ *and a set of constraints* C, *an ABA framework* $\mathcal{F}_C$ *equipped with constraints is defined as*

$$\mathcal{F}_C = \langle \mathcal{L}, \mathcal{R} \cup \mathcal{R}_C, \mathcal{A}, ^- \rangle, \text{ where}$$

$$\mathcal{R}_C = \{\neg a_i \leftarrow a_1, \ldots, a_{i-1}, a_{i+1}, \ldots, a_m | \leftarrow a_1, \ldots, a_m \in C, \ \bar{a}_i = \neg a_i, 1 \le i \le m\}.$$

Constrains defined in Definition 7 help users in expressing knowledge. And furthermore they are useful to eliminate undesirable *Sname* extensions in ABA just as integrity constraints are used to eliminate undesirable answer sets in answer set programming [19,20]. (Details are omitted due to limitations of space.)

The following properties hold for $\mathcal{F}_c$.

Theorem 2. *$\mathcal{A}sms \subseteq \mathcal{A}$ is conflict-free in $\mathcal{F}_C = \langle \mathcal{L}, \mathcal{R} \cup \mathcal{R}_C, \mathcal{A}, ^- \rangle$ if and only if $\mathcal{A}sms$ is conflict-free in $\mathcal{F} = \langle \mathcal{L}, \mathcal{R}, \mathcal{A}, ^- \rangle$ and satisfies a set of constraints* C, *namely $\{a_1, \ldots, a_m\} \not\subseteq \mathcal{A}sms$ for $\forall \leftarrow a_1, \ldots, a_m \in C$.*

Proof. See Appendix. □

Theorem 3. *A conflict-free set $\mathcal{A}sms$ in $\mathcal{F}_C = \langle \mathcal{L}, \mathcal{R} \cup \mathcal{R}_C, \mathcal{A}, ^- \rangle$ satisfies a set of constraints* C.

Proof. See Appendix. □

Proposition 1. *In $\mathcal{F}_C = \langle \mathcal{L}, \mathcal{R} \cup \mathcal{R}_C, \mathcal{A}, ^- \rangle$, every Sname assumption extension $\mathcal{A}sms \subseteq \mathcal{A}$ satisfies a set of constraints* C.

Proof. This is obviously proved based on Theorem 3. □

Example 4 (Cont. Example 2). Let us express the knowledge of *Cakes* except preferences (i.e. the game rule no. 3) in ABA. Suppose that a_i (resp. b_i, c_i) $(1 \le i \le 2)$ stands for the solution of the problem such that Almond cake (resp. Brownie cake, Cheesecake) is taken at i-th round. Let $\mathcal{A} = \{a_1, b_1, c_1, a_2, b_2, c_2\}$ and $\bar{a}_i = \neg a_i$, $\bar{b}_i = \neg b_i$, $\bar{c}_i = \neg c_i$, where $\neg a_i$ (resp. $\neg b_i$, $\neg c_i$) denotes the negation of a_i (resp. b_i, c_i). Then

- the game rule no. 1 (a) is expressed by four constraints as follows:
   $\leftarrow a_1, b_1, c_1 \quad \leftarrow a_1, a_2 \quad \leftarrow b_1, b_2 \quad \leftarrow c_1, c_2$
- The game rule no. 1 (b) is expressed by three constraints as follows:
   $\leftarrow a_2, b_2 \quad \leftarrow b_2, c_2 \quad \leftarrow c_2, a_2$
- The game rule no. 2 is expressed by the rule as follows:
   $\neg b_2 \leftarrow c_1, a_1$

Cakes except preferences is modeled in ABA $\mathcal{F}_C = \langle \mathcal{L}, \mathcal{R} \cup \mathcal{R}_C, \mathcal{A}, {}^- \rangle$, where

- $\mathcal{R} = \{\neg b_2 \leftarrow c_1, a_1\}$
- $\mathcal{R}_C = \{\neg c_1 \leftarrow a_1, b_1, \quad \neg a_1 \leftarrow b_1, c_1, \quad \neg b_1 \leftarrow c_1, a_1, \quad \neg a_2 \leftarrow a_1,$
  $\neg a_1 \leftarrow a_2, \quad \neg b_2 \leftarrow b_1, \quad \neg b_1 \leftarrow b_2, \quad \neg c_2 \leftarrow c_1, \quad \neg c_1 \leftarrow c_2,$
  $\neg b_2 \leftarrow a_2, \quad \neg a_2 \leftarrow b_2, \quad \neg c_2 \leftarrow b_2, \quad \neg b_2 \leftarrow c_2, \quad \neg a_2 \leftarrow c_2, \quad \neg c_2 \leftarrow a_2\}$
- $\mathcal{A} = \{a_1, b_1, c_1, a_2, b_2, c_2\}$, and $\bar{a}_i = \neg a_i$, $\bar{b}_i = \neg b_i$, $\bar{c}_i = \neg c_i$ $(i = 1, 2)$.

22 arguments are constructed in $\mathcal{F}_C$ as follows.

- $A_1 : \{a_1\} \vdash \neg a_2$ $\bullet$ $A_2 : \{a_2\} \vdash \neg a_1$ $\bullet$ $B_1 : \{b_1\} \vdash \neg b_2$
- $B_2 : \{b_2\} \vdash \neg b_1$ $\bullet$ $C_1 : \{c_1\} \vdash \neg c_2$ $\bullet$ $C_2 : \{c_2\} \vdash \neg c_1$
- $AB : \{a_1, b_1\} \vdash \neg c_1$ $\bullet$ $BC : \{b_1, c_1\} \vdash \neg a_1$ $\bullet$ $CA : \{c_1, a_1\} \vdash \neg b_1$
- $\alpha_1 : \{a_1\} \vdash a_1$ $\bullet$ $\alpha_2 : \{a_2\} \vdash a_2$ $\bullet$ $\beta_1 : \{b_1\} \vdash b_1$ $\bullet$ $\beta_2 : \{b_2\} \vdash b_2$
- $\gamma_1 : \{c_1\} \vdash c_1$ $\bullet$ $\gamma_2 : \{c_2\} \vdash c_2$ $\bullet$ $CA2 : \{c_1, a_1\} \vdash \neg b_2$
- $P_1 : \{a_2\} \vdash \neg b_2$ $\bullet$ $P_2 : \{a_2\} \vdash \neg c_2$ $\bullet$ $Q_1 : \{b_2\} \vdash \neg a_2$
- $Q_2 : \{b_2\} \vdash \neg c_2$ $\bullet$ $R_1 : \{c_2\} \vdash \neg b_2$ $\bullet$ $R_2 : \{c_2\} \vdash \neg a_2$

Thus $\mathcal{F}_C$ has three preferred and stable argument extensions E_i (resp. assumption extensions $asms_i = \mathtt{Args2Asms}(E_i)$ satisfying constraints) $(1 \le i \le 3)$ as follows:

$$E_1 = \{A_1, C_1, \alpha_1, \gamma_1, CA, CA2\}, \qquad \text{with } asms_1 = \{a_1, c_1\}$$
$$E_2 = \{A_2, B_1, C_1, \alpha_2, \beta_1, \gamma_1, BC, P_1, P_2\}, \quad \text{with } asms_2 = \{a_2, b_1, c_1\}$$
$$E_3 = \{A_1, B_1, C_2, \alpha_1, \beta_1, \gamma_2, AB, R_1, R_2\}, \quad \text{with } asms_3 = \{a_1, b_1, c_2\}$$

Note that E_1 (resp. E_2, E_3) as well as $asms_1$ (resp. $asms_2, asms_3$) denote that a pair of Almond and Cheesecake cakes (resp. a pair of Brownie cake and Cheesecake, a pair of Almond and Brownie cakes) is chosen in the first round.

Definition 10 (ABA equipped with preferences and constraints). *Given an ABA framework $\mathcal{F} = \langle \mathcal{L}, \mathcal{R}, \mathcal{A}, {}^- \rangle$, a set of constraints C and a sentence ordering $\preceq \subseteq \mathcal{L} \times \mathcal{L}$, an ABA framework $\mathcal{F}_{PC}$ equipped with preferences $\preceq$ and constraints C is defined as*

$$\mathcal{F}_{PC} = \langle \mathcal{L}, \mathcal{R} \cup \mathcal{R}_C, \mathcal{A}, {}^-, \preceq \rangle, \text{ where}$$

$\mathcal{R}_C = \{\neg a_i \leftarrow a_1, \ldots, a_{i-1}, a_{i+1}, \ldots, a_m \mid \leftarrow a_1, \ldots, a_m \in \mathsf{C}, \ \bar{a}_i = \neg a_i, 1 \le i \le m\}$.

Example 5 (Cont. Example 4). By incorporating the preferences given in *Cakes*, i.e. $a_i \preceq b_j$ $(1 \le i, j \le 2)$ into the ABA $\mathcal{F}_C$ shown in Example 4, we obtain p_ABA $\mathcal{F}_{PC} = \langle \mathcal{L}, \mathcal{R} \cup \mathcal{R}_C, \mathcal{A}, {}^-, \preceq \rangle$, where $\mathcal{F}_C = \langle \mathcal{L}, \mathcal{R} \cup \mathcal{R}_C, \mathcal{A}, {}^- \rangle$ has three preferred (resp. stable) argument extensions E_1, E_2, E_3 as shown in Example 4.

Since $E_1 \sqsubseteq_{ex} E_2$ is derived due to $\alpha_i \le \beta_j$ or $a_i \preceq b_j$ $(1 \le i, j \le 2)$, both E_2 and E_3 (resp. $\{a_2, b_1, c_1\}$ and $\{a_1, b_1, c_2\}$) are obtained as preferred and stable $\mathcal{P}$-argument extensions (resp. $\mathcal{P}$-assumption extensions) in $\mathcal{F}_{PC}$. Accordingly we again obtain the solution of *Cakes*.

4.4 Prioritized Logic Programming As Argumentation Equipped with Preferences

In [23], we showed that p_ABA can capture Sakama and Inoue's preferred answer sets of a prioritized logic program (PLP) [19]. Hence we show that the PLP expressing the knowledge of *Cakes* as well as the p_ABA instantiated with the PLP enable us to obtain its solution based on the respective semantics as follows.

Example 6 (Cont. Example 2). Let a_i (resp. b_i, c_i) $(1 \leq i \leq 2)$ be a propositional atom which means that Almond cake (resp. Brownie cake, Cheesecake) is taken at i-th round. Then

- The game rule no. 1 (a) is expressed by rules of a normal logic program as follows:
  $b_1 \leftarrow not\ a_1, \quad c_1 \leftarrow not\ a_1, \quad c_1 \leftarrow not\ b_1, \quad a_1 \leftarrow not\ b_1, \quad a_1 \leftarrow not\ c_1,$
  $b_1 \leftarrow not\ c_1$
- The game rules no. 1 (b) and no. 2 are expressed by rules as follows:
  $a_2 \leftarrow not\ a_1, \quad b_2 \leftarrow not\ a_1,\ not\ c_1, \quad c_2 \leftarrow not\ c_1$
- The game rule no. 3 is expressed by $a_i \preceq b_j$ $(1 \leq i, j \leq 2)$.

These lead to PLP (P, Φ) as follows:
$P = \{b_1 \leftarrow not\ a_1, \quad c_1 \leftarrow not\ a_1, \quad c_1 \leftarrow not\ b_1, \quad a_1 \leftarrow not\ b_1, \quad a_1 \leftarrow not\ c_1,$
$\qquad b_1 \leftarrow not\ c_1, \quad a_2 \leftarrow not\ a_1, \quad b_2 \leftarrow not\ a_1, not\ c_1, \quad c_2 \leftarrow not\ c_1\}$
$\Phi = \{(a_i, b_j) | 1 \leq i, j \leq 2)\}.$
P has three answer sets (i.e. stable models) S_i $(1 \leq i \leq 3)$ as follows:
$$S_1 = \{a_1, c_1\}, \quad S_2 = \{b_1, c_1, a_2\}, \quad S_3 = \{a_1, b_1, c_2\}.$$
$S_1 \sqsubseteq_{as} S_2$ is derived due to Φ^* [19]. Hence S_2 and S_3 corresponding to $asms_2$ and $asms_3$ in Example 5 are obtained as preferred answer sets of the PLP (P, Φ).

On the other hand, according to [23, Corollary 2] (i.e. [22, Theorem 2]), we can construct the p_ABA $\mathcal{F}_{\mathrm{PLP}} = \langle \mathcal{L}_P, P, \mathcal{A}, ^-, \Phi^* \rangle$ instantiated with this PLP, where $\mathcal{A} = HB_{not} = \{not\ p \mid p \in HB_P\}$ for $HB_P = \{a_1, b_1, c_1, a_2, b_2, c_2\}$, $\mathcal{L}_P = HB_P \cup HB_{not}$, $\overline{not\ p} = p$ for $p \in HB_P$ and Φ^* is the reflexive and transitive closure of $\Phi = \{(a_i, b_j) | 1 \leq i, j \leq 2)\}$. Then as indicated by [23, Corollary 2], we can obtain two stable $\mathcal{P}$-argument extensions E_2 and E_3 of $\mathcal{F}_{\mathrm{PLP}}$ with
$$\mathrm{Concs}(E_2) = \{b_1, c_1, a_2, not\ a_1, not\ b_2, not\ c_2\},$$
$$\mathrm{Concs}(E_3) = \{a_1, b_1, c_2, not\ c_1, not\ a_2, not\ b_2\}$$

corresponding to S_2 and S_3. As a result, we again obtain the solution of *Cakes*.

5 Related Work

Čyras and Toni [5] proposed an ABA$^+$ framework: $\langle \mathcal{L}, \mathcal{R}, \mathcal{A}, ^-, \leqslant \rangle$, where $\langle \mathcal{L}, \mathcal{R}, \mathcal{A}, ^- \rangle$ is an ABA framework and $\leqslant$ is a preorder on $\mathcal{A}$. They newly introduced $<$-attacks $\subseteq \mathcal{P}(\mathcal{A}) \times \mathcal{P}(\mathcal{A})$ consisting of two types depending on $\leqslant$. Its semantics is given by a $<$-*Sname* extension $E \subseteq \mathcal{A}$ as defined by replacing the notion of *attacks* with <-*attacks* in standard ABA. Compared ABA$^+$ with p_ABA, the form of ABA$^+$ is a special case of p_ABA $\langle \mathcal{L}, \mathcal{R}, \mathcal{A}, \mathcal{C}, \preceq \rangle$ as far as its underlying ABA is flat because $\leqslant \subseteq \mathcal{A} \times \mathcal{A}$ is a subset of $\preceq \subseteq \mathcal{L} \times \mathcal{L}$ and $^-$ is a special case of $\mathcal{C}$ s.t. $|\mathcal{C}(\alpha)|=1$ for $\forall \alpha \in \mathcal{A}$. Hence none of preferences over hypotheses (e.g. (1), (2)), preferences over goals $G \subseteq \mathcal{L} \setminus \mathcal{A}$ which are often required in decision-making and practical reasoning, and preferences on (defeasible) rules for epistemic reasoning can be expressed in ABA$^+$. In contrast, p_ABA has a mechanism to represent and reason with all of these preferences in its framework [22,23]. Therefore p_ABA has far much more expressive power than ABA$^+$.

Prakken proposed $ASPIC^+$ for structured argumentation with preferences [17,18]. Comparison between $ASPIC^+$ and p_ABA is discussed in detail in [23].

Dung [11,12] proposed a new approach of structured argumentation with priorities for $ASPIC^+$-type argumentation formalisms. A novel attack relation (assignment) called *regular* [12] (resp. *normal* [11]) which takes account of priorities over defeasible rules is defined without constructing argument orderings.

Coste-Marquis et al. proposed constrained argumentation frameworks [3] where constraints on admissible arguments in abstract argumentation are considered. Instead in our approach, constraints on assumptions expressed by rules without head can be treated in ABA and p_ABA as shown in Subsect. 4.3.

6   Discussion and Conclusion

Čyras and Toni claimed that p_ABA cannot solve two examples (i.e. [5, Example 1], [4, Example 1]) since the given preferences do not work in their p_As whose underlying ABAs have a unique extension, and proposed ABA$^+$. Against their claim, it is shown in Sects. 3 and 4 that p_ABAs in which we encoded the respective knowledge give us solutions of them without any difficulties. In conclusion, they could not obtain the solutions of these examples not due to the property of p_ABA but due to their incorrect knowledge encodings in p_ABA.

In what follows, we show that the semantics of ABA$^+$ has a serious problem as to treating preferences. As addressed in Example 1, Čyras and Toni presented ABA$^+$ consisting of $\mathcal{R} = \{\overline{\alpha} \leftarrow \beta\}$, $\mathcal{A} = \{\alpha, \beta\}$, $\beta \leqslant \alpha$. Based on its semantics, $\{\alpha\}$ is selected as a unique <-complete extension due to $\beta \leqslant \alpha$ though $\langle \mathcal{L}, \mathcal{R}, \mathcal{A}, ^- \rangle$ has a unique extension $\{\beta\}$ [5]. Now consider a real world problem as follows.

Example 7. Usually the famous legal principle: "innocent until proven guilty" is applied to the suspect under no evidence.

It is expressed by ABA $\mathcal{F}$ consisting of $\mathcal{R} = \{innocent \leftarrow not\ guilty\}$, $\mathcal{A} = \{not\ innocent,\ not\ guilty\}$, $\overline{not\ innocent} = innocent$, $\overline{not\ guilty} = guilty$.

Hereupon suppose that someone prefers "not innocent" to "not guilty" (since he prefers "guilty" to "innocent") though there is no evidence proving the suspect is guilty. Obviously under this situation, the human legal reasoning result in court is "innocent" regardless of any preference for innocence or guilt because there is no evidence of "guilty". On the other hand, given the preference s.t.

"*not guilty* $\leqslant$ *not innocent*" along with $\mathcal{F}$, ABA$^+$ has {*not innocent*} as its unique extension, whereas p_ABA has the unique $\mathcal{P}$-argument extension E with $\text{Concs}(E) = \{innocent, not\ guilty\}$ regardless of preferences since the underlying ABA has the unique extension E. Thus p_ABA yields "innocent", while ABA$^+$ yields "not innocent". Hence thanks to the property discussed in the introduction, p_ABA gives us the human legal reasoning result, i.e. "innocent, whereas ABA$^+$ cannot perform such typical non-monotonic reasoning with preferences.

In contrast, according to the correspondence between $ASPIC^+$ and ABA [17,18], this ABA$^+$ can be faithfully mapped to $ASPIC^+$ consisting of $\mathcal{R}_s = \{\beta \rightarrow \neg\alpha\}$, $\mathcal{K}_p = \{\alpha, \beta\}$ (or $\mathcal{K}_a = \{\alpha, \beta\}$), $\overline{\alpha} = \neg\alpha$ and $\beta \leq' \alpha$. As for the case $\mathcal{K}_p = \{\alpha, \beta\}$, $defeat = \emptyset$ is derived. Then the mapped $ASPIC^+$ has a unique complete extension E^+ with $\text{concs}(E^+) = \{\alpha, \beta, \neg\alpha\}$, which is not directly consistent [17,18]. Similarly this ABA$^+$ may be also mapped to Dung's rule-based system [11,12], say $\mathcal{R}_{\text{dung}}$, which consists of $d_0: \Rightarrow \beta$ $d_1: \Rightarrow \alpha\ r : \beta \rightarrow \neg\alpha$ and $d_0 \prec d_1$. Surprisingly when we replace the symbol β (resp. α) with a (resp. b), $\mathcal{R}_{\text{dung}}$ coincides with the rule-based system shown in [12, Example 7] in which no regular attack relation assignment exists as discussed by Dung [12].

We are the first to show *prioritized logic programming as argumentation equipped with preferences* (cf. Subsect. 4.4) [23] as Dung showed logic programming as argumentation [7]. Nevertheless our future work is to explore the other types of the semantics for p_ABA so that it can capture the other types of prioritized logic programming such as Brewka and Eiter's preferred answer sets, Delgrande, Schaub and Tompits' preferred answer sets and so on [6].

Acknowledgments. This work was supported by KAKENHI (Grant-in-Aid for Scientific Research(S)17H06103).

Appendix

Proof of Theorem 2 ($\Longleftarrow$). Let $\mathcal{A}sms$ be conflict-free in $\mathcal{F}$ and satisfies $\{a_1, \ldots, a_m\} \not\subseteq \mathcal{A}sms$ for $\forall \leftarrow a_1, \ldots, a_m \in \mathsf{C}$. Since $\mathcal{A}sms$ is conflict-free in $\mathcal{F}$, $\forall \alpha \in \mathcal{A}sms$ is not attacked by arguments constructed by using only rules from $\mathcal{R}$ in $\mathcal{F}_{\mathsf{C}}$. Now suppose that $\mathcal{A}sms$ is not conflict-free in $\mathcal{F}_{\mathsf{C}}$. Then for some $\{a_1, \ldots, a_{k-1}, a_{k+1}, \ldots, a_m\} \subseteq \mathcal{A}sms$, there exists an argument $\{a_1, \ldots, a_{k-1}, a_{k+1}, \ldots, a_m\} \vdash \neg a_k$ constructed by the rule from $\mathcal{R}_{\mathcal{C}}$ that attacks $\mathcal{A}sms$, which denotes that a_k ($1 \leq k \leq m$) is in $\mathcal{A}sms$. Thus $\{a_1, \ldots, a_{k-1}, a_k, a_{k+1}, \ldots, a_m\} \subseteq \mathcal{A}sms$ is derived. This contradicts that $\mathcal{A}sms$ satisfies $\{a_1, \ldots, a_m\} \not\subseteq \mathcal{A}sms$ for $\forall \leftarrow a_1, \ldots, a_m \in \mathsf{C}$. Thus it is derived that $\mathcal{A}sms$ is conflict-free in $\mathcal{F}_{\mathsf{C}}$.

($\Longrightarrow$) Let $\mathcal{A}sms$ be conflict-free in $\mathcal{F}_{\mathsf{C}} = \langle \mathcal{L}, \mathcal{R} \cup \mathcal{R}_{\mathsf{C}}, \mathcal{A}, ^- \rangle$. Then $\mathcal{A}sms$ is also conflict-free in $\mathcal{F} = \langle \mathcal{L}, \mathcal{R}, \mathcal{A}, ^- \rangle$ due to $\mathcal{R} \subseteq (\mathcal{R} \cup \mathcal{R}_{\mathsf{C}})$. Now suppose that for this $\mathcal{A}sms$ which is conflict-free in $\mathcal{F}$, there exists some constraint $\exists \leftarrow a_1, \ldots, a_m \in \mathsf{C}$ which satisfies $\{a_1, \ldots, a_m\} \subseteq \mathcal{A}sms$. Then in $\mathcal{F}_{\mathsf{C}}$, there exists some argument $\{a_1, \ldots, a_{k-1}, a_{k+1}, \ldots, a_m\} \vdash \neg a_k$ built from $\mathcal{R}_{\mathsf{C}}$ that attacks $\mathcal{A}sms$. This contradicts that $\mathcal{A}sms$ is conflict-free in $\mathcal{F}_{\mathsf{C}}$. Hence it holds $\{a_1, \ldots, a_m\} \not\subseteq \mathcal{A}sms$ for $\forall \leftarrow a_1, \ldots, a_m \in \mathsf{C}$ w.r.t. $\mathcal{A}sms$ which is the conflict-free in $\mathcal{F}$. $\qquad\square$

Proof of Theorem 3. Suppose that in $\mathcal{F}_C$, there is some conflict-free set $\mathcal{A}sms \subseteq \mathcal{A}$ which does not satisfy some constraint in C, that is, $\{a_1, \ldots, a_m\} \subseteq \mathcal{A}sms$ holds for $\exists \leftarrow a_1, \ldots, a_m \in C$. Then using rules from $\mathcal{R}_C$, it is possible to construct the argument $\{a_1, \ldots, a_{k-1}, a_{k+1}, \ldots, a_m\} \vdash \neg a_k$ that attacks $a_k \in \mathcal{A}sms$ $(1 \leq k \leq m)$. This contradicts that $\mathcal{A}sms$ is conflict-free. □

References

1. Bondarenko, A., Dung, P.M., Kowalski, R.A., Toni, F.: An abstract, argumentation-theoretic approach to default reasoning. Artif. Intell. **93**, 63–101 (1997)
2. Caminada, M., Sá, S., Alcântara, J., Dvořák, W.: On the difference between assumption-based argumentation and abstract argumentation. In: BNAIC 2013, pp. 25–32 (2013). IFCoLog J. Logic Appl. **2**(1), 15–34 (2015)
3. Coste-Marquis, S., Devred, C., Marquis, P.: Constrained argumentation frameworks. In: Proceedings of KR 2006, pp. 112–122 (2006)
4. Čyras, K.: Argumentation-based reasoning with preferences. In: Proceedings of PAAMS 2016, pp. 199–210 (2016)
5. Čyras, K., Toni, F.: ABA$^+$: assumption-based argumentation with preferences. In: Proceedings of KR 2016, pp. 553–556 (2016)
6. Delgrande, J.P., Schaub, T., Tompits, H., Wang, K.: A classification and survey of preference handling approaches in nonmonotonic reasoning. J. Comput. Intell. **20**(2), 308–334 (2004). Wiley
7. Dung, P.M.: On the acceptability of arguments and its fundamental role in nonmonotonic reasoning, logic programming and n-person games. Artif. Intell. **77**, 321–357 (1995)
8. Dung, P.M., Mancarella, P., Toni, F.: Computing ideal sceptical argumentation. Artif. Intell. **171**(10–15), 642–674 (2007)
9. Dung, P.M., Kowalski, R.A., Toni, F.: Assumption-based argumentation. In: Simari G., Rahwan I. (eds.) Argumentation in Artificial Intelligence, pp. 199–218, Springer, Boston (2009). doi:10.1007/978-0-387-98197-0_10
10. Dung, P.M., Thang, P.M.: Closure and consistency rationalities in logic-based argumentation. In: Balduccini, M., Son, T.C. (eds.) Logic Programming, Knowledge Representation, and Nonmonotonic Reasoning. LNCS, vol. 6565, pp. 33–43. Springer, Heidelberg (2011). doi:10.1007/978-3-642-20832-4_3
11. Dung, P.M.: An axiomatic analysis of structured argumentation with priorities. Artif. Intell. **231**, 107–150 (2016)
12. Dung, P.M.: A Canonical semantics for structured argumentation with priorities. In: Proceedings of COMMA-2016, pp. 263–274 (2016)
13. Fan, X., Toni, F.: A general framework for sound assumption-based argumentation dialogues. Artif. Intell. **216**, 20–54 (2014)
14. Kakas, A.C., Kowalski, R.A., Toni, F.: Abductive logic programming. J. Logic Comput. **2**(6), 719–770 (1992)
15. Lifschitz, V.: Computing circumscription. In: Proceedings of IJCAI 1985, pp. 121–127 (1985)
16. McCarthy, J.: Applications of circumscription to formalizing commonsense knowledge. Artif. Intell. **28**, 89–116 (1986)
17. Modgil, S.J., Prakken, H.: The $ASPIC+$ framework for structured argumentation: a tutorial. Argum. Comput. **5**, 31–62 (2014)

18. Prakken, H.: An abstract framework for argumentation with structured arguments. Argum. Comput. **1**, 93–124 (2010)
19. Sakama, C., Inoue, K.: Prioritized logic programming and its application to commonsense reasoning. Artif. Intell. **123**, 185–222 (2000)
20. Sakama, C., Inoue, K.: An abductive framework for computing knowledge base updates. TPLP **3**(6), 671–713 (2003)
21. Wakaki, T., Nitta, K., Sawamura, H.: Computing abductive argumentation in answer set programming. In: McBurney, P., Rahwan, I., Parsons, S., Maudet, N. (eds.) ArgMAS 2009. LNCS, vol. 6057, pp. 195–215. Springer, Heidelberg (2010). doi:10.1007/978-3-642-12805-9_12
22. Wakaki, T.: Assumption-based argumentation equipped with preferences. In: Dam, H.K., Pitt, J., Xu, Y., Governatori, G., Ito, T. (eds.) PRIMA 2014. LNCS, vol. 8861, pp. 116–132. Springer, Cham (2014). doi:10.1007/978-3-319-13191-7_10
23. Wakaki, T.: Assumption-based argumentation equipped with preferences and its application to decision-making, practical reasoning, and epistemic reasoning. J. Comput. Intell,. doi:10.1111/coin.12111, Accepted 17 October 2016. Published early view article (online version) on 20 March 2017. Wiley (2017)

Semantic Change and Extension Enforcement in Abstract Argumentation

Sylvie Doutre[1] and Jean-Guy Mailly[2(✉)]

[1] IRIT, Université Toulouse 1 Capitole, Toulouse, France
doutre@irit.fr
[2] LIPADE, Université Paris Descartes, Paris, France
jean-guy.mailly@parisdescartes.fr

Abstract. Change in argumentation frameworks has been widely studied in the recent years. Most of the existing works on this topic are concerned with change of the structure of the argumentation graph (addition or removal of arguments and attacks), or change of the outcome of the framework (acceptance statuses of arguments). Change on the acceptability semantics that is used in the framework has not received much attention so far. Such a change can be motivated by different reasons, especially it is a way to change the outcome of the framework. In this paper, it is shown how semantic change can be used as a way to reach a goal about acceptance statuses in a situation of extension enforcement.

1 Introduction

Recently, the dynamics of argumentation frameworks (AFs) has received much attention [5–7,10–12,15–17,19,25]. Essentially, we can distinguish between two kinds of approaches for change in AFs: some of them deal with the structure of the AF (the set of arguments and the attack relation), while the other ones deal with the statuses of arguments (extensions, labellings, skeptically accepted arguments,...). However, a third component of the argumentation process has received almost no attention: the semantics which links the structure of the AF and the arguments statuses. Even if some approaches allow to change the semantics during the process (see for instance [6]), it is not explained *why* the semantics has to change, nor *how* the new semantics is selected. In this paper, we study these questions by focusing on extension-based semantics, that is, semantics that, when applied to an AF, produce a set of acceptable sets of arguments called extensions.

Two main reasons may motivate a change of the semantics. First, it may be required by some practical considerations. Indeed, an issue with some argumentation semantics is their high complexity. This theoretical complexity is not a practical problem if we consider some particular classes of AFs, or if the size of the AF is not too large. However, if at some point, for an agent, using some high complexity semantics is the best choice for some reason – for instance, because it guarantees the existence of at least one extension, or a number of extensions smaller than with another potential semantics –, the evolution of the AF may

S. Moral et al. (Eds.): SUM 2017, LNAI 10564, pp. 194–207, 2017.
DOI: 10.1007/978-3-319-67582-4_14

justify a change of the semantics. If the agent interacts with other agents in the context of a debate for instance, arguments and attacks may be added to the AF. Such additions increase the size of the AF, and they may cause the AF to leave the structural class it belongs to; this may make the computation of the extensions, and of related decision problems, not efficient anymore. A change of semantics may then be suitable.

A second reason that may motivate a change of the semantics, is as an alternative way to enforce some constraint on the acceptance statuses of arguments, or on sets of arguments. Actually, there may be limitations in given applications, which prevent to modify the attack relation and to modify the set of arguments (*e.g.* the debate the arguments and the attacks come from has ended; nothing can be added any longer). Then, if the agent has to enforce a constraint about acceptance statuses, the only component which may be modified is the semantics (that is, the way to reason about the AF). In fact, whether or not a change of the structure of the AF is possible, we show that a change of semantics can be a way to reach this goal with less change on the structure of the AF.

Main Contributions

1. We give a unified abstract framework to describe change of AFs, which encompasses all existing approaches for modifying AFs. This allows to use the same tools to analyze and extend these different approaches.
2. We extend existing work on the characteristics of extension enforcement [5], *i.e.* we provide new results about the minimal change to make on an AF to ensure that a set of arguments is (included in) an extension, w.r.t. a specific semantics.
3. We study the success rate of semantic change for extension enforcement, *i.e.* the percentage of AFs for which the result is better (w.r.t. minimal change on the AF structure) when semantic change is used. This contribution relies on the abstract framework defined in 1., and benefits from the new characteristics given in 2.

Organization of the Paper. Section 2 presents background notions about abstract argumentation. Section 3 proposes a very general way to define change in argumentation frameworks, which encompasses all existing approaches. In Sect. 4, we show how semantic change can be used to enforce an acceptability constraint in an argumentation framework. Section 5 describes our experimental analysis of the semantic change success rate. The last section concludes the paper and describes some research tracks for future work.

2   Background Notions

[22] considers argumentation as the study of relations between arguments, without taking into account the origin of arguments or their internal structure. In this context, an argumentation framework (AF) is a directed graph $\langle A, R \rangle$ where the nodes in A are the *arguments* and the edges in R represent *attacks* between arguments. We consider only finite AFs, *i.e.* the set of arguments A is finite. $(a_i, a_j) \in$

R means that a_i attacks a_j; a_i is called an *attacker* of a_j. An argument a_i (resp. a set of arguments S) *defends* an argument a_j against its attacker a_k if a_i (resp. some argument in S) attacks a_k. The *range* of a set of arguments S w.r.t. R, denoted S_R^+, is the subset of A which contains S and the arguments attacked by S; formally $S_R^+ = S \cup \{a_j \mid \exists a_i \in S \text{ s.t. } (a_i, a_j) \in R\}$. Different methods allow to evaluate the arguments. A common approach is to compute *extensions*, which are sets of jointly acceptable arguments. Different semantics have been defined, which yield different kinds of extensions [2,22].

Definition 1. Let $F = \langle A, R \rangle$ be an AF. A set $S \subseteq A$ is

- conflict-free w.r.t. F if $\nexists a_i, a_j \in S$ s.t. $(a_i, a_j) \in R$;
- admissible w.r.t. F if S is conflict-free and S defends each $a_i \in S$;
- a naive extension of F if S is a maximal conflict-free set (w.r.t. $\subseteq$);
- a complete extension of F if S is admissible and S contains all the arguments that it defends;
- a preferred extension of F if S is a maximal complete extension (w.r.t. $\subseteq$);
- a stable extension of F if S is conflict-free and $S_R^+ = A$;
- a grounded extension of F if S is a minimal complete extension (w.r.t. $\subseteq$);

As shortcuts, we write respectively $cf, ad, na, co, pr, st, gr$ for these semantics. For each semantics σ, the σ-extensions of F are denoted $Ext_\sigma(F)$.

We introduce the notion of defense function[1] of a set of arguments in an AF.

Definition 2. Given an AF $F = \langle A, R \rangle$ and a set of arguments $E \subseteq A$, the *defense function of E in F* is the mapping from E and F to the set of arguments $f(E, F)$ defined by:

$$f(E, F) = \{a \in A \mid E \text{ defends } a \text{ against all its attackers}\}$$

Example 1. Let us consider the argumentation framework F_1 given at Fig. 1, and let us illustrate some of the semantics.
$Ext_{ad} = \{\emptyset, \{a_1\}, \{a_4\}, \{a_4, a_6\}, \{a_1, a_3\}, \{a_1, a_4\}, \{a_1, a_4, a_6\}\}$, $Ext_{st}(F) = \{\{a_1, a_4, a_6\}\}$, $Ext_{pr}(F) = \{\{a_1, a_3\}, \{a_1, a_4, a_6\}\}$, $Ext_{co}(F) = \{\{a_1\}, \{a_1, a_3\}, \{a_1, a_4, a_6\}\}$, $Ext_{gr}(F) = \{\{a_1\}\}$.

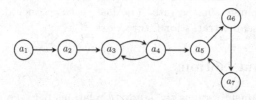

Fig. 1. The AF F_1

---

[1] This function is called *characteristic function* by [22]. We call it defense function to avoid confusion with the characteristics from [5].

Let us introduce a way to measure the difference between argumentation semantics. This distance between semantics has been proposed by [20]. It relies upon the relationships between the sets of extensions that the semantics produce.

Definition 3. Let $\Sigma = \{\sigma_1, \ldots, \sigma_n\}$ be a set of semantics, the *extension inclusion graph* of Σ is defined by $Inc(\Sigma) = \langle \Sigma, D \rangle$ with $D \subseteq \Sigma \times \Sigma$ such that $(\sigma_i, \sigma_j) \in D$ if and only if

- for each AF F, $Ext_{\sigma_i}(F) \subseteq Ext_{\sigma_j}(F)$;
- there is no $\sigma_k \in \Sigma$ ($k \neq i, k \neq j$) such that $Ext_{\sigma_i}(F) \subseteq Ext_{\sigma_k}(F)$ and $Ext_{\sigma_k}(F) \subseteq Ext_{\sigma_j}(F)$.

Given $\sigma_i, \sigma_j \in \Sigma$, the *$\Sigma$-inclusion difference measure* between semantics is the length of the shortest non-oriented path between σ_i and σ_j in $Inc(\Sigma)$, denoted $\delta_{Inc,\Sigma}(\sigma_i, \sigma_j)$.

Example 2. Figure 2 describes the extension inclusion graph of $\Sigma = \{cf, ad, na, st, pr, co, gr\}$. We observe, for instance, that $\delta_{Inc,\Sigma}(st, ad) = 3$, $\delta_{Inc,\Sigma}(pr, gr) = 2$, and $\delta_{Inc,\Sigma}(co, pr) = 1$.

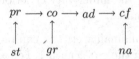

Fig. 2. Extension inclusion graph $Inc(\Sigma)$

3   Abstracting Change in Argumentation

We propose here an abstract definition which encompasses all methods for change in argumentation into a global family.

Definition 4. A *change operator* is a mapping χ from a multiset of AFs $\mathcal{F} = \{\!\{F_1, \ldots, F_n\}\!\}$, a formula φ from a logical language and a semantics σ, to a multiset $\mathcal{F}' = \{\!\{F'_1, \ldots, F'_k\}\!\}$ and a semantics σ'. Formally,

$$\chi(\mathcal{F}, \varphi, \sigma) = (\mathcal{F}', \sigma')$$

Most of existing operations on change in argumentation consider a single AF in the input and the output, which are obviously special cases of multiset. It is similar for approaches which consider a set as the outcome. [18] considers a profile of AFs as the input, which can be equivalently defined as a multiset since the order of the AFs in the tuple is not considered. Except [6], existing works do not consider semantic change, which means that $\sigma' = \sigma$ for these approaches. The formula represents a piece of information which is at the origin of the change (for instance in a context of belief revision [15,16] or update [19,25]). More generally, it is a constraint which has to be satisfied by the result of the operation, like an integrity constraint in a belief merging context [18]. The language of the formula is not the same depending the approach (*e.g.* each of [15, 16, 19, 25] has its own language). Some approaches also do not use directly a formula from a logical language, but can be mapped to a formula from a given language. For instance, adding or removing attacks and arguments [10, 12] are equivalent to formulae from the language defined in [16]. Similarly, sets of arguments considered for extension enforcement [5, 6, 17] are special cases of the formulae defined in [15, 18].

Among these approaches, some of them consider some notion of minimality, like minimal change on the attack relation [5, 16, 19, 25], minimal change on the acceptance statuses of arguments [15, 16, 18], or minimal cardinality [15, 18]. We can give a general definition of minimality in the change process.

Definition 5. A *minimality criterion* is a mapping from a tuple $\langle \mathcal{F}, \varphi, \sigma, \mathcal{F}', \sigma' \rangle$ to a tuple of positive real numbers $d(\langle \mathcal{F}, \varphi, \sigma, \mathcal{F}', \sigma' \rangle)$.
Given two such tuples t_1, t_2, we define $t_1 < t_2$ if the ith element of t_1 is smaller than the ith element of t_2, when i is the smallest index such that t_1 and t_2 are different.

Given a multiset of AFs $\mathcal{F} = \{\!\{F_1, \ldots, F_n\}\!\}$, a formula φ and a semantics σ, a change operator χ satisfies the minimality criterion d iff $\chi(\mathcal{F}, \varphi, \sigma) = (\mathcal{F}', \sigma')$ and $d(\langle \mathcal{F}, \varphi, \sigma, \mathcal{F}', \sigma' \rangle)$ is minimal.

Obviously, the simplest minimality criteria can be defined with a single number, so $d(\langle \mathcal{F}, \varphi, \sigma, \mathcal{F}', \sigma' \rangle)$ is a tuple of length 1. For instance, we instantiate this definition with extension enforcement operators [5, 6, 17].

Definition 6. Given an AF $F = \langle A, R \rangle$ and a set of arguments $E \subseteq A$, a *strict* (resp. *non-strict*) *enforcement operator* is a change operator which maps $\mathcal{F} = \{\!\{F\}\!\}$, a formula $\varphi_E = \bigwedge_{a_i \in E} a_i$ and a semantics σ to $\mathcal{F}' = \{\!\{F'\}\!\}$ and σ' such that $E \in Ext_{\sigma'}(F')$ (resp. $\exists \epsilon \in Ext_{\sigma'}(F')$ with $E \subseteq \epsilon$).
An enforcement is *minimal* iff if satisfies the minimality criterion

$$d(\langle \mathcal{F}, \varphi, \sigma, \mathcal{F}', \sigma' \rangle) = \langle d_H(\mathcal{F}, \mathcal{F}') \rangle$$

where d_H is the Hamming distance between graphs[2] [3].

We say that F' is an enforcement of E in F. We use $\varphi_E = \bigwedge_{a_i \in E} a_i$ to specify that the set E is the enforcement request; this is reminiscent of the logical encodings used in [17, 26].

Some change operators use more complex minimality criteria, which combine m simple criteria. In this case, we can represent it with a m-length tuple; this is the case of *e.g.* [15, 16, 18].

---

[2] The Hamming distance between two graphs $F_1 = \langle A_1, R_1 \rangle$ and $F_2 = \langle A_2, R_2 \rangle$ is the cardinality of the symmetric difference between R_1 and R_2; in other words, in the present case, it is the number of attacks that it is necessary to add/remove from one graph to get the other.

[3] Since here $\mathcal{F}, \mathcal{F}'$ are singletons, the Hamming distance between graphs can be directly used. For other kinds of change operators, it should be generalized to multisets.

4 Extension Enforcement and Semantic Change

In this section, we study how semantic change can be useful for extension enforcement. We first recall the definition of the five existing enforcement approaches. Then we show on intuitive examples that changing the semantics can permit to enforce an extension with fewer change on the structure (or even *without* any structural change). Finally, we extend Baumann's study on minimal change depending on the semantics, and we define a more general class of enforcement operators which reach our goal: perform extension enforcement with minimal structural change *by semantic change*.

4.1 Extension Enforcement Operators

In the first work on extension enforcement [6], it is considered that everything which appears in the current AF cannot be changed. The authorized changes are the addition of arguments, and possibly of attacks concerning at least one new argument. This kind of change is called a *normal expansion*. Special cases of normal expansion are called *strong expansion* and *weak expansion*. A strong expansion (resp. weak expansion) is an expansion which adds only strong arguments (resp. weak arguments), which are arguments that cannot be attacked by (resp. cannot attack) the previous arguments.

Definition 7. Let F, F' be two AFs such that F' is a strict (resp. non-strict) enforcement of a set of arguments E in F.

- If F' is a normal expansion of F, then the change from F to F' is a *strict* (resp. *non-strict*) *normal enforcement*.
- If F' is a strong expansion of F, then the change from F to F' is a *strict* (resp. *non-strict*) *strong enforcement*.
- If F' is a weak expansion of F, then the change from F to F' is a *strict* (resp. *non-strict*) *weak enforcement*.

Then, [17] considers new approaches which, on the opposite, question the attack relation between existing arguments. Two operators are proposed.

Definition 8. Let $F = \langle A, R \rangle, F' = \langle A', R' \rangle$ be two AFs such that F' is a strict (resp. non-strict) enforcement of the set of arguments E in F.

- If $A = A'$ and $R \neq R'$, then the change from F to F' is a *strict* (resp. *non-strict*) *argument-fixed enforcement*.
- If $A \subseteq A'$, then the change from F to F' is a *strict* (resp. *non-strict*) *general enforcement*.

In all these approaches, it is considered that

- either the semantics does not change in the enforcement;
- or the new semantics is given as a parameter of the operator: it is not specified why the semantics should change, nor why this particular semantics should be the new one.

We use $Nor_x, Str_x, Weak_x, Fix_x$ and Gen_x to denote these enforcement methods, with $x \in \{s, ns\}$ corresponding to strict and non-strict.

4.2   Minimal Structural Change Through Semantic Change

Example 3. Let us consider again the AF F_1 given at Fig. 1. We want to enforce the set $E = \{a_1, a_3\}$ as an extension. We consider that the agent is currently using the stable semantics. Obviously, structural change is required if the agent does not change the semantics. But we have seen previously that E is already an extension of F if we consider, for instance, the preferred or the complete semantics. So if the agent considers a change of semantics, the enforcement can be realized without any change on the structure.

Of course, in some situations, only switching the semantics may not be sufficient to reach the goal, if none of the possible semantics leads to build extensions which are consistent with this goal. In this case, and even if structural change is permitted, then the semantic change can still be a means to minimize the structural change required to reach the goal. Indeed, even if structural changes are permitted (or required), it can be costly for the agents to perform such changes. Such modifications of the set of arguments and of the set of attacks may then have to be limited.

The minimal change problem for extension enforcement has already been studied in [5], for a subset of the possible enforcement approaches. First, it only considers some particular target semantics (stable, preferred, complete, admissible). Also, the argument-fixed enforcement operators is not considered. Finally, only non-strict enforcement is characterized. For each pair of these semantics and enforcement operators, the minimal number of changes (addition or removal of attacks) to reach an enforcement is called the *characteristic*. This characteristic is a natural number when the enforcement is possible; $+\infty$ means that the enforcement is impossible under the given semantics.

We continue this study of characteristics and we give here some results for argument-fixed enforcement. We first need to introduce some notations.

Definition 9. Given an AF $F = \langle A, R \rangle$, and $X \subseteq A$,

- $R_\downarrow(F, X) = R \cap (X \times X)$ for any $X \subseteq A$;
- $na(F, X) = \{a_i \in A \setminus X \mid \forall a_j \in X, (a_i, a_j) \notin R \text{ and } (a_j, a_j) \notin R\}$
- $ad(F, X) = \{a_i \in A \setminus X \mid \exists a_j \in X, (a_i, a_j) \in R \text{ and } \forall a_j \in X, (a_j, a_i) \notin R\}$
- $st(F, X) = \{a_i \in A \setminus X \mid \forall a_j \in X, (a_j, a_i) \notin R\}$.

Proposition 1. Let $F = \langle A, R \rangle$ be an AF, and $E \subseteq A$. The characteristic of strict argument-fixed enforcement for $\sigma \in \{cf, ad, st, co, pr, na\}$ is defined by the function $V_{\sigma, Fix_s}^F(E)$:

$$V_{cf, Fix_s}^F(E) = |R_\downarrow(F, E)|$$
$$V_{na, Fix_s}^F(E) = |R_\downarrow(F, E)| + |na(F, E)|$$
$$V_{ad, Fix_s}^F(E) = |R_\downarrow(F, E)| + |ad(F, E)|$$
$$V_{st, Fix_s}^F(E) = |R_\downarrow(F, E)| + |st(F, E)|$$
$$V_{co, Fix_s}^F(E) = \min\{|R'\Delta R| + |R_\downarrow(F', E)| \mid f(E, F') = E, F' = \langle A, R' \rangle\}$$
$$V_{pr, Fix_s}^F(E) = \min\{|R'\Delta R| + |R_\downarrow(F', E)| \mid E \subseteq f(E, F'), \forall E \subset E' \subseteq A,$$
$$E' \not\subseteq f(E', F'), F' = \langle A, R' \rangle\}$$

We observe that these results are in line with the complexity results from [26]. Indeed, these characteristics suggest polynomial-time algorithm to compute the minimal enforcement of E under cf, na, ad and st semantics. Obtaining a better formulation for the other characteristics is still challenging.

Proposition 2. *Let $F = \langle A, R \rangle$ be an AF, and $E \subseteq A$. The characteristic of non-strict argument-fixed enforcement for $\sigma \in \{cf, ad, st, co, pr, na\}$ is defined by the function $V_{\sigma,Fix_{ns}}^F(E)$:*

$$V_{na,Fix_{ns}}^F(E) = V_{cf,Fix_{ns}}^F(E) = |R_\downarrow(F, E)|$$
$$V_{ad,Fix_{ns}}^F(E) = \min(\{|R_\downarrow(F, E')| + |ad(F, E')| \mid E \subseteq E' \subseteq A\})$$
$$V_{st,Fix_{ns}}^F(E) = \min(\{|R_\downarrow(F, E')| + |st(F, E')| \mid E \subseteq E' \subseteq A\})$$
$$V_{pr,Fix_{ns}}^F(E) = V_{co,Fix_{ns}}^F(E) = V_{ad,Fix_{ns}}^F(E)$$

We notice that these results are reminiscent of the characteristics for general enforcement [5].

Observation 1. For $Op \in \{Nor, Str, Weak\}$, the characteristic is trivial for conflict-free and naive semantics: either the set E is conflict-free, then the characteristic is 0; or E is not conflict-free, then the characteristic is $+\infty$.

Now, we generalize the definition of enforcement operators to take into account semantic change.

Definition 10. Let $F = \langle A, R \rangle$ be an AF, σ a semantics, Σ be a set of semantics, and $E \subseteq A$. Given $Op \in \{Nor, Str, Weak, Fix, Gen\}$ and $x = s$ (resp. $x = ns$), the minimal change enforcement of E in F w.r.t. Op_x is defined as $\chi(\{\!\{F\}\!\}, \varphi_E, \sigma) = (\{\!\{F'\}\!\}, \sigma')$ with $\sigma' \in \Sigma$, such that $E \in Ext_{\sigma'}(F')$ (resp. $\exists \epsilon \in Ext_{\sigma'}(F')$ s.t. $E \subseteq \epsilon$), and the criterion $\langle V_{\sigma,Op_x}^F(E), \delta_{Inc,\Sigma}(\sigma, \sigma') \rangle$ is satisfied.

This means that contrary to previous works on extension enforcement, the target semantics is not a parameter of the enforcement operator. It is chosen to guarantee that:

- the characteristic (*i.e.* the structural change) is minimal;
- in the case when several semantics have the same characteristic, the chosen one should minimize the semantic change.

Example 4. Let us come back to the AF F_1 described at Fig. 1. We want to enforce the set $E = \{a_1, a_3\}$ as an extension, with $\sigma = st$ the semantics currently used by the agent. E is not a stable extension, neither the grounded extension or a naive extension. However, it is a preferred, complete, admissible and conflict-free extension. This means that

- for every $\sigma' \in \{pr, co, ad, cf\}$, $V_{\sigma,Op_x}^F(E) = 0$ for every Op_x;
- for every $\sigma' \in \{st, gr, na\}$, $V_{\sigma,Op_x}^F(E) > 0$ for every Op_x.

This guarantees that the result of the enforcement (whatever the operator Op_x) is the AF F_1 itself, with one of the semantics $\{pr, co, ad, cf\}$. We observe that

$\delta_{Inc,\Sigma}(st,pr) = 1$, $\delta_{Inc,\Sigma}(st,co) = 2$, $\delta_{Inc,\Sigma}(st,ad) = 3$ and $\delta_{Inc,\Sigma}(st,cf) = 4$, so the new semantics is the preferred semantics. Formally, the result of enforcing E in F_1 is

$$Op_x(\{\{F_1\}\}, \bigwedge_{a_i \in E} a_i, st) = (\{\{F_1\}\}, pr)$$

We use here $\delta_{Inc,\Sigma}$ to illustrate our approach, but other difference measures between semantics could be used to define minimal semantic change. The inclusion graph that we use here is a particular case of relation graph as defined in [20]. Some other interesting notions of relation graphs could be used to define distances between semantics, like intertranslatability graphs [23] or skepticism relations [3]. [20] also mentions other approaches, based on the properties satisfied by the semantics, or based on the actual set of extensions of an AF w.r.t. the different semantics. This offers a wide range of possibilities to define minimal semantic change.

Observation 2. Our approach cannot give a worse result, w.r.t. structural change, than the classical enforcement approaches (by "classical", we mean approaches without semantic change, or with a given target semantics). Moreover, we can identify some basic cases for which our approach is sure to give a better result than classical approaches. For instance, as illustrated by Example 4, when the set E to be enforced is not a σ-extension of the considered AF F (with σ the current semantics), but E is known to be a σ'-extension of F, with σ' one of the possible alternative semantics. In this situation, it is guaranteed that enforcing E in F with our semantic change-based approach is possible without any structural change, while classical approaches do not permit this.

5   Empirical Study

In this section, we present an empirical study of the success of semantic change for extension enforcement. We have computed the result of some enforcement requests for a large set of AFs (using the strict argument-fixed enforcement approach), w.r.t. different semantics ($\Sigma = \{ad, st, co, na\}$), and for each pair $(\sigma_1, \sigma_2) \in \Sigma \times \Sigma$, we have compared $V^F_{\sigma_1, Fix_s}$ and $V^F_{\sigma_2, Fix_s}$. When $V^F_{\sigma_1, Fix_s}$ is significantly higher than $V^F_{\sigma_2, Fix_s}$ for a given AF F, this means that semantic change is relevant for this AF, w.r.t. this pair of semantics and enforcement operator. Indeed, in this case, changing the semantics from σ_1 to σ_2 allows to reach one's goal (enforcing a set of arguments E) with a lower cost (w.r.t. change of the graph). In the following subsections, we first present in detail our experimental protocol, then we provide an analysis of our results.

5.1   Protocol

We have used the AFs and enforcement requests from [26], which are available online. They provide AFs with different size of arguments $|A| \in \{50, 100, 150, 200, 250, 300\}$. The AFs are generated following the Erdös-Rényi model [24].

For $p \in \{0.05, 0.1, 0.2, 0.3\}$, each pair $(a_i, a_j) \in A \times A$ has a probability p to belong to the attack relation R. For each $|A|$ and each p, five AFs have been generated. Finally, for each AF, five sets of arguments $E \subset A$ have been randomly generated for each $|E|/|A| \in \{0.05, 0.1, 0.2, 0.3\}$. This means that for each $|A|$, 400 enforcement problem instances $(F = \langle A, R \rangle, E \subset A)$ have been generated.

For all these enforcement requests, we have computed the result of the argument-fixed strict enforcement for $\sigma \in \{na, ad, stb, co\}$. Enforcement under the naive semantics has been done through a software that we have developed in Java. For the other semantics, we have used Pakota, the enforcement solver provided by [26].[4]

The experiments have been done on a 64 bits Ubuntu 16.04 system, equipped with 8 Gio of RAM and a CPU Intel Core i5 with 3.20 GHz. The time limit was set to 10 min.

5.2   Analysis of the Results

Figure 3 presents our results for a subset of the instances, namely the AFs with $|A| = 50$ and the associated enforcement requests $E \subset A$. We only present the results for this class of AFs for a matter of readability. Indeed, for the other values of $|A|$, the results appear to be remarkably similar. Also, we only present 3 of the 6 possible combinations of semantics: (ad, st) (represented by $\triangle$), (ad, na) (represented by $\times$) and (co, st) (represented by $\square$). For each of these combinations (σ_1, σ_2), each point represents an instance (*i.e.* a pair $(F = \langle A, R \rangle, E \subseteq A)$), such that the point abscissa is the minimal change to enforce E in F w.r.t. σ_1, and its ordinate is the value for the enforcement w.r.t. σ_2. So, a point situated under the diagonal represents an instance for which the minimal change to perform the enforcement w.r.t. σ_1 is higher than the minimal change to perform the enforcement w.r.t. σ_2 (and vice-versa for the points above the diagonal). We observe that semantic change actually brings something to extension enforcement. Indeed for most of the instances, the points are situated far from the diagonal, which means that they can benefit from semantic change. On the opposite, the points situated on the diagonal represent instances for which semantic change does not improve the "quality" of enforcement.

Let us mention the fact that we have similar results for the pairs of semantics (st, na) and (co, na). Only the pair (ad, co) results in points close to the diagonal for a high proportion of the instances. For $|A| \in \{100, 150, 200, 250, 300\}$, we observe similar results. Let us still mention that the higher the value of $|A|$, the higher the proportion of instances with a ratio close to 1. But even for $|A| = 300$, there is still a significant amount of instances which benefit from semantic change (*i.e.* instance with a significative difference between $V^F_{\sigma_1, Fix_s}$

---

[4] Pakota also provides the possibility to execute enforcement under the preferred semantics. Because of the higher complexity of the enforcement problem under the preferred semantics, our experiment has encountered a high number of timeouts. For this reason, we exclude preferred semantics of our empirical analysis for now.

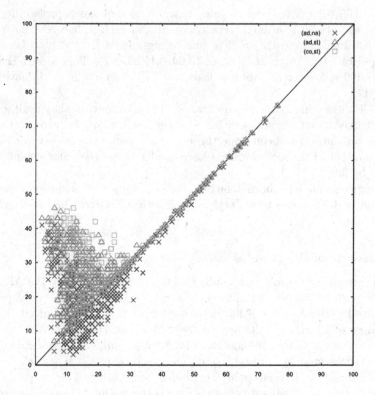

Fig. 3. Comparing minimal change depending on the semantics, for AFs with 50 arguments

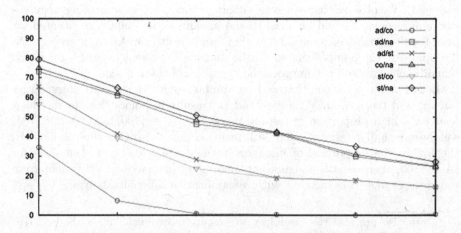

Fig. 4. Success percentage for different semantic change situations

and $V^F_{\sigma_2,Fix_s}$).[5] Figure 4 presents, for each $|A|$ and each pair of semantics, the percentage of instances for which the ratio $V^F_{\sigma_1,Fix_s}/V^F_{\sigma_2,Fix_s}$ is smaller than 0.9 or greater than 1.1, *i.e.* the percentage of instance for which semantic change is successful.

6 Conclusion

This paper addresses particular aspects of the *dynamics of argumentation frameworks*. Most of the existing approaches in this domain concern either a change of the structure of an AF, or a change on the acceptance statuses of arguments (both being related). We argue that it makes sense in some applications to permit the agent to change her reasoning process, which is represented by the acceptance semantics. This change can be motivated by a need of computational efficiency (requirement of a lower complexity), or by properties to be enforced on the set of extensions (e.g. requirement of some arguments to be accepted), with a minimal change of the graph structure.

Such a change in the reasoning process is related to what is discussed in [8,9]. Roughly speaking, the idea is that an agent can be able to use different reasoning processes, such as one which is harder to compute and probably more rational, and another one which is easier to compute and based on some less rational concepts (for instance, there can be some bias due to the agent's perception of the source of information). Semantic change in argumentation can be conducted by similar ideas.

In this paper, we have first defined a very abstract framework to describe change in argumentation. This framework is useful to describe and analyze the different approaches for argumentation dynamics with the same tools. Then we have instantiated this framework for a specific (and well-studied) family of change operators for AFs: extension enforcement. We show that allowing an agent to change the semantics when performing an extension enforcement is useful in some situations, since this semantic change cannot provide a worse result (w.r.t. the number of modifications of the graph) than "classical" enforcement, and can even provide better results. This claim is grounded on the new study of characteristics. We have conducted an experimental study which shows the impact of semantic change on a large set of instances.

Several interesting questions have arisen from this work. Naturally, we want to complete our study of characteristics and our experiments with more semantics. The ideal semantics [21], the prudent semantics [14] or the SCC-recursive semantics [4] are good candidates. Determining the missing characteristics (for instance, the characteristics of the strict versions of operators studied by Baumann in [5]) is also an important future work. Since the difference between semantics is here evaluated in the setting of the well-known extension-based semantics, the extension of our approach to labelling-based semantics seems to

---

[5] A complete description and analysis of our experiments, including the instances, the enforcement system, and the curves for every value of $|A|$ and every pair (σ_1, σ_2) is available online: http://www.math-info.univ-paris5.fr/~jmailly/expSemChange.

be quite immediate. On the contrary, semantic change for ranking-based semantics [1] requires a deeper investigation. Regarding our experimental study, we want to explore more in depth the impact of the different parameters on the semantic change, for instance the size of the AF, the size of the set of arguments to be enforced, and the probability of attacks. We have considered here the Erdös-Rényi model, which captures an interesting graph structure, and which has already been the object of other studies [26]. We plan to conduct similar studies with other families of graphs [13] to determine whether the impact of semantic change is different for these families. Also, we want to extend extension enforcement systems to benefit from the study of characteristics: computing the characteristics for a list of enforcement operators and a list of semantics, we can choose the best operator and semantics to enforce a set with minimal change of the graph.

Finally, we want to study the impact of semantic change on some operations which return a set [15,18]. In these papers, the outcome of the operation represents some uncertain result (intuitively, the set is interpreted as a "disjunction" of AFs). Our goal is to determine whether semantic change can help to reduce the cardinality of the set (i.e. reduce the uncertainty of the result).

References

1. Amgoud, L., Ben-Naim, J.: Ranking-based semantics for argumentation frameworks. In: Liu, W., Subrahmanian, V.S., Wijsen, J. (eds.) SUM 2013. LNCS (LNAI), vol. 8078, pp. 134–147. Springer, Heidelberg (2013). doi:10.1007/978-3-642-40381-1_11

2. Baroni, P., Caminada, M., Giacomin, M.: An introduction to argumentation semantics. Knowl. Eng. Rev. **26**, 365–410 (2011)

3. Baroni, P., Giacomin, M.: Skepticism relations for comparing argumentation semantics. Int. J. Approx. Reason. **50**(6), 854–866 (2009)

4. Baroni, P., Giacomin, M., Guida, G.: SCC-recursiveness: a general schema for argumentation semantics. Artif. Intell. **168**, 162–210 (2005)

5. Baumann, R.: What does it take to enforce an argument? minimal change in abstract argumentation. In: Proceedings of ECAI 2012, pp. 127–132 (2012)

6. Baumann, R., Brewka, G.: Expanding argumentation frameworks: enforcing and monotonicity results. In: Proceedings of COMMA 2010, pp. 75–86 (2010)

7. Bisquert, P., Cayrol, C., Saint-Cyr, F.D., Lagasquie-Schiex, M.-C.: Change in argumentation systems: exploring the interest of removing an argument. In: Benferhat, S., Grant, J. (eds.) SUM 2011. LNCS (LNAI), vol. 6929, pp. 275–288. Springer, Heidelberg (2011). doi:10.1007/978-3-642-23963-2_22

8. Bisquert, P., Croitoru, M., Dupin de Saint-Cyr, F.: Four ways to evaluate arguments according to agent engagement. In: Guo, Y., Friston, K., Aldo, F., Hill, S., Peng, H. (eds.) BIH 2015. LNCS (LNAI), vol. 9250, pp. 445–456. Springer, Cham (2015). doi:10.1007/978-3-319-23344-4_43

9. Bisquert, P., Croitoru, M., Saint-Cyr, F.D.: Towards a dual process cognitive model for argument evaluation. In: Beierle, C., Dekhtyar, A. (eds.) SUM 2015. LNCS (LNAI), vol. 9310, pp. 298–313. Springer, Cham (2015). doi:10.1007/978-3-319-23540-0_20

10. Boella, G., Kaci, S., van der Torre, L.: Dynamics in argumentation with single extensions: attack refinement and the grounded extension (Extended Version). In: McBurney, P., Rahwan, I., Parsons, S., Maudet, N. (eds.) ArgMAS 2009. LNCS, vol. 6057, pp. 150–159. Springer, Heidelberg (2010). doi:10.1007/978-3-642-12805-9_9

11. Booth, R., Kaci, S., Rienstra, T., van der Torre, L.: A logical theory about dynamics in abstract argumentation. In: Liu, W., Subrahmanian, V.S., Wijsen, J. (eds.) SUM 2013. LNCS, vol. 8078, pp. 148–161. Springer, Heidelberg (2013). doi:10.1007/978-3-642-40381-1_12

12. Cayrol, C., Dupin de Saint-Cyr, F., Lagasquie-Schiex, M.C.: Change in abstract argumentation frameworks: adding an argument. J. Artif. Intell. Res. **38**, 49–84 (2010)

13. Cerutti, F., Giacomin, M., Vallati, M.: Generating structured argumentation frameworks: AFBenchGen2. In: Proceedings of COMMA 2016, pp. 467–468 (2016)

14. Coste-Marquis, S., Devred, C., Marquis, P.: Prudent semantics for argumentation frameworks. In: Proceedings of ICTAI 2005, pp. 568–572 (2005)

15. Coste-Marquis, S., Konieczny, S., Mailly, J.G., Marquis, P.: On the revision of argumentation systems: minimal change of arguments statuses. In: Proceedings of KR 2014 (2014)

16. Coste-Marquis, S., Konieczny, S., Mailly, J.-G., Marquis, P.: A translation-based approach for revision of argumentation frameworks. In: Fermé, E., Leite, J. (eds.) JELIA 2014. LNCS, vol. 8761, pp. 397–411. Springer, Cham (2014). doi:10.1007/978-3-319-11558-0_28

17. Coste-Marquis, S., Konieczny, S., Mailly, J.G., Marquis, P.: Extension enforcement in abstract argumentation as an optimization problem. In: Proceedings of IJCAI 2015 (2015)

18. Delobelle, J., Haret, A., Konieczny, S., Mailly, J.G., Rossit, J., Woltran, S.: Merging of abstract argumentation frameworks. In: Proceedings of KR 2016, pp. 33–42 (2016)

19. Doutre, S., Herzig, A., Perrussel, L.: A dynamic logic framework for abstract argumentation. In: Proceedings of KR 2014, pp. 62–71 (2014)

20. Doutre, S., Mailly, J.G.: Quantifying the difference between argumentation semantics. In: Proceedings of COMMA 2016 (2016)

21. Dung, P.M., Mancarella, P., Toni, F.: Computing ideal sceptical argumentation. Artif. Intell. **171**(10–15), 642–674 (2007)

22. Dung, P.M.: On the acceptability of arguments and its fundamental role in nonmonotonic reasoning, logic programming, and n-person games. Artif. Intell. **77**(2), 321–357 (1995)

23. Dvořák, W., Spanring, C.: Comparing the expressiveness of argumentation semantics. In: Proceedings of COMMA 2012, pp. 261–272 (2012)

24. Erdös, P., Rényi, A.: On random graphs I. Publicationes Mathematicae, pp. 290–297 (1959)

25. de Saint-Cyr, F.D., Bisquert, P., Cayrol, C., Lagasquie-Schiex, M.: Argumentation update in YALLA (yet another logic language for argumentation). Int. J. Approx. Reason. **75**, 57–92 (2016)

26. Wallner, J.P., Niskanen, A., Järvisalo, M.: Complexity results and algorithms for extension enforcement in abstract argumentation. In: Proceedings of AAAI 2016, pp. 1088–1094 (2016)

Measuring Disagreement
in Argumentation Graphs

Leila Amgoud[⊠] and Jonathan Ben-Naim

IRIT – CNRS, Toulouse, France
amgoud@irit.fr

Abstract. The aim of this paper is to evaluate to what extent an *argumentation graph* (a set of *arguments* and *attacks* between them) is conflicting. For that purpose, we introduce the novel notion of *disagreement measure* as well as a set of principles that such a measure should satisfy. We propose some intuitive measures and show that they fail to satisfy some of the principles. Then, we come up with a more discriminating measure which satisfies them all. Finally, we relate some measures to those quantifying inconsistency in knowledge bases.

1 Introduction

An *argumentation framework* is a graph whose nodes are *arguments* and edges are *attacks* between pairs of arguments. The graph may be extracted from a knowledge base (e.g., in [1]), or from a dialogue between agents (e.g., [2]), etc. Whatever the source of the graph, the presence of attacks means existence of *disagreements* and three questions raise quite naturally: (1) how to *model* disagreements? (2) what is their *amount*? and how to *solve* them? Works in computational argumentation focused mainly on questions (1) and (3). They assume that disagreements in an argumentation graph are nothing more than the attacks of the graph, and represent them either as abstract relations between pairs of arguments (e.g., in [3]), or as logical relations between arguments (e.g., undercut [4], rebuttal [5]). An impressive amount of work has also been done on *solving* disagreements using the so-called *acceptability semantics*, of which extension semantics [3] are some examples.

The question of measuring the *amount* of disagreements in an argumentation graph has never been studied. Consider the six argumentation graphs below. There is no method in the literature that evaluates the amount of disagreement in each of them.

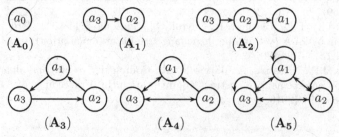

© Springer International Publishing AG 2017
S. Moral et al. (Eds.): SUM 2017, LNAI 10564, pp. 208–222, 2017.
DOI: 10.1007/978-3-319-67582-4_15

Existing semantics solve disagreements without bothering about their amount. Nevertheless, quantifying disagreement is relevant for various purposes. Namely, in the context of inconsistency handling, an argumentation graph is extracted from a (inconsistent) knowledge base (KB). Quantifying disagreements in the graph allows evaluating at what extent the knowledge base is inconsistent. Motivated by important applications like software specifications, quantifying inconsistency in a KB has become a hot topic the last six years (e.g. [6–11]). Since the number of applications of argumentation grows steadily, it is important that the approach has its own tools for answering various needs of the applications including measuring inconsistency. Hence, argumentation not only handles inconsistency in KBs, but it will also be used for measuring inconsistency in those KBs.

The contribution of this paper is fourfold: First, we introduce the novel notion of *disagreement measure*, that is a real-valued function that assigns to each argumentation framework a value representing its amount of disagreements. Second, we propose principles that a disagreement measure should satisfy. These principles serve as theoretical criteria for judging and comparing disagreement measures. Third, we define five intuitive disagreement measures, one of which satisfies all the principles. Finally, we make a first bridge with works on inconsistency measures by showing that some of our measures return the same result as an existing inconsistency measure.

The paper is structured as follows: Sect. 2 recalls basic concepts. Section 3 defines disagreement measures and proposes principles they should satisfy. Section 4 introduces six measures and discusses their properties. Section 5 shows how some measures evaluate inconsistency in KBs.

2   Basic Concepts

An argumentation framework (or argumentation graph) $\mathbf{A}$ is a graph consisting of a *non-empty* set $\mathcal{A}$ of nodes representing *arguments*, and a set $\mathcal{R}$ of *links* (or edges). A link $r \in \mathcal{R}$ is an *ordered* pair (a_1, a_2) representing a *direct attack* from argument a_1 to argument a_2 $(a_1, a_2 \in \mathcal{A})$. Throughout the paper, we write $\mathbf{A} = \langle \mathcal{A}, \mathcal{R} \rangle$. A graph is finite iff its set of arguments is finite.

- A *path* from argument a to argument b in $\mathbf{A}$ is a sequence $\langle a_0, \ldots, a_n \rangle$ of arguments of $\mathcal{A}$ such that $a_0 = a$, $a_n = b$, for any $0 \leq i < n$, $(a_i, a_{i+1}) \in \mathcal{R}$, and for all $i \neq j$, $a_i \neq a_j$. We say that b is *reachable* from a when there is a path from a to b. If $n = 2m + 1$ and $m > 0$, then the pair (a_0, a_n) is an *indirect attack* on a_n.
- A *cycle* is a path $\langle a_0, \ldots, a_n \rangle$ such that $(a_n, a_0) \in \mathcal{R}$. It is *elementary* iff there does not exist a cycle $\langle b_0, \ldots, b_m \rangle$ such that $\{b_0, \ldots, b_m\} \subset \{a_0, \ldots, a_n\}$. A graph is *acyclic* if it does not contain any cycle.
- An argumentation graph $\mathbf{A} = \langle \{a_0, \ldots, a_n\}, \mathcal{R} \rangle$ is a cycle iff $\mathcal{R} = \{(a_i, a_{i+1}) \mid 0 \leq i < n\} \cup \{(a_n, a_0)\}$. The graph $\mathbf{A} = \langle \{a_0, \ldots, a_n\}, \mathcal{R} \rangle$ is a *simple path* iff $\mathcal{R} = \{(a_i, a_{i+1}) \mid 0 \leq i < n\}$.

- The *length* of a path (resp. cycle) $\langle a_0, \ldots, a_n \rangle$ is n (resp. $n+1$).
- An *isomorphism* from $\mathbf{A} = \langle \mathcal{A}, \mathcal{R} \rangle$ to $\mathbf{A}' = \langle \mathcal{A}', \mathcal{R}' \rangle$ is a bijective function f from $\mathbf{A}$ to $\mathbf{A}'$ such that $\forall a, b \in \mathcal{A}, (a, b) \in \mathcal{R}$ iff $(f(a), f(b)) \in \mathcal{R}'$.

Notations: We denote by `Args` an infinite set of all possible arguments, and by $\mathcal{U}$ the universe of finite argumentation graphs built from `Args`. For any argumentation graph $\mathbf{A} = \langle \mathcal{A}, \mathcal{R} \rangle$, $\texttt{Arg}(\mathbf{A}) = \mathcal{A}$, $\texttt{Att}(\mathbf{A}) = \mathcal{R}$, and $\texttt{SelfAtt}(\mathbf{A}) = \{a \in \mathcal{A} \mid (a, a) \in \mathcal{R}\}$.

3   Principles for Disagreement Measures

Our aim is to evaluate the amount of disagreements contained in an argumentation graph. This is done by a *disagreement measure*, that is a real-valued function that assigns a *disagreement value* to every argumentation graph.

Definition 1 (Disagreement Measure). *A disagreement measure is a function* $\mathcal{K} : \mathcal{U} \to [0, +\infty)$. *For an argumentation graph* $\mathbf{A} = \langle \mathcal{A}, \mathcal{R} \rangle \in \mathcal{U}$, $\mathcal{K}(\mathbf{A})$ *is called the* disagreement value *of* $\mathbf{A}$.

For two argumentation graphs $\mathbf{A}$ and $\mathbf{A}'$, we say that $\mathbf{A}$ is *more conflicting* than $\mathbf{A}'$ if $\mathcal{K}(\mathbf{A}) > \mathcal{K}(\mathbf{A}')$. The value 0 stands for absence of disagreements.

We propose next a set of *principles* that any disagreement measure should satisfy. The first principle states that the disagreement value of an argumentation graph does not depend on the identity of its arguments. Note that this axiom is used in most axiomatic approaches including game theory (e.g., Shapley value [12]).

Principle 1 (Anonymity). *For all argumentation graphs* $\mathbf{A} = \langle \mathcal{A}, \mathcal{R} \rangle$ *and* $\mathbf{A}' = \langle \mathcal{A}', \mathcal{R}' \rangle$ *in* $\mathcal{U}$, *if* $\mathbf{A}$ *and* $\mathbf{A}'$ *are isomorphic, then* $\mathcal{K}(\mathbf{A}) = \mathcal{K}(\mathbf{A}')$.

The second principle states that attacks are the only source of disagreements. Thus, any argumentation graph that has an empty attack relation receives the value 0. This axiom is somehow similar to the consistency axiom proposed in [6] for measuring inconsistency in knowledge bases.

Principle 2 (Agreement). *For any argumentation graph* $\mathbf{A} = \langle \mathcal{A}, \mathcal{R} \rangle \in \mathcal{U}$, *if* $\mathcal{R} = \emptyset$, *then* $\mathcal{K}(\mathbf{A}) = 0$.

The third principle concerns "harmless" arguments (i.e., arguments which neither attack nor are attacked by other arguments). The principle states that adding such arguments to an argumentation graph will not modify its disagreement value. This axiom is also in the same spirit as the "free formula independence" axiom proposed in [6].

Principle 3 (Dummy). *For any argumentation graph* $\mathbf{A} = \langle \mathcal{A}, \mathcal{R} \rangle \in \mathcal{U}$, *for any* $a \in \texttt{Args} \setminus \mathcal{A}$, $\mathcal{K}(\mathbf{A}) = \mathcal{K}(\mathbf{A}')$, *where* $\mathbf{A}' = \langle \mathcal{A} \cup \{a\}, \mathcal{R} \rangle$.

The next principle states that if new attacks are added to an argumentation graph, its disagreement value increases. This axiom is in the spirit of monotony axiom in [6] which states that if a knowledge base is extended by formulas, its inconsistency degree cannot decrease.

Principle 4 (Monotony). *For any argumentation graph* $\mathbf{A} = \langle \mathcal{A}, \mathcal{R} \rangle \in \mathcal{U}$, *for any* $r \in (\mathcal{A} \times \mathcal{A}) \setminus \mathcal{R}$, $\mathcal{K}(\mathbf{A}) < \mathcal{K}(\mathbf{A}')$, *where* $\mathbf{A}' = \langle \mathcal{A}, \mathcal{R} \cup \{r\} \rangle$.

So far, we have seen that disagreements contained in an argumentation graph are due to direct attacks (i.e., elements of $\mathcal{R}$). It is also well-known that the role of such attacks is to *weaken* their targets (see the weakening property in [13]). Indeed, whatever the semantics that is used for evaluating arguments, it should satisfy the weakening property since it defines the essence of attacks. However, the effect of weakening may propagate in the graph, giving birth to *indirect attacks*. Consider the following graph.

$$\boxed{a_0} \rightarrow \boxed{a_1} \rightarrow \boxed{a_2} \rightarrow \boxed{a_3}$$

Under stable semantics [3], the graph has one extension $\{a_0, a_2\}$, and the argument a_3 is rejected. If we remove the attack from a_0 to a_1, the new graph has $\{a_0, a_1, a_3\}$ as stable extension, and a_3 becomes accepted. Thus, the attack (a_0, a_1) has a negative effect on a_3. The same phenomenon occurs under the h-categorizer semantics proposed by Besnard and Hunter [1]. The argument a_3 has an acceptability degree 0.60 in the initial graph and 0.66 in the modified one. Thus, a_3 looses weight in presence of the attack (a_0, a_1). The argument a_0 is then considered as an indirect attacker of a_3. This shows that indirect attacks are also source of disagreement in argumentation graphs since they are not only harmful for their direct targets (a_1 in the example), but also to the indirect ones (a_3).

The next principle states that an acyclic graph containing indirect attacks is more conflicting than an acyclic graph containing only direct ones. This holds for graphs that have the same number of arguments and the same number of attacks.

Principle 5 (Reinforcement) *For argumentation graphs* $\mathbf{A} = \langle \mathcal{A}, \mathcal{R} \rangle$ *and* $\mathbf{A}' = \langle \mathcal{A}', \mathcal{R}' \rangle$ *in* $\mathcal{U}$ *such that:*

- $\mathcal{A} = \mathcal{A}' = \{a_0, \ldots, a_n, b_0, \ldots, b_n\}$ *with* $n \geq 3$,
- $\mathcal{R} = \{(a_i, b_i) \mid i \in \{0, \ldots, n-1\}\}$,
- $\mathcal{R}' = \{(a_i, a_{i+1}) \mid i \in \{0, \ldots, n-1\}\}$,

it holds that $\mathcal{K}(\mathbf{A}') > \mathcal{K}(\mathbf{A})$.

The two graphs $\mathbf{A}$ and $\mathbf{A}'$ have $n-1$ direct attacks. In addition, $\mathbf{A}'$ contains at least one indirect attack (e.g. (a_0, a_n) when $n = 3$). So, $\mathbf{A}$ is less conflicting than $\mathbf{A}'$. Note that due to the Anonymity principle, Reinforcement holds also for argumentation graphs that contain different sets of arguments.

The two argumentation graphs of Reinforcement are acyclic. Assume now an acyclic graph with 100 direct attacks and a 10-length elementary cycle.

The latter contains thus 10 attacks and several indirect attacks. Which of the two graphs is more conflicting? There are two possible (but incompatible) answers to this question: (i) to give more weight to disagreements generated by direct attacks, (ii) to give an overwhelming weight to cycle since it represents a *deadlock* situation while conflicts are *open* in an acyclic graph. This second choice is captured by the following *optional* principle.

Principle 6 (Cycle Precedence) *For all graphs* $\mathbf{A} = \langle \mathcal{A}, \mathcal{R} \rangle$ *and* $\mathbf{A}' = \langle \mathcal{A}', \mathcal{R}' \rangle$ *in* $\mathcal{U}$, *if* $\mathbf{A}$ *is acyclic and* $\mathbf{A}'$ *is an elementary cycle, then* $\mathcal{K}(\mathbf{A}) < \mathcal{K}(\mathbf{A}')$.

The last and optional principle says that a disagreement measure could take the size of cycles into account. The idea is that the larger the size of a cycle is, the less severe the disagreement; said differently, the less arguments are needed to produce a cycle, the more "obvious" and strong the disagreement. For instance, a cycle of length 2 is more conflicting than a cycle of length 1000. The latter is less visible than the former.

Principle 7 (Size Sensitivity). *For all elementary cycles* $\mathbf{A} = \langle \mathcal{A}, \mathcal{R} \rangle$, $\mathbf{A}' = \langle \mathcal{A}', \mathcal{R}' \rangle$ *in* $\mathcal{U}$, *if* $|\mathcal{A}'| < |\mathcal{A}|$, *then* $\mathcal{K}(\mathbf{A}) < \mathcal{K}(\mathbf{A}')$.

The seven principles are independent (none of them follows from the others). They are also compatible (they can be satisfied all together by a disagreement measure).

Theorem 1. *The principles are independent and compatible.*

4   Five Disagreement Measures

This section introduces disagreement measures and analytically evaluates them against the proposed principles, especially the five mandatory ones. We introduce them from the most naive to the most elaborated one.

4.1   Connectance Measure

The first measure that comes in mind for evaluating disagreements in an argumentation graph is the one that counts the number of attacks in a graph. Such a measure is very natural since disagreements come from attacks.

Definition 2 (Connectance measure). *Let* $\mathbf{A} = \langle \mathcal{A}, \mathcal{R} \rangle$ *be an argumentation graph.* $\mathcal{K}_c(\mathbf{A}) = |\mathcal{R}|$.

Let us illustrate the measure with a running example.

Example 1. *Consider the six argumentation graphs from the introduction. It can be checked that* $\mathcal{K}_c(\mathbf{A_0}) = 0$, $\mathcal{K}_c(\mathbf{A_1}) = 1$, $\mathcal{K}_c(\mathbf{A_2}) = 2$, $\mathcal{K}_c(\mathbf{A_3}) = 3$, $\mathcal{K}_c(\mathbf{A_4}) = 5$, *and* $\mathcal{K}_c(\mathbf{A_5}) = 9$.

The measure $\mathcal{K}_c$ satisfies four out of seven principles.

Theorem 2. *Connectance measure satisfies Anonymity, Agreement, Dummy, and Monotony. It violates Reinforcement, Size sensitivity and Cycle Precedence.*

The fact that $\mathcal{K}_c$ violates Reinforcement means that it does not take into account indirect attacks, which is a real weakness of a disagreement measure. This shows also that the amount of disagreement is not the simple number of attacks.

4.2 In-Degree Measure

The second candidate measure counts the number of arguments that are attacked in an argumentation graph.

Definition 3 (In-degree measure). *Let* $\mathbf{A} = \langle \mathcal{A}, \mathcal{R} \rangle$ *be an argumentation graph.* $\mathcal{K}_i(\mathbf{A}) = |\{a \in \mathcal{A} \mid \exists (x, a) \in \mathcal{R}\}|.$

Let us illustrate the measure with the six graphs given in the introduction.

Example 1 (cont): According to the In-degree measure, $\mathcal{K}_i(\mathbf{A_0}) = 0$, $\mathcal{K}_i(\mathbf{A_1}) = 1$, $\mathcal{K}_i(\mathbf{A_2}) = 2$, and $\mathcal{K}_i(\mathbf{A_3}) = \mathcal{K}_i(\mathbf{A_4}) = \mathcal{K}_i(\mathbf{A_5}) = 3$. Thus, $\mathbf{A_3}$ is more conflicting than $\mathbf{A_2}$ which is more conflicting than $\mathbf{A_1}$.

This measure satisfies only three out of seven principles.

Theorem 3. *In-degree measure satisfies Anonymity, Agreement, and Dummy. It violates Monotony, Reinforcement, Cycle Precedence, and Size Sensitivity.*

This measure has two weaknesses: it does not distinguish an elementary cycle from a complete graph (see graphs $\mathbf{A_3}$ and $\mathbf{A_5}$ in Example 1). Moreover, like Connectance measure, it does not take into account indirect attacks.

Remark: In-degree measure focuses on attacked arguments. One may define another measure which rather evaluates the number of "aggressive" arguments, that is, arguments which attack other arguments. Such a measure satisfies (respectively violates) exactly the same principles as In-degree measure. Thus, it is not a good candidate for assessing disagreement in an argumentation graph.

4.3 Extension-Based Measures

We now define two measures that are based on acceptability semantics, namely on extension-based semantics proposed in [3]. Those semantics were introduced for solving disagreements in an argumentation graph. Before introducing the measures, let us first recall the semantics we will consider. Let $\mathbf{A} = \langle \mathcal{A}, \mathcal{R} \rangle$ be an argumentation graph and $\mathcal{E} \subseteq \mathcal{A}$.

- $\mathcal{E}$ is *conflict-free* iff $\nexists a, b \in \mathcal{E}$ such that $(a, b) \in \mathcal{R}$.
- $\mathcal{E}$ *defends* an argument $a \in \mathcal{A}$ iff $\forall b \in \mathcal{A}$, if $(b, a) \in \mathcal{R}$, then $\exists c \in \mathcal{E}$ such that $(c, b) \in \mathcal{R}$.

Definition 4 (Acceptability semantics). *Let $\mathbf{A} = \langle \mathcal{A}, \mathcal{R} \rangle$ be an argumentation graph, and $\mathcal{E} \subseteq \mathcal{A}$ be conflict-free.*

- *$\mathcal{E}$ is a* naive extension *iff it is a maximal (w.r.t. set $\subseteq$) conflict-free set.*
- *$\mathcal{E}$ is a* preferred extension *iff it is a maximal (w.r.t. set $\subseteq$) set that defends all its elements.*

Notations: $\text{Ext}_x(\mathbf{A})$ denotes the set of all extensions of $\mathbf{A}$ under semantics x where $x \in \{n, p\}$ and n (respectively p) stands for naive (respectively preferred).

The basic idea behind extension-based measures is that the existence of multiple extensions means presence of disagreements in the graph. Furthermore, the greater the number of extensions of an argumentation graph, the greater the amount of disagreements in the graph. However, a disagreement measure which counts only the number of extensions (under a given semantics) may miss disagreements. Consider the following argumentation graph:

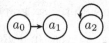

This graph has two naive extensions ($\{a_0\}$ and $\{a_1\}$), which are mainly due to the conflict between a_0 and a_1 neglecting thus the self-attack. Similarly, the graph has a single preferred extension $\{a_0\}$ and the self-attack is again neglected. In what follows, we propose two measures (one for each of the two semantics recalled above) which take into account both the number of extensions and the number of self-attacking arguments in an argumentation graph.

Definition 5. (Extension-based measure) *Let $\mathbf{A} = \langle \mathcal{A}, \mathcal{R} \rangle$ be an argumentation graph and $x \in \{n, p\}$.*

$$\mathcal{K}_e^x(\mathbf{A}) = |\text{Ext}_x(\mathbf{A})| + |\text{SelfAtt}(\mathbf{A})| - 1.$$

The subtraction of 1 in the above equation is required in order to ensure agreement in case of empty attack relations.

Example 1 (cont): Under naive semantics, $\mathcal{K}_e^n(\mathbf{A_0}) = 0$, $\mathcal{K}_e^n(\mathbf{A_1}) = \mathcal{K}_e^n(\mathbf{A_2}) = 1$, $\mathcal{K}_e^n(\mathbf{A_3}) = \mathcal{K}_e^n(\mathbf{A_4}) = 2$, and $\mathcal{K}_e^n(\mathbf{A_5}) = 3$.

Under preferred semantics, $\mathcal{K}_e^p(\mathbf{A_0}) = \mathcal{K}_e^p(\mathbf{A_1}) = \mathcal{K}_e^p(\mathbf{A_2}) = \mathcal{K}_e^p(\mathbf{A_3}) = 0$, $\mathcal{K}_e^p(\mathbf{A_4}) = 1$, $\mathcal{K}_e^p(\mathbf{A_5}) = 3$.

These two measures are clearly not powerful since they are not discriminating as shown in Example 1. For instance, under preferred semantics, the corresponding measure does not make any difference between graphs with empty attack relations ($\mathbf{A_0}$) and those that have one preferred (resp. stable) extension ($\mathbf{A_1}$ and $\mathbf{A_2}$). The measure is also unable to make a difference between a graph which has one non-empty extension and a graph which has a single empty extension ($\mathbf{A_2}$ and $\mathbf{A_3}$). The following result confirms these observations. Indeed, the two measures satisfy only three principles out of seven.

Theorem 4. *Extension-based measures satisfy Anonymity, Agreement and Dummy. They violate Monotony, Reinforcement, Cycle Precedence, and Size Sensitivity.*

Despite the fact that these measures satisfy (respectively violate) the same principles as $\mathcal{K}_i$, they may return different results. Indeed, $\mathcal{K}_i$ assigns the same value to $\mathbf{A_3}$ and $\mathbf{A_4}$ while $\mathcal{K}_e^p$ assigns to them different values. Similarly, $\mathcal{K}_i$ assigns different values to $\mathbf{A_1}$ and $\mathbf{A_2}$ while naive measure assigns to both graphs the same value 1.

Remark: It is worth mentioning that it is possible to define other measures using other extension semantics like complete, stable, semi-stable, etc. However, they will all satisfy the same set of principles as the two discussed above.

4.4   Distance-Based Measure

The previous disagreement measures are unable to take into account indirect attacks. Our last measure escapes this limitation. It satisfies thus reinforcement as well as all the other principles. The basic idea for capturing indirect attacks (and of course direct attacks) is to check the existence of a path between any pair of arguments of an argumentation graph. Since two arguments may be related by several paths, we consider the shortest one. Then, we compute a global distance for the graph which is the sum of the lengths of those paths. Before defining formally the new measure, let us first recall the notion of distance in graphs.

Table 1. Satisfaction of principles by the measures (the symbol ● stands for satisfaction and ○ for violation.)

	$\mathcal{K}_e^x$	$\mathcal{K}_c$	$\mathcal{K}_i$	$\mathcal{K}_d$
Anonymity	●	●	●	●
Agreement	●	●	●	●
Dummy	●	●	●	●
Monotony	○	●	○	●
Reinforcement	○	○	○	●
Cycle precedence	○	○	○	●
Size sensitivity	○	○	○	●

Definition 6 (Distance). *Let $\mathbf{A} = \langle \mathcal{A}, \mathcal{R} \rangle$ be an argumentation graph and $a, b \in \mathcal{A}$. If $a \neq b$, then the distance between a and b in $\mathbf{A}$, $d(a, b)$, is the length of the shortest path from a to b if b is reachable from a, and $d(a, b) = k$ if b is not reachable from a. If $a = b$, $d(a, b)$ is the length of the shortest elementary cycle in which a is involved, and $d(a, b) = k$ if a is not involved in cycles. Throughout the paper, we assume that $k = |\mathcal{A}| + 1$.*

Note that k is set to $|\mathcal{A}| + 1$ because the longest path in an argumentation graph is $|\mathcal{A}| - 1$ and the length of the longest cycle is $|\mathcal{A}|$.

Example 1 (cont): In argumentation graph $\mathbf{A_3}$, $d(a_1, a_1) = 3$, $d(a_1, a_2) = 2$ and $d(a_1, a_3) = 1$. In graph $\mathbf{A_2}$, $d(a_1, a_1) = 4$ and $d(a_1, a_3) = 4$ (here $k = 4$).

The domain of the distance function is delimited as follows.

Proposition 1. *Let* $\mathbf{A} = \langle \mathcal{A}, \mathcal{R} \rangle$ *be an argumentation graph. For all* $a, b \in \mathcal{A}$, $d(a, b) \in [1, k]$.

The global distance of an argumentation graph is the sum of lengths of the shortest paths between any pair of arguments.

Definition 7 (Global distance). *For any argumentation graph* $\mathbf{A} = \langle \mathcal{A}, \mathcal{R} \rangle$,

$$\mathcal{D}(\mathbf{A}) = \sum_{a_i \in \mathcal{A}} \sum_{a_j \in \mathcal{A}} d(a_i, a_j)$$

Example 1 (cont): $\mathcal{D}(\mathbf{A_0}) = 2$, $\mathcal{D}(\mathbf{A_1}) = 10$, $\mathcal{D}(\mathbf{A_2}) = 28$, $\mathcal{D}(\mathbf{A_3}) = 18$, $\mathcal{D}(\mathbf{A_4}) = 13$ and $\mathcal{D}(\mathbf{A_5}) = 9$.

Let us now delimit the upper and lower bounds of the global distance of an argumentation graph.

Proposition 2. *For any argumentation graph* $\mathbf{A} = \langle \mathcal{A}, \mathcal{R} \rangle$,

$$min \leq \mathcal{D}(\mathbf{A}) \leq max$$

where $max = n^2 \times (n + 1)$, $min = n^2$ *and* $n = |\mathcal{A}|$.

We show next that the upper bound is reached by an argumentation graph in case its attack relation is empty, while the lower bound is reached when the graph is complete.

Proposition 3. *For any argumentation graph* $\mathbf{A} = \langle \mathcal{A}, \mathcal{R} \rangle$,

- $\mathcal{D}(\mathbf{A}) = max$ *iff* $\mathcal{R} = \emptyset$
- $\mathcal{D}(\mathbf{A}) = min$ *iff* $\mathcal{R} = \mathcal{A} \times \mathcal{A}$

Distance-based measure evaluates to what extent the global distance of an argumentation graph is close to the upper bound. The more it is close to max, the less disagreements are in the graph. The closer the global distance is to min, the more the graph contains a lot of conflicts.

Definition 8 (Distance-based measure). *For any argumentation graph* $\mathbf{A} = \langle \mathcal{A}, \mathcal{R} \rangle$,

$$\mathcal{K}_d(\mathbf{A}) = \frac{max - \mathcal{D}(\mathbf{A})}{max - min}$$

where $max = n^2 \times (n + 1)$, $min = n^2$ *and* $n = |\mathcal{A}|$.

Let us illustrate this measure with the running example.

Example 1 (cont): $\mathcal{K}_d(\mathbf{A_0}) = 0$, $\mathcal{K}_d(\mathbf{A_1}) = 0.25$, $\mathcal{K}_d(\mathbf{A_2}) = 0.29$, $\mathcal{K}_d(\mathbf{A_3}) = 0.66$, $\mathcal{K}_d(\mathbf{A_4}) = 0.88$, and $\mathcal{K}_d(\mathbf{A_5}) = 1$.

This measure computes somehow the *degree of connectivity* of an argumentation graph. Indeed, a high disagreement value means that the graph is highly connected, and small disagreement value means that the graph is not very connected. It makes thus fine grained comparisons of argumentation graphs, namely of various forms of cyclic graphs. In Example 1, $\mathbf{A_5}$ is more conflicting than $\mathbf{A_4}$ which is itself more conflicting than $\mathbf{A_3}$.

In what follows, we introduce the notion of *connectivity degree* of an argumentation graph. It is the proportion of pairs of arguments which are related by at least one path.

Definition 9 (Connectivity degree). *The* connectivity degree *of an argumentation graph* $\mathbf{A} = \langle \mathcal{A}, \mathcal{R} \rangle$ *is*

$$\mathsf{Co}(\mathbf{A}) = \frac{|\{(a,b) \in \mathcal{A}^2 \mid d(a,b) < k\}|}{|\mathcal{A}|^2}.$$

The next result shows that the upper bound of the disagreement value of an argumentation graph is exactly the connectivity degree of the graph.

Theorem 5. *For any argumentation graph* $\mathbf{A} = \langle \mathcal{A}, \mathcal{R} \rangle$, $\mathcal{K}_d(\mathbf{A}) \in [0, \mathsf{Co}(\mathbf{A})]$.

Proof. Let $\mathbf{A} = \langle \mathcal{A}, \mathcal{R} \rangle$ be an argumentation graph such that $|\mathcal{A}| = n$. Let $\mathcal{B} = \{(a_i, a_j) \in \mathcal{A} \times \mathcal{A} \mid d(a_i, a_j) < k\}$. From Proposition 4, $\mathcal{K}_d(\mathbf{A}) = \mathsf{Co}(\mathbf{A}) + \frac{\mathsf{Co}(\mathbf{A})}{n} - \frac{\sum_{(a_i,a_j) \in \mathcal{B}} d(a_i,a_j)}{n^3} = \frac{\mathsf{Co}(\mathbf{A})(n+1)}{n} - \frac{\sum_{(a_i,a_j) \in \mathcal{B}} d(a_i,a_j)}{n^3}$. It holds that $\sum_{(a_i,a_j) \in \mathcal{B}} d(a_i, a_j) \geq n^2 \mathsf{Co}(\mathbf{A})$ (since $|\mathcal{B}| = n^2 \mathsf{Co}(\mathbf{A}$ and $d(a,b) \in [1,k]$). Thus, $\frac{\sum_{(a_i,a_j) \in \mathcal{B}} d(a_i,a_j)}{n^3} \geq \frac{n^2 \mathsf{Co}(\mathbf{A})}{n^3}$. Consequently, $\frac{\mathsf{Co}(\mathbf{A})}{n} - \frac{\sum_{(a_i,a_j) \in \mathcal{B}} d(a_i,a_j)}{n^3} \leq 0$.

So, $\mathsf{Co}(\mathbf{A}) + \frac{\mathsf{Co}(\mathbf{A})}{n} - \frac{\sum_{(a_i,a_j) \in \mathcal{B}} d(a_i,a_j)}{n^3} \leq \mathsf{Co}(\mathbf{A})$. Thus, $\mathcal{K}_d(\mathbf{A}) \leq \mathsf{Co}(\mathbf{A})$.

Let us now characterize the disagreement values of elementary cycles. The shorter an elementary cycle, the more conflicting it is. The maximal value (1) is given to cycles of length 1, that is graphs that contain only one argument, moreover it is self-attacking. This value decreases when the length of cycles increases. However, we show that it cannot be less than 0.5. This means that the distance-based measure considers cycles as very conflicting even when they are very long, which is very natural.

Proposition 4. *For any elementary cycle* $\mathbf{A} = \langle \mathcal{A}, \mathcal{R} \rangle$, $\mathcal{K}_d(\mathbf{A}) \in (\frac{1}{2}, 1]$.

Proof. Let $\mathbf{A} = \langle \mathcal{A}, \mathcal{R} \rangle$ be an elementary cycle, and let $n = |\mathcal{A}|$. For any $a \in \mathcal{A}$, $\sum_{b_i \in \mathcal{A}} d(a, b_i) = 1 + 2 + 3 + \ldots + n = \frac{n \times (n+1)}{2}$. Thus, $\mathcal{D}(\mathbf{A}) = \frac{n^2 \times (n+1)}{2}$. $\mathcal{K}_d(\mathbf{A}) = \frac{n^2 \times (n+1) - \mathcal{D}(\mathbf{A})}{n^2 \times (n+1) - n^2}$, thus $\mathcal{K}_d(\mathbf{A}) = \frac{n^2 \times (n+1)}{2n^3} = \frac{n+1}{2n} = \frac{1}{2} + \frac{1}{2n}$. $\mathcal{K}_d(\mathbf{A}) = 1$ in case $n = 1$, i.e., $\mathbf{A}$ is made of a self attacking argument. Since $\mathbf{A}$ is finite, then $\mathcal{A}$ is finite. Consequently, $\mathcal{K}_d(\mathbf{A}) > \frac{1}{2}$.

The next result delimits the disagreement values of acyclic argumentation graphs.

Proposition 5. *For any acyclic argumentation graph* $\mathbf{A} = \langle \mathcal{A}, \mathcal{R} \rangle$, $\mathcal{K}_d(\mathbf{A}) \in [0, \frac{1}{2})$.

Proof. Let $\mathbf{A} = \langle \mathcal{A}, \mathcal{R} \rangle$ be an acyclic argumentation graph such that $|\mathcal{A}| = n$. Since $\mathbf{A}$ is acyclic, then $\forall a \in \mathcal{A}$, $d(a,a) = k$. Moreover, $\forall a,b \in \mathcal{A}$, if $d(a,b) < k$ then $d(b,a) = k$ (since there is no cycle in the graph). Thus, $|\{(a,b) \in \mathcal{A}^2 \mid d(a,b) < k\}| \leq \frac{n^2-n}{2}$. Consequently, $\frac{|\{(a,b)\in\mathcal{A}^2 \mid d(a,b)<k\}|}{n^2} \leq \frac{n^2-n}{2n^2}$. We get $\mathrm{Co}(\mathbf{A}) \leq \frac{1}{2} - \frac{1}{2n}$. Since $\mathcal{A}$ is finite, then $\mathrm{Co}(\mathbf{A}) < \frac{1}{2}$. From Theorem 5, $\mathcal{K}_d(\mathbf{A}) \leq \mathrm{Co}(\mathbf{A})$. So, $\mathcal{K}_d(\mathbf{A}) < \frac{1}{2}$.

The two previous results show that the measure $\mathcal{K}_d$ considers any acyclic graph as strictly less conflicting than any elementary cycle. Moreover, the ratio of disagreement in an acyclic graph is always not very high and can never reach the maximal value 1. On the contrary, the ratio of disagreement in an elementary cycle is always high.

Proposition 6. *Let* $\mathbf{A} = \langle \mathcal{A}, \mathcal{R} \rangle$ *and* $\mathbf{A}' = \langle \mathcal{A}', \mathcal{R}' \rangle$ *be simple paths. If* $|\mathcal{A}| < |\mathcal{A}'|$, *then* $\mathcal{K}_d(\mathbf{A}) < \mathcal{K}_d(\mathbf{A}')$.

Proof. Let $\mathbf{A} = \langle \mathcal{A}, \mathcal{R} \rangle$ and $\mathbf{A}' = \langle \mathcal{A}', \mathcal{R}' \rangle$ be two simple paths. Let $n = |\mathcal{A}|$ and $n' = |\mathcal{A}'|$. Assume that $n < n'$. Thus, $n^2 < n'^2$ and $\frac{1}{n^2} > \frac{1}{n'^2}$. Consequently, $1 - \frac{1}{n^2} < 1 - \frac{1}{n'^2}$ and then $\mathcal{K}_d(\mathbf{A}) < \mathcal{K}_d(\mathbf{A}')$.

The distance-based measure satisfies all our principles.

Theorem 6. $\mathcal{K}_d$ *satisfies all the seven principles.*

Proof. Let $\mathbf{A} = \langle \mathcal{A}, \mathcal{R} \rangle$ be an argumentation graph. Anonymity is obviously satisfied. From Proposition 3, if $\mathcal{R} = \emptyset$, $\mathcal{D}(\mathbf{A}) = max$, thus $\mathcal{K}_d(\mathbf{A}) = 0$ which ensures Agreement.

Let $\mathbf{A} = \langle \mathcal{A}, \mathcal{R} \rangle$ and $\mathbf{A}' = \langle \mathcal{A}', \mathcal{R}' \rangle$ be two elementary cycles such that $|\mathcal{A}| = n$, $|\mathcal{A}'| = m$ and $m > n$. $\mathcal{D}(\mathbf{A}) = \frac{n^2(n+1)}{2}$, thus $\mathcal{K}_d(\mathbf{A}) = \frac{n+1}{2n}$, and $\mathcal{D}(\mathbf{A}') = \frac{m^2(m+1)}{2}$ and $\mathcal{K}_d(\mathbf{A}') = \frac{m+1}{2m} = \frac{1}{2} + \frac{1}{2m}$. Since $m > n$ then $2m > 2n$ and $\frac{1}{2m} < \frac{1}{2n}$. Consequently, $\mathcal{K}_d(\mathbf{A}') < \mathcal{K}_d(\mathbf{A})$. This shows that Size Sensitivity is satisfied. Let us now show that $\mathcal{K}_d$ satisfies Dummy principle. Assume that $|\mathcal{A}| = n$ Let $a \in \mathrm{Args} \setminus \mathcal{A}$ and $\mathbf{A}' = \langle \mathcal{A} \cup \{a\}, \mathcal{R} \rangle$. We denote by k the maximal distance in graph $\mathbf{A}$ and by k' the maximal distance in graph $\mathbf{A}'$. From definition, $k' = n + 2$ since $|\mathrm{Arg}(\mathbf{A}')| = n + 1$. Since the new arguments does not attack and is not attacked by other arguments, then the original distances in graph $\mathbf{A}$ will not change except those that got value k which will be incremented by 1 each. Thus, $\mathcal{D}(\mathbf{A}') = \mathcal{D}(\mathbf{A}) + (2n+1)k' + x$ where $x \geq 0$ is the number of pairs (a_i, a_j) of arguments for which the length of the shortest path from a_i to a_j is equal to k in graph $\mathbf{A}$. We get $\mathcal{D}(\mathbf{A}') = \mathcal{D}(\mathbf{A}) + 2n^2 + 5n + 2 + x$. Moreover, $\mathcal{K}_d(\mathbf{A}) = 1 + \frac{1}{n} - \frac{\mathcal{D}(\mathbf{A})}{n^3}$ and $\mathcal{K}_d(\mathbf{A}') = 1 + \frac{1}{n+1} - \frac{\mathcal{D}(\mathbf{A})}{(n+1)^3} - \frac{x+2n^2+5n+2}{(n+1)^3}$. Thus, $\mathcal{K}_d(\mathbf{A}') < \mathcal{K}_d(\mathbf{A})$.

Let us now show that monotony is also satisfied. Let $\mathcal{R}' \subseteq (\mathcal{A} \times \mathcal{A}) \setminus \mathcal{R}$ and $\mathbf{A}' = \langle \mathcal{A}, \mathcal{R} \cup \mathcal{R}' \rangle$. Both $\mathbf{A}$ and $\mathbf{A}'$ have the same min and max distances since they have the same number of arguments. Consequently, $\mathcal{K}_d(\mathbf{A}) = 1 + \frac{1}{n} - \frac{\mathcal{D}(\mathbf{A})}{n^3}$ and $\mathcal{K}_d(\mathbf{A}') = 1 + \frac{1}{n} - \frac{\mathcal{D}(\mathbf{A}')}{n^3}$, with $n = |\mathcal{A}|$. Assume that $\mathcal{D}(\mathbf{A}') > \mathcal{D}(\mathbf{A})$. This means that there exists $a, b \in \mathcal{A}$ such $d(a,b) = x$ in graph $\mathbf{A}$, $d(a,b) = y$ in graph $\mathbf{A}'$ and $y > x$. This is impossible since the shortest path in $\mathbf{A}$ between a and b still exists in $\mathbf{A}'$. Thus, in $\mathbf{A}'$, the shortest path between a and b is either the same as in $\mathbf{A}$ or a path with $y < x$ because of the additional attacks of $\mathcal{R}'$.

Cycle Precedence follows from Propositions 4 and 5.

Reinforcement is also satisfied. Since the two graphs in the principle are assumed to have the same number of arguments, then both graphs have the same max and min values. It is thus sufficient to compare the global distances of the graphs. We can easily compute the following values: $\mathcal{D}(\mathbf{A}) = 8n^3 + 26n^2 + 32n + 14$, and $\mathcal{D}(\mathbf{A}') = \frac{20}{3}n^3 + 25n^2 + \frac{91}{3}n + 12$. $\mathcal{D}(\mathbf{A}) > \mathcal{D}(\mathbf{A}')$, thus $\mathcal{K}(\mathbf{A}) < \mathcal{K}(\mathbf{A}')$.

Distance-based measure satisfies all the principles. Thus, it takes into account both the direct attacks in an argumentation graph as well as the indirect ones. All these features make it the best candidate for measuring disagreement in argumentation graphs. Table 1 recalls for each measure, the principles it satisfies and those it violates.

5 Links Between Disagreement Measures and Inconsistency Measures

In this section, we consider argumentation graphs $\langle \mathcal{A}, \mathcal{R} \rangle$ that are generated from a propositional knowledge base Σ. The arguments of $\mathcal{A}$ are defined as follows:

Definition 10 (Argument). *Let Σ be a propositional knowledge base. An argument is a pair (X, x) s.t. $X \subseteq \Sigma$, X is consistent, $X \vdash x$[1], and $\nexists X' \subset X$ such that $X' \vdash x$.*

Regarding the attack relation $\mathcal{R}$, we consider *assumption-attack* defined in [5].

Definition 11 (Assumption-Attack). *An argument (X, x) attacks (Y, y) $(((X, x), (Y, y)) \in \mathcal{R})$ iff $\exists y' \in Y$ such that $x \equiv \neg y'$[2].*

In [14], the authors proposed a measure ($\mathcal{I}$) that quantifies the amount of inconsistency in a propositional knowledge base Σ. That amount is equal to the number of maximal (for set inclusion) consistent subsets of Σ and the number of inconsistent formulas in Σ minus 1.

---

[1] The symbol $\vdash$ stands for propositional inference relation.
[2] The symbol $\equiv$ stands for logical equivalence.

Definition 12 (Inconsistency Measure). *For any propositional knowledge base Σ,*

$$\mathcal{I}(\Sigma) = |\text{Max}(\Sigma)| + |\text{Inc}(\Sigma)| - 1.$$

$\text{Max}(\Sigma)$ *is the set of maximal (for set $\subseteq$) consistent subsets of Σ, and $\text{Inc}(\Sigma)$ is the set of inconsistent formulae in Σ. $\mathcal{I}(\Sigma)$ is called the inconsistency value of Σ.*

Given a knowledge base Σ, we show that its inconsistency value (as computed by measure $\mathcal{I}$) is equal to the disagreement values of the corresponding argumentation graph using the three extension-based measures $\mathcal{K}_e^x$ with $x \in \{n, p\}$.

Theorem 7. *Let Σ be a propositional knowledge base such that $\text{Inc}(\Sigma) = \emptyset$. Let $\mathbf{A} = \langle \mathcal{A}, \mathcal{R} \rangle$ be the argumentation graph built over Σ. The following holds:*

$$\mathcal{K}_e^n(\mathbf{A}) = \mathcal{K}_e^p(\mathbf{A}) = \mathcal{I}(\Sigma)$$

Proof. Let Σ be a propositional knowledge base such that $\text{Inc}(\Sigma) = \emptyset$. Let $\mathbf{A} = \langle \mathcal{A}, \mathcal{R} \rangle$ be the argumentation graph built over Σ. From Theorem 8 in [15], $\text{Ext}_n(\mathbf{A}) = \text{Ext}_p(\mathbf{A})$. Furthermore, there is a full correspondence between the naive extensions of $\mathbf{A}$ and the maximal (for set inclusion) consistent subsets of Σ. Hence, $|\text{Ext}_n(\mathbf{A})| = |\max(\Sigma)|$. Since $\text{Inc}(\Sigma) = \emptyset$, then $\mathcal{I}(\Sigma) = |\max(\Sigma)| - 1$. Since by definition of arguments, $\text{SelfAtt}(\mathbf{A}) = \emptyset$, then $\mathcal{K}_e^x(\mathbf{A}) = |\text{Ext}_x(\mathbf{A})| - 1$. Thus, $\mathcal{K}_e^x(\mathbf{A}) = \mathcal{I}(\Sigma)$.

This result shows that not only the two extension-based measures return the same result in case of propositional knowledge bases, but also they are equivalent to the inconsistency measure proposed in [14].

6   Related Work

Despite the great amount of work on argumentation, there is no work on computing the amount of disagreements in argumentation graphs. Our paper presented the first attempt in this direction. In [16], the authors studied to what extent the extensions (under a given semantics) of an argumentation graph are different. The problem they addressed is thus completely different from the purpose of our paper.

Several measures were proposed in the literature for quantifying inconsistency in propositional knowledge bases (e.g., [6–8]). Our extension-based measures are equivalent to one of them, namely the one proposed in [14].

7   Conclusion

This paper studied for the first time how to quantify disagreements in an argumentation graph. It showed that disagreements is more than direct attacks. It proposed principles which serve as theoretical criteria for validating and comparing disagreement measures. It defined six intuitive measures and investigated their properties. The distance-based measure is the most powerful one. It not only satisfies all the proposed principles, but it is also very discriminating, that is, it provides a fine grained evaluation of argumentation graphs. Moreover, it captures very well the two sources of disagreement: direct and indirect attacks. Furthermore, the paper made a first bridge with works on inconsistency measures. It showed that extension-based measures return the same amount of conflict as one proposed in [14].

This work can be extended in several ways. First, we plan to investigate more deeply the relationship between the disagreement value of an argumentation graph and existing inconsistency degree of the knowledge base over which the graph is built. A particular focus will be put on distance-based measure since it captures well indirect attacks in an argumentation graph. Another line of research consists of evaluating the contribution of each argument to the disagreement value of a graph. Such information may be useful in a dialogues for identifying the culprit that should be attacked.

Acknowledgments. This work was supported by ANR-13-BS02-0004 and ANR-11-LABX-0040-CIMI.

References

1. Besnard, P., Hunter, A.: A logic-based theory of deductive arguments. Artif. Intell. J. **128**(1–2), 203–235 (2001)
2. Bonzon, E., Maudet, N.: On the outcomes of multiparty persuasion. In: 10th International Conference on Autonomous Agents and Multiagent Systems (AAMAS 2011), pp. 47–54 (2011)
3. Dung, P.M.: On the acceptability of arguments and its fundamental role in nonmonotonic reasoning, logic programming and n-person games. Artif. Intell. **77**(2), 321–358 (1995)
4. Pollock, J.: How to reason defeasibly. Artif. Intell. **57**(1), 1–42 (1992)
5. Elvang-Gøransson, M., Krause, P.J., Fox, J.: Acceptability of arguments as 'logical uncertainty'. In: Clarke, M., Kruse, R., Moral, S. (eds.) ECSQARU 1993. LNCS, vol. 747, pp. 85–90. Springer, Heidelberg (1993). doi:10.1007/BFb0028186
6. Hunter, A., Konieczny, S.: On the measure of conflicts: shapley inconsistency values. Artif. Intell. **174**(14), 1007–1026 (2010)
7. Grant, J., Hunter, A.: Distance-based measures of inconsistency. In: Gaag, L.C. (ed.) ECSQARU 2013. LNCS (LNAI), vol. 7958, pp. 230–241. Springer, Heidelberg (2013). doi:10.1007/978-3-642-39091-3_20
8. Jabbour, S., Ma, Y., Raddaoui, B.: Inconsistency measurement thanks to MUS decomposition. In: International Conference on Autonomous Agents and Multi-Agent Systems, AAMAS 2014, pp. 877–884 (2014)
9. Thimm, M.: Towards large-scale inconsistency measurement. In: Lutz, C., Thielscher, M. (eds.) KI 2014. LNCS (LNAI), vol. 8736, pp. 195–206. Springer, Cham (2014). doi:10.1007/978-3-319-11206-0_19
10. Jabbour, S., Ma, Y., Raddaoui, B., Sais, L., Salhi, Y.: On structure-based inconsistency measures and their computations via closed set packing. In: Proceedings of the 2015 International Conference on Autonomous Agents and Multiagent Systems, AAMAS 2015, pp. 1749–1750 (2015)

11. Jabbour, S., Sais, L.: Exploiting MUS structure to measure inconsistency of knowledge bases. In: 22nd European Conference on Artificial Intelligence, ECAI 2016, pp. 991–998 (2016)
12. Shapley, L.: A values for n-person games. In: Kuhn, H.W., Tucker, A.W. (eds.) Contributions to the Theory of Games, vol. II. Annal of Mathematics Studies (AM-28) (1953)
13. Amgoud, L., Ben-Naim, J.: Axiomatic foundations of acceptability semantics. In: Proceedings of the Fifteenth International Conference Principles of Knowledge Representation and Reasoning, KR 2016, pp. 2–11 (2016)
14. Grant, J., Hunter, A.: Measuring consistency gain and information loss in stepwise inconsistency resolution. In: Liu, W. (ed.) ECSQARU 2011. LNCS, vol. 6717, pp. 362–373. Springer, Heidelberg (2011). doi:10.1007/978-3-642-22152-1_31
15. Amgoud, L., Besnard, P.: Logical limits of abstract argumentation frameworks. J. Appl. Non-Class. Logics **23**(3), 229–267 (2013)
16. Booth, R., Caminada, M., Podlaszewski, M., Rahwan, I.: Quantifying disagreement in argument-based reasoning. In: International Conference on Autonomous Agents and Multiagent Systems, AAMAS 2012, pp. 493–500 (2012)

Belief in Attacks in Epistemic Probabilistic Argumentation

Sylwia Polberg[1]([✉]), Anthony Hunter[1], and Matthias Thimm[2]

[1] University College London, London, UK
sylwia.polberg@gmail.com
[2] University of Koblenz-Landau, Koblenz, Germany

Abstract. The epistemic approach to probabilistic argumentation assigns belief to arguments. This is valuable in dialogical argumentation where one agent can model the beliefs another agent has in the arguments and this can be harnessed to make strategic choices of arguments to present. In this paper, we extend this epistemic approach by also representing the belief in attacks. We investigate properties of this proposal and compare it to the constellations approach showing neither subsumes the other.

Keywords: Abstract argumentation · Probabilistic argumentation · Epistemic argumentation

1 Introduction

Abstract argumentation as proposed by Dung [8] provides an important formalism for representing and evaluating arguments and counterarguments. Proposals for probabilistic argumentation extend this to address aspects of uncertainty arising in argumentation. The two main approaches to probabilistic argumentation are the constellations and the epistemic approaches [14]. In the **constellations approach**, the uncertainty is in the topology of the graph. This approach is useful when one agent is not sure what arguments and attacks another agent is aware of, and so this can be captured by a probability distribution over the space of possible argument graphs. In the **epistemic approach**, the topology of the argument graph is fixed, but there is uncertainty as to the degree to which each argument is believed.

In this paper, we extend the epistemic approach with a probability distribution over the power set of attacks which we use to represent the uncertainty in each attack. To illustrate, we consider a listener to a political discussion on the radio. This is a situation where the listener acquires all the arguments and attacks that are presented, but does not add or delete arguments or attacks. The argument graph is given in Fig. 1. Often the listener would evaluate the

---

This research is funded by EPSRC Project EP/N008294/1 "Framework for Computational Persuasion".

S. Moral et al. (Eds.): SUM 2017, LNAI 10564, pp. 223–236, 2017.
DOI: 10.1007/978-3-319-67582-4_16

arguments and attacks. For instance, she may have a low belief in A_3 because she has found World Bank predictions to be unreliable in the past, and she may have a high belief in argument A_2, but a low belief in the attack by A_2 on A_1. As a result, she may have a high belief in A_1. Note, in the constellations approach, it is not possible to represent all the arguments and attacks in one graph, and then assign belief to them.

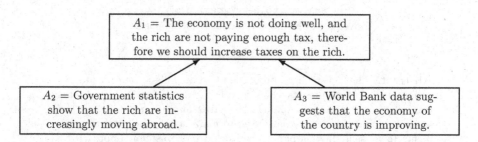

Fig. 1. Example of an argument graph acquired by a listener to a debate

Often uncertainty in attacks arises because "real-world" arguments are normally enthymemes (i. e. some or all of the premises and/or the claim are implicit). When an agent posits an enthymeme, the recipient decodes it to recover the intended argument. This creates a risk that the recipient decodes it differently to the way intended (as illustrated by the attack of A_2 on A_1 in Fig. 1).

A potentially valuable role for the extended epistemic approach is in supporting an agent (X) when arguing with another agent (Y). Agent X can model agent Y to reflect the arguments *and* attacks that X thinks Y believes. This extends proposals for using the epistemic approach for user modelling in persuasion dialogues [15,16].

The contributions of this paper are: (1) a set of constraints on probability distributions that take into account uncertainty of arguments and attacks; (2) results on the constraints showing inter-relationships between them; (3) results showing how with certain combinations of constraints recover and generalize Dung's dialectical semantics; and (4) a comparison with the constellations approach showing how neither subsumes the other. All proofs are available online[1].

2   Preliminaries

We start with a brief review of abstract argumentation as proposed by [8]. An **argument graph** (or a framework) is a directed graph $G = (\mathcal{A}, \mathcal{R})$, where $\mathcal{A}$ is the set of arguments and $\mathcal{R} \subseteq \mathcal{A} \times \mathcal{A}$ is the set of attacks. The way we decide which arguments can be accepted or rejected (or neither) is called a semantics.

---

[1] http://www0.cs.ucl.ac.uk/staff/a.hunter/papers/extended_epistemic_full.pdf

We focus on the argument-based approach [1,5] and the adaptation of the attack-based approach from [2].

An **argument labeling** is a total function $L : \mathcal{A} \rightarrow \{\text{in}, \text{out}, \text{und}\}$ [1,5]. By $\text{in}(L), \text{out}(L)$ and $\text{und}(L)$ we denote the arguments mapped respectively to in, out and und(ecided) by L. We will often write a labeling as a triple (I, O, U), where I, O and U are sets of arguments mapped to in, out and und. We say that a set of elements attacks another element if it contains an appropriate attacker. We can now introduce the notion of legality, on which our semantics are based.

Definition 1. *An argument $A \in \mathcal{A}$ is an* **attacker** *of $B \in \mathcal{A}$ iff $(A, B) \in \mathcal{R}$. Let $L : \mathcal{A} \rightarrow \{\text{in}, \text{out}, \text{und}\}$ be a labeling:*

- $X \in \text{in}(L)$ *is* **legally in** *iff all its attackers are in $\text{out}(L)$.*
- $X \in \text{out}(L)$ *is* **legally out** *iff it has an attacker in $\text{in}(L)$.*
- $X \in \text{und}(L)$ *is* **legally und** *iff not all of its attackers are in $\text{out}(L)$ and it does not have an attacker in $\text{in}(L)$.*

Definition 2. *Let $L : \mathcal{A} \rightarrow \{\text{in}, \text{out}, \text{und}\}$ be a labeling:*

(cf) *L is* **conflict-free** *iff it holds that if $A \in \text{in}(L)$, then none of its attackers is in $\text{in}(L)$, and every $A \in \text{out}(L)$ is legally out*

(ad) *L is* **admissible** *iff every $A \in \text{in}(L)$ is legally in and every $A \in \text{out}(L)$ is legally out.*

(co) *L is* **complete** *if it is admissible and every $A \in \text{und}(L)$ is legally und.*

Additionally, a complete labeling is stable (st) if $\text{und}(L) = \emptyset$, it is preferred (pr) if $\text{in}(L)$ is maximal wrt. $\subseteq$, and it is grounded (gr) if $\text{in}(L)$ is minimal wrt. $\subseteq$.

In Dung's semantics, attacks are seen as secondary to arguments. For example, ensuring that no attack on a given argument is accepted is equivalent to making sure that no argument carrying out an attack is accepted. However, this correspondence does not always hold for various generalizations of Dung's graph, in which a given conflict may not be successful due to preferences, probabilities, or when it is a target of an attack as well [4]. Therefore, the status assigned to a given attack is not necessarily the same as assigned to its source. We will now adapt the approach from [2] and focus on the **extended labelings**, which are total functions $L^\star : \mathcal{A} \cup \mathcal{R} \rightarrow \{\text{in}, \text{out}, \text{und}\}$. We introduce the notion of an extended attacker (attacker*), which is now a conflict, not an argument, and the attackee can be both an argument and a relation.

Definition 3. *For an attack $\alpha = (A, B) \in \mathcal{R}$, the source of α is $src(\alpha) = A$ and the target of α is $trg(\alpha) = B$. An attack $\alpha \in \mathcal{R}$ is an* **attacker** * *(1) of $B \in \mathcal{A}$ iff $B = trg(\alpha)$, and (2) of $\beta \in \mathcal{R}$ iff $trg(\alpha) = src(\beta)$.*

By replacing attacker with attacker* in the previous definitions and maximizing/minimizing attacks as well as arguments, we obtain the attack-based semantics, further distinguished with *. We have the following correspondence between these two families of semantics [2]. Please observe that that extended labelings can, in general, only be projected to their corresponding ordinary labelings if they are at least complete.

Proposition 1. *If $L = (I, O, U)$ is a σ–labeling, where $\sigma \in \{$cf, ad, co, pr, gr, st$\}$, then $L^\star = (I \cup \{\alpha \mid src(\alpha) \in I\}, O \cup \{\alpha \mid src(\alpha) \in O\}, U \cup \{\alpha \mid src(\alpha) \in U\})$ is a $\sigma^\star$–labeling. If $L^\star = (I^\star, O^\star, U^\star)$ is a $\delta^\star$–labeling, where $\delta \in \{$co, pr, gr, st$\}$, then $L = (I^\star \cap \mathcal{A}, O^\star \cap \mathcal{A}, U^\star \cap \mathcal{A})$ is a δ–labeling.*

We use $\sigma(G)$ to denote the set of labelings of G according to the semantics $\sigma \in \{$cf, ad, co, pr, gr, st, cf*, ad*, co*, pr*, gr*, st$^\star\}$. We will say that a set of arguments $\mathcal{S}$ is a σ–extension iff there exists a σ–labeling L s.t. in(L) $= \mathcal{S}$.

Example 1. Consider the graph G_1 below. The admissible labelings are $L_1 = (\emptyset, \emptyset, \{A, B, C\})$, $L_2 = (\{A\}, \{B\}, \{C\})$, $L_3 = (\{B\}, \{A\}, \{C\})$ and $L_4 = (\{B\}, \{A, C\}, \emptyset)$. Apart from L_3, all of them are complete. L_1 is grounded, L_2 and L_4 are preferred, and L_4 is stable.

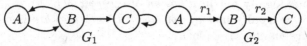

$$G_1 \qquad\qquad G_2$$

Now consider the graph G_2 above. The admissible labelings are $L_1 = (\emptyset, \emptyset, \{A, B, C\})$, $L_2 = (\{A\}, \emptyset, \{B, C\})$, $L_3 = (\{A\}, \{B\}, \{C\})$, and $L_4 = (\{A, C\}, \{B\}, \emptyset)$. L_4 is the single complete, preferred, stable and grounded labeling. The admissible* labelings of G_2 are $L_1^\star = (\emptyset, \emptyset, \{A, B, C, r_1, r_2\})$, $L_2^\star = (\{A\}, \emptyset, \{B, C, r_1, r_2\})$, $L_3^\star = (\{r_1\}, \emptyset, \{A, B, C, r_2\})$, $L_4^\star = (\{r_1\}, \{B\}, \{A, C, r_2\})$, $L_5^\star = (\{r_1\}, \{r_2\}, \{A, B, C\})$, $L_6^\star = (\{r_1\}, \{B, r_2\}, \{A, C\})$, $L_7^\star = (\{A, r_1\}, \emptyset, \{B, C, r_2\})$, $L_8^\star = (\{A, r_1\}, \{B\}, \{C, r_2\})$, $L_9^\star = (\{A, r_1\}, \{r_2\}, \{B, C\})$, $L_{10}^\star = (\{A, r_1\}, \{B, r_2\}, \{A, C\})$, $L_{11}^\star = (\{r_1, C\}, \{r_2\}, \{A, B\})$, $L_{12}^\star = (\{r_1, C\}, \{B, r_2\}, \{A\})$, $L_{13}^\star = (\{A, r_1, C\}, \{r_2\}, \{B\})$ and $L_{14}^\star = (\{A, r_1, C\}, \{B, r_2\}, \emptyset)$. $L_{14}^\star$ is the single complete*, preferred*, stable* and grounded* labeling.

We can observe that even though every argument-based labeling has a corresponding extended one, the removal of attacks from the extended labeling does not necessarily give us a standard one (e.g. $L_{11}^\star$ and $L_{12}^\star$).

3   Constellations Approach

We review the constellations approach [13] which extends [7,19]. It allows representation of the uncertainty over the topology of the graph: Each subgraph of the original graph is assigned a probability to represent the chances of it being the actual argument graph of the agent. It can be used to model what arguments and attacks an agent is aware of.

Definition 4. *For $G = (\mathcal{A}, \mathcal{R})$ and $G' = (\mathcal{A}', \mathcal{R}')$, the subgraph relation, denoted $\sqsubseteq$, is defined as $G' \sqsubseteq G$ iff $\mathcal{A}' \subseteq \mathcal{A}$ and $\mathcal{R}' \subseteq (\mathcal{A}' \times \mathcal{A}') \cap \mathcal{R}$. The set of subgraphs of G is $\mathsf{Sub}(G) = \{G' \mid G' \sqsubseteq G\}$. A subgraph $(\mathcal{A}', \mathcal{R}')$ is **full** iff $\mathcal{A}' \subseteq \mathcal{A}$ and $\mathcal{R}' = (\mathcal{A}' \times \mathcal{A}') \cap \mathcal{R}$. A subgraph $(\mathcal{A}', \mathcal{R}')$ is **spanning** iff $\mathcal{A}' = \mathcal{A}$ and $\mathcal{R}' \subseteq \mathcal{R}$.*

If our uncertainty is about which arguments appear in the graph, then only the full (induced) subgraphs of the argument graph have non-zero probability. If we are only uncertain about which attacks appear, then it is the spanning subgraphs of the argument graph that can have non-zero probability.

Definition 5. *A* **subgraph distribution** *is a function* $P^c : \mathsf{Sub}(G) \to [0,1]$ *with* $\sum_{G' \in \mathsf{Sub}(G)} P^c(G') = 1$. *A subgraph distribution* P^c *is a* **full subgraph distribution** *iff if* $(\mathcal{A}', \mathcal{R}')$ *is not a full subgraph, then* $P^c((\mathcal{A}', \mathcal{R}')) = 0$. *A subgraph distribution* P^c *is a* **spanning subgraph distribution** *iff iff if* $(\mathcal{A}', \mathcal{R}')$ *is not a spanning subgraph,* $P^c((\mathcal{A}', \mathcal{R}')) = 0$.

Determining the probability that a set of arguments is an extension (labeling) of a particular type (e. g. grounded, preferred, etc.) is is done by collecting the probabilities of the subgraphs producing the desired labelings. In a similar fashion, we can derive the probability of an argument being accepted in a labeling of a given type.

Definition 6. *For* $\mathcal{S} \subseteq \mathcal{A}$ *and* $\sigma \in \{\mathsf{cf}, \mathsf{ad}, \mathsf{co}, \mathsf{pr}, \mathsf{gr}, \mathsf{st}\}$, *the probability that* $L : \mathcal{S} \to \{\mathsf{in}, \mathsf{out}, \mathsf{und}\}$ *is a* σ*-labeling is:*

$$P_\sigma(L) = \sum_{G' \in \mathsf{Sub}(G)\ s.t.\ L \in \sigma(G')} P^c(G')$$

Definition 7. *Given a semantics* $\sigma \in \{\mathsf{ad}, \mathsf{co}, \mathsf{pr}, \mathsf{gr}, \mathsf{st}\}$, *the probability that* $A \in \mathcal{A}$ *is* in *in a* σ*-labeling is*

$$P_\sigma(A) = \sum_{G' \in \mathsf{Sub}(G)\ s.t.\ L \in \sigma(G')\ and\ A \in \mathsf{in}(L)} P(G')$$

Example 2. Consider the graph $G = (\{A, B\}, \{(A, B)\})$. Its subgraphs are $G_1 = (\{A, B\}, \{(A, B)\})$, $G_2 = (\{A, B\}, \emptyset)$, $G_3 = (\{A\}, \emptyset)$, $G_4 = (\{B\}, \emptyset)$ and $G_5 = (\emptyset, \emptyset)$. Out of them, G_1, G_3, G_4 and G_5 are full, and G_1 and G_2 are spanning. Consider the following subgraph distribution P^c: $P^c(G_1) = 0.09$, $P^c(G_2) = 0.81$, $P^c(G_3) = 0.01$ and $P^c(G_4) = 0.09$ and $P^c(G_5) = 0$. The probability of a given set being a grounded extension is as follows: $P_{\mathsf{gr}}(\{A, B\}) = P^c(G_2) = 0.81$; $P_{\mathsf{gr}}(\{A\}) = P^c(G_1) + P^c(G_3) = 0.1$; $P_{\mathsf{gr}}(\{B\}) = P^c(G_4) = 0.09$; and $P_{\mathsf{gr}}(\{\}) = P^c(G_5) = 0$. Therefore $P_{\mathsf{gr}}(A) = 0.91$ and $P_{\mathsf{gr}}(B) = 0.9$.

4 Extended Epistemic Approach

In the original version of the epistemic approach [3,14,17,18,22], an argument graph has an associated probability distribution over the sets of arguments. From this, we derive the probability of a single argument and interpret it as the belief that an agent has in it (i. e. the degree to which the agent believes the premises and the conclusion drawn from those premises). We say that an agent believes an argument A to some degree when $P(A) > 0.5$, disbelieves an argument to some degree when $P(A) < 0.5$, and neither believes nor disbelieves an argument when $P(A) = 0.5$. Here we extend the approach with uncertainty over attacks. For this, we introduce the probability of attack (i.e. the degree of belief that the attacker does indeed attack the attackee). We use two functions in the definition because we want to investigate the interplay between them.

Definition 8. *An* **epistemic bidistribution** *is a pair* (P^a, P^r) *where*

- P^a *is a function* $P^a : 2^{\mathcal{A}} \rightarrow [0,1]$ *with* $\sum_{\mathcal{S} \subseteq \mathcal{A}} P^a(\mathcal{S}) = 1$ *(***argument belief distribution***)*.
- P^r *is a function* $P^r : 2^{\mathcal{R}} \rightarrow [0,1]$ *with* $\sum_{\mathcal{S} \subseteq \mathcal{R}} P^r(\mathcal{S}) = 1$ *(***attack belief distribution***)*.

The **probability of an argument** A *is* $P^a(A) = \sum_{\mathcal{S} \subseteq \mathcal{A} \ s.t. \ A \in \mathcal{S}} P^a(\mathcal{S})$. *The* **probability of an attack** α *is* $P^r(\alpha) = \sum_{\mathcal{S} \subseteq \mathcal{R} \ s.t. \ \alpha \in \mathcal{S}} P^r(\mathcal{S})$. *Finally, let* $P^b(X)$ *denote* $P^a(X)$ *(resp.* $P^r(X)$*) when* $X \in \mathcal{A}$ *(resp.* $X \in \mathcal{R}$*)*.

In order to simplify the notation, we drop the brackets for representing the probability of an attack relation, i.e., for $(A,B) \in \mathcal{R}$, instead of $P^r((A,B))$ we write $P^r(A,B)$.

The epistemic probability distributions can be constrained by imposing rationality postulates. In what follows we will build up on some of the postulates from [18] and introduce some new ones. The previous results can be retrieved by considering bidistributions in which all attacks are believed. We separate our new approaches into two families of postulates.

We start with the independent family of postulates in Definition 9 and give results on inter-relationships in Fig. 2 where $\mathcal{P}_\mu$ is the set of bidistributions satisfying postulate μ in G. The family is called independent because there is no dependence imposed between belief in attacks and belief in attackers, i.e. the probabilities assigned to an attack α and to $src(\alpha)$ are not necessarily related. RAT*, TER*, COH*, and OPT* require that both the attacker and the attack itself need to be believed in order to affect the attackee, or that either of them can be disbelieved in order for belief in the target. For RAT* (resp. STC*), if an attacker and its attack are believed, then the attackee is not believed (resp. disbelieved). As a dual for STC*, PRO* ensures that if an attack and attackee are believed, the attacker is not believed. TRU* requires that an argument is believed when there is no evidence to the contrary. By DIS*, an argument can only be disbelieved for a reason. TER* simply limits beliefs to three values corresponding precisely to the in, out and und statuses from the standard semantics. The ABIN* postulate prohibits being undecided about beliefs. Finally, while all the previous properties consider belief and disbelief, the COH* and OPT* properties give margins for probability assignments—one focuses on the upper, the other on the lower bound. By varying the use of also undecided attacks, we can specialize our axioms further, as seen in the case of RPRO* and RCOH*.

Definition 9. *(The* **independent family** *of postulates). An epistemic bidistribution* (P^a, P^r) *is:*

- **(RAT*)** **rational*** *if for all* $A, B \in \mathcal{A}$ *s.t.* $(A,B) \in \mathcal{R}$ *and* $P^r(A,B) > 0.5$, $P^a(A) > 0.5$ *implies* $P^a(B) \leq 0.5$.
- **(STC*)** **strict*** *if for all* $A, B \in \mathcal{A}$, *s.t.* $(A,B) \in \mathcal{R}$ *and* $P^r(A,B) > 0.5$, $P^a(A) > 0.5$ *implies* $P^a(B) < 0.5$.
- **(PRO*)** **protective*** *if for all* $A, B \in \mathcal{A}$ *s.t.* $(A,B) \in \mathcal{R}$ *and* $P^r(A,B) > 0.5$, $P^a(B) > 0.5$ *implies* $P^a(A) < 0.5$.

(RPRO*) **restricted protective*** *if for all $A, B \in \mathcal{A}$ s.t. $(A, B) \in \mathcal{R}$ and $P^r(A, B) \geq 0.5$, $P^a(B) > 0.5$ implies $P^a(A) < 0.5$.*

(TRU*) **trusting*** *if for every $B \in \mathcal{A}$, it holds that if for all $C \in \mathcal{A}$ s.t. $(C, B) \in \mathcal{R}$, either $P^a(C) < 0.5$, or $P^r(C, B) < 0.5$, then $P^a(B) > 0.5$.*

(DIS*) **discharging*** *if for every $B \in \mathcal{A}$, if $P^a(B) < 0.5$, then there exists $C \in \mathcal{A}$ s.t. $(C, B) \in \mathcal{R}$, $P^r(C, B) > 0.5$ and $P^a(C) > 0.5$.*

(TER*) **ternary*** *if for all $X \in \mathcal{A} \cup \mathcal{R}$, $P^b(X) \in \{0, 0.5, 1\}$.*

(ABIN*) **attack binary*** *if for all $X \in \mathcal{R}$, $P^r(X) \neq 0.5$.*

(COH*) **coherent*** *if for all $A, B \in \mathcal{A}$ s.t. $(A, B) \in \mathcal{R}$ and $P^r(A, B) > 0.5$, $P^a(A) \leq 1 - P^a(B)$.*

(RCOH*) **restricted coherent*** *if for all $A, B \in \mathcal{A}$, s.t. $(A, B) \in \mathcal{R}$ and $P^r(A, B) \geq 0.5$, $P^a(A) \leq 1 - P^a(B)$.*

(OPT*) **optimistic*** *if for every $A \in \mathcal{A}$, it holds that*
$$P^a(A) \geq 1 - \sum_{B \ s.t. \ (B,A) \in \mathcal{R}, P^r(B,A) > 0.5 \ and \ P^a(B) > 0.5} P^a(B).$$

In the independent family, the belief we have in an attacker does not constrain the belief we may have in its attack. We consider it an intuitive modeling, as we do not have to believe two arguments in order to acknowledge a conflict between them. Imagine two people witnessing a robbery, one claiming that the criminal ran away in a car, the other that he used a bike. The statements are clearly conflicting and we can believe the attacks between them independently of the belief we have in the witnesses. Similarly, we do not need to believe a given attack even if we believe the arguments participating in it, as exemplified in the introduction.

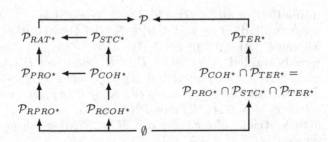

Fig. 2. Relationships for the independent family of postulates where $\mathcal{P}_{\mu_1} \rightarrow \mathcal{P}_{\mu_2}$ denotes $\mathcal{P}_{\mu_1}(G) \subseteq \mathcal{P}_{\mu_2}(G)$

We present a second family of postulates in Definition 10, called the dependent family, and give results on inter-relationships in Fig. 3. This second family is motivated by the observation that in some situations (e.g. when argument graphs are obtained from logical knowledge bases), it is natural to expect that there is a dependence between belief in an attacker and its attack. Moreover, in many approaches that explicitly include the attacks in extensions and labelings,

230   S. Polberg et al.

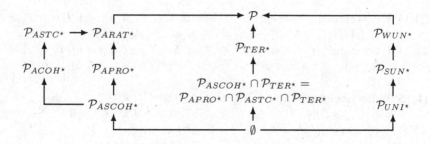

Fig. 3. Relationships for the dependent family of postulates where $\mathcal{P}_{\mu_1} \to \mathcal{P}_{\mu_2}$ denotes $\mathcal{P}_{\mu_1}(G) \subseteq \mathcal{P}_{\mu_2}(G)$

the conflicts need to conform to the same semantics as the arguments. Consequently, we can demand that the belief in an argument affects the belief in its attacks and vice versa. For this, we introduce the UNI*, SUN* and WUN* postulates below. Similarly, we also present the attack postulates, which constrain the belief both in the attacked argument and the conflict whose source is attacked, thus implicitly acknowledging the dependency between the two. Moreover, while in the independent family the beliefs in the attack and the attacker had to be mentioned explicitly due to their independence, in this family we consider just the attack itself. This also reflects the intuition behind the attack–based approach.

Definition 10. *(The **dependent family** of postulates). An epistemic bidistribution (P^a, P^r) is:*

- **(UNI*)** **unified*** *if for all $(A, B) \in \mathcal{R}$, $P^r(A, B) = P^a(A)$*
- **(SUN*)** **semi–unified*** *if for all $(A, B) \in \mathcal{R}$, $P^a(A) > 0.5$ iff $P^r(A, B) > 0.5$ and $P^a(A) < 0.5$ iff $P^r(A, B) < 0.5$.*
- **(WUN*)** **weakly unified*** *if for all $(A, B) \in \mathcal{R}$, either both $P^r(A, B) \geq 0.5$ and $P^a(A) \geq 0.5$ or both $P^r(A, B) \leq 0.5$ and $P^a(A) \leq 0.5$.*
- **(ARAT*)** **attack rational*** *iff for every $\alpha \in \mathcal{R}$, if $P^r(\alpha) > 0.5$ and α is an attacker* of $X \in \mathcal{A} \cup \mathcal{R}$, then $P^b(X) \leq 0.5$.*
- **(ASTC*)** **attack strict*** *iff for every $\alpha \in \mathcal{R}$, if $P^r(\alpha) > 0.5$ and α is an attacker* of $X \in \mathcal{A} \cup \mathcal{R}$, then $P^b(X) < 0.5$.*
- **(APRO*)** **attack protective*** *iff for every $X \in \mathcal{A} \cup \mathcal{R}$ and $\alpha \in \mathcal{R}$ s.t. α is an attacker* of X, if $P^b(X) > 0.5$, then $P^r(\alpha) < 0.5$.*
- **(ATRŮ*)** **attack trusting*** *iff for every $X \in \mathcal{A} \cup \mathcal{R}$, it holds that if for every attacker* $\beta \in \mathcal{R}$ of X it is the case that $P^r(\beta) < 0.5$, then $P^b(X) > 0.5$.*
- **(ADIS*)** **attack discharging*** *iff for every $X \in \mathcal{A} \cup \mathcal{R}$, if $P^b(X) < 0.5$, then there exists an attacker* $\beta \in \mathcal{R}$ of X s.t. $P^r(\beta) > 0.5$.*
- **(ACOH*)** **attack coherent*** *iff for every $X \in \mathcal{A} \cup \mathcal{R}$ and $\alpha \in \mathcal{R}$ s.t. α is an attacker* of X, if $P^r(\alpha) > 0.5$, then $P^r(\alpha) \leq 1 - P^b(X)$.*
- **(ASCOH*)** **attack strongly coherent*** *iff for every $X \in \mathcal{A} \cup \mathcal{R}$ and $\alpha \in \mathcal{R}$ s.t. α is an attacker* of X, if $P^r(\alpha) \geq 0.5$, then $P^r(\alpha) \leq 1 - P^b(X)$.*

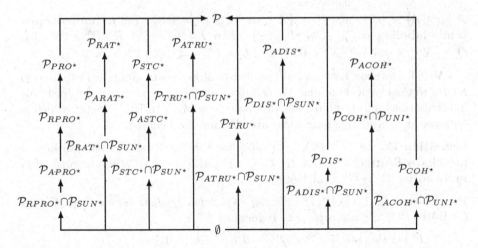

Fig. 4. Classes of probability functions where $\mathcal{P}_{\mu_1} \rightarrow \mathcal{P}_{\mu_2}$ denotes $\mathcal{P}_{\mu_1}(G) \subseteq \mathcal{P}_{\mu_2}(G)$

Some relationships between the two postulate families are given in Fig. 4.

Although the independent family is more argument-driven, while the dependent family is more attack-driven, there is a meeting point between them. In particular, by the use of postulates that tie the belief we have in a conflict to the belief we have in its source, we can up to a certain degree replace one family with the other.

Example 3. Consider again G_2 from Example 1 and an epistemic bidistribution (P^a, P^r) (partially) defined through the following constraints:

$$P^a(A) = 0.9 \qquad P^a(B) = 0.5 \qquad P^a(C) = 0.6$$
$$P^r(r_1) = 0.7 \qquad P^r(r_2) = 0.6$$

Then (P^a, P^r) is (among others) rational*, weakly unified*, and (trivially) attack trusting*. It is, for example, not strict*, not protective*, and not attack strongly coherent*.

5   Relationship with Classical Semantics

In order to compare the extended epistemic approach with Dung's classical approach, we introduce the notions of epistemic and extended epistemic labelings. Elements that are believed (disbelieved or neither) are simply assigned in (respectively, out and und). We will show which postulates need to be satisfied in order for the (extended) epistemic labelings to conform to the desired classical semantics and vice versa.

Definition 11. *Let (P^a, P^r) be an epistemic bidistribution. The **epistemic labeling** is $L_{P^a} = (I, O, U)$, where $I = \{A \in \mathcal{A} \mid P^a(A) > 0.5\}$, $O = \{A \in$*

$\mathcal{A} \mid P^a(A) < 0.5\}$, and $U = \{A \in \mathcal{A} \mid P^a(A) = 0.5\}$. The **extended epistemic labeling** is $L^*_{P^a,P^r} = (I, O, U)$, where $I = \{X \in \mathcal{A} \cup \mathcal{R} \mid P^b(X) > 0.5\}$, $O = \{X \in \mathcal{A} \cup \mathcal{R} \mid P^b(X) < 0.5\}$, and $U = \{X \in \mathcal{A} \cup \mathcal{R} \mid P^b(X) = 0.5\}$.

We will first show how our independent family of postulates (i. e. Definition 9) relates to the classical notions. We can draw a connection between the epistemic bidistributions and the labelings of a subgraph of the original graph, which is obtained by considering only those attacks that are believed:

Definition 12. *Let (P^a, P^r) be an epistemic bidistribution. The set of* **believed attacks** *is* $\mathsf{BAtts}(P^r) = \{(A, B) \in \mathcal{R} \mid P^r(A, B) > 0.5\}$. *The subgraph of G* **induced by** P^r *is* $G' = (\mathcal{A}, \mathsf{BAtts}(P^r))$.

Proposition 2. *Let (P^a, P^r) be an epistemic bidistribution and $G' = (\mathcal{A}, \mathsf{BAtts}(P^r))$ the subgraph of G induced by P^r.*

- $(P^a, P^r) \in \mathcal{P}_{RAT^*}(G) \cap \mathcal{P}_{DIS^*}(G)$ iff $L_{P^a} \in \mathsf{cf}(G')$.
- $(P^a, P^r) \in \mathcal{P}_{PRO^*}(G) \cap \mathcal{P}_{DIS^*}(G)$ iff $L_{P^a} \in \mathsf{ad}(G')$.
- $(P^a, P^r) \in \mathcal{P}_{PRO^*}(G) \cap \mathcal{P}_{STC^*}(G) \cap \mathcal{P}_{DIS^*}(G) \cap \mathcal{P}_{TRU^*}(G)$ iff $L_{P^a} \in \mathsf{co}(G')$.

We can observe that although one attack distribution induces only one subgraph, a single subgraph can be induced by multiple distributions. This is due to the fact that the removal of the attacks depends on whether an attack is believed at all, not on the degree of this belief. Moreover, there can be infinitely many argument distributions associated with a single complete labeling due to the fact that any of the values from $[0, 0.5)$ (or $(0.5, 1]$) can lead to an out (or in) assignment of a given labeling. Although this is to be expected taking into account the fact that probabilistic semantics carry more information than the classical ones, we would also like to distinguish those probability functions that can be uniquely associated with a given subgraph and its complete labelings. We thus propose the definition of complete probability bidistributions; a given subgraph can be induced only by a single ternary and attack binary distribution, while ternary, trusting, disapproving and coherent postulates lead to a tighter relation with the labelings:

Definition 13. *An epistemic bidistribution (P^a, P^r) is* **complete** *iff $(P^a, P^r) \in \mathcal{P}_{COH^*}(G) \cap \mathcal{P}_{DIS^*}(G) \cap \mathcal{P}_{TRU^*}(G) \cap \mathcal{P}_{TER^*}(G) \cap \mathcal{P}_{ABIN^*}(G)$.*

From this, we can further define the preferred, complete and stable bidistributions that lead to appropriate labelings in the associated subgraph by maximizing or minimizing particular assignments, similarly as in the classical semantics. In this case, instead of focusing on in, out and und assignments, we look for probabilities 1, 0 and 0.5.

Although the extended epistemic approach is quite general, the epistemic labelings without any constraints on the attack distributions are connected to the labellings of the subgraphs of a given framework, not necessarily the framework itself. However, if we apply the dependency postulates from the dependent family (Definition 10)— in particular, the semi-unified one—we can observe that we can focus on the original graph again. The only difference wrt. the previous results is the use of the restricted, not standard protectiveness.

Proposition 3. *The following holds:*

- If $L \in \mathsf{cf}(G)$, then there exists $(P^a, P^r) \in \mathcal{P}_{RAT^\star}(G) \cap \mathcal{P}_{DIS^\star}(G) \cap \mathcal{P}_{SUN^\star}(G)$ s.t. $L = L_{P^a}$.
- If $L \in \mathsf{ad}(G)$, then there exists $(P^a, P^r) \in \mathcal{P}_{RPRO^\star}(G) \cap \mathcal{P}_{DIS^\star}(G) \cap \mathcal{P}_{SUN^\star}(G)$ s.t. $L = L_{P^a}$.
- If $L \in \mathsf{co}(G)$, then there exists $(P^a, P^r) \in \mathcal{P}_{RPRO^\star}(G) \cap \mathcal{P}_{STC^\star}(G) \cap \mathcal{P}_{DIS^\star}(G) \cap \mathcal{P}_{TRU^\star}(G) \cap \mathcal{P}_{SUN^\star}(G)$ s.t. $L = L_{P^a}$.
- If $(P^a, P^r) \in \mathcal{P}_{RAT^\star}(G) \cap \mathcal{P}_{DIS^\star}(G) \cap \mathcal{P}_{SUN^\star}(G)$, then $L_{P^a} \in \mathsf{cf}(G)$.
- If $(P^a, P^r) \in \mathcal{P}_{RPRO^\star}(G) \cap \mathcal{P}_{DIS^\star}(G) \cap \mathcal{P}_{SUN^\star}(G)$, then $L_{P^a} \in \mathsf{ad}(G)$.
- If $(P^a, P^r) \in \mathcal{P}_{RPRO^\star}(G) \cap \mathcal{P}_{STC^\star}(G) \cap \mathcal{P}_{DIS^\star}(G) \cap \mathcal{P}_{TRU^\star}(G) \cap \mathcal{P}_{SUN^\star}(G)$, then $L_{P^a} \in \mathsf{co}(G)$.

This leads to the following complete probability bidistribution, which can uniquely describe the complete labelings of the underlying framework. Using this, we can also retrieve the preferred, grounded and stable labellings as for the classical case (as discussed in Sect. 2).

Definition 14. *An epistemic bidistribution (P^a, P^r) is **jointly complete** iff $(P^a, P^r) \in \mathcal{P}_{SUN^\star}(G) \cap \mathcal{P}_{RCOH^\star}(G) \cap \mathcal{P}_{DIS^\star}(G) \cap \mathcal{P}_{TRU^\star}(G) \cap \mathcal{P}_{TER^\star}(G)$.*

Let us now focus on the extended classical semantics. As we could have already observed in Example 1, the admissible* labelings were not necessarily corresponding to the admissible ones. However, we can easily grasp it with our attack epistemic postulates.

Proposition 4. *The following holds:*

- $(P^a, P^r) \in \mathcal{P}_{WUN^\star}(G) \cap \mathcal{P}_{ARAT^\star}(G) \cap \mathcal{P}_{ADIS^\star}(G)$ iff $L^\star_{P^a, P^r} \in \mathsf{cf}^\star(G)$.
- $(P^a, P^r) \in \mathcal{P}_{WUN^\star}(G) \cap \mathcal{P}_{APRO^\star}(G) \cap \mathcal{P}_{ADIS^\star}(G)$ iff $L^\star_{P^a, P^r} \in \mathsf{ad}^\star(G)$.
- $(P^a, P^r) \in \mathcal{P}_{SUN^\star}(G) \cap \mathcal{P}_{APRO^\star}(G) \cap \mathcal{P}_{ASTC^\star}(G) \cap \mathcal{P}_{ADIS^\star}(G) \cap \mathcal{P}_{ATRU^\star}(G)$ iff $L^\star_{P^a, P^r} \in \mathsf{co}^\star(G)$.

The fact that the complete* labelings correspond to bidistributions satisfying the SUN* postulate gives us one more important result. In particular, under the SUN* postulate we can replace the other postulates from the dependent family with their counterparts from the independent family. This also means that we can use the jointly complete bidistributions in order to uniquely retrieve the extended labelings of G that are at least complete*.

Theorem 1. *Let (P^a, P^r) be an epistemic bidistribution. Then $L^\star_{P^a, P^r} \in \mathsf{co}^\star(G)$ iff $(P^a, P^r) \in \mathcal{P}_{SUN^\star}(G) \cap \mathcal{P}_{RPRO^\star}(G) \cap \mathcal{P}_{STC^\star}(G) \cap \mathcal{P}_{DIS^\star}(G) \cap \mathcal{P}_{TRU^\star}(G)$.*

These results show that our new proposal for epistemic probabilities can generalize a wider range of argumentation semantics than the original one [18]. Moreover, what we have presented can be easily extended to handle the attack-based semantics from [24] and recursive attacks from [2].

6   Comparison with Constellations Approach

The reasoning behind the epistemic and constellations approaches is different, with the former intended to reflect the belief in arguments and attacks, and the latter expressing the uncertainty concerning the topology of the graph, e. g., as to which arguments and attacks are known about or what elements should appear in the graph. Nevertheless, we can still draw some connections between them. We can observe that in a subgraph distribution assigning non-zero probability only to subgraphs without attacks, the grounded extension of each subgraph would consist of all of its arguments. These extensions and their probabilities produce an argument distribution. Thus, the constellations approach can up to some extent mimic the epistemic approach:

Proposition 5. *For each argument-belief distribution P^a over G, there is a constellations distribution P^c over $\mathsf{Sub}(G)$ s.t. for all arguments A in $\mathcal{A}$, $P^a(A) = P^c_{\mathrm{gr}}(A)$.*

In turn, a spanning or full subgraph distribution can be simulated with the attack or argument belief distribution due to the fact that part of a subgraph becomes "fixed" and not directly subject to any uncertainty.

Proposition 6. *For each spanning subgraph distribution P^c over G, there is an attack belief distribution P^r s.t. for all subgraphs $G' \sqsubseteq G$, and for all sets of attacks $\mathcal{S} \subseteq \mathcal{R}$, if $\mathcal{R}' = \mathcal{S}$, then $P^c(G') = P^r(\mathcal{S})$.*

Proposition 7. *For each full subgraph distribution P^c over G, there is an argument belief distribution P^a s.t. for all subgraphs $G' \sqsubseteq G$, and for all sets of arguments $\mathcal{S} \subseteq \mathcal{A}$, if $\mathcal{A}' = \mathcal{S}$, then $P^c(G') = P^a(\mathcal{S})$.*

However, we can observe that if a subgraph distribution is neither a full subgraph distribution nor a spanning subgraph distribution, then the constellations approach cannot be captured by the epistemic approach. Moreover, in the constellations approach, the marginal value of a given argument (i. e. the total probability of subgraphs containing this argument) is never less than the marginal for any attack involving that argument. In contrast, the belief in an attacker can be greater than then belief in the attack or attackee. This shows that, in general, the epistemic approach cannot be captured by the constellations method.

Definition 15. *Let P^c be a subgraph distribution. The **argument marginal function** is $P^m(A) = \sum_{G' \in \mathsf{Sub}(G)\ s.t.\ A \in \mathcal{A}'} P(G')$. The **attack marginal function** is $P^m(A, B) = \sum_{G' \in \mathsf{Sub}(G)\ s.t.\ (A,B) \in \mathcal{R}'} P(G')$.*

Proposition 8. *Let P^c be a subgraph distribution. For all $(A, B) \in \mathcal{R}$, $P^c(A) \geq P^c(A, B)$.*

Example 4. Consider the graph $G_1 = (\{A, B, C\}, \{(A, B), (B, A), (C, B)\})$ and its subgraphs $G_2 = (\{A, B, C\}, \{(A, B), (B, A))\})$ and $G_3 = (\{A, B\}, \{(A, B)\})$.

For this graph, we consider the subgraph distribution $P^c(G_1) = 0.3$, $P^c(G_2) = 0.5$ and $P^c(G_3) = 0.2$, which is neither a full subgraph nor a spanning subgraph distribution. We cannot use P^a or P^r to represent P^c.

We can now consider an epistemic bidistribution (P^a, P^r) s.t. $P^r(\{A, B\}) = 1$ and for every set $\mathcal{S} \subseteq \mathcal{A}$ s.t. $A \in \mathcal{S}$, $P^a(\mathcal{S}) = 0$ (the remaining assignments are arbitrary as long as we obtain a distribution). Therefore, $P^a(A) < P^a((A, B))$. But there cannot be any subgraph distribution P^c for G_1 s.t. $P^m(A) < P^m((A, B))$.

These results show that extended epistemic and constellations approach, although related, do not subsume each other.

7 Conclusions

In this paper, we extend the epistemic approach to account for belief in attacks as well as arguments. We do this by introducing the notion of an epistemic bidistribution. We then provide two families of postulates that offer a variety of ways of constraining the bidistributions according to different notions of rational behaviour, give some relationships between these two families, and show how these postulates relate to classical semantics for abstract argumentation, and we show how the extended epistemic and constellations approaches do not subsume each other.

Important dimensions for probabilistic argumentation include the constellations approach to abstract argumentation (e.g. [6,7,9,19,20]), the equational approach to abstract argumentation [10], and probabilistic structured argumentation (e.g. [7,21,23]). The extended epistemic approach is complementary to these existing approaches (see Sect. 6, for differences with the constellations approach, and see [10], for differences between the epistemic and equational approaches).

The epistemic approach is a promising approach to user modelling in persuasion where a persuader can model the beliefs in arguments of the persuadee and update the model during a dialogue [11,15,16], and the user model can be harnessed to make strategic choices of move in a dialogue using decision theory [12]. The extended epistemic approach offers richer user models, and pontentially more effective decisions about moves (as indicated by our example in Sect. 1).

References

1. Baroni, P., Caminada, M., Giacomin, M.: An introduction to argumentation semantics. Knowl. Eng. Rev. **26**(4), 365–410 (2011)
2. Baroni, P., Cerutti, F., Giacomin, M., Guida, G.: AFRA: argumentation framework with recursive attacks. Int. J. Approximate Reasoning **52**(1), 19–37 (2011)
3. Baroni, P., Giacomin, M., Vicig, P.: On rationality conditions for epistemic probabilities in abstract argumentation. In: Proceedings of COMMA 2014. FAIA, vol. 266, pp. 121–132. IOS Press (2014)

4. Brewka, G., Polberg, S., Woltran, S.: Generalizations of Dung frameworks and their role in formal argumentation. IEEE Intell. Syst. **29**(1), 30–38 (2014)
5. Caminada, M., Gabbay, D.M.: A logical account of formal argumentation. Stud. Logica. **93**, 109–145 (2009)
6. Dondio, P.: Multi-valued and probabilistic argumentation frameworks. In: Proceedings of COMMA 2014. FAIA, vol. 266, pp. 253–260. IOS Press (2014)
7. Dung, P., Thang, P.: Towards (probabilistic) argumentation for jury-based dispute resolution. In: Proceedings of COMMA 2010. FAIA, vol. 216, pp. 171–182. IOS Press (2010)
8. Dung, P.M.: On the acceptability of arguments and its fundamental role in non-monotonic reasoning, logic programming and n-person games. Artif. Intell. **77**(2), 321–358 (1995)
9. Fazzinga, B., Flesca, S., Parisi, F.: On the complexity of probabilistic abstract argumentation frameworks. ACM Trans. Comput. Logic **16**(3), 22:1–22:39 (2015)
10. Gabbay, D., Rodrigues, O.: Probabilistic argumentation: an equational approach. Log. Univers. **9**(3), 345–382 (2015)
11. Hadoux, E., Hunter, A.: Computationally viable handling of beliefs in arguments for persuasion. In: Proceedings of ICTAI 2016, pp. 319–326. IEEE (2016)
12. Hadoux, E., Hunter, A.: Strategic sequences of arguments for persuasion using decision trees. In: Proceedings of AAAI 2017, pp. 1128–1134. AAAI Press (2017)
13. Hunter, A.: Some foundations for probabilistic abstract argumentation. In: Proceedings of COMMA 2012. FAIA, vol. 245, pp. 117–128. IOS Press (2012)
14. Hunter, A.: A probabilistic approach to modelling uncertain logical arguments. Int. J. Approximate Reasoning **54**(1), 47–81 (2013)
15. Hunter, A.: Modelling the persuadee in asymmetric argumentation dialogues for persuasion. In: Proceedings of IJCAI 2015, pp. 3055–3061. AAAI Press (2015)
16. Hunter, A.: Persuasion dialogues via restricted interfaces using probabilistic argumentation. In: Schockaert, S., Senellart, P. (eds.) SUM 2016. LNCS, vol. 9858, pp. 184–198. Springer, Cham (2016). doi:10.1007/978-3-319-45856-4_13
17. Hunter, A., Thimm, M.: Probabilistic argument graphs for argumentation lotteries. In: Proceedings of COMMA 2014. FAIA, vol. 266, pp. 313–324. IOS Press (2014)
18. Hunter, A., Thimm, M.: Probabilistic argumentation with epistemic extensions and incomplete information. Technical report, ArXiv, May 2014
19. Li, H., Oren, N., Norman, T.J.: Probabilistic argumentation frameworks. In: Modgil, S., Oren, N., Toni, F. (eds.) TAFA 2011. LNCS, vol. 7132, pp. 1–16. Springer, Heidelberg (2012). doi:10.1007/978-3-642-29184-5_1
20. Polberg, S., Doder, D.: Probabilistic abstract dialectical frameworks. In: Fermé, E., Leite, J. (eds.) JELIA 2014. LNCS, vol. 8761, pp. 591–599. Springer, Cham (2014). doi:10.1007/978-3-319-11558-0_42
21. Riveret, R., Rotolo, A., Sartor, G., Prakken, H., Roth, B.: Success chances in argument games: a probabilistic approach to legal disputes. In: Proceedings of JURIX 2007, pp. 99–108. IOS Press (2007)
22. Thimm, M.: A probabilistic semantics for abstract argumentation. In: Proceedings of ECAI 2012. FAIA, vol. 242, pp. 750–755. IOS Press (2012)
23. Timmer, S.T., Meyer, J.-J.C., Prakken, H., Renooij, S., Verheij, B.: Explaining Bayesian networks using argumentation. In: Destercke, S., Denoeux, T. (eds.) ECSQARU 2015. LNCS, vol. 9161, pp. 83–92. Springer, Cham (2015). doi:10.1007/978-3-319-20807-7_8
24. Villata, S., Boella, G., van der Torre, L.: Attack semantics for abstract argumentation. In: Proceedings of IJCAI 2011, pp. 406–413. AAAI Press (2011)

A Parametrized Ranking-Based Semantics for Persuasion

Elise Bonzon[1], Jérôme Delobelle[2](✉), Sébastien Konieczny[2], and Nicolas Maudet[3]

[1] LIPADE, Université Paris Descartes, Paris, France
bonzon@parisdescartes.fr
[2] CRIL, CNRS - Université d'Artois, Lens, France
{delobelle,konieczny}@cril.fr
[3] Sorbonne Universités, UPMC Univ Paris 06,
CNRS - LIP6, UMR 7606, Paris, France
nicolas.maudet@lip6.fr

Abstract. In this paper we question the ability of the existant ranking semantics for argumentation to capture persuasion settings, emphasizing in particular the phenomena of protocatalepsis (the fact that it is often efficient to anticipate the counter-arguments of the audience), and of fading (the fact that long lines of argumentation become ineffective). It turns out that some widely accepted principles of ranking-based semantics are incompatible with a faithful treatment of these phenomena. We thus propose a parametrized semantics based on propagation of values, which allows to control the scope of arguments to be considered for evaluation. We investigate its properties (identifying in particular threshold values guaranteeing that some properties hold), and report experimental results showing that the family of rankings that may be returned have a high coherence rate.

Keywords: Argumentation · Persuasion · Ranking semantics

1 Introduction

Recently, the quest for a principled method to analyse networks of contradictory arguments has stimulated a number of work. Taken in their abstract form, such networks are *argumentation frameworks*, as defined by Dung [10]. Sharing the view that identifying sets of mutually acceptable arguments (extensions) is sometimes not sufficient, many "gradual" (returning a value) [3,9,16,17] or "ranking" (returning an order) semantics have been proposed [1,2,6,7,13,19,20]. Each of these proposals has some merit, and nicely designed examples convince indeed that, in some situations at least, they should be the method of choice. When it comes to comparing these approaches (beyond their formal properties like convergence or uniqueness of solution), things become difficult. This is so because the basis of comparison is not so clear in the first place, different proposals emphasizing different properties. In [5], many existing semantics were

© Springer International Publishing AG 2017
S. Moral et al. (Eds.): SUM 2017, LNAI 10564, pp. 237–251, 2017.
DOI: 10.1007/978-3-319-67582-4_17

compared on the basis on all the axioms mentioned in the literature. However, even the relevance of some axioms may be very much dependent on the context of application. What is often missing to compare these approaches is thus a clear indication of the applications they target.

In this paper, we aim at defining a good ranking semantics for persuasion. In this context, what constitutes an efficient argumentation has been rather extensively studied, and may constitute an interesting basis for comparaison. We shall concentrate on two well documented phenomena:

- *Procatalepsis*: anticipating the counter-arguments of an audience [22] is often a way to strengthen his own arguments, and many phenomena are well documented. To illustrate this, we extend an example from Besnard and Hunter [4, p. 85]: a (made-up) sales pitch intended to persuade to buy a specific car.

  (a1) The car x is a high performance family car with a diesel engine and a price of 32000
  (a2) In general, diesel engines have inferior performance compared with gazoline engines
  (a3) But, with these new engines, the difference in performance [...] is negligible.
  (a4) You may find that the price is high
  (a5) But it will be amortized because Diesel engines last longer than other engines.

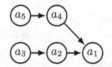

Apart from the fact that the example predates recent diesel scandals, what is striking is that it blatantly contradicts an axiom "Void Precedence" satisfied (to the best of our knowledge) by *all* gradual semantics and which considers non-attacked arguments as the most acceptable arguments. In this kind of persuasion contexts, it is clearly more convincing to state the more plausible counter-argument to (a_1) in order to provide some convincing defenses against them, that simply state (a_1) alone.

- *Fading*: Long lines of argumentation become ineffective in practice, because the audience easily looses track of the relation between the arguments. This is supported by recent evidence [21] which shows (in the context of their study, an extensive analysis of debates which took place on the subreddit "ChangeMyView"), that the arguments located at a distance of 10 from an other argument is about the limit. While some ranking semantics incorporate features which can be used to *discount* the strength of arguments relatively to their distance, this is not the case of all semantics.

We conclude that current ranking semantics are poorly equipped to be used in a context of persuasion. Our research question is thus to design a ranking-based semantics suited for persuasion, catering in particular for the fading effect and the fact void precedence might not be satisfied.

Our contribution is a gradual semantics which permits to account for these phenomena. Most importantly, some parameter allows to regulate how these principles are respected. Our vision is that, equipped with our ranking semantics, a seller facing different sales pitch might decide which is more likely to make the case. In general, this contribution could thus be used as an ingredient for developing strategies for computation persuasion techniques [14].

The remainder of this paper is as follows. In Sect. 2 we recall the necessary background in formal argumentation. Section 3.1 presents the basic principle of propagation on which our proposal, detailed in Sect. 3.2, is built. This semantics takes a parameter, the attenuation factor δ, which allows to control the convergence speed and the obtained rankings. More importantly, we study in detail the relation between this factor and the "Void Precedence" axiom. In order to allow proper comparison with other proposals in the literature, the paper concludes with a study of the axiomatic properties of our semantics, as well as an extended example.

2 Preliminaries

Following Dung, we define an *argumentation framework* (AF) as a binary attack-relation over a (finite) set of abstracts arguments.

Definition 1 ([10]). *An* **argumentation framework (AF)** *is a pair* $F = \langle \mathcal{A}, \mathcal{R} \rangle$ *where* $\mathcal{A}$ *is a set of* **arguments***, and* $\mathcal{R} \subseteq \mathcal{A} \times \mathcal{A}$ *is a binary relation called the* **attack relation***. Notation* $(a, b) \in \mathcal{R}$ *means that a attacks b. Let* $Arg(F) = \mathcal{A}$.

One of the main goals of argumentation theory is to identify which arguments are rationally *acceptable* according to different notions of acceptability. In [10], the acceptability of an argument depends on its membership to some extensions, whereas ranking-based semantics aim to rank arguments from the most to the least acceptable ones.

Definition 2. *A* **ranking semantics** σ *associates to any argumentation framework* $F = \langle \mathcal{A}, \mathcal{R} \rangle$ *a ranking* $\succeq_F^\sigma$ *on* $\mathcal{A}$*, where* $\succeq_F^\sigma$ *is a preorder (a reflexive and transitive relation) on* $\mathcal{A}$*. $a \succeq_F^\sigma b$ means that a is at least as acceptable as b ($a \simeq_F^\sigma b$ is a shortcut for $a \succeq_F^\sigma b$ and $b \succeq_F^\sigma a$, and $a \succ_F^\sigma b$ is a shortcut for $a \succeq_F^\sigma b$ and $b \not\succeq_F^\sigma a$).*

When the ranking semantics σ and the graph F is clear from the context, we will use $\succeq$ instead of $\succeq_F^\sigma$.

Let us introduce some notations that help us to define our ranking semantics in the next section.

Notation 1. *Let* $F = \langle \mathcal{A}, \mathcal{R} \rangle$ *and* $a, b \in \mathcal{A}$*. A* **path** *from a to b, denoted by $p(a, b)$, is a sequence of nodes $s = \langle a_0, \ldots, a_n \rangle$ such that from each node there is an edge to the next node in the sequence: $a_0 = a$, $a_n = b$ and $\forall i < n, (a_i, a_{i+1}) \in \mathcal{R}$. Its* **length** *is denoted by $|p(a, b)|$ and is equal to the number of edges it is composed of.*

Notation 2. *Let $\Delta_n(a) = \{b \mid \exists p(b,a), \ with \ |p(b,a)| = n\}$ be the set of arguments that are bound by a path of length n to the argument a. An argument $b \in \Delta_n(a)$ is a* **defender** *(resp.* **attacker***) of a if n is even (resp. odd). A path from b to a is a* **branch** *if b is not attacked, i.e. if $\Delta_1(b) = \emptyset$. It is a* **defense branch** *(resp.* **attack branch***) if b is a defender (resp. attacker) of a. $\Delta^{B^+}(a)$ (resp. $\Delta^{B^-}(a)$) denotes the set of all the defense (resp. attack) branches of a.*

While our method is general, in the context of this paper we shall also pay special attention to tree shaped argumentation frameworks where an argument a, called *root argument*, has only defense branches (*i.e.* $\Delta^{B^-}(a) = \emptyset$ and $\Delta^{B^+}(a) \neq \emptyset$). Such frameworks will be called **persuasion pitches**. The AF in the introduction is an example of persuasion pitch with a_1 as root argument.

Table 1. Computation of the valuation P of each argument from F_1 when $\epsilon = 0.5$ and $\delta = 0.4$

$P_i^{0.5,0.4}$	a	b	c	d	e	f
0	0.5	1	0.5	0.5	0.5	0.5
1	-0.1	1	0.1	0.1	0.3	0.3
2	-0.02	1	0.1	0.34	0.38	0.46
3	-0.052	1	0.1	0.308	0.316	0.364
$\vdots$	$\vdots$	$\vdots$	$\vdots$	$\vdots$	$\vdots$	$\vdots$
14	-0.0402	1	0.1	0.3161	0.3506	0.3736

3   Variable-Depth Propagation

3.1   The Propagation Principle

The semantics we propose in this paper follows the principle of *propagation* already used by some ranking semantics [6,20]. In short, the idea is to assign a positive initial value to each argument in the AF (arguments may start with the same initial value [20] or start with distinct values like in [6], where non-attacked arguments have greater value than attacked ones). Then each argument propagates its value into the argumentation framework, alternating the polarity according to the considered path (negatively if it is an attack path, positively if it is a defense one).

Inspired by these definitions, we formally define this propagation principle, including in addition a damping factor δ which allows to decrease the impact of attackers situated further away along a path (the longer the path length i, the smaller the δ^i). Among other things, such a damping factor will allow to guarantee the convergence of the computation of the arguments' values, as also proposed in [18,20].

Definition 3. *Let* $\langle \mathcal{A}, \mathcal{R} \rangle$ *be an argumentation framework. The valuation P of* $a \in \mathcal{A}$*, at step* i*, is given by:*

$$P_i^{\epsilon,\delta}(a) = \begin{cases} v_\epsilon(a) & if \; i = 0 \\ P_{i-1}^{\epsilon,\delta}(a) + (-1)^i \delta^i \sum_{b \in \Delta_i(a)} v_\epsilon(b) & otherwise \end{cases}$$

with $\delta \in \,]0,1[$ *be an attenuation factor and* $v_\epsilon : \mathcal{A} \to \mathbb{R}^+$ *a valuation function that assigns an initial weight to each argument, with* $\epsilon \in [0,1]$ *such that* $\forall b \in \mathcal{A}$*,* $v_\epsilon(b) = 1$ *if* $\Delta_1(b) = \emptyset$*;* $v_\epsilon(b) = \epsilon$ *otherwise.*

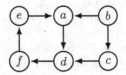

Fig. 1. The argumentation framework F_1

Example 1. *Let us compute the valuation* P *of each argument in* F_1 *(see Fig. 1) when* $\epsilon = 0.5$ *and* $\delta = 0.4$ *and give the results in the Table 1.*
In focusing on the argument f*, we can see that it begins with an initial weight of* 0.5 $(P_0^{0.5,0.4}(f) = 0.5)$ *because it is attacked. Then, it receives negatively the value sent by its direct attacker* d *which is also attacked:* $P_1^{0.5,0.4}(f) = P_0^{0.5,0.4}(f) -$ $0.4 \times v_{0.5}(d) = 0.3$*. Then, during the second step* $(i = 2)$*, it receives positively the weights from* a *and* c *attenuated by* δ^2*:* $P_2^{0.5,0.4}(f) = P_1^{0.5,0.4}(f) + 0.4^2 \times$ $(v_{0.5}(a) + v_{0.5}(c)) = 0.46$*. When* $i = 3$*, it receives negatively the weight of 1 from* b *and the weight of 0.5 from* e *attenuated by* δ^3*:* $P_3^{0.5,0.4}(f) = P_2^{0.5,0.4}(f) - 0.4^3 \times$ $(v_{0.5}(b) + v_{0.5}(e)) = 0.364$*. And so on and so forth.*

The following proposition answers to the question of convergence of the valuation P for each argument for every AF.

Proposition 1. *Let* $\langle \mathcal{A}, \mathcal{R} \rangle$ *be an argumentation framework and* $\delta \in \,]0,1[$*. For all* $a \in \mathcal{A}$*, the sequence* $\{P_i^{\epsilon,\delta}(a)\}_{i=0}^{+\infty}$ *converges.*

The **propagation number** of an argument a the value $P^{\epsilon,\delta}(a) = \lim_{i \to +\infty} P_i^{\epsilon,\delta}(a)$.

Example 1. (cont.) *The propagation number of each argument from* F_1 *(see Fig. 1) is represented in the shaded cell in the Table 1.*

3.2 The Two-Phase Propagation Method

The non-attacked arguments play a key role for assessing the acceptability of arguments in Dung's classical semantics. Although, as explained before, we do not necessarily want them ranked above any other, these arguments must keep a specific role, at least to distinguish attack and defense branches (as also suggested in the global valuation approach of Cayrol and Lagasquie-Schiex [7]). Our solution is a two-phase process. In the first phase, non-attacked arguments propagate their weights (=1) in the argumentation graph, while attacked arguments have weight 0. Any pairwise strict comparison (based on propagation number) resulting from this process is fixed. In the second phase (that is, to break ties among arguments equally valued in the first phase), we re-run the propagation phase, this time setting an initial weight $\epsilon \neq 0$. Formally:

Definition 4. *Let $\epsilon \in \,]0,1]$ and $\delta \in \,]0,1[$. The ranking-based semantics Variable-Depth Propagation $vdp^{\epsilon,\delta}$ associates to any argumentation framework $\langle \mathcal{A}, \mathcal{R} \rangle$ a ranking $\succeq$ on $\mathcal{A}$ such that $\forall a, b \in \mathcal{A}$:*

$$a \succeq b \text{ iff } P^{0,\delta}(a) > P^{0,\delta}(b) \text{ or } (P^{0,\delta}(a) = P^{0,\delta}(b) \text{ and } P^{\epsilon,\delta}(a) \geq P^{\epsilon,\delta}(b))$$

Example 1 (cont.) *According to the previous definition, we need first to compute the propagation number of each argument with $\epsilon = 0$. We obtain the following propagation numbers: $P^{0,0.4}(a) = -0.4105$, $P^{0,0.4}(b) = 1$, $P^{0,0.4}(c) = -0.4$, $P^{0,0.4}(d) = 0.1642$, $P^{0,0.4}(e) = 0.0263$ and $P^{0,0.4}(f) = -0.0656$. Thus, we can already obtain the following ranking:*

$$b \succ d \succ e \succ f \succ c \succ a$$

Note that no arguments are equally acceptable here, so it is not necessary to perform the second phase.

A concern might be that the value of ϵ used in the second phase might change the ranking obtained. We show that this is not the case:

Proposition 2. *Let $F = \langle \mathcal{A}, \mathcal{R} \rangle$ be an argumentation framework and $\delta \in \,]0,1[$. $\forall \epsilon, \epsilon' \in \,]0,1]$,*

$$vdp^{\epsilon,\delta} = vdp^{\epsilon',\delta}$$

Please note that even if different values of ϵ do not change the preorder, it is necessary to keep it in the process in order to make a distinction between non-attacked and attacked arguments (see Definition 3 about the valuation function v_ϵ). However, this is a purely internal artefact without any effect on the outcome of the method. To make this clear, we note vdp^δ instead of $vdp^{\epsilon,\delta}$ to describe our parametrized ranking semantics in general.

Now regarding δ, two different values can produce different preorders. Indeed, this parameter allows to choose the scope of influence of the arguments in the system in addition to allow the convergence of the valuation P. For instance, with a value of δ close to 0, only the nearest arguments (so a little part of the AF) are taken into consideration to compute the different propagation numbers whereas with a value of δ close to 1, (almost) all the AF will be inspected. Following the principle of the fading effect, it is natural to assume that arguments located at a long distance from another argument become ineffective. In terms of design, it seems very interesting to have the ability to control this parameter so as to

specify a maximal depth after which arguments see their influence on the value
of others vanish.

To better understand how to take this principle into account, let us inspect
the algorithm used to compute the propagation numbers. First, a positive num-
ber is assigned to each argument ($\forall a \in \mathcal{A}$, $P_0^{\epsilon,\delta}(a) = 1$ if a is non-attacked or
$P_0^{\epsilon,\delta}(a) = \epsilon$ otherwise). Then at step $i \in \mathbb{N}$, we add (or remove) the accumulated
score until the previous step ($P_{i-1}^{\epsilon,\delta}(a)$) and the attenuated weights (v_ϵ and δ^i)
received from arguments at the beginning of a path with a length of i (Δ_i):
$P_i^{\epsilon,\delta}(a) = P_{i-1}^{\epsilon,\delta}(a) + (\text{-}1)^i \delta^i \sum_{b \in \Delta_i(a)} v_\epsilon(b)$. We stop the process when, between two
steps, the difference with the previous step for all the valuations P is smaller
than a fixed precision threshold μ, i.e. $\forall a \in \mathcal{A}, |P_i^{\epsilon,\delta}(a) - P_{i-1}^{\epsilon,\delta}(a)| < \mu$. Thus,
given a precision, one can choose δ according to the maximal expected depth.

Proposition 3. *Let F be an argumentation framework, $i \in \mathbb{N} \backslash \{0\}$ be the max-
imal depth and μ be the precision threshold. If $\delta < \sqrt[i]{\dfrac{\mu}{\max\limits_{a \in Arg(F)} (|\Delta_i(a)|)}}$ then the*
sequence $\{P_i^{\epsilon,\delta}(a)\}_{i=0}^{+\infty}$ converges before step $i + 1$.

Example 1 (cont.) *Suppose that one considers that the maximal depth should
be 5. In using the previous formula with a precision $\mu = 0.0001$, then δ should
be smaller than $\sqrt[5]{\frac{0.0001}{3}} \simeq 0.127$. Thus, a value close to this limit, for instance
$\delta = 0.12$, ensures that only the arguments until a depth of 5 (included) are
considered.*

We can also find a computational advantage to represent the fading effect.
Indeed, as the number of steps needed to find the propagation number of each
argument is smaller as if we need to browse all the AF, the ranking is computed
faster.

4   Ranking-Based Properties

We now investigate the properties satisfied by vdp. We start by inspecting the
case of Void Precedence, before checking other properties discussed in the literature. These results give us some invariants, *i.e.* constraints on the resulting
rankings. As we have seen, by tuning the parameter δ, different rankings can still
be returned, it is why we report experimental results showing that the diversity
of rankings remains small.

4.1   Void Precedence

One of the very distinctive feature of vdp is that an attacked argument can have
a better score (and so a better rank) than a non-attacked argument. Indeed,
when a given argument has many defense branches and few attack branches, it
receives many positive weights. Thus, according to the choice of δ, this argument
can obtain a greater score than the score of non-attacked arguments.

Void Precedence (VP). A non-attacked argument is ranked strictly higher than any attacked argument: $\Delta_1(a) = \emptyset$ and $\Delta_1(b) \neq \emptyset \Rightarrow a \succ b$

Let us illustrate that with the persuasion pitch used by the salesman in the introduction.

Example 2. *Consider the AF illustrated in the introduction, where the argument a_1 has two defense branches. In computing the propagation number of each argument, with $\delta = 0.95$ and $\epsilon = 0$, we obtain $P^{0,0.95}(a_1) = 1.805$, $P^{0,0.95}(a_2) = P^{0,0.95}(a_4) = -0.95$ and $P^{0,0.95}(a_3) = P^{0,0.95}(a_5) = 1$. With a non-zero $\epsilon = 0.5$, $P^{0.5,0.95}(a_1) = 1.36$, $P^{0.5,0.95}(a_2) = P^{0.5,0.95}(a_4) = -0.45$ and $P^{0.5,0.95}(a_3) = P^{0.5,0.95}(a_5) = 1$. So, one can infer the following preorder where a_1 is better ranked than non-attacked arguments:*

$$a_1 \succ a_3 \simeq a_5 \succ a_2 \simeq a_4$$

In fact, there exists a threshold for the parameter δ below which VP is satisfied.

Proposition 4. *Let $\langle \mathcal{A}, \mathcal{R} \rangle$ be an argumentation framework and $\delta^M = \sqrt{\dfrac{1}{\max_{a \in \mathcal{A}}(|\Delta_2(a)|)}}$.*

If $\delta < \delta^M$ then vdp^δ satisfies VP

Let us check which values of δ are needed to satisfy VP for the argumentation framework in the introduction:

Example 2 (cont.) *The argument a_1 has the highest number of direct defenders with $|\Delta_2(a_1)| = 2$. The value of δ should be now: $\delta < \delta^M = \sqrt{1/2} \simeq 0.7071$. So if $\delta = 0.7$, we obtain $P^{0,0.7}(a_1) = 0.98$, $P^{0,0.7}(a_2) = P^{0,0.7}(a_4) = -0.7$ and $P^{0,0.7}(a_3) = P^{0,0.7}(a_5) = 1$ when $\epsilon = 0$ and $P^{0.5,0.7}(a_1) = 0.78$, $P^{0.5,0.7}(a_2) = P^{0.5,0.7}(a_4) = -0.2$ and $P^{0.5,0.7}(a_3) = P^{0.5,0.7}(a_5) = 1$ when $\epsilon = 0.5$. These results allow to obtain the following preorder: $a_3 \simeq a_5 \succ a_1 \succ a_2 \simeq a_4$.*

The question to know if VP should hold or not relates to the status of the missing information in argumentation systems. If all the information are available, then "really unattacked" arguments should be better that any attacked argument, as it is the case with the other semantics. But there are cases where the argumentation systems encode the information currently available, and that is susceptible to be completed. This is this case that we attack in this paper with procatalepsis. Non-attacked argument can be seen as an argument which has *not yet* been debated whereas it is more difficult to find counter-arguments to an argument already attacked but defended thereafter.

Thus, our method departs from other approaches in its treatment of the VP property, but to a certain extent only. For instance, in a persuasion pitch a single line of defense is not enough to be more convincing than a non-attacked argument. On the other hand, when this condition is met a simple condition for the violation of VP in persuasion pitches can be stated:

Proposition 5 *Let $PP = \langle \mathcal{A}, \mathcal{R} \rangle$ be a persuasion pitch with $a \in \mathcal{A}$ as the root argument. Then,*
(i) if $|\Delta^{B^+}(a)| < 2$ then vdp^δ satisfies VP;
(ii) if $|\Delta^{B^+}(a)| \geq 2$ and $\delta > \sqrt[m]{\frac{1}{|\Delta^{B^+}(a)|}}$ with m the length of the longest defense branch of a then vdp^δ violates VP.

Let us discuss about the link between the two principles concerning δ. Indeed, the value of δ should not be too small in order to take into account enough arguments (and not only the direct attackers for example) to obtain a significant result. But, in the same way, it also should not be too high if one wants to capture the procatalepsis principle. Understanding this interplay can provide valuable information, in particular in the persuasion context. Suppose for instance that the persuader knows that a given δ value is expected, corresponding to the profile of a specific audience. Then, this value being fixed, it is possible to infer that a certain number of defense branches will be required. Hence, instead of developing, say, two long lines of persuasion, the persuader will instead favor the deployment of a number of alternative lines in her persuasion pitch.

Interestingly, it turns out that in the context of our method, the property VP is related with another property studied in the literature, namely *defense precedence*:

Defense Precedence (DP). For two arguments with the same number of direct attackers, a defended argument is ranked higher than a non-defended argument:

$$|\Delta_1(a)| = |\Delta_1(b)|, \Delta_2(a) \neq \emptyset \text{ and } \Delta_2(b) = \emptyset \Rightarrow a \succ b$$

Proposition 6. *If vdp^δ satisfies VP then it satisfies DP.*

Note that this is not the case in general (some semantics satisfy VP but not DP).

4.2   Other Properties

Several other properties have been proposed, and studied in the literature (see [5] for an overview). Below we study how our method stands with respect to these properties. We give their informal definition and point the reader to [5] for the complete versions. Basic general properties are the fact that a ranking on a set of arguments should only depend on the attack relation (*Abstraction*, **Abs**); that the ranking between two arguments should be independent of arguments that are not connected to either of them (*Independence*, **In**); that all arguments can be compared (*Total*, **Tot**); and that all non-attacked arguments should be equally acceptable (*Non-attacked Equivalence*, **NaE**).

Local properties (like the already introduced DP) confine themselves to the level of direct attackers: (*Counter-Transitivity*, **CT**) states that if the direct attackers of b are (i) at least as numerous and (ii) acceptable as those of a, then a should be at least as acceptable as b, while in its strict version (**SCT**) either (i) or (ii) must be strict, implying a strict comparison between a and b.

Global properties specify how the ranking should be affected on the basis of the comparison of attack and defense branches. More precisely: adding a defense branch to an attacked argument should increase its acceptability (*Addition of Defense Branch*, **+DB**); increasing the length of an attack branch of an argument should increase its acceptability (*Increase of Attack Branch*, ↑**AB**); adding an attack branch to an argument should decrease its acceptability (*Addition of Attack Branch*, **+AB**); and increasing the length of a defense branch of an argument should decrease its acceptability (*Increase of Defense Branch*, ↑**DB**). Note that +DB is indeed restricted to *attacked arguments*, otherwise its incompatibility with VP is obvious. In the same spirit, (*Attack vs Full Defense*, **AvsFD**), *i.e.* the fact that an argument with only defense branches and no attack branch should be strictly more acceptable than an argument attacked once by a non-attacked argument. For persuasion pitches, this property can be simply reformulated as "a persuasion pitch for x should make it more acceptable than stating x with an attacking argument". This seems compelling in our context, thus providing further evidence of the inability of many of the existing semantics to properly capture persuasion settings.

Let us now check which properties are satisfied by vdp:

Proposition 7. *Let* $\delta \in]0,1[$. *vdp$^\delta$ satisfies Abs, In, Tot, NaE, +AB and AvsFD.*

Some global properties like +DB, ↑DB and ↑AB are not satisfied because of the fading effect. Indeed, when the branch, which is added or extended, is too long, the arguments at the end of this branch have no impact on the targeted argument. It is why, we propose to define the corresponding properties (+DB$_i$, ↑DB$_i$ and ↑AB$_i$) which capture the same idea but with the additional condition that the maximal length of the branch is i.

Proposition 8. *With* $\delta \in]\delta^m, 1[$ *s.t.* $\delta^m = \sqrt[i]{\dfrac{\mu}{\max\limits_{a \in Arg(F)} \left(|\Delta_i(a)|\right)}}$ *where i represents the length of the branch which is added or extended then vdp$^\delta$ satisfies also +DB$_i$, ↑DB$_i$ and ↑AB$_i$.*

These results are reported in Table 2. For comparison purpose, we also include in this table the results of some semantics from the literature where the same set [5] of properties has been already checked. Namely, these semantics are: the semantics based on Social Argumentation Frameworks SAF [8,11,16] restricted to Dung's argumentation framework, the semantics Categoriser Cat [3,19], the Discussion-based semantics Dbs and the Burden-based semantics Bbs [1], the global semantics based on tuple values $Tuples^*$ [7], and the semantics $M\&T$ [17].

We first remark that vdp satisfy the "basic" properties according to [5] (Abs, In, +AB, NaE and Tot), at the exception of VP as intended by design and discussed earlier.

We can also note that vdp always satisfies property AvsFD, and for a specific δ the property +DB. Indeed, the possibility to rank arguments with various different defenders higher than arguments which are defended only once seems

Table 2. Summary of the properties satisfied by vdp ($\forall \delta$, and with $\delta^m < \delta' < \delta^M$) and some existing ranking semantics from the literature where the same set of properties has been already checked. A cross $\times$ means that the property is not satisfied, symbol $\checkmark$ means that the property is satisfied and $\checkmark_i$ means that the i-version of the property (cf Proposition 8) is satisfied. Shaded cells are results proved in this paper.

Properties	SAF	Cat	Dbs	Bbs	$Tuples^*$	M&T	vdp$^\delta$	vdp$^{\delta'}$
Abs	$\checkmark$	$\checkmark$	$\checkmark$	$\checkmark$	$\checkmark$	$\checkmark$	$\checkmark$	$\checkmark$
In	$\checkmark$	$\checkmark$	$\checkmark$	$\checkmark$	$\checkmark$	$\checkmark$	$\checkmark$	$\checkmark$
Tot	$\checkmark$	$\checkmark$	$\checkmark$	$\checkmark$	$\times$	$\checkmark$	$\checkmark$	$\checkmark$
NaE	$\checkmark$	$\checkmark$	$\checkmark$	$\checkmark$	$\checkmark$	$\checkmark$	$\checkmark$	$\checkmark$
+AB	$\checkmark$	$\checkmark$	$\checkmark$	$\checkmark$	$\checkmark$	$\checkmark$	$\checkmark$	$\checkmark$
AvsFD	$\times$	$\times$	$\times$	$\times$	$\checkmark$	$\checkmark$	$\checkmark$	$\checkmark$
+DB	$\times$	$\times$	$\times$	$\times$	$\checkmark$	$\times$	$\times$	$\checkmark_i$
↑AB	$\checkmark$	$\checkmark$	$\checkmark$	$\checkmark$	$\checkmark$	$\times$	$\times$	$\checkmark_i$
↑DB	$\checkmark$	$\checkmark$	$\checkmark$	$\checkmark$	$\checkmark$	$\times$	$\times$	$\checkmark_i$
VP	$\checkmark$	$\checkmark$	$\checkmark$	$\checkmark$	$\checkmark$	$\checkmark$	$\times$	$\checkmark$
DP	$\checkmark$	$\checkmark$	$\checkmark$	$\checkmark$	$\times$	$\times$	$\times$	$\checkmark$
CT / SCT	$\checkmark$	$\checkmark$	$\checkmark$	$\checkmark$	$\times$	$\times$	$\times$	$\times$

very interesting. For instance, in persuasion scenarios, a claim defended with various different arguments may be more credible than a claim only defended once.

In the end, the only properties which are never satisfied are CT and SCT. In fact, it is easy to show that these properties are incompatible with +DB. Intuitively, whereas +DB considers that adding a defense is positive for an argument, SCT says that adding any branch (so including defense branch) to an argument should decrease its acceptability.

4.3   On the Diversity of Rankings

A nice feature of our semantics is thus that the designer can choose whether VP holds or not, giving rise to different rankings. However, one may be worried that the diversity of rankings is so high that the semantics becomes too sensitive to small modifications of the parameter δ. To check this, we applied our semantics on 1000 randomly generated AFs[1] for different values of $\delta \in \{0.001, 0.2, 0.4, 0.6, 0.8, 0.9\}$. Then, we computed the similarity degree between two rankings from two different values of δ in using the Kendall tau distance [15] which returns a value between 0 and 1.

---

[1] The generation algorithms are based on the three algorithms used for producing the benchmarks of the competition ICCMA'15 (see http://argumentationcompetition. org/2015/results.html).

Definition 5. *Let* $\langle A, R \rangle$ *and* $\tau_{\sigma_1}, \tau_{\sigma_2}$ *the orders returned by the ranking semantics* σ_1 *and* σ_2 *respectively. The Kendall tau distance between* τ_{σ_1} *and* τ_{σ_2} *is calculated as follow:*

$$K(\tau_{\sigma_1}, \tau_{\sigma_2}) = \frac{\sum_{\{i,j\} \in A} \overline{K_{i,j}}(\tau_{\sigma_1}, \tau_{\sigma_2})}{0.5 \times |A| \times (|A| - 1)}$$

with:

- $\overline{K_{i,j}}(\tau_{\sigma_1}, \tau_{\sigma_2}) = 1$ *if* $i \succ^{\sigma_1} j$ *and* $i \succ^{\sigma_2} j$, *or* $i \prec^{\sigma_1} j$ *and* $i \prec^{\sigma_2} j$, *or* $i \simeq^{\sigma_1} j$ *and* $i \simeq^{\sigma_2} j$,
- $\overline{K_{i,j}}(\tau_{\sigma_1}, \tau_{\sigma_2}) = 0$ *if* $i \succ^{\sigma_1} j$ *and* $i \prec^{\sigma_2} j$ *or vice versa*,
- $\overline{K_{i,j}}(\tau_{\sigma_1}, \tau_{\sigma_2}) = 0.5$ *if* $i \succ^{\sigma_1} j$ *or* $i \prec^{\sigma_1} j$ *and* $i \simeq^{\sigma_2} j$ *or vice versa*.

Thus, two rankings with a Kendall tau distance of 1 are fully similar whereas a score of 0 means that they are totally reversed. The Table 3 contains, for each pair of semantics, the average Kendall tau distance ($\times 100$) computed on the 1000 generated AFs. The results show that the obtained rankings stay pretty close because the smallest observed similarity between the smallest and largest value of δ is 86.26%. This similarity remains overall very high, showing that the semantics remains quite stable as the parameter varies.

Table 3. Average Kendall tau distance on 1000 randomly generated AFs for different values of δ

$P_i^{0.5,0.4}$	a	b	c	d	e	f
0	0.5	1	0.5	0.5	0.5	0.5
1	-0.1	1	0.1	0.1	0.3	0.3
2	-0.02	1	0.1	0.34	0.38	0.46
3	-0.052	1	0.1	0.308	0.316	0.364
⋮	⋮	⋮	⋮	⋮	⋮	⋮
14	-0.0402	1	0.1	0.3161	0.3506	0.3736

5   Comparison with Related Work

Now, let us show that, in general, the different semantics proposed in the literature may return a large variety of rankings. To show this, we will use the example of Fig. 1.

The range of semantics considered here is more important than the previous section because we include recent semantics for which the axiomatic properties have not (to the best of our knowledge) been yet studied. Two kinds of existing semantics are excluded from this study, because of specificities which make the comparison difficult. The first ones are the semantics which return a partial preorder between arguments (*i.e.* some arguments could be incomparable) like the global semantics based on tuple values [7] and the semantics proposed in [13]. The second category is the semantics that return a set of rankings for a same AF, like [12]. Thus, we consider the semantics *Cat*, *M&T*, *SAF*, *Dbs* and *Bbs* that have already been mentioned on Sect. 4.2. We will also consider the semantics using the fuzzy label *FL* [9], the α-Burden-based semantics α-*BBS* [2], the counting semantics *CS* [20] and the propagation semantics $Propa_\epsilon, Propa_{1+\epsilon}, Propa_{1\to\epsilon}$ from [6]. All these rankings are represented in Fig. 2.

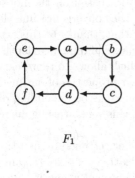

F_1

Semantics	Order between arguments
M&T	$b \succ d \simeq e \succ a \simeq c \simeq f$
FL	$b \succ d \succ e \succ f \succ a \simeq c$
SAF	
α-BBS ($\alpha = 5$)	$b \succ d \succ e \succ f \succ c \succ a$
$Propa_{1\to\epsilon}$	
$vdp^\delta (\delta = 0.3)$	
Dbs/Bbs	
CS	
α-BBS ($\alpha = 0.5$)	$b \succ f \succ e \succ c \succ d \succ a$
$Propa_\epsilon (\epsilon = 0.8)$	
Cat	
$Propa_\epsilon (\epsilon = 0.3)$	$b \succ f \succ e \succ d \succ c \succ a$
$Propa_{1+\epsilon}$	
$vdp^\delta (\delta = 0.8)$	$d \succ b \succ e \succ c \succ f \succ a$

Fig. 2. Orders obtained with the different ranking semantics applied to F_1

The particularity of our ranking semantics is clearly visible in this table because only vdp with $\delta \geq \frac{\sqrt{2}}{2}$ consider d as strictly more acceptable than b. All the other semantics satisfy VP, so consider b, which is non-attacked, as the best argument in this AF. Indeed, thanks to the even cycle, d receives only positive weights from the only non-attacked argument b and is also defended by e. However, d is not always among the best arguments as it is the case with the semantics which consider that a new defense branch can reinforce an argument (*i.e.* the semantics that satisfy +DB and AvsFD). Indeed, the semantics which consider a defense as a weak attack (*i.e.* the semantics that satisfy SCT) judge that even if d is defended, it is still directly attacked once more than c, e and f. The reverse reasoning hold with the argument f which is one of the best arguments for the semantics that satisfy SCT whereas is stay quite acceptable for the semantics that satisfy +DB and AvsFD. The worst arguments is always a which is attacked twice including once by b which is non-attacked. It is why it is almost always worst than c which is directly attacked by b only.

Finally, it is important to note that even if our proposal is related to Pu et al. [20] and Bonzon et al. [6] concerning the propagation method, it is substantially different, and has clearly different properties and behaviors (see Fig. 2).

6   Conclusion

Many ranking-based semantics have been proposed recently in the literature. Despite detailed studies of their properties, it remains hard to see which is more appropriate for a given application context. In this paper, we took the problem the other way and challenged these semantics in the context of persuasion, emphasizing in particular two well-documented phenomena occurring in practice: protocatalepsis and fading. It turns out that none of the proposed semantics is really appropriate – all of them commit for instance to the "Void Precedence" property which is incompatible with the procatalepsis principle. This motivated us to introduce a new parametrized ranking semantics based on the notion of propagation. An attenuation factor is used to allow the convergence but also to decrease the impact of further arguments. We show that, thanks to this attenuation factor, fading can be captured by selecting a maximal influence depth. For some values of this parameter VP is not satisfied, which allows to represent protocatalepsis in persuasion pitches. We also study other properties of our method, and we experimentally study how diverse can the rankings be depending of the value of the parameter. Future work include testing this semantics on currently developed computation persuasion tools.

Our methodology may also prove inspiring in other settings: by questioning the relevance of the existing semantics in other application contexts (*e.g.* negotiation), we may find out that some specific phenomena are not properly captured, and that other adjustments are required.

Acknowledgements. This work benefited from the support of the project AMANDE ANR-13-BS02-0004 of the French National Research Agency (ANR).

References

1. Amgoud, L., Ben-Naim, J.: Ranking-based semantics for argumentation frameworks. In: Liu, W., Subrahmanian, V.S., Wijsen, J. (eds.) SUM 2013. LNCS, vol. 8078, pp. 134–147. Springer, Heidelberg (2013). doi:10.1007/978-3-642-40381-1_11
2. Amgoud, L., Ben-Naim, J., Doder, D., Vesic, S.: Ranking arguments with compensation-based semantics. In: Proceedings of the 15th International Conference on Principles of Knowledge Representation and Reasoning (KR 2016), pp. 12–21 (2016)
3. Besnard, P., Hunter, A.: A logic-based theory of deductive arguments. Artif. Intell. **128**(1–2), 203–235 (2001)
4. Besnard, P., Hunter, A.: Elements of Argumentation. MIT Press (2008)
5. Bonzon, E., Delobelle, J., Konieczny, S., Maudet, N.: A comparative study of ranking-based semantics for abstract argumentation. In: Proceedings of the 30th AAAI Conference on Artificial Intelligence (AAAI 2016), pp. 914–920 (2016)
6. Bonzon, E., Delobelle, J., Konieczny, S., Maudet, N.: Argumentation ranking semantics based on propagation. In: Proceedings of the 6th International Conference on Computational Models of Argument (COMMA 2016), pp. 139–150 (2016)
7. Cayrol, C., Lagasquie-Schiex, M.-C.: Graduality in argumentation. J. Artif. Intell. Res. **23**, 245–297 (2005)

8. Correia, M., Cruz, J., Leite, J.: On the efficient implementation of social abstract argumentation. In: Proceedings of the 21st European Conference on Artificial Intelligence (eCAI 2014), pp. 225–230 (2014)
9. da Costa Pereira, C., Tettamanzi, A., Villata, S.: Changing one's mind: erase or rewind? In: Proceedings of the 22nd International Joint Conference on Artificial Intelligence, (IJCAI 2011), pp. 164–171 (2011)
10. Dung, P.H.: On the acceptability of arguments and its fundamental role in nonmonotonic reasoning, logic programming and n-person games. Artif. Intell. **77**(2), 321–358 (1995)
11. Eğilmez, S., Martins, J., Leite, J.: Extending social abstract argumentation with votes on attacks. In: Black, E., Modgil, S., Oren, N. (eds.) TAFA 2013. LNCS, vol. 8306, pp. 16–31. Springer, Heidelberg (2014). doi:10.1007/978-3-642-54373-9_2
12. Gabbay, D.M.: Equational approach to argumentation networks. Argument Comput. **3**(2–3), 87–142 (2012)
13. Grossi, D., Modgil, S.: On the graded acceptability of arguments. In: Proceedings of the 24th International Joint Conference on Artificial Intelligence (IJCAI 2015), pp. 868–874 (2015)
14. Hunter, A.: Opportunities for argument-centric persuasion in behaviour change. In: Fermé, E., Leite, J. (eds.) JELIA 2014. LNCS, vol. 8761, pp. 48–61. Springer, Cham (2014). doi:10.1007/978-3-319-11558-0_4
15. Kendall, M.G.: A new measure of rank correlation. Biometrika **30**(1/2), 81–93 (1938)
16. Leite, J., Martins, J.: Social abstract argumentation. In: Proceedings of the 22nd International Joint Conference on Artificial Intelligence, (IJCAI 2011), pp. 2287–2292 (2011)
17. Matt, P.-A., Toni, F.: A game-theoretic measure of argument strength for abstract argumentation. In: Hölldobler, S., Lutz, C., Wansing, H. (eds.) JELIA 2008. LNCS, vol. 5293, pp. 285–297. Springer, Heidelberg (2008). doi:10.1007/978-3-540-87803-2_24
18. Pu, F., Luo, J., Luo, G.: Some supplementaries to the counting semantics for abstract argumentation. In: Proceedings of the 27th IEEE International Conference on Tools with Artificial Intelligence (ICTAI 2015), pp. 242–249 (2015)
19. Pu, F., Luo, J., Zhang, Y., Luo, G.: Argument ranking with categoriser function. In: Buchmann, R., Kifor, C.V., Yu, J. (eds.) KSEM 2014. LNCS, vol. 8793, pp. 290–301. Springer, Cham (2014). doi:10.1007/978-3-319-12096-6_26
20. Pu, F., Luo, J., Zhang, Y., Luo, G.: Attacker and defender counting approach for abstract argumentation. In: Proceedings of the 37th Annual Meeting of the Cognitive Science Society (CogSci 2015) (2015)
21. Tan, C., Niculae, V., Danescu-Niculescu-Mizil, C., Lee, L.: Winning arguments: interaction dynamics and persuasion strategies in good-faith online discussions. In: Proceedings of the 25th International Conference on World Wide Web (WWW 2016), pp. 613–624 (2016)
22. Walton, D.: Dialog Theory for Critical Argumentation. John Benjamins Publishing (2007)

A Probabilistic Programming Language for Influence Diagrams

Steven D. Prestwich[✉], Federico Toffano, and Nic Wilson

Insight Centre for Data Analytics, Department of Computer Science,
University College Cork, Cork, Ireland
{steven.prestwich,federico.toffano,nic.wilson}@insight-centre.org

Abstract. Probabilistic Programming (PP) extends the expressiveness
and scalability of Bayesian networks via programmability. Influence Diagrams (IDs) extend Bayesian Networks with decision variables and utility
functions, allowing them to model sequential decision problems. Limited-
Memory IDs (LIMIDs) further allow some earlier events to be ignored or
forgotten. We propose a generalisation of PP and LIMIDs called IDLP,
implemented in Logic Programming and with a solver based on Reinforcement Learning and sampling. We show that IDLP can model and
solve LIMIDs, and perform PP tasks including inference, finding most
probable explanations, and maximum likelihood estimation.

1 Introduction

Probabilistic Programming (PP) [5,11] is a tool for statistical modelling that
facilitates the modeling of large Bayesian networks, and typical tasks include
the computation of posterior probability distributions, finding most probable
explanations, and maximum likelihood estimation. It allows a user to define
complex models with few lines of code and has a growing list of applications.
A recent DARPA project in the USA (*Probabilistic Programming for Advancing
Machine Learning*) explores new applications to machine learning.

PP has greatly extended the expressiveness and flexibility of traditional
graphical approaches, by unifying Turing complete programming languages with
probabilistic modeling. There are many PP languages (PPLs), some based on
existing programming languages and others self-contained: we refer the reader to
[12] for publications, systems and PP news. Using a PPL users can completely
specify large, complex probabilistic models to which inference can be applied
automatically.

All PPLs allow the user to define *random variables* but very few also provide
decision variables. Allowing both random and decision variables could extend
the advantages of PP to many more applications of a type encountered in fields
such as Reinforcement Learning, Approximate Dynamic Programming, Stochastic Programming, Stochastic Dynamic Programming and Simulation Optimisation: *sequential decision problems under uncertainty*. However, we know of no PP
system able to tackle such problems. Only the DTProbLog language [2] provides
decision variables, in a limited way (see Sect. 6).

© Springer International Publishing AG 2017
S. Moral et al. (Eds.): SUM 2017, LNAI 10564, pp. 252–265, 2017.
DOI: 10.1007/978-3-319-67582-4_18

In this paper we propose an extension of PP to sequential decision problems. Our modelling approach is based on Influence Diagrams (IDs) [7], a graphical model that generalises Bayesian networks by adding decision variables and utility functions. Hence, our approach combines the advantages of two others: the features of IDs that enable them to model decision problems, and the expressiveness and pragmatism of PP (see Fig. 1). We present a language called IDLP based on logic programming, but we are also developing a Python-based implementation for wider useability. To solve the models we combine Reinforcement Learning algorithms with sampling to find optimal policies to some well-known IDs. We also show that our IDLP can tackle other PP problems besides inference.

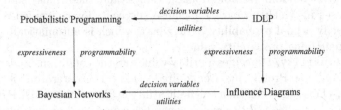

Fig. 1. Relationship of IDLP to other approaches

The paper is organised as follows: Sect. 2 describes our new PPL and shows its use for inference; Sect. 3 introduces decision variables and utilities, and shows the connection with IDs; Sect. 4 considers other PP tasks; Sect. 5 describes our solution method; and Sect. 6 discusses related work and concludes the paper.

2 The IDLP Language

We call our language IDLP (see Sect. 3 for an explanation of the name). Some PPLs are newly created languages, while others are existing programming languages plus a few primitives for common operations such as declaring a variable with a specified probability distribution. IDLP is of the latter type and is based on Prolog.

There already exist several Prolog-based PPLs which are surveyed in [13]. Most follow Sato's *distribution semantics* [16], but we find the representation of non-binary choices slightly unnatural in these languages (via *multi-valued annotated disjunctions* in the ProbLog language [14]) so we do not use the distribution semantics. Instead we simply use Prolog as a convenient symbolic programming language. (Readers unfamiliar with Prolog might find some details obscure, but unfortunately space is too limited to provide a Prolog introduction.)

To introduce IDLP's primitives we apply it to an example of a central PP task: *probabilistic inference*, that is inferring a conditional probability from a probabilistic program, and as an example we use Pearl's *alarm* example [10]. The IDLP model is shown in Fig. 2. The primitives are:

- Predicate `variables/1`[1] specifies a list of *IDLP variables* in a strict order, with their discrete domain values contained in a list `[...]`. In this example all variables are random (denoted by `r`) but later we shall add decision variables (denoted by `d`). Unlike Prolog variables whose names start with an uppercase letter, IDLP variables are ground terms and may have arguments.
- An infix operator $= ^\sim$ is used to specify *IDLP variable assignments*: `alarm = ~yes` means that variable `alarm` takes value `yes`.
- Predicate `p/2` describes discrete probability distributions for each IDLP variable assignment. No Prolog cuts (`!`) are needed to make this predicate deterministic, as the first successful p-clause fixes the probability (following standard Prolog operational semantics). IDLP requires probabilities to be provided for all random variable domain values, and these must sum to 1. To catch cases in which no earlier p-clause succeeds, for convenience the user may specify a final probability by `default`, which is automatically replaced by the remaining probability value.
- Predicate `utility/1` specifies a utility value and the conditions under which it occurs. Again no Prolog cuts are required to make this deterministic, and if no `utility`-clause succeeds then a default utility of 0 is assumed. IDLP computes *conditional expected utilities* which generalise conditional probabilities: in this example a utility value of 1 occurs when `earthquake` is assigned to `yes`, otherwise the default value of 0 occurs. This conditional expected utility is equal to the conditional probability

  `p(earthquake=~yes|alarm=~yes)`

- Predicate `evidence/0` specifies the condition used in the conditional expected utility. In this example the evidence is true if `alarm=~yes`. (It is of course possible to specify evidence that is always false.)
- Finally, to take (say) 1 million samples we call the goal `?-samples(1000000)`.

The user need only provide Prolog code for these predicates, and any Prolog techniques can be used to do this: this is the advantage of a PPL over a graphical approach. Given these predicates and the `samples` goal, IDLP computes conditional expected utilities by repeated simulation and rejection sampling (see Sect. 5). The conditional probability in this example is correctly computed by IDLP to be approximately 0.23.

As described so far, IDLP is a Prolog-based PPL with discrete (binary or nonbinary) random variables, slightly generalised to compute conditional expected utilities instead of conditional probabilities. It uses simulation and rejection sampling to compute conditional expectations. Next we introduce decision variables.

3   Decision Variables

We now add decision variables to IDLP to enable it to model IDs, hence the name IDLP: *Influence Diagrams in Logic Programming*. In fact IDLP can model a slight generalisation of IDs, which do not typically contain evidence.

---

[1] In standard Prolog notation a predicate `P/A` has name `P` and arity `A`.

```
variables([ r(burglary,[yes,no]), r(earthquake,[yes,no]),
            r(alarm,[yes,no]) ]).

p(burglary=~yes,0.7).      p(burglary=~no,0.3).
p(earthquake=~yes,0.2).    p(earthquake=~no,0.8).

p(alarm=~yes,0.9) :- burglary=~yes, earthquake=~yes.
p(alarm=~yes,0.8) :- burglary=~yes, earthquake=~no.
p(alarm=~yes,0.1) :- burglary=~no,  earthquake=~yes.
p(alarm=~no,default).

evidence :- alarm=~yes.

utility(1) :- earthquake=~yes.
```

Fig. 2. IDLP model for Pearl's alarm example

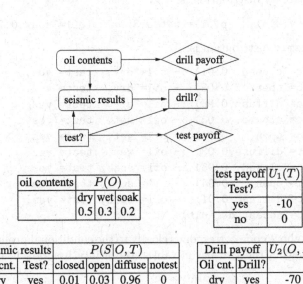

oil contents	P(O)		
	dry	wet	soak
	0.5	0.3	0.2

test payoff	$U_1(T)$
Test?	
yes	-10
no	0

Seismic results	$P(S|O,T)$			
Oil cnt.	Test?	closed	open	diffuse	notest
dry	yes	0.01	0.03	0.96	0
dry	no	0	0	0	1
wet	yes	0.03	0.94	0.03	0
wet	no	0	0	0	1
soak	yes	0.95	0.04	0.01	0
soak	no	0	0	0	1

Drill payoff	$U_2(O,D)$	
Oil cnt.	Drill?	
dry	yes	-70
dry	no	0
wet	yes	50
wet	no	0
soak	yes	200
soak	no	0

Fig. 3. Oil wildcatter ID

3.1   Influence Diagrams

IDLP's decision variables also have discrete (binary or non-binary) domains and can occur anywhere in the ordered `variables` list. Decision variables are automatically assigned values that maximise the expected utility. For example consider the well-known *Oil Wildcatter* problem [15] shown in Fig. 3. We use the following list of variables:

```
variables([ r(oil,[dry,wet,soak]), d(test,[yes,no]),
            r(seismic,[closed,open,diffuse,notest]),
            d(drill,[yes,no])]).
```

which conforms to the chronological order of events:

- because of ancient geological events the oil state is either `dry`, `wet` or `soak`
- we decide whether or not to `test` for oil
- the result of the test is either `closed`, `open` or `diffuse`, or the special value `notest` if a test was not performed.
- if we applied the test, then based on the result (but not on the unobservable variable `oil`) we decide whether or not to drill

The `oil` probability distribution is:

```
p(oil=~dry,0.5).    p(oil=~wet,0.3).    p(oil=~soak,0.2).
```

while the `seismic` distribution is:

```
p(seismic=~closed, 0.01) :- oil=~dry,  test=~yes.
p(seismic=~open,   0.03) :- oil=~dry,  test=~yes.
p(seismic=~diffuse,0.96) :- oil=~dry,  test=~yes.
p(seismic=~closed, 0.03) :- oil=~wet,  test=~yes.
p(seismic=~open,   0.94) :- oil=~wet,  test=~yes.
p(seismic=~diffuse,0.03) :- oil=~wet,  test=~yes.
p(seismic=~closed, 0.95) :- oil=~soak, test=~yes.
p(seismic=~open,   0.04) :- oil=~soak, test=~yes.
p(seismic=~diffuse,0.01) :- oil=~soak, test=~yes.
p(seismic=~notest,default).
```

Note that the above code corresponds closely to the conditional probability tables in Fig. 3. Following the two payoff tables, the utility function has two components:

```
utility(R) :- drill_payoff(R1), test_payoff(R2), R is R1+R2.

drill_payoff(-10) :- test=~yes.
drill_payoff(0).

test_payoff(-70) :- oil=~dry,  drill=~yes.
test_payoff(50)  :- oil=~wet,  drill=~yes.
test_payoff(200) :- oil=~soak, drill=~yes.
test_payoff(0).
```

There is no evidence so we simply write:

```
evidence.
```

For this problem the `test` decision does not depend on any other variable, but the `drill` decision depends on whether a test was made, and if so on its result. To model this aspect of IDs we introduce a new primitive predicate `depends/2`, whose first argument is a decision variable, and whose second argument is a list of the decision and/or random variables on which it depends (the ordering of this list is arbitrary as it represents a set). In this example the decision `test` does not depend on anything, while the decision `drill` depends on the `test` decision and on the random `seismic` result:

```
depends(test,[]).
depends(drill,[test,seismic]).
```

The `oil` variable is unobservable so no decision depends on it. Solving this model by simulation (see Sect. 5) we get an expected utility of 42.73 which is close to the known optimal expected utility of 42.75. The policy it finds is correct: apply the seismic test, and drill if the test result is open or closed.

3.2   LIMIDs

Limited-Memory IDs (LIMIDs) [9] model situations in which some events may be forgotten or ignored, and might not be strictly ordered. We still list variables in a total order, choosing any that is consistent with the actual partial order. Forgotten or ignored variables are treated in the same way as the unobservable variable in the wildcatter example: decisions do not depend on them.

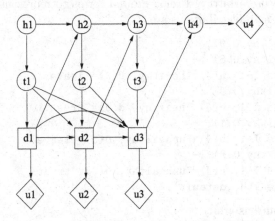

Fig. 4. Pig breeding ID

We take another well-known example: the pig breeding problem [9]. The details of the problem are not given via tables but in a verbal description, which we paraphrase from [9]:

A pig breeder grows pigs for four months then sells them. During this period the pig may or may not develop a disease. If it has the disease when it must be sold, then it must be sold for slaughtering and its expected market price is 300 (Danish kroner). If it is disease-free its expected market price is 1000. Once a month, a veterinary surgeon test the pig for the disease. If it is ill then the test indicates this with probability 0.80, and if it is healthy then the test indicates this with probability 0.90. At each monthly visit the surgeon may or may not treat the pig, and the treatment costs 100. A pig has the disease in month 1 with probability 0.10. A healthy pig develops the disease in the next month with probability 0.20 without treatment and 0.10 with treatment. An unhealthy pig remains unhealthy in the next month with probability 0.90 without treatment, and 0.50 with treatment.

Further details come in two versions, modelled by an ID and a LIMID respectively.

ID Version. Here the pig breeder knows which pigs have been treated each month and their test results, and bases treatment decisions on this information. The ID is shown in Fig. 4. The IDLP variables are:

```
variables([
    r(h(1),[healthy,unhealthy]), r(t(1),[pos,neg]),
    d(d(1),[treat,leave]), r(h(2),[healthy,unhealthy]),
    r(t(2),[pos,neg]), d(d(2),[treat,leave]),
    r(h(3),[healthy,unhealthy]), r(t(3),[pos,neg]),
    d(d(3),[treat,leave]), r(h(4),[healthy,unhealthy]) ]).
```

Note that we parameterise the t-, d- and h-variables using an argument denoting time: we can use arbitrary Prolog ground terms to represent variables and domain values. The h distributions are:[2]

```
p(h(1)=~healthy,0.9).
p(h(I)=~healthy,0.8) :-
    I>1, J is I-1, h(J)=~healthy, d(J)=~leave.
p(h(I)=~healthy,0.9) :-
    I>1, J is I-1, h(J)=~healthy, d(J)=~treat.
p(h(I)=~healthy,0.1) :-
    I>1, J is I-1, h(J)=~unhealthy, d(J)=~leave.
p(h(I)=~healthy,0.5) :-
    I>1, J is I-1, h(J)=~unhealthy, d(J)=~treat.
p(h(_)=~unhealthy,default).
```

and the t distributions are:

```
p(t(I)=~pos,0.8) :- h(I)=~unhealthy.
p(t(I)=~pos,0.1) :- h(I)=~healthy.
p(t(_)=~neg,default).
```

---

[2] The underscore _ character is a Prolog *anonymous variable* that matches any term and indicates a "don't care" value.

We have exploited the fact that variables at different times have similar distributions, to obtain a more compact description. The utility function has 4 components: profit u(4) from the sale, and possible payments u(1), u(2) and u(3) for treatment:

```
utility(R) :- u(1,T1), u(2,T2), u(3,T3), u4(S),
              R is S+T1+T2+T3.

u(4,300)   :- h(4)=~unhealthy.
u(4,1000)  :- h(4)=~healthy.

u(I,-100)  :- d(I)=~treat.
u(I,0)     :- d(I)=~leave.
```

There is no evidence:

```
evidence.
```

Using the incoming arcs to each decision in the ID (the h are unobservable so they do not appear in any list):

```
depends(d(1),[t(1)]).
depends(d(2),[t(1),t(2),d(1)]).
depends(d(3),[t(1),t(2),t(3),d(1),d(2)]).
```

LIMID Version. Here the pig breeder does not keep detailed records and bases treatment decisions only on the previous test result for each pig. The LIMID is shown in Fig. 5. The IDLP model is as above except for the **depends** predicate:

```
depends(d(I),[t(I)]).
```

The optimal policy is not to treat in month 1 whatever the first test result, treat in month 2 if tests 1 and 2 are positive, and treat in month 3 if tests 2 or 3 are positive, with expected utility 729.255.[3] Solving the LIMID model we almost always find the optimal policy with expected utility of approximately 727: ignore the first test (do not treat) and follow the other two (treat if the test was positive). On the few occasions that we do not find the optimal policy, we find other policies with near-optimal expected utility (less than 1% optimality gap).

4   Probabilistic Programming Revisited

We now show that as well as computing conditional probabilities, IDLP can perform other PP tasks.

---

[3] A different policy is given in [9]: treat in month 3 if tests 1 and 2, or 3, are positive. We find that their policy has expected utility 725.884 while ours is optimal. They cite our expected utility so we believe this was simply a typographical error. To compute the expected value of a policy we use the variable elimination algorithm where, instead of maximising over the decision variables, we set their values according to the policy.

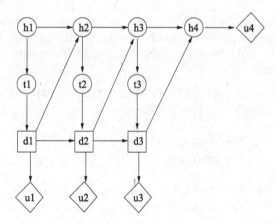

Fig. 5. Pig breeding LIMID

4.1 Most Probable Explanations

IDLP can be used to solve another PP problem via its decision variables: finding *most probable explanations* (MPE). In particular: find the assignment to a set of variables that maximises a conditional probability. For example suppose we wish to find the MPE of an alarm in Pearl's alarm example. The possible explanations and their probabilities are given in the model:

```
burglary=~yes earthquake=~yes 0.90
burglary=~no  earthquake=~yes 0.10
burglary=~yes earthquake=~no  0.80
burglary=~no  earthquake=~no  0.00
```

so we should get the explanation `burglary=~yes` and `earthquake=~yes`. But we really want only one most probable cause: if we forbid the case in which both are true then we should get `burglary=~yes` and `earthquake=~no`.

We can solve this type of problem by introducing decision variables b and e which respectively indicate the truth or falsity of `burglary` and `earthquake` in the MPE:

```
variables([ d(b,[yes,no]), d(e,[yes,no]),
            r(burglary,[yes,no]), r(earthquake,[yes,no]),
            r(alarm,[yes,no]) ]).
```

and making `evidence` true if

$$(b \oplus e) \wedge (b \leftrightarrow burglary) \wedge (e \leftrightarrow earthquake)$$

(where $\oplus$ denotes exclusive-or) as follows:

```
evidence :-
    reify(b=~yes,B), reify(e=~yes,E),
    reify(burglary=~yes,Bu), reify(earthquake=~yes,Ea),
    B+E=:=1, B==Bu, E==Ea.
```

where `reify(G,B)` assigns 1 to B if goal G succeeds, otherwise 0; this additional primitive predicate is useful for constructing logical combinations of assignments as in this example. The utility is:

```
utility(1) :- true(alarm=~yes).
```

so we are maximising `p(alarm=~yes|evidence)`. Again no decision variable depends on another:

```
depends(_,[]).
```

We get expected utility 0.8 with decisions `b=~yes` and `e=~no`, indicating that a burglary (and no earthquake) is a more probable explanation of the alarm than an earthquake (and no burglary).

4.2 Maximum Likelihood Estimation

Another PP task is *maximum likelihood estimation* (MLE) in which we must estimate probability distributions: see for example [6] which solves the problem in the ProbLog PPL via a BDD. Suppose we want to choose random variable distributions to maximise a conditional probability. As an example we modify the Alarm example so that the probability of an alarm occurring, given both a burglary and an earthquake, is unknown. Suppose that we must choose probability P to maximise `p(alarm=~yes|burglary=~yes)`.

To model this problem we approximate the continuous P by a discrete variable q with 5 possible values, the largest being the original probability 0.8. The model is shown in Fig. 6. We correctly find expected utility 0.8 with decision `q=~0.8`.

```
p(burglary=~yes,0.7).      p(burglary=~no,0.3).
p(earthquake=~yes,0.2).    p(earthquake=~no,0.8).

p(alarm=~yes,P) :- burglary=~yes, earthquake=~yes, q=~P.
p(alarm=~yes,0.8) :- burglary=~yes, earthquake=~no.
p(alarm=~yes,0.1) :- burglary=~no,  earthquake=~yes.
p(alarm=~no,default).

variables([ d(q,[0.0,0.2,0.4,0.6,0.8]), r(burglary,[yes,no]),
            r(earthquake,[yes,no]), r(alarm,[yes,no])]).

utility(1) :- alarm=~yes.

evidence :- burglary=~yes.

depends(_,[]).
```

Fig. 6. The alarm example with an unknown probability

4.3 Hybrid Problems

Most probable explanations and maximum likelihood estimations can both be computed efficiently by specialised PP algorithms, and our prototype rejection sampling approach is probably not competitive. However, an advantage of being able to solve different types of problem with a single algorithm is that we can also solve hybrid problems combining their features. For example we can tackle problems containing elements of both MPE and MLE. Suppose we modify the MPE Alarm ID in Sect. 4.1 by replacing the probability

```
p(alarm=~yes,0.8) :- burglary=~yes, earthquake=~yes.
```

by the one from the MLE Alarm ID in Sect. 4.2:

```
p(alarm=~yes,P) :- burglary=~yes, earthquake=~yes, q=~P.
```

Then we obtain a problem that is neither pure MPE nor MLE but has features of both. In this hybrid problem we must choose a probability P and an explanation e that together maximise the conditional expected utility $p(alarm=\sim yes \mid e)$. Solving the problem we find burglary=~yes, earthquake=~no and q=~0.8, giving conditional expected utility 0.8.

5 Solution by Reinforcement Learning

We learn optimal policies by combining techniques from Reinforcement Learning (RL) and sampling. RL is a fundamental machine learning problem, distinct from supervised and unsupervised learning. It tackles the problem of how a software agent should choose actions in an environment in order to maximise an expected reward. Temporal Difference (TD) Learning is a class of algorithms for solving such problems, and can be viewed as a hybrid of Monte Carlo and Dynamic Programming methods with roots in Behavioural Psychology [18].

We implemented a Monte Carlo RL algorithm from [18] with a variety of ϵ-greedy action selection methods. In this algorithm an *episode* (a simulation) proceeds from an *initial state* to a *terminal state* by taking *actions*. The *state values* $V(s)$ for states s are stored in a table (assuming a finite number of possible states) and initialised arbitrarily (for example by setting them to 0), and eventually converge to the expected total reward assuming we follow an optimal policy from s, via *bootstrapping*. To achieve convergence, the *learning rate* α and the ϵ parameter decay from 1 to 0 as the number of simulations performed increases. Any action may incur a reward r, but in our application the only rewards occur at the end of an episode. Updating a state value is referred to as *backup*. The discount rate parameter γ is set to 1 as we do not consider infinite horizon problems.

During a simulation or episode, we assign random and decision variables to domain values in the specified order: random variables are assigned values according to their distributions as specified by the Bayesian network expressed by the p predicate, as before; decision variables are assigned by our algorithm using ϵ-greedy action selection. The subset of assignments that are observable at any state is specified by the depends predicate. The only rewards occur at the end of a simulation (if the evidence is true in that simulation), and may be any function of the variable assignments as specified by the utility predicate. The RL algorithms are slightly modified here to handle evidence via rejection sampling: backup only occurs at the end of each episode if evidence is true. Though rejection sampling is also used in some other PPLs it can be costly, and we are developing another version of our language using Gibbs sampling.

If all information is available at any state then some RL algorithms are guaranteed to converge to an optimum policy because the problem is a Markov Decision Problem (MDP). Our adaptation has the same guarantee for simulations in which the evidence is true, if all previous variables are observable to each decision variable. However, if some information is invisible then the problem is a Partially Observable MDP (POMDP) and there is no such guarantee. So far we have obtained optimal policies but more sophisticated RL methods might be required for other IDs.

Note that applying our RL-based solver to pure PP problems (without decision variables) is equivalent to using simple rejection sampling for learning conditional probabilities, as long as the learning rate α decays appropriately (inversely with time). This close connection shows that IDLP is a strict generalisation of a simple PP approach, and makes RL algorithms a natural choice for extending PP with decision variables.

6 Conclusion and Related Work

We described a Prolog-based PPL called IDLP, extended to include decision variables and utilities. It can model multistage decision problems represented by IDs and (more generally) LIMIDs. IDLP solves these problems using a simple Reinforcement Learning algorithm combined with rejection sampling, and we showed that it can also perform PP tasks including inference, finding most probable explanations, and maximum likelihood estimation, as well as hybrids of these tasks. IDLP allows a user to model complex IDs with few lines of code, and to find approximate solutions for large IDs that are intractable for exact algorithms. Note that our Monte-Carlo sampling is based on an epsilon-greedy action selection that avoids wasting samples on policies with low rewards. This enables it to find a good solution without computing a precise expected utility for all possible policies.

Monte Carlo methods have been used before to solve IDs. [4] solve IDs by decision and random variable reduction, using Monte Carlo methods to sample the state of each random variable in a subset of relevant variables. The relevant variables are those random and decision variables whose values are required to take a particular decision. The aim is to determine a decision function for each decision variable, one by one, based on the relative estimated maximum conditional expected utility. So, as in our method, they do not compute posterior distributions explicitly, unlike exact methods such as [17] in which posterior distributions are computed by arc reversal. [3] use Monte Carlo sampling and approximation functions to handle large IDs, breaking them down into several single-stage decision problems. Our approach is different to these, as we apply Monte Carlo Reinforcement Learning and sampling to all the random variables of the ID, and take decisions according to epsilon-greedy action selection. Reinforcement Learning methods have of course been applied to many ID-like problems modelled as MDPs and POMDPs [18], but to the best of our knowledge this has not previously been proposed in the context of a PPL.

IDLP is not the only PPL with decision variables and utilities: DTProbLog (Decision-Theoretic ProbLog) [2], another Prolog-based PPL, also supports them. However, DTProbLog is not designed to solve *sequential* or *multistage* decision problems: in Stochastic Programming terminology it solves only *single-stage* problems in which all decisions are made before any random events occur [1]. IDLP allows decision and random variables to occur in any order in the variable list, enabling it to model multistage problems whose solution is not a fixed decision variable assignment but a policy tree.

We see the main contribution of this work as the convergence of tools and ideas from three areas: Probabilistic Programming, Influence Diagrams and Reinforcement Learning. We have used only simple techniques and examples from each of these fields, but our approach can potentially benefit from advances in each. This is a work in progress and we intend to improve it in several ways. Firstly, we are developing a Python-based system that should be easier for non-Prolog programmers to use. Secondly, a performance bottleneck is caused by our use of rejection sampling, which prevents us from tackling PP problems in which the evidence occurs in a vanishingly small proportion of scenarios. Though several PP systems also provide rejection sampling, the most powerful systems rely on Monte Carlo Markov Chain (MCMC) sampling algorithms. In future work we shall combine MCMC with Reinforcement Learning, perhaps using techniques such as those in [8]. Thirdly, another potential performance bottleneck is the occurrence of an exponential number of states in some IDs. This can be alleviated by discarding information to obtain simpler policies as in LIMIDs, but for some problems we will need more sophisticated *state aggregation* techniques such as tile coding or neural networks [18]. These choices may affect convergence so a great deal of work remains to be done, drawing on the literature from several research areas.

Acknowledgements. This work was supported in part by Science Foundation Ireland (SFI) under Grant Number SFI/12/RC/2289.

References

1. Birge, J.R., Louveaux, F.V.: Introduction to Stochastic Programming. Springer, New York (2011). doi:10.1007/978-1-4614-0237-4
2. van den Broeck, G., Thon, I., van Otterlo, M., de Raedt, L.: DTProbLog: a decision-theoretic probabilistic prolog. In: 24th AAAI Conference on Artificial Intelligence (2010)

3. Cano, A., Gómez, M., Moral, S.: A forward-backward Monte Carlo method for solving influence diagrams. Int. J. Approximate Reasoning **42**, 119–135 (2006)

4. Charnes, J.M., Shenoy, P.P.: Multistage Monte Carlo Method for solving influence diagrams using local computation. Manage. Sci. **50**(3), 405–418 (2004)

5. Gordon, A.D., Henzinger, T.A., Nori, A.V., Rajamani, S.K.: Probabilistic programming. in: International Conference on Software Engineering (2014)

6. Gutmann, B., Thon, I., De Raedt, L.: Learning the parameters of probabilistic logic programs from interpretations. In: Gunopulos, D., Hofmann, T., Malerba, D., Vazirgiannis, M. (eds.) ECML PKDD 2011. LNCS, vol. 6911, pp. 581–596. Springer, Heidelberg (2011). doi:10.1007/978-3-642-23780-5_47

7. Howard, R.A., Matheson, J.E.: Influence Diagrams. Readings in Decision Analysis, Strategic Decisions Group, Menlo Park, CA, Chap. 38, pp. 763–771 (1981)

8. Kimura, H.: Reinforcement learning in multi-dimensional state-action space using Random Rectangular Coarse Coding and Gibbs Sampling. In: International Conference on Intelligent Robots and Systems. IEEE (2007)

9. Lauritzen, S.L., Nilsson, D.: Representing and solving decision problems with limited information. Manage. Sci. **47**, 1238–1251 (2001)

10. Pearl, J.: Probabilistic Reasoning in Intelligent Systems: Networks of Plausible Inference. Series in Representation and Reasoning. Morgan Kaufman Publishers, San Mateo (1988)

11. Pfeffer, A.: Practical Probabilistic Programming. Manning Publications, Greenwich (2016)

12. http://probabilistic-programming.org/wiki/Home

13. De Raedt, L., Kimmig, A.: Probabilistic (logic) programming concepts. Mach. Learn. **100**(1), 5–47 (2015). Springer New York LLC

14. De Raedt, L., Kimmig, A., Toivonen, H.: ProbLog: a probabilistic prolog and its application in link discovery. In: IJCAI 2007, Proceedings of the 20th International Joint Conference on Artificial Intelligence, pp. 2462–2467 (2007)

15. Raiffa, H.: Decision Analysis. Addison-Wesley, Reading (1968)

16. Sato, T.: A statistical learning method for logic programs with distribution semantics. In: Proceedings of the 12th International Conference on Logic Programming, pp. 715–729. MIT Press (1995)

17. Shachter, R.D.: Evaluating influence diagrams. Oper. Res. **34**(6), 871–882 (1986)

18. Sutton, R.S., Barto, A.G.: Reinforcement Learning: an Introduction. MIT Press, Cambridge (1998)

First-Order Typed Model Counting
for Probabilistic Conditional Reasoning
at Maximum Entropy

Marco Wilhelm[1(✉)], Marc Finthammer[2], Gabriele Kern-Isberner[1],
and Christoph Beierle[2]

[1] Department of Computer Science, TU Dortmund, Dortmund, Germany
`marco.wilhelm@tu-dortmund.de`
[2] Department of Computer Science, University of Hagen, Hagen, Germany

Abstract. First-order typed model counting extends first-order model counting by the ability to distinguish between different types of models. In this paper, we exploit this benefit in order to calculate weighted conditional impacts (WCIs) which play a central role in nonmonotonic reasoning based on conditionals. More precisely, WCIs store information about the verification and the falsification of conditionals with respect to a possible worlds semantics, and therefore serve as sufficient statistics for maximum entropy (ME) distributions as models of probabilistic conditional knowledge bases. Formally, we annotate formulas with algebraic types that encode concisely all structural information needed to compute WCIs, while allowing for a systematic and efficient counting of models. In this way, our approach to typed model counting for ME-reasoning integrates both structural and counting aspects in the same framework.

1 Introduction

In recent years, *relational probabilistic logics* [6,9,10,15] became the focus of attention as they provide a strong means to model uncertain knowledge about interactions between individual objects. Reasoning in these logics, however, requires elaborate strategies to deal with large numbers of objects in order to tractably draw inferences, such as exploiting symmetries and making use of the indistinguishability of certain objects. Some encouraging results in this research field of *lifted probabilistic inference* were presented in [1,2,14]. These works basically capitalize on their efficient first-order model counting techniques that build a firm basis for further investigations; the core idea is to compile first-order sentences into sd-DNNF-circuits which allow for recursive model counting.

In this paper, we lift the abovementioned techniques to a *conditional* relational probabilistic logic based on conditionals of the form $(B|A)[p]$, stating that "if A holds, then B follows with probability p". To this end, we introduce a structural enhancement of sd-DNNF-circuits which we call sd-DNNF^S-circuits. These circuits allow one to label the edges of the circuit with elements of a commutative monoid S. By choosing generators of conditional structures [8] as these elements,

© Springer International Publishing AG 2017
S. Moral et al. (Eds.): SUM 2017, LNAI 10564, pp. 266–279, 2017.
DOI: 10.1007/978-3-319-67582-4_19

which encode the trivalent evaluation of conditionals, sd-DNNF^S-circuits are able to capture the whole logical information provided by a conditional knowledge base in a very comprised way. This information can be utilized by *(first-order) typed model counting on* sd-DNNF^S-*circuits*, which is the equivalent to ordinary model counting on sd-DNNF-circuits. More precisely, the typed model counting task (TMC) extends model counting by simultaneously group models into different *types* that are determined by the elements of the commutative monoid associated to the satisfied parts of the sd-DNNF^S-circuit (see also our previous work on *propositional* typed model counting in [17]). The main result of this paper eventually is that typed model counting is expedient for determining the *weighted conditional impacts* (WCIs; cf. [4]) relevant to a knowledge base $\mathcal{R}$. Basically, WCIs are conditional structures together with their frequencies of occurrence. Once these WCIs are known, the maximum entropy distribution (ME-distribution) as a preferable model of $\mathcal{R}$ (cf. [8,12]) can be computed efficiently using a highly optimized variant of *generalized iterative scaling* (GIS; cf. [4]), which principally allows for lifted inferences. However, until now, computing the WCIs forms a bottle-neck of ME-reasoning since no efficient algorithms for these computations are known except for very limited problem classes. In this regard, typed model counting provides a promising framework for efficient WCI-computations.

After recalling some basics of reasoning at maximum entropy, we discuss a GIS-algorithm that calculates ME-distributions based on WCIs, present typed model counting for structured first-order sentences, especially on sd-DNNF^S-circuits, exploit it for WCI-calculations, discuss related work and conclude.

2   Preliminaries

We consider a *first-order language* FOL over a signature $\Sigma = (\mathsf{Pred}, \mathsf{Const})$ consisting of a finite set of predicate symbols Pred and a finite set of constants Const without functions of arity greater than zero. An *atom* $P(t_1, \ldots, t_n)$ is a predicate P of arity n followed by *terms* t_i, $i = 1, \ldots, n$, where a term is either a variable or a constant. A *literal* is an atom or its negation, and a variable is called *free* iff it is not bound by a quantifier. A formula without variables is *grounded*. Every formula $A \in$ FOL can be grounded by substituting every free variable in A with a constant and by carrying out all quantifications. With $\mathsf{Gr}(A)$ we denote the set of all *ground instances* of A built this way. Further, $\mathsf{Gr}(\mathsf{Atoms})$ denotes the set of all ground instances of all atoms (= ground atoms). To shorten formulas, we abbreviate $A \wedge B$ as AB, $\neg A$ as $\overline{A}$, and $A \vee \overline{A}$ as $\top$.

A *(probabilistic) conditional* $(B|A)[p]$ with $A, B \in$ FOL and $p \in [0,1]$ is a formal representation of the statement "if A holds, then B follows with probability p", albeit we have to clarify its meaning if A or B contains free variables. With $\mathsf{Gr}((B|A)[p])$ we denote the set of all ground instances of $(B|A)[p]$ (e.g., $(R(a,b)|Q(a))[p]$ and $(R(a,a)|Q(a))[p]$ are ground instances of $(R(x,y)|Q(x))[p]$ if $a, b \in$ Const, but $(R(a,b)|Q(b))[p]$ is not). Further, $\mathsf{var}(r)$ denotes the set of all free variables in r, where r is a conditional or a formula.

The semantics of conditionals is given by probability distributions over *possible worlds*. Here, a possible world is simply a Herbrand interpretation and thus a subset of Gr(Atoms). We denote the set of all possible worlds with Ω. A possible world ω *entails* a ground atom $A \in$ Gr(Atoms), written $\omega \models A$, iff $A \in \omega$. We extend the entailment relation $\models$ to arbitrary grounded formulas in the usual way. The *aggregating semantics* [10] now allows us to define the concept of models for conditionals even if they contain free variables. It is inspired by statistical approaches, but sums up probabilities instead of just counting instances.

Definition 1 (Aggregating Semantics). *Let* $\mathcal{P} : \Omega \to [0,1]$ *be a probability distribution, and let* $r = (B|A)[p]$ *be a conditional.* $\mathcal{P}$ *is a model of* r, *written*

$$\mathcal{P} \models r, \; \text{iff} \; \frac{\sum_{(B'|A')[p] \in \mathsf{Gr}(r)} \mathcal{P}(A'B')}{\sum_{(B'|A')[p] \in \mathsf{Gr}(r)} \mathcal{P}(A')} = p,$$

where $\mathcal{P}(A) = \sum_{\omega \in \Omega: \, \omega \models A} \mathcal{P}(\omega)$ *for grounded formulas* A.

A *knowledge base* $\mathcal{R}$ is a finite set of conditionals[1], and a probability distribution $\mathcal{P}$ is a *model of* $\mathcal{R}$, written $\mathcal{P} \models \mathcal{R}$, iff it is a model of every conditional in $\mathcal{R}$. A knowledge base is *consistent* iff it has at least one model. Among all models of a consistent knowledge base $\mathcal{R}$, the ME-*distribution* $\mathcal{P}_{\mathsf{ME}}(\mathcal{R})$ shows especially good properties from a logical point of view [7], and complies with commonsense reasoning in an optimal way [13]. As $\mathcal{P}_{\mathsf{ME}}(\mathcal{R})$ follows the *principle of maximum entropy (ME-principle)* [12,16], it is the unique distribution with maximal entropy that models $\mathcal{R}$.

Definition 2 (ME-Distribution). *Let* $\mathcal{R}$ *be a consistent knowledge base. The* ME-distribution *of* $\mathcal{R}$ *is defined by*

$$\mathcal{P}_{\mathsf{ME}}(\mathcal{R}) = \arg \max_{\mathcal{P} \models \mathcal{R}} - \sum_{\omega \in \Omega} \mathcal{P}(\omega) \log \mathcal{P}(\omega).$$

As a central prerequisite for our further investigations, we observe that possible worlds with the same *conditional structure* [8] also have the same ME-probability, which is a gratifying property of the ME-distribution. The conditional structure basically states how often a possible world ω verifies ($\omega \models A'_i B'_i$) and falsifies ($\omega \models A'_i \overline{B'_i}$) the (ground instances of the) conditionals in $\mathcal{R}$.

Definition 3 (Conditional Structure). *Let* $\mathcal{R} = \{r_1, \ldots, r_n\}$ *be a consistent knowledge base with* $r_i = (B_i|A_i)[p_i]$, $i = 1, \ldots, n$, *and let* $\omega \in \Omega$. *The conditional structure of* ω *with respect to* $\mathcal{R}$ *is defined by*

$$\sigma_{\mathcal{R}}(\omega) = \prod_{i=1,\ldots,n} \; \prod_{(B'_i|A'_i)[p_i] \in \mathsf{Gr}(r_i)} \begin{cases} \mathbf{a}_i^+ & \text{iff } \omega \models A'_i B'_i \\ \mathbf{a}_i^- & \text{iff } \omega \models A'_i \overline{B'_i} \, , \\ 1 & \text{iff } \omega \models \overline{A'_i} \end{cases}$$

---

[1] Non-conditional statements of the form "$A \in$ FOL holds with probability p" can be accommodated into $\mathcal{R}$ by adding the conditional $(A|\top)[p]$ to $\mathcal{R}$.

where $\mathbf{a_i^+}$ and $\mathbf{a_i^-}$ for $i = 1, \ldots, n$ are the generators of a free commutative monoid $(\mathcal{G_R}, \cdot, 1)$.[2] Using the functions

$$\mathsf{ver}_i(\omega) = |\{(B_i'|A_i')[p_i] \in \mathsf{Gr}(r_i) \mid \omega \models A_i'B_i'\}|,$$
$$\mathsf{fal}_i(\omega) = |\{(B_i'|A_i')[p_i] \in \mathsf{Gr}(r_i) \mid \omega \models A_i'\overline{B_i'}\}|,$$

the conditional structure can also be written as $\sigma_{\mathcal{R}}(\omega) = \prod\limits_{i=1}^{n} (\mathbf{a_i^+})^{\mathsf{ver}_i(\omega)}(\mathbf{a_i^-})^{\mathsf{fal}_i(\omega)}.$

As possible worlds with the same conditional structure verify and falsify the same conditionals equally often, they are *indifferent* from the knowledge base's perspective and the notion of conditional structure induces an equivalence relation on Ω: Two possible worlds ω and ω' are equivalent with respect to $\mathcal{R}$, written $\omega \sim_{\mathcal{R}} \omega'$, iff $\sigma_{\mathcal{R}}(\omega) = \sigma_{\mathcal{R}}(\omega')$, i.e., iff for all $i = 1, \ldots, n$ both $\mathsf{ver}_i(\omega) = \mathsf{ver}_i(\omega')$ and $\mathsf{fal}_i(\omega) = \mathsf{fal}_i(\omega')$ hold. The following proposition holds.

Proposition 1. *Let $\mathcal{R}$ be a consistent knowledge base, and let $\omega \sim_{\mathcal{R}} \omega'$. Then, $\mathcal{P}_{\mathsf{ME}}(\omega) = \mathcal{P}_{\mathsf{ME}}(\omega')$.*

As a consequence, it is sufficient to calculate the equivalence classes $[\omega]_{\sim_{\mathcal{R}}}$ and the ME-probability $\mathcal{P}_{\mathsf{ME}}(\omega)$ for only one representative per equivalence class in order to determine $\mathcal{P}_{\mathsf{ME}}(\mathcal{R})$, as we will discuss in the next section.

3   Computing the ME-Distribution

Our algorithm $\mathsf{GIS^{WCI}}$ computes the ME-distribution $\mathcal{P}_{\mathsf{ME}}(\mathcal{R})$ for a consistent $\mathcal{R}$. It is based on the well-known *generalized iterative scaling* approach [3], but avoids iterations over possible worlds. Instead, it works on the so-called *weighted conditional impacts* (WCIs) that comprise, in a computer processible way, all the information about the equivalence classes $[\omega]_{\sim_{\mathcal{R}}}$ which is necessary to calculate $\mathcal{P}_{\mathsf{ME}}(\mathcal{R})$. Hence, WCIs serve as sufficient statistics for ME-distributions.

Definition 4 (Weighted Conditional Impact). *Let $\mathcal{R}$ be a consistent knowledge base consisting of the conditionals $r_i = (B_i|A_i)[p_i]$, $i = 1, \ldots, n$. The conditional impact is a mapping $\gamma_{\mathcal{R}} : \Omega/\sim_{\mathcal{R}} \to (\mathbb{N}_0 \times \mathbb{N}_0)^n$ defined by*

$$\gamma_{\mathcal{R}}([\omega_{\sim_{\mathcal{R}}}]) = \langle \gamma_1, \ldots, \gamma_n \rangle \quad with \quad \gamma_i = \langle \mathsf{ver}_i(\omega), \mathsf{fal}_i(\omega) \rangle, \quad i = 1, \ldots, n.$$

The weighted conditional impact *(WCI) maps each equivalence class $[\omega]_{\sim_{\mathcal{R}}}$ to the tuple consisting of its conditional impact and its cardinality and is given by*

$$\mathsf{WCI}([\omega_{\sim_{\mathcal{R}}}]) = \langle \gamma_{\mathcal{R}}([\omega]_{\sim_{\mathcal{R}}}), |[\omega]_{\sim_{\mathcal{R}}}| \rangle.$$

The weighted conditional impact matrix $\mathsf{WCI}(\mathcal{R}) \in (\mathbb{N}_0)^{d \times 2n+1}$ *of $\mathcal{R}$ is given by*

$$\mathsf{WCI}(\mathcal{R})_{j,k} = \begin{cases} ((\mathsf{WCI}([\omega_j]_{\sim_{\mathcal{R}}})_1)_i)_1 & \text{iff } k = 2i - 1 \\ ((\mathsf{WCI}([\omega_j]_{\sim_{\mathcal{R}}})_1)_i)_2 & \text{iff } k = 2i \\ \mathsf{WCI}([\omega_j]_{\sim_{\mathcal{R}}})_2 & \text{iff } k = 2n + 1 \end{cases},$$

where $\{[\omega_1]_{\sim_{\mathcal{R}}}, \ldots, [\omega_d]_{\sim_{\mathcal{R}}}\} = \Omega/\sim_{\mathcal{R}}$, thus $d = |\Omega/\sim_{\mathcal{R}}|$.

---

[2] We usually omit the operation symbol ".".

Table 1. Parameters describing the logical aspects of $\mathcal{R}$ from Ex. 1

$[\omega]_{\sim_{\mathcal{R}}}$	$\sigma_{\mathcal{R}}(\omega)$	$|[\omega]_{\sim_{\mathcal{R}}}|$	$\mathsf{WCI}([\omega]_{\sim_{\mathcal{R}}})$
$[\omega_1]_{\sim_{\mathcal{R}}}$	$\mathbf{a_1^+}^2 \quad \mathbf{a_2^+}$	1	$\langle\langle\langle 2,0\rangle, \langle 1,0\rangle\rangle, 1\rangle$
$[\omega_2]_{\sim_{\mathcal{R}}}$	$\mathbf{a_1^+}\,\mathbf{a_1^-}\,\mathbf{a_2^+}$	2	$\langle\langle\langle 1,1\rangle, \langle 1,0\rangle\rangle, 2\rangle$
$[\omega_3]_{\sim_{\mathcal{R}}}$	$\mathbf{a_1^+} \quad \mathbf{a_2^+}$	2	$\langle\langle\langle 1,0\rangle, \langle 1,0\rangle\rangle, 2\rangle$
$[\omega_4]_{\sim_{\mathcal{R}}}$	$\mathbf{a_1^+} \quad\quad \mathbf{a_2^-}$	2	$\langle\langle\langle 1,0\rangle, \langle 0,1\rangle\rangle, 2\rangle$
$[\omega_5]_{\sim_{\mathcal{R}}}$	$\mathbf{a_1^-}^2\,\mathbf{a_2^+}$	1	$\langle\langle\langle 0,2\rangle, \langle 1,0\rangle\rangle, 1\rangle$
$[\omega_6]_{\sim_{\mathcal{R}}}$	$\mathbf{a_1^-}\,\mathbf{a_2^+}$	2	$\langle\langle\langle 0,1\rangle, \langle 1,0\rangle\rangle, 2\rangle$
$[\omega_7]_{\sim_{\mathcal{R}}}$	$\mathbf{a_1^-} \quad \mathbf{a_2^-}$	2	$\langle\langle\langle 0,1\rangle, \langle 0,1\rangle\rangle, 2\rangle$
$[\omega_8]_{\sim_{\mathcal{R}}}$	$\mathbf{a_2^-}$	4	$\langle\langle\langle 0,0\rangle, \langle 0,1\rangle\rangle, 4\rangle$

Our definition of the weighted conditional impact based on the equivalence classes of possible worlds is in coincidence with the definition presented in [4], where WCIs are defined for possible worlds. The weighted conditional impact matrix $\mathsf{WCI}(\mathcal{R})$ is an even more comprised representation of all weighted conditionals impacts with respect to $\mathcal{R}$ and will simplify notations in Proposition 2.

Example 1. We consider the knowledge base $\mathcal{R} = \{r_1, r_2\}$ with

$$r_1 = (\underline{F}amous(x)|\underline{M}illionaire(x))[0.7], \quad r_2 = (\underline{M}illionaire(\underline{a}lice)|\top)[0.9].$$

The equivalence classes of the possible worlds with respect to $\sim_{\mathcal{R}}$ (and the set of constants $\mathsf{Const} = \{\underline{a}lice, \underline{b}ob\}$) are

$$
\begin{aligned}
[\omega_1]_{\sim_{\mathcal{R}}} &= \{\ \{F(a), F(b), M(a), M(b)\} & \},\\
[\omega_2]_{\sim_{\mathcal{R}}} &= \{\ \{F(a), M(a), M(b)\}, \{F(b), M(a), M(b)\}\ \},\\
[\omega_3]_{\sim_{\mathcal{R}}} &= \{\ \{F(a), M(a)\}, & \{F(a), F(b), M(a)\}\ \},\\
[\omega_4]_{\sim_{\mathcal{R}}} &= \{\ \{F(b), M(b)\}, & \{F(a), F(b), M(b)\}\ \},\\
[\omega_5]_{\sim_{\mathcal{R}}} &= \{\ \{M(a), M(b)\} & \},\\
[\omega_6]_{\sim_{\mathcal{R}}} &= \{\ \{M(a)\}, & \{F(b), M(a)\} & \},\\
[\omega_7]_{\sim_{\mathcal{R}}} &= \{\ \{M(b)\}, & \{F(a), M(b)\} & \},\\
[\omega_8]_{\sim_{\mathcal{R}}} &= \{\ \emptyset, \ \{F(a)\}, \ \{F(b)\}, \{F(a), F(b)\} & \}.
\end{aligned}
$$

Their cardinalities, conditional structures, and weighted conditional impacts are shown in Table 1. The weighted conditional impact matrix is

$$
\mathsf{WCI}(\mathcal{R}) =
\begin{pmatrix}
2 & 0 & 1 & 0 & 1 \\
1 & 1 & 1 & 0 & 2 \\
1 & 0 & 1 & 0 & 2 \\
1 & 0 & 0 & 1 & 2 \\
0 & 2 & 1 & 0 & 1 \\
0 & 1 & 1 & 0 & 2 \\
0 & 1 & 0 & 1 & 2 \\
0 & 0 & 0 & 1 & 4
\end{pmatrix}.
$$

Once $\text{WCI}(\mathcal{R})$ is given, the algorithm GIS^{WCI} proceeds to calculate the ME-distribution $\mathcal{P}_{\text{ME}}(\mathcal{R})$ as shown in Fig. 1. The algorithm was presented in [4] first and is recalled in a more informal way here. Calculating the required input $\text{WCI}(\mathcal{R})$ is a difficult task, at least if one wants to avoid iterations over Ω, which is necessary when having lifted inference in mind. Until now, efficient strategies to compute $\text{WCI}(\mathcal{R})$ are known for very restricted classes of knowledge bases only [5]. In the rest of the paper, we present *(first-order) typed model counting* which proves to be a fruitful framework for further investigations on calculating $\text{WCI}(\mathcal{R})$. In the course of this, the knowledge base $\mathcal{R}$ is compiled into a so-called structured formula $\phi(\mathcal{R})$ whose models are counted and classified into different types by the typed model counting task. The outcome of this approach is isomorphic to $\text{WCI}(\mathcal{R})$. The whole process of calculating the ME-distribution of a probabilistic knowledge base $\mathcal{R}$ by exploiting WCIs is illustrated in Fig. 2.

Input $\text{WCI}(\mathcal{R})$ and probabilities $p_1, \ldots, p_n$ of the n conditionals in $\mathcal{R}$

Output   Values $\alpha_0, \alpha_1, \ldots, \alpha_n$ determining $\mathcal{P}_{\text{ME}}(\mathcal{R})$ via
$$\mathcal{P}_{\text{ME}}(\mathcal{R})(\omega) = \alpha_0 \prod_{i=1}^{n} \alpha_i^{\text{ver}_i(\omega)(1-p_i)-\text{fal}_i(\omega)p_i}$$

1. Initialize iteration counter $k = 0$ and uniform probabilities $P_{(k)}([\omega]_{\sim_{\mathcal{R}}})$ for equivalence classes in $\Omega/\sim_{\mathcal{R}}$
2. Repeatedly calculate an approximation to $\alpha_1, \ldots, \alpha_n$ until an abortion condition holds based on the quality of the approximation
   (a) Increase iteration counter k
   (b) Update scaling factors $\beta_{(k),i}$ based on probabilities $P_{(k-1)}([\omega]_{\sim_{\mathcal{R}}})$ by employing $\text{WCI}(\mathcal{R})$ and p_i
   (c) Scale probabilities $P_{(k)}([\omega]_{\sim_{\mathcal{R}}})$ and values $\alpha_{(k),i}$ by updated $\beta_{(k),i}$
   (d) Normalize scaled probabilities $P_{(k)}([\omega]_{\sim_{\mathcal{R}}})$
   ↻ End loop
3. Set values α_i to $\alpha_{(k),i}$
4. Determine normalization value α_0 wrt. α_i and p_i

Fig. 1. Algorithm GIS^{WCI} to compute $\mathcal{P}_{\text{ME}}(\mathcal{R})$ from $\text{WCI}(\mathcal{R})$ without iterating over Ω

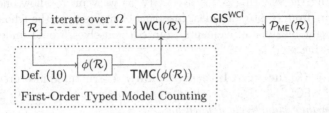

Fig. 2. First-order typed model counting and its relevance for calculating $\mathcal{P}_{\text{ME}}(\mathcal{R})$

4   First-Order Typed Model Counting (TMC)

(First-order) typed model counting (TMC) extends first-order model counting by the ability to classify models into different *types* (see also [17] for a definition in the propositional case). These types are represented by elements of a commutative monoid $(\mathcal{S}, \otimes, \mathbf{1}_{\mathcal{S}})$ that are directly incorporated into the formulas. Therewith, TMC allows for a more fine-grained evaluation of the formula. When using typed model counting to calculate WCIs in Sect. 5, we will instantiate $(\mathcal{S}, \otimes, \mathbf{1}_{\mathcal{S}})$ with $(\mathcal{G}_{\mathcal{R}}, \cdot, 1)$ such that the different model types represent conditional structures.

As a formal basis for TMC, we define the *structured language* $\mathsf{FOL}^{\mathcal{S}}$ which consists of all formulas in FOL and additionally allows to concatenate elements from $\mathcal{S}$ to the left of any part of a formula as long as they are not in the scope of negations (e.g., $\forall X.(\mathbf{s} \circ \neg A(X) \vee B(X))$ is in $\mathsf{FOL}^{\mathcal{S}}$ if $\mathbf{s} \in \mathcal{S}$ and $A, B \in \mathsf{Pred}$, but $\forall X.(\neg(\mathbf{s} \circ A(X)) \vee B(X))$ is not).

Definition 5 (Structured Language). *Let* FOL *be defined as before (see Sect. 2), and let* $(\mathcal{S}, \otimes, \mathbf{1}_{\mathcal{S}})$ *be a commutative monoid. The structured language* $\mathsf{FOL}^{\mathcal{S}}$ *is the smallest set such that*

$$A, \ B \wedge C, \ B \vee C, \ \exists X.B, \ \forall X.B, \ (\mathbf{s} \circ B) \ \in \mathsf{FOL}^{\mathcal{S}},$$

where $A \in \mathsf{FOL}$, $B, C \in \mathsf{FOL}^{\mathcal{S}}$, $X \in \mathsf{var}(B)$, $\mathbf{s} \in \mathcal{S}$, *and* $\circ : \mathcal{S} \times \mathsf{FOL}^{\mathcal{S}} \to \mathsf{FOL}^{\mathcal{S}}$ *is an outer operation between* $\mathcal{S}$ *and* $\mathsf{FOL}^{\mathcal{S}}$.[3] *Further, we claim that formulas in* $\mathsf{FOL}^{\mathcal{S}}$ *have no free variables, i.e., they are sentences.*

In order to be able to interpret and count the typed models of formulas in $\mathsf{FOL}^{\mathcal{S}}$, we enrich $(\mathcal{S}, \otimes, \mathbf{1}_{\mathcal{S}})$ with a second binary operation $\oplus$. More precisely, we add an additional element $\mathbf{0}_{\mathcal{S}}$ to $\mathcal{S}$ which will serve as the identity element with respect to $\oplus$. Then, we build the closure $\mathcal{S}^{\oplus}$ of $\mathcal{S} \cup \{\mathbf{0}_{\mathcal{S}}\}$ under application of $\oplus$ while ensuring that $\otimes$ and $\oplus$ behave distributively. Eventually, we obtain the commutative semiring $(\mathcal{S}^{\oplus}, \oplus, \otimes, \mathbf{0}_{\mathcal{S}}, \mathbf{1}_{\mathcal{S}})$. We further abbreviate $n \cdot \mathbf{s} = \bigoplus_{i=1}^{n} \mathbf{s}$ and $\mathbf{s}^{n} = \bigotimes_{i=1}^{n} \mathbf{s}$ for $\mathbf{s} \in \mathcal{S}$ and $n \in \mathbb{N}$. Note that the differentiation between the monoid $\mathcal{S}$ and the semiring $\mathcal{S}^{\oplus}$ is necessary, as we want to allow one to insert only elements from $\mathcal{S}$ into formulas. Inserting elements from the whole semiring $\mathcal{S}^{\oplus}$, that might be sums of elements from $\mathcal{S}$ themselves, would undermine the idea of counting models.

Definition 6 (Structured Interpretation). *A structured interpretation is a mapping* $\mathcal{I}^{\mathcal{S}} : \mathsf{FOL}^{\mathcal{S}} \to \mathcal{S} \cup \{\mathbf{0}_{\mathcal{S}}\}$ *that maps every ground atom to either* $\mathbf{0}_{\mathcal{S}}$ *or* $\mathbf{1}_{\mathcal{S}}$ *and is further inductively defined by*

1. $\quad \mathcal{I}^{\mathcal{S}}(\neg A) = \begin{cases} \mathbf{1}_{\mathcal{S}} & \text{iff } \mathcal{I}^{\mathcal{S}}(A) = \mathbf{0}_{\mathcal{S}} \\ \mathbf{0}_{\mathcal{S}} & \text{iff } \mathcal{I}^{\mathcal{S}}(A) = \mathbf{1}_{\mathcal{S}} \end{cases}$,

2. $\quad \mathcal{I}^{\mathcal{S}}(B \wedge C) = \mathcal{I}^{\mathcal{S}}(B) \otimes \mathcal{I}^{\mathcal{S}}(C)$,

---

[3] The operation $\circ$ shall be the one that binds weakest.

$$3. \quad \mathcal{I}^{\mathcal{S}}(B \vee C) = \begin{cases} \mathcal{I}^{\mathcal{S}}(B) & \text{iff } \mathcal{I}^{\mathcal{S}}(C) = \mathbf{0}_{\mathcal{S}} \\ \mathcal{I}^{\mathcal{S}}(C) & \text{iff } \mathcal{I}^{\mathcal{S}}(B) = \mathbf{0}_{\mathcal{S}}, \\ \mathcal{I}^{\mathcal{S}}(B) \otimes \mathcal{I}^{\mathcal{S}}(C) & \text{otherwise} \end{cases}$$

4. $\quad \mathcal{I}^{\mathcal{S}}(\mathbf{s} \circ B) = \mathbf{s} \otimes \mathcal{I}^{\mathcal{S}}(B),$

5. $\quad \mathcal{I}^{\mathcal{S}}(\exists X.B) = \mathcal{I}^{\mathcal{S}}(\bigvee_{c \in \mathsf{Const}} B[X/c]),$

6. $\quad \mathcal{I}^{\mathcal{S}}(\forall X.B) = \mathcal{I}^{\mathcal{S}}(\bigwedge_{c \in \mathsf{Const}} B[X/c]),$

where $A \in \mathsf{FOL}$, $B, C \in \mathsf{FOL}^{\mathcal{S}}$, $\mathbf{s} \in \mathcal{S}$, $X \in \mathsf{var}(B)$, and $B[X/c]$ is B after substituting every occurrence of X by the constant c.

Note that the structured interpretation of a formula $A \in \mathsf{FOL}$ (i.e., A is free of elements from $\mathcal{S}$) is exactly the same as the ordinary interpretation of A when mapping $\mathbf{1}_{\mathcal{S}}$ to 1 (true) and $\mathbf{0}_{\mathcal{S}}$ to 0 (false).

Example 2. Let $\mathbf{s} \in \mathcal{S}$, and let $\mathcal{I}^{\mathcal{S}}$ be a structured interpretation such that $\mathcal{I}^{\mathcal{S}}(M(c)) = \mathbf{1}_{\mathcal{S}}$ for all $c \in \mathsf{Const}$. Then,

$$\mathcal{I}^{\mathcal{S}}(\mathbf{s} \circ \forall X.M(X)) = \mathbf{s} \quad \text{and} \quad \mathcal{I}^{\mathcal{S}}(\forall X.\mathbf{s} \circ M(X)) = \mathbf{s}^n \quad \text{where } n = |\mathsf{Const}|.$$

This difference is caused by the fact that elements in $\mathcal{S}$ do not need to be idempotent, which is in contrast to ordinary interpretations of formulas in FOL.

There is a one-to-one correspondence between structured interpretations and possible worlds obtained by the *structured interpretation induced by* $\omega \in \Omega$:

$$\mathcal{I}^{\mathcal{S}}_{\omega}(A) = \begin{cases} \mathbf{1}_{\mathcal{S}} & \text{iff } A \in \omega \\ \mathbf{0}_{\mathcal{S}} & \text{iff } A \notin \omega \end{cases}, \quad A \in \mathsf{Gr}(\mathsf{Atoms}).$$

The set of all structured interpretations then is $\{\mathcal{I}^{\mathcal{S}}_{\omega} \mid \omega \in \Omega\}$, and we may build our further investigations upon those structured interpretations that are induced by possible worlds.

Definition 7 (Typed Model Counting). *The structured interpretation* $\mathcal{I}^{\mathcal{S}}_{\omega}$ *is a typed model of* $A \in \mathsf{FOL}^{\mathcal{S}}$, *written* $\mathcal{I}^{\mathcal{S}}_{\omega} \models A$, *iff* $\mathcal{I}^{\mathcal{S}}_{\omega}(A) \in \mathcal{S}$, *i.e., iff* $\mathcal{I}^{\mathcal{S}}_{\omega}(A) \neq \mathbf{0}_{\mathcal{S}}$. *It is a* model of type $\mathbf{s}$ *iff* $\mathcal{I}^{\mathcal{S}}_{\omega}(A) = \mathbf{s}$. *The* typed model counting *task is calculating*

$$\mathsf{TMC}(A) = \bigoplus_{\mathcal{I}^{\mathcal{S}}_{\omega} \models A} \mathcal{I}^{\mathcal{S}}_{\omega}(A).$$

Example 3. Let $\mathsf{Const} = \{a, b, c\}$ and $\mathbf{s_1}, \mathbf{s_2} \in \mathcal{S}$. The structured formula

$$A = (\mathbf{s_1} \circ M(a)) \vee (\mathbf{s_2} \circ M(b)M(c))$$

has five typed models (when M is the only predicate): Three of type $\mathbf{s_1}$ (where $M(a)$ is true, i.e., $\mathcal{I}^{\mathcal{S}}_{\omega}(M(a)) = \mathbf{1}_{\mathcal{S}}$, but not both $M(b)$ and $M(c)$ are true), one of type $\mathbf{s_2}$ (where $M(b)$ and $M(c)$ are true, but $M(a)$ is not), and one of type $\mathbf{s_1} \otimes \mathbf{s_2}$ (where $M(a), M(b)$, and $M(c)$ are true). Hence,

$$\mathsf{TMC}(A) = 3 \cdot \mathbf{s_1} \oplus \mathbf{s_2} \oplus (\mathbf{s_1} \otimes \mathbf{s_2}).$$

In order to perform typed model counting, we seize on and extend the approach to first-order model counting presented in [1,2] that compiles clausal theories into so-called sd-DNNF-circuits on which first-order model counting can be performed in time polynomial in the size of the circuit. Here, we compile *structured* formulas into sd-DNNF^S-*circuits* for which we have to define when structured formulas are semantically equivalent.

Definition 8 (S-Equivalence). *Let $A, B \in$ FOL^S be structured formulas. A is S-equivalent to B, written $A \equiv_S B$, iff $\mathcal{I}^S_\omega(A) = \mathcal{I}^S_\omega(B)$ for every structured interpretation $\mathcal{I}^S_\omega$.*

The definition of S-equivalence is more restrictive than the ordinary equivalence of formulas in FOL since S-equivalent formulas not only need to have the same models, but the models also have to be of the same type.

Definition 9 (sd-DNNF^S-Circuit). *An sd-DNNF^S-circuit is a rooted directed acyclic graph whose edges are labeled with elements of S^4, whose leaf nodes represent ground literals, and whose inner nodes represent one of the following operators:[5]*

1. **decomposable conjunction $A \bigwedge B$**: *representing the formula $A \wedge B$ in FOL^S with the constraint that A and B do not share any ground atoms,*
2. **smooth deterministic disjunction $A \bigvee B$**: *representing the formula $A \vee B$ in FOL^S with the constraints that A and B contain the same ground atoms and are mutually exclusive, i.e., $\mathcal{I}^S_\omega(A) \otimes \mathcal{I}^S_\omega(B) = \mathbf{0}_S$ for every structured interpretation $\mathcal{I}^S_\omega$,*
3. **set conjunction $\bigwedge C$**: *representing a decomposable conjunction over isomorphic operands, i.e., operands that are S-equivalent up to a permutation of constants,*
4. **set disjunction $\bigvee C$**: *representing a smooth deterministic disjunction over isomorphic operands.*

Every structured formula in FOL^S can be compiled into an S-equivalent sd-DNNF^S-circuit, which can be proved as follows: As we consider a finite signature, every quantification in a formula $A \in$ FOL^S can be executed until there is no more variable in A. As a consequence, A is S-equivalent to a propositional structured formula. In [17], we have proved that every propositional structured formula is equivalent to a formula in sd-DNNF^S which is a normal form that is in one-to-one correspondence to sd-DNNF^S-circuits. The great advantage of sd-DNNF^S-circuits is that counting their typed models can be performed recursively by interpreting them as algebraic circuits (here, A and B are subgraphs):[6]

---

[4] We write edge labels in dashed frames and omit the label "$\mathbf{1}_S$".

[5] Note that sd-DNNF^S-circuits principally represent grounded formulas, and so are $A, B, C \in$ FOL^S in the following definitions. The power of sd-DNNF^S-circuits lies in the capability of consolidating isomorphic formulas via set conjunction respectively set disjunction.

[6] In TMC(A) resp. TMC(B), those structured interpretations are considered that are restricted to the ground atoms in A resp. B.

Example 4. Let $|\mathsf{Const}| \geq 2$ and $\mathbf{s} \in \mathcal{S}$. We consider the structured formula

$$\psi = \exists X.\mathbf{s} \circ \underline{M}illionaire(X).$$

The single instances of the existential quantification are not mutually exclusive as there may exist distinct ground atoms $M(a)$ and $M(b)$ which both are mapped to $\mathbf{1}_\mathcal{S}$ by the same structured interpretation (i.e., there may exist more than one millionaire within a possible world). In order to compile ψ into an sd-DNNF$^\mathcal{S}$-circuit, it is expedient to distinguish structured interpretations by the number of ground instances of $M(X)$ they map to $\mathbf{1}_\mathcal{S}$. This can be achieved via set disjunction and set conjunction as shown in Fig. 3.

Now, one possible application of TMC is as follows: For a fixed size of Const, typed model counting of ψ not only tells us how many possible worlds exist with at least one millionaire, but we can read out of $\mathsf{TMC}(\psi)$ how many possible worlds exist with exactly k millionaires: Let $|\mathsf{Const}| = 3$. Then,

$$\mathsf{TMC}(\psi) = 3 \cdot \mathbf{s} \oplus 3 \cdot \mathbf{s}^2 \oplus \mathbf{s}^3,$$

and, e.g., $3 \cdot \mathbf{s}^2$ says that there are three possible worlds with exactly two millionaires.

5 Typed Model Counting for **ME-Reasoning**

We now apply typed model counting (TMC) to ME-reasoning, in particular to calculating WCIs. To this end, we convert knowledge bases into structured formulas that are interpreted as the conditional structures of exactly those possible worlds which induced the respective structured interpretations. Thus, conditional structures will serve as the semantics of the formulas in this section.

Definition 10 (Knowledge Base Compilation). *Let $\mathcal{R}$ be a consistent knowledge base consisting of the conditionals $r_i = (B_i|A_i)[p_i]$, $i = 1,\ldots,n$. Further, let $\mathsf{FOL}^{\mathcal{G}_\mathcal{R}}$ be the structured language $\mathsf{FOL}^\mathcal{S}$ after instantiating $(\mathcal{S}, \otimes, \mathbf{1}_\mathcal{S})$*

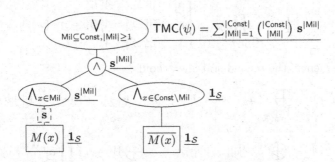

$$\mathsf{TMC}(\psi) = \sum_{|\mathsf{Mil}|=1}^{|\mathsf{Const}|} \binom{|\mathsf{Const}|}{|\mathsf{Mil}|} \mathbf{s}^{|\mathsf{Mil}|}$$

Fig. 3. sd-DNNF$^\mathcal{S}$-circuit for ψ from Ex. 4. Annotated typed model counts are underlined and can be obtained by instantiating the leave nodes with $\mathbf{1}_\mathcal{S}$ and incrementally counting the typed models of the parent nodes until one reaches the root node with the free commutative monoid $(\mathcal{G}_\mathcal{R}, \cdot, 1)$. We define[7]

$$\phi(\mathcal{R}) = \bigwedge_{i=1}^{n} \bigvee_{X \in \mathsf{var}(r_i)} X.\phi_i(r_i) \quad \text{with}$$

$$\phi_i(r_i) = (\mathbf{a_i^+} \circ A_i B_i) \vee (\mathbf{a_i^-} \circ A_i \overline{B_i}) \vee \overline{A_i}, \ i = 1, \ldots, n.$$

The formula $\phi(\mathcal{R})$ is a conjunction over all ground instances of all conditionals in $\mathcal{R}$, and every conjunct consists of three mutually exclusive disjuncts: every ground instantiation of $(\mathbf{a_i^+} \circ A_i B_i)$ refers to the verification of the respective ground instance of the i-th conditional, $(\mathbf{a_i^-} \circ A_i \overline{B_i})$ refers to the conditional's falsification, and $\overline{A_i}$ covers the case in which the conditional is not applicable. In total, $\phi(\mathcal{R})$ comprises all the information about the logical part of the knowledge base $\mathcal{R}$ within a single structured formula.

The following proposition states how the weighted conditional impact matrix $\mathsf{WCI}(\mathcal{R})$ of $\mathcal{R}$ can be read out of the typed models of $\phi(\mathcal{R})$. Hence, it bridges the gap between $\mathcal{R}$ and $\mathsf{WCI}(\mathcal{R})$ as illustrated in Fig. 2 using typed model counting, and therefore it constitutes the central result of this paper.

Proposition 2. *Let $\mathcal{R} = \{r_1, \ldots, r_n\}$ be a consistent knowledge base, and let $W = \mathsf{WCI}(\mathcal{R}) \in (\mathbb{N}_0)^{d \times 2n+1}$ be the weighted conditional impact matrix of $\mathcal{R}$. Further let $\omega \in \Omega$. Then,*

1. $\mathcal{I}_\omega^{\mathcal{G}_\mathcal{R}}(\phi(\mathcal{R})) = \sigma_\mathcal{R}(\omega)$,
2. $\mathsf{TMC}(\phi(\mathcal{R})) = \bigoplus_{[\omega]_{\sim_\mathcal{R}} \in \Omega/\sim_\mathcal{R}} |[\omega]_{\sim_\mathcal{R}}| \cdot \sigma_\mathcal{R}(\omega)$

$$= \bigoplus_{j=1}^{d} W_{j,2n+1} \prod_{i=1}^{n} \mathbf{a_i^+}^{W_{j,2i-1}} \mathbf{a_i^-}^{W_{j,2i}}.$$

Proof. We prove the first statement of the proposition:

$$\mathcal{I}_\omega^{\mathcal{G}_\mathcal{R}}(\phi(\mathcal{R})) = \mathcal{I}_\omega^{\mathcal{G}_\mathcal{R}}\left(\bigwedge_{i=1}^{n} \bigvee_{X \in \mathsf{var}(r_i)} X.\phi_i(r_i)\right) = \prod_{i=1}^{n} \mathcal{I}_\omega^{\mathcal{G}_\mathcal{R}}\left(\bigvee_{X \in \mathsf{var}(r_i)} X.\phi_i(r_i)\right)$$

$$= \prod_{i=1}^{n} \prod_{(B_i'|A_i')[p_i] \in \mathsf{Gr}(r_i)} \mathcal{I}_\omega^{\mathcal{G}_\mathcal{R}}((\mathbf{a_i^+} \circ A_i' B_i') \vee (\mathbf{a_i^-} \circ A_i' \overline{B_i'}) \vee \overline{A_i'})$$

$$= \prod_{i=1}^{n} \prod_{(B_i'|A_i')[p_i] \in \mathsf{Gr}(r_i)} \begin{cases} \mathbf{a_i^+} & \text{iff } \omega \models A_i' B_i' \\ \mathbf{a_i^-} & \text{iff } \omega \models A_i' \overline{B_i'} \\ 1 & \text{iff } \omega \models \overline{A_i'} \end{cases} = \sigma_\mathcal{R}(\omega).$$

As a consequence, the second statement holds:

$$\mathsf{TMC}(\phi(\mathcal{R})) = \bigoplus_{\mathcal{I}_\omega^{\mathcal{G}_\mathcal{R}} \models \phi(\mathcal{R})} \mathcal{I}_\omega^{\mathcal{G}_\mathcal{R}}(\phi(\mathcal{R})) = \bigoplus_{\mathcal{I}_\omega^{\mathcal{G}_\mathcal{R}} \models \phi(\mathcal{R})} \sigma_\mathcal{R}(\omega) \overset{(*)}{=} \bigoplus_{\omega \in \Omega} \sigma_\mathcal{R}(\omega)$$

$$= \bigoplus_{[\omega]_{\sim_\mathcal{R}} \in \Omega/\sim_\mathcal{R}} |[\omega]_{\sim_\mathcal{R}}| \cdot \sigma_\mathcal{R}(\omega) \overset{(\dagger)}{=} \bigoplus_{j=1}^{d} W_{j,2n+1} \prod_{i=1}^{n} \mathbf{a_i^+}^{W_{j,2i-1}} \mathbf{a_i^-}^{W_{j,2i}}.$$

---

[7] Here, $\forall_{X \in \mathsf{var}(r_i)} X.\phi_i(r_i) = \forall X_1.(\ldots \forall X_m.\phi_i(r_i))$ where $\mathsf{var}(r_i) = \{X_1, \ldots, X_m\}$. Note that $\mathsf{var}(r_i) = \mathsf{var}(\phi_i(r_i))$ for $i = 1, \ldots, n$, which can be proved easily.

The Equation ($\star$) holds, since $\mathcal{I}_\omega^{\mathcal{G}_\mathcal{R}}(\phi(\mathcal{R})) \in \mathcal{G}_\mathcal{R}$ for every $\omega \in \Omega$ due to the definition of $\phi(\mathcal{R})$, and (†) follows from Definition 4.

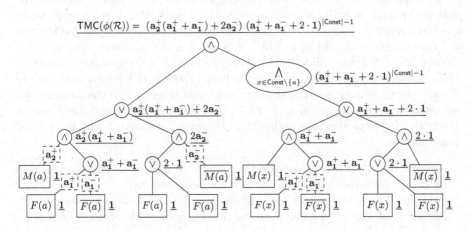

Fig. 4. sd-DNNF$^\mathcal{S}$-circuit wrt. Ex. 5. Annotated typed model counts are underlined

Example 5. We recall the knowledge base from Example 1. Its compilation as per Definition 10 results in

$$\phi(\mathcal{R}) = [\forall X.(\mathbf{a_1^+} \circ M(X)F(X)) \vee (\mathbf{a_1^-} \circ M(X)\overline{F(X)})$$
$$\vee (\overline{M(X)})] \wedge [(\mathbf{a_2^+} \circ M(a)) \vee (\mathbf{a_1^-} \circ \overline{M(a)})].$$

An sd-DNNF$^\mathcal{S}$-circuit for $\phi(\mathcal{R})$ is shown in Fig. 4. The annotated counts are given with respect to an arbitrary number of constants $|\mathsf{Const}| \geq 1$. When setting $|\mathsf{Const}| = 2$, the outcome of the typed model counting task is

$$\mathsf{TMC}(\phi(\mathcal{R})) = (\mathbf{a_2^+}(\mathbf{a_1^+} + \mathbf{a_1^-}) + 2\mathbf{a_2^-})\,(\mathbf{a_1^+} + \mathbf{a_1^-} + 2 \cdot 1)$$
$$= \mathbf{a_1^+}^2\mathbf{a_2^+} + 2\mathbf{a_1^+}\mathbf{a_1^-}\mathbf{a_2^+} + 2\mathbf{a_1^+}\mathbf{a_2^+} + 2\mathbf{a_1^+}\mathbf{a_2^-}$$
$$+ \mathbf{a_1^-}^2\mathbf{a_2^+} + 2\mathbf{a_1^-}\mathbf{a_2^+} + 2\mathbf{a_1^-}\mathbf{a_2^-} + 4\mathbf{a_2^-}$$

which is in accordance with the data given in Table 1. When $\mathsf{TMC}(\phi(\mathcal{R}))$ is calculated with respect to a parameterized number of constants $|\mathsf{Const}|$ as in this example, it is an easy task to determine $\mathsf{WCI}(\mathcal{R})$ for an arbitrary number of constants. Note that this is not possible when, for instance, iterating over all possible worlds in Ω in order to calculate $\mathsf{WCI}(\mathcal{R})$, as the size of Ω heavily depends on $|\mathsf{Const}|$.

6   Related Work

There is a strong connection between typed model counting (TMC) and algebraic model counting (AMC) [11] as both approaches combine the task of counting the models of a formula with algebraic aspects. Whereas in AMC it is allowed to map literals to arbitrary elements of a commutative semiring and not just to elements of a commutative monoid as in TMC, it is not possible to insert elements from the algebraic structure directly into formulas, which is an essential feature in TMC. To some extent, this lack of expressivity of AMC can be overcome: Instead of inserting an element s of the algebraic structure directly into the formula, one can introduce a novel ground atom which is interpreted as the certain element s during the algebraic model counting task. However, this workaround has several disadvantages:

1. The novel ground atoms increase the number of interpretations that have to be checked whether they model a formula unnecessarily.
2. Additional constraints have to be made, in order to ensure that certain combinations of the novel ground atoms do not hold in the same model (e.g., think about the case that a novel ground atom V indicates that a conditional is verified, and another ground atom F indicates that it is falsified. One has to ensure that V and F do not hold at the same time, for example by adding the constraint $V \Leftrightarrow \overline{F}$.),
3. While conjunctions and disjunctions are idempotent operations, elements in algebraic structures do not have to be idempotent. As a consequence, it might happen that several distinct novel ground atoms for the same element of the algebraic structure have to be introduced (e.g., for every single ground instance of a conditional the novel ground atoms indicating the verification, falsification respectively the non-applicability of the conditional have to be distinct).

As a consequence, TMC constitutes a framework that fits better than AMC to the requirements that WCI-calculations impose.

7   Conclusion and Future Work

We extended the concept of typed model counting (TMC), which was primarily defined for propositional logics in [17], to a first-order setting. To this end, we introduced the language of structured formulas which are classical first-order formulas equipped with elements of a commutative monoid. Structured formulas allow for a more fine-grained evaluation than classical formulas do, which results in the ability to distinguish between different types of models. We utilized this ability for calculating weighted conditional impacts (WCIs) that play a central role in nonmonotonic reasoning with conditionals. These WCIs we made use of in our algorithm GIS^{WCI} that efficiently computes the ME-distribution as a model of a probabilistic conditional knowledge base. In future work, we aim to develop an algorithm that capitalizes on both TMC an GIS^{WCI} and allows for lifted inferences at maximum entropy. For this purpose, we want to build on the results presented in [1] by formulating rules for compiling structured formulas into sd-DNNF^S-circuits and by investigating their complexity.

Acknowledgements. This research was supported by the German National Science Foundation (DFG) Research Unit FOR 1513 on Hybrid Reasoning for Intelligent Systems.

References

1. Van den Broeck, G.: Lifted inference and learning in statistical relational models. Ph.D. thesis, KU Leuven (2013)
2. Van den Broeck, G., Taghipour, N., Meert, W., Davis, J., De Raedt, L.: Lifted probabilistic inference by first-order knowledge compilation. In: Proceedings of 22nd IJCAI Conference, IJCAI/AAAI, pp. 2178–2185 (2011)
3. Darroch, J.N., Ratcliff, D.: Generalized iterative scaling for log-linear models. Ann. Math. Stat. **43**, 1470–1480 (1972). Institute of Mathematical Statistics
4. Finthammer, M., Beierle, C.: A two-level approach to maximum entropy model computation for relational probabilistic logic based on weighted conditional impacts. In: Straccia, U., Calì, A. (eds.) SUM 2014. LNCS, vol. 8720, pp. 162–175. Springer, Cham (2014). doi:10.1007/978-3-319-11508-5_14
5. Finthammer, M., Beierle, C.: Towards a more efficient computation of weighted conditional impacts for relational probabilistic knowledge bases under maximum entropy semantics. In: Hölldobler, S., Krötzsch, M., Peñaloza, R., Rudolph, S. (eds.) KI 2015. LNCS, vol. 9324, pp. 72–86. Springer, Cham (2015). doi:10.1007/978-3-319-24489-1_6
6. Getoor, L., Taskar, B. (eds.): Introduction to Statistical Relational Learning. MIT Press, Cambridge (2007)
7. Kern-Isberner, G.: A note on conditional logics and entropy. Int. J. Approx. Reason. **19**(3–4), 231–246 (1998)
8. Kern-Isberner, G.: Conditionals in Nonmonotonic Reasoning and Belief Revision. Springer, Heidelberg (2001)
9. Kern-Isberner, G., Lukasiewicz, T.: Combining probabilistic logic programming with the power of maximum entropy. Artif. Intell. **157**(1–2), 139–202 (2004)
10. Kern-Isberner, G., Thimm, M.: Novel semantical approaches to relational probabilistic conditionals. In: Proceedings of 12th KR Conference. AAAI Press (2010)
11. Kimmig, A., Van den Broeck, G., de Raedt, L.: Algebraic model counting. Int. J. Appl. Logic **22**(C), 46–62 (2017)
12. Paris, J.B.: The Uncertain Reasoner's Companion - A Mathematical Perspective. Cambridge University Press, Cambridge (1994)
13. Paris, J.B.: Common sense and maximum entropy. Synthese **117**(1), 75–93 (1999)
14. Poole, D.: First-order probabilistic inference. In: Proceedings of 18th IJCAI Conference, pp. 985–991. Morgan Kaufmann (2003)
15. Richardson, M., Domingos, P.M.: Markov logic networks. Mach. Learn. **62**(1–2), 107–136 (2006)
16. Shore, J.E., Johnson, R.W.: Axiomatic derivation of the principle of maximum entropy and the principle of minimum cross-entropy. IEEE Trans. Inf. Theor. **26**(1), 26–37 (1980)
17. Wilhelm, M., Kern-Isberner, G.: Typed model counting and its application to probabilistic conditional reasoning at maximum entropy. In: Proceedings of 30th FLAIRS Conference, pp. 748–753. AAAI Press (2017)

Towards Statistical Reasoning in Description Logics over Finite Domains

Rafael Peñaloza[1]$^{(\boxtimes)}$ and Nico Potyka[2]

[1] KRDB Research Centre, Free University of Bozen-Bolzano, Bolzano, Italy
rafael.penaloza@unibz.it
[2] University of Osnabrück, Osnabrück, Germany
npotyka@uni-osnabrueck.de

Abstract. We present a probabilistic extension of the description logic $\mathcal{ALC}$ for reasoning about statistical knowledge. We consider conditional statements over proportions of the domain and are interested in the probabilistic-logical consequences of these proportions. After introducing some general reasoning problems and analyzing their properties, we present first algorithms and complexity results for reasoning in some fragments of Statistical $\mathcal{ALC}$.

1 Introduction

Probabilistic logics enrich classical logics with probabilities in order to incorporate uncertainty. In [5], probabilistic logics have been classified into three types that differ in the way how they handle probabilities. Type 1 logics enrich classical interpretations with probability distributions over the domain and are well suited for reasoning about statistical probabilities. This includes proportional statements like "2% of the population suffer from a particular disease." Type 2 logics consider probability distributions over possible worlds and are better suited for expressing subjective probabilities or degrees of belief. For instance, a medical doctor might say that she is 90% sure about her diagnosis. Type 3 logics combine type 1 and type 2 logics allow to reason about both kinds of uncertainty.

One basic desiderata of probabilistic logics is that they generalize a classical logic. That is, the probabilistic interpretation of formulas with probability 1 should agree with the classical interpretation. However, given that first-order logic is undecidable, a probabilistic first-order logic that satisfies our basic desiderata will necessarily be undecidable. In order to overcome the problem, we can, for instance, restrict to Herbrand interpretations over a fixed domain [2,9,13] or consider decidable fragments like description logics [3,8,10].

Probabilistic type 2 extensions of description logics have been previously studied in [11]. In the unpublished appendix of this work, a type 1 extension of $\mathcal{ALC}$ is presented along with a proof sketch for ExpTime-completeness of the corresponding satisfiability problem. This type 1 extension enriches classical interpretations with probability distributions over the domain as suggested in [5].

S. Moral et al. (Eds.): SUM 2017, LNAI 10564, pp. 280–294, 2017.
DOI: 10.1007/978-3-319-67582-4_20

We consider a similar, but more restrictive setting here. We are interested in an $\mathcal{ALC}$ extension that allows statistical reasoning. However, we do not impose a probability distribution over the domain. Instead, we are only interested in reasoning about the proportions of a population satisfying some given properties. For instance, given statistical information about the relative frequency of certain symptoms, diseases and the relative frequency of symptoms given diseases, one can ask the relative frequency of a disease given a particular combination of symptoms. Therefore, we consider only classical $\mathcal{ALC}$ interpretations with finite domains and are interested in the relative proportions that are true in these interpretations.

Hence, interpretations in our framework can be regarded as a subset of the interpretations in [11], namely those with finite domains and a uniform probability distribution over the domain. These interpretations are indeed sufficient for our purpose. In particular, by considering strictly less interpretations, we may be able to derive tighter answer intervals for some queries. Our approach bears some resemblance to the random world approach from [4]. However, the authors in [4] consider possible worlds with a fixed domain size N and are interested in the limit of proportions as N goes to infinity. We are interested in all finite possible worlds that satisfy certain proportions and ask what statistical statements must be true in all these worlds.

We begin by introducing Statistical $\mathcal{ALC}$ in Sect. 2 together with three relevant reasoning problems. Namely, the Satisfiability Problem, the l-Entailment problem and the p-Entailment problem. In Sect. 3, we will then discuss some logical properties of Statistical $\mathcal{ALC}$. In Sects. 4 and 5, we present first computational results for fragments of Statistical $\mathcal{ALC}$. We had to omit several proofs in order to meet space restrictions. All proofs can be found in the full version of the paper [16].

2   Statistical $\mathcal{ALC}$

We start by revisiting the classical description logic $\mathcal{ALC}$. Given two disjoint sets N_C of *concept names* and N_R of *role names*, $\mathcal{ALC}$ concepts are built using the grammar rule $C ::= \top \mid A \mid \neg C \mid C \sqcap C \mid \exists r.C$, where $A \in N_C$ and $r \in N_R$. One can express disjunction, universal quantification and subsumption through the usual logical equivalences like $C_1 \sqcup C_2 \equiv \neg(\neg C_1 \sqcap \neg C_2)$. For the semantics, we focus on finite interpretations. An $\mathcal{ALC}$ *interpretation* $\mathcal{I} = (\Delta^{\mathcal{I}}, \cdot^{\mathcal{I}})$ consist of a non-empty, finite domain $\Delta^{\mathcal{I}}$ and an interpretation function $\cdot^{\mathcal{I}}$ that maps concept names $A \in N_C$ to sets $A^{\mathcal{I}} \subseteq \Delta^{\mathcal{I}}$ and roles names $r \in N_R$ to binary relations $r^{\mathcal{I}} \subseteq \Delta^{\mathcal{I}} \times \Delta^{\mathcal{I}}$. Two $\mathcal{ALC}$ concepts C_1, C_2 are *equivalent* ($C_1 \equiv C_2$) iff $C_1^{\mathcal{I}} = C_2^{\mathcal{I}}$ for all interpretations $\mathcal{I}$.

Here, we consider a probabilistic extension of $\mathcal{ALC}$. Statistical $\mathcal{ALC}$ knowledge bases consist of probabilistic conditionals that are built up over $\mathcal{ALC}$ concepts.

Definition 1 (Conditionals, Statistical KB). *A probabilistic $\mathcal{ALC}$ conditional is an expression of the form $(C \mid D)[\ell, u]$, where C, D are $\mathcal{ALC}$ concepts* *and* $\ell, u \in \mathbb{Q}$ *are rational numbers such that* $0 \le \ell \le u \le 1$. *A statistical* $\mathcal{ALC}$ knowledge base *(KB) is a set* $\mathcal{K}$ *of probabilistic* $\mathcal{ALC}$ *conditionals.*

For brevity, we usually call probabilistic $\mathcal{ALC}$ conditionals simply *conditionals*.

Example 2. Let $\mathcal{K}_{flu} = \{(\exists \mathsf{has.fever} \mid \exists \mathsf{has.flu})[0.9, 0.95], (\exists \mathsf{has.flu} \mid \top)[0.01, 0.03]\}$. $\mathcal{K}_{flu}$ states that 90 to 95 percent of patients who have the flu have fever, and that only 1 to 3 percent of patients have the flu.

Intuitively, a conditional $(C \mid D)[\ell, u]$ expresses that the relative proportion of elements of D that also belong to C is between ℓ and u. In order to make this more precise, consider a finite $\mathcal{ALC}$ interpretation $\mathcal{I}$, and an $\mathcal{ALC}$ concept X. We denote the cardinality of $X^{\mathcal{I}}$ by $[X]^{\mathcal{I}}$, that is, $[X]^{\mathcal{I}} := |X^{\mathcal{I}}|$. The interpretation $\mathcal{I}$ satisfies $(C \mid D)[\ell, u]$, written as $\mathcal{I} \models (C \mid D)[\ell, u]$, iff either $[D]^{\mathcal{I}} = 0$ or

$$\frac{[C \sqcap D]^{\mathcal{I}}}{[D]^{\mathcal{I}}} \in [\ell, u]. \tag{1}$$

$\mathcal{I}$ satisfies a statistical $\mathcal{ALC}$ knowledge base $\mathcal{K}$ iff it satisfies all conditionals in $\mathcal{K}$. In this case, we call $\mathcal{I}$ a model of $\mathcal{K}$ and write $\mathcal{I} \models \mathcal{K}$. We denote the set of all models of $\mathcal{K}$ by $\mathrm{Mod}(\mathcal{K})$. As usual, $\mathcal{K}$ is *consistent* if $\mathrm{Mod}(\mathcal{K}) \ne \emptyset$ and inconsistent otherwise. We call two knowledge bases $\mathcal{K}_1, \mathcal{K}_2$ equivalent and write $\mathcal{K}_1 \equiv \mathcal{K}_2$ iff $\mathrm{Mod}(\mathcal{K}_1) = \mathrm{Mod}(\mathcal{K}_2)$.

Example 3. Consider again the KB $\mathcal{K}_{flu}$ from Example 2. Let $\mathcal{I}$ be an interpretation with 1000 individuals. 10 of these have the flu and 9 have both the flu and fever. Then $\mathcal{I} \in \mathrm{Mod}(\mathcal{K}_{flu})$.

In classical $\mathcal{ALC}$, knowledge bases are defined by a set of general concept inclusions (GCIs) $C \sqsubseteq D$ that express that C is a subconcept of D. An interpretation $\mathcal{I}$ satisfies $C \sqsubseteq D$ iff $C^{\mathcal{I}} \subseteq D^{\mathcal{I}}$. As shown next, GCIs can be seen as a special kind of conditionals, and hence statistical $\mathcal{ALC}$ KBs are a generalization of classical $\mathcal{ALC}$ KBs.

Proposition 4. *For all statistical* $\mathcal{ALC}$ *interpretations* $\mathcal{I}$, *we have* $\mathcal{I} \models C \sqsubseteq D$ *iff* $\mathcal{I} \models (D \mid C)[1, 1]$.

Proof. If $\mathcal{I} \models C \sqsubseteq D$ then $C^{\mathcal{I}} \subseteq D^{\mathcal{I}}$ and $C^{\mathcal{I}} \cap D^{\mathcal{I}} = C^{\mathcal{I}}$. If $C^{\mathcal{I}} = \emptyset$, we have $[C]^{\mathcal{I}} = 0$. Otherwise $\frac{[C \sqcap D]^{\mathcal{I}}}{[C]^{\mathcal{I}}} = 1$. Hence, $\mathcal{I} \models (D \mid C)[1, 1]$.

Conversely, assume $\mathcal{I} \models (D \mid C)[1, 1]$. If $[C]^{\mathcal{I}} = 0$, then $C^{\mathcal{I}} = \emptyset$ and $\mathcal{I} \models C \sqsubseteq D$. Otherwise, $\frac{[C \sqcap D]^{\mathcal{I}}}{[C]^{\mathcal{I}}} = 1$, that is, $[C \sqcap D]^{\mathcal{I}} = [C]^{\mathcal{I}}$. If there was a $d \in C^{\mathcal{I}} \setminus D^{\mathcal{I}}$, we had $[C \sqcap D]^{\mathcal{I}} < [C]^{\mathcal{I}}$, hence, we have $C^{\mathcal{I}} \subseteq D^{\mathcal{I}}$ and $\mathcal{I} \models C \sqsubseteq D$. □

Given a statistical $\mathcal{ALC}$ knowledge base $\mathcal{K}$, the first problem that we are interested in is deciding consistency of $\mathcal{K}$. We define the satisfiability problem for statistical $\mathcal{ALC}$ knowledge bases as usual.

Satisfiability Problem: Given a knowledge base $\mathcal{K}$, decide whether $\mathrm{Mod}(\mathcal{K}) \ne \emptyset$.

Example 5. Consider again the knowledge base $\mathcal{K}_{flu}$ from Example 2. The conditional $(\exists\mathsf{has.flu} \mid \top)[0.01, 0.03]$ implies that $[\exists\mathsf{has.flu}]^{\mathcal{I}} \geq 0.01$ for all models $\mathcal{I} \in \mathrm{Mod}(\mathcal{K}_{flu})$. $(\exists\mathsf{has.fever} \mid \exists\mathsf{has.flu})[0.9, 0.95]$ implies $[\exists\mathsf{has.fever} \sqcap \exists\mathsf{has.flu}]^{\mathcal{I}} \geq 0.9[\exists\mathsf{has.flu}]^{\mathcal{I}}$. Therefore, $[\exists\mathsf{has.fever}]^{\mathcal{I}} \geq [\exists\mathsf{has.fever} \sqcap \exists\mathsf{has.flu}]^{\mathcal{I}} \geq 0.9[\exists\mathsf{has.flu}]^{\mathcal{I}} \geq 0.009$. Hence, adding the conditional $(\exists\mathsf{has.fever} \mid \top)[0, 0.005]\}$ renders $\mathcal{K}_{flu}$ inconsistent.

If $\mathcal{K}$ is consistent, we are interested in deriving (implicit) probabilistic conclusions. We can think of different reasoning problems in this context. First, we can define an entailment relation analogously to logical entailment. Then, the probabilistic conditional $(C \mid D)[\ell, u]$ is an *l-consequence* of the KB $\mathcal{K}$ iff $\mathrm{Mod}(\mathcal{K}) \subseteq \mathrm{Mod}(\{(C \mid D)[\ell, u]\})$. In this case, we write $\mathcal{K} \models_l (C \mid D)[\ell, u]$. In the context of type 2 probabilistic conditionals, this entailment relation has also been called just *logical consequence* [9].

l-Entailment Problem: Given a knowledge base $\mathcal{K}$ and a conditional $(C \mid D)$ $[\ell, u]$, decide whether $\mathcal{K} \models_l (C \mid D)[\ell, u]$.

Example 6. Consider again the KB $\mathcal{K}_{flu}$ from Example 2. As explained in Example 5, $[\exists\mathsf{has.fever}]^{\mathcal{I}} \geq 0.009$ holds for all models $\mathcal{I} \in \mathrm{Mod}(\mathcal{K})$. Therefore, it follows that $\mathcal{K}_{flu} \models_l (\exists\mathsf{has.fever} \mid \top)[0.009, 1]$. That is, our statistical information suggests that at least 9 out of 1,000 of our patients have fever.

Example 7. Consider a domain with birds (B), penguins (P) and flying animals (F). We let $\mathcal{K}_{birds} = \{(B \mid \top)[0.5, 0.6], (F \mid B)[0.85, 0.9], (F \mid P)[0, 0]\}$. Note that the conditional $(F \mid B)[0.85, 0.9]$ is actually equivalent to $(\neg F \mid B)[0.1, 0.15]$. Furthermore, for all $\mathcal{I} \in \mathrm{Mod}(\mathcal{K}_{birds})$, $(F \mid P)[0, 0]$ implies $[P \sqcap F]^{\mathcal{I}} = 0$. Therefore, we have $[P \sqcap B]^{\mathcal{I}} = [B \sqcap P \sqcap F]^{\mathcal{I}} + [B \sqcap P \sqcap \neg F]^{\mathcal{I}} \leq 0 + [B \sqcap \neg F]^{\mathcal{I}} \leq 0.15[B]^{\mathcal{I}}$. Hence, $\mathcal{K}_{birds} \models_l (P \mid B)[0, 0.15]$. That is, our statistical information suggests that at most 15 out of 100 birds in our population are penguins.

As usual, the satisfiability problem can be reduced to the l-entailment problem.

Proposition 8. $\mathcal{K}$ *is inconsistent iff* $\mathcal{K} \models_l (\top \mid \top)[0, 0]$.

Proof. If $\mathcal{K}$ is inconsistent, then $\mathrm{Mod}(\mathcal{K}) = \emptyset$ and so $\mathcal{K} \models_l (\top \mid \top)[0, 0]$.

Conversely, assume $\mathcal{K} \models_l (\top \mid \top)[0, 0]$. We have $[\top]^{\mathcal{I}} > 0$ and $\frac{[\top \sqcap \top]^{\mathcal{I}}}{[\top]^{\mathcal{I}}} = 1$ for all interpretations $\mathcal{I}$. Hence, $\mathrm{Mod}(\{(\top \mid \top)[0, 0]\}) = \emptyset$ and since $\mathcal{K} \models_l (\top \mid \top)$ $[0, 0]$, we must have $\mathrm{Mod}(\mathcal{K}) = \emptyset$ as well. □

Often, we do not want to check whether a specific conditional is entailed, but rather deduce tight probabilistic bounds for a statement. This problem is often referred to as the *probabilistic entailment problem* in other probabilistic logics, see [6,9,13] for instance. Consider a *query* of the form $(C \mid D)$, where C, D are $\mathcal{ALC}$ concepts. We define the p-Entailment problem similar to the probabilistic entailment problem for type 2 probabilistic logics.

p-Entailment Problem: Given knowledge base $\mathcal{K}$ and a query $(C \mid D)$, find minimal and maximal solutions of the optimization problems

$$\inf_{\mathcal{I} \in \text{Mod}(\mathcal{K})} \Big/ \sup_{\mathcal{I} \in \text{Mod}(\mathcal{K})} \frac{[C \sqcap D]^{\mathcal{I}}}{[D]^{\mathcal{I}}}$$

$$\textit{subject to} \qquad [D]^{\mathcal{I}} > 0$$

Since the objective function $\frac{[C \sqcap D]^{\mathcal{I}}}{[D]^{\mathcal{I}}}$ is bounded from below by 0 and from above by 1, the infimum m and the maximum M are well-defined whenever there is a model $\mathcal{I} \in \text{Mod}(\mathcal{K})$ such that $[D]^{\mathcal{I}} > 0$. In this case, we say that $\mathcal{K}$ p-entails $(C \mid D)[m, M]$ and write $\mathcal{K} \models_p (C \mid D)[m, M]$. In the context of type 2 probabilistic conditionals, this entailment relation has also been called *tight logical consequence* [9]. If $[D]^{\mathcal{I}} = 0$ for all $\mathcal{I} \in \text{Mod}(\mathcal{K})$, the p-Entailment problem is infeasible, that is, there exists no solution.

Example 9. In Example 7, we found that $\mathcal{K}_{birds} \models_l (P \mid B)[0, 0.15]$. This bound is actually tight. Since 0 is always a lower bound and we showed that 0.15 is an upper bound, it suffices to give examples of interpretations that take these bounds. For the lower bound, let $\mathcal{I}_0$ be an interpretation with 200 individuals. 100 of these individuals are birds and 85 are birds that can fly. There are no penguins. Then $\mathcal{I}_0$ is a model of $\mathcal{K}_{birds}$ with $[B]^{\mathcal{I}_0} > 0$ that satisfies $(P \mid B)[0, 0]$. Construct $\mathcal{I}_1$ from $\mathcal{I}_0$ by letting the 15 non-flying birds be penguins. Then $\mathcal{I}_1$ is another model of $\mathcal{K}_{birds}$ and $\mathcal{I}_1$ satisfies $(P \mid B)[0.15, 0.15]$. Hence, we also have $\mathcal{K}_{birds} \models_p (P \mid B)[0, 0.15]$.

If $\mathcal{K} \models_p (C \mid D)[m, M]$, one might ask whether the values between m and M are actually taken by some model of $\mathcal{K}$ or whether there can be large gaps in between. For the probabilistic entailment problem for type 2 logics, we can show that the models of $\mathcal{K}$ do indeed yield a dense interval by noting that each convex combination of models is a model and applying the Intermediate Value Theorem from Real Analysis. However, in our framework, we do not consider probability distributions over possible worlds, but the worlds themselves, which are discrete in nature. We therefore cannot apply the same tools here. However, for each two models that yield different probabilities for a query, we can find another model that takes the probability in the middle of these probabilities.

Lemma 10 (Bisection Lemma). *Let C, D be two arbitrary $\mathcal{ALC}$ concepts. If there exist $\mathcal{I}_0, \mathcal{I}_1 \in \text{Mod}(\mathcal{K})$ such that $r_0 = \frac{[C \sqcap D]^{\mathcal{I}_0}}{[D]^{\mathcal{I}_0}} < \frac{[C \sqcap D]^{\mathcal{I}_1}}{[D]^{\mathcal{I}_1}} = r_1$, then there is an $\mathcal{I}_{0.5} \in \text{Mod}(\mathcal{K})$ such that $\frac{[C \sqcap D]^{\mathcal{I}_{0.5}}}{[D]^{\mathcal{I}_{0.5}}} = \frac{r_0 + r_1}{2}$.*

We can now show that for each value between the lower and upper bound given by p-entailment, we can find a model that gives a probability arbitrarily close to this value.

Proposition 11 (Intermediate Values). *Let $\mathcal{K} \models_p (C \mid D)[m, M]$. Then for every $x \in (m, M)$ (where (m, M) denotes the open interval between m and M) and for all $\epsilon > 0$, there is a $\mathcal{I}_{x,\epsilon} \in \text{Mod}(\mathcal{K})$ such that $\left| \frac{[C \sqcap D]^{\mathcal{I}_{x,\epsilon}}}{[D]^{\mathcal{I}_{x,\epsilon}}} - x \right| < \epsilon$.*

3   Logical Properties

We now discuss some logical properties of Statistical $\mathcal{ALC}$. We already noted that Statistical $\mathcal{ALC}$ generalizes classical $\mathcal{ALC}$ in Proposition 4. Furthermore, p-entailment yields a tight and dense (Proposition 11) answer interval for all queries whose condition can be satisfied by models of the knowledge base. Let us also note that statistical $\mathcal{ALC}$ is *language invariant*. That is, increasing the language by adding new concept or role names does not change the semantics of $\mathcal{ALC}$. This can be seen immediately by observing that the interpretation of conditionals in (1) depends only on the concept and role names that appear in the conditional.

Statistical $\mathcal{ALC}$ is also *representation invariant* in the sense that for all concepts C_1, D_1 and C_2, D_2, if $C_1 \equiv C_2$ and $D_1 \equiv D_2$ then $(C_1 \mid D_1)[l, u] \equiv (C_2 \mid D_2)[l, u]$. Hence, changing the syntactic representation of conditionals does not change their semantics. In particular, entailment results are independent of such changes.

Both l- and p-entailment satisfy the following *independence* property: whether or not $\mathcal{K} \models_l (C \mid D)[\ell, u]$ ($\mathcal{K} \models_p (C \mid D)[m, M]$) depends only on the conditionals in $\mathcal{K}$ that are connected with the query. This may simplify answering the query by reducing the size of the KB. In order to make this more precise, we need some additional definitions. For an arbitrary $\mathcal{ALC}$ concept C, $\mathrm{Sig}(C)$ denotes the set of all concept and role names appearing in C. The conditionals $(C_1 \mid D_1)[\ell_1, u_1]$ and $(C_2 \mid D_2)[\ell_2, u_2]$ are *directly connected* (written $(C_1 \mid D_1)[\ell_1, u_1] \rightleftharpoons (C_2 \mid D_2)[\ell_2, u_2]$) if and only if $(\mathrm{Sig}(C_1) \cup \mathrm{Sig}(D_1)) \cap (\mathrm{Sig}(C_2) \cup \mathrm{Sig}(D_2)) \neq \emptyset$. That is, two conditionals are directly connected iff they share concept or role names. Let $\rightleftharpoons^*$ denote the transitive closure of $\rightleftharpoons$. We say that $(C_1 \mid D_1)[\ell_1, u_1]$ and $(C_2 \mid D_2)[\ell_2, u_2]$ are *connected* iff $(C_1 \mid D_1)[\ell_1, u_1] \rightleftharpoons^* (C_2 \mid D_2)[\ell_2, u_2]$. The *restriction of $\mathcal{K}$ to conditionals connected to $(C \mid D)[\ell, u]$* is the set $\{\kappa \in \mathcal{K} \mid \kappa \rightleftharpoons^* (C \mid D)[\ell, u]\}$. Using an analogous definition for queries (qualitative conditionals) $(C_1 \mid D_1)$ and $(C_2 \mid D_2)$, we get the following result.

Proposition 12 (Independence). *If $\mathcal{K}$ is consistent, we have*

1. $\mathcal{K} \models_l (C \mid D)[\ell, u]$ *iff* $\{\kappa \in \mathcal{K} \mid \kappa \rightleftharpoons^* (C \mid D)[\ell, u]\} \models_l (C \mid D)[\ell, u]$.
2. $\mathcal{K} \models_p (C \mid D)[m, M]$ *iff* $\{\kappa \in \mathcal{K} \mid \kappa \rightleftharpoons^* (C \mid D)\} \models_p (C \mid D)[m, M]$.

Another interesting property of probabilistic logics is *continuity*. Intuitively, continuity states that minor changes in the knowledge base do not yield major changes in the derived probabilities. However, as demonstrated by Courtney and Paris, this condition is too strong when reasoning with the maximum entropy model of the knowledge base [14, p. 90]. The same problem arises for the probabilistic entailment problem [17, Example 4]. While these logics considered subjective probabilities, the same problem occurs in our setting for statistical probabilities as we demonstrate now.

Example 13. Consider the knowledge base

$$\mathcal{K} = \{(B \mid A)[0.4, 0.5], (C \mid A)[0.5, 0.6], (B \mid C)[1, 1], (C \mid B)[1, 1]\}.$$

The interpretation $\mathcal{I} = (\{a, b\}, \cdot^{\mathcal{I}})$ with $A^{\mathcal{I}} = \{a, b\}$, $B^{\mathcal{I}} = C^{\mathcal{I}} = \{b\}$ is a model of $\mathcal{K}$, i.e., $\mathcal{K}$ is consistent. In particular, since A is interpreted by the whole domain of $\mathcal{I}$ we know that

$$\mathcal{K} \models_p (A \mid \top)[m, 1]$$

for some $m \in [0, 1]$. As explained in Proposition 4, deterministic conditionals correspond to concept inclusions and so $(B \mid C)[1, 1]$ and $(C \mid B)[1, 1]$ imply that $B^{\mathcal{I}'} = C^{\mathcal{I}'}$ for all models $\mathcal{I}'$ of $\mathcal{K}$. Therefore, $\frac{[B \sqcap A]^{\mathcal{I}'}}{[A]^{\mathcal{I}'}} = \frac{[C \sqcap A]^{\mathcal{I}'}}{[A]^{\mathcal{I}'}}$. Let $\mathcal{K}'$ denote the knowledge base that is obtained from $\mathcal{K}$ by decreasing the upper bound of the first conditional in $\mathcal{K}$ by an arbitrarily small $\epsilon > 0$. That is,

$$\mathcal{K}' = \{(B \mid A)[0.4, 0.5 - \epsilon], (C \mid A)[0.5, 0.6], (B \mid C)[1, 1], (C \mid B)[1, 1]\}.$$

Then the only way to satisfy the first two conditionals in $\mathcal{K}'$ is by interpreting A by the empty set. Indeed, the interpretation $\mathcal{I}_\emptyset$ that interprets all concept names by the empty set is a model of $\mathcal{K}'$. So $\mathcal{K}'$ is consistent and

$$\mathcal{K}' \models_p (A \mid \top)[0, 0].$$

Hence, a minor change in the probabilities in the knowledge base can yield a severe change in the entailed probabilities. This means that the p-entailment relation that we consider here is not continuous in this way either.

As an alternative to this strong notion of continuity, Paris proposed to measure the difference between KBs by the Blaschke distance between their models. Blaschke continuity says that if KBs are close with respect to the Blaschke distance, the entailed probabilities are close. Blaschke continuity is satisfied by some probabilistic logics under maximum entropy and probabilistic entailment [14, 17]. In [14, 17], probabilistic interpretations are probability distributions over a finite number of classical interpretations and the distance between two interpretations is the distance between the corresponding probability vectors. We cannot apply this definition here because we interpret conditionals by means of classical interpretations. It is not at all clear what a reasonable definition for the distance between two classical interpretations is. We leave the search for a reasonable topology on the space of classical interpretations for future work.

4   Statistical $\mathcal{EL}$

Proposition 4 and the fact that reasoning in $\mathcal{ALC}$ is ExpTime-complete, show that our reasoning problems are ExpTime-hard. However, we did not find any upper bounds on the complexity of reasoning in $\mathcal{ALC}$ so far. We will therefore focus on some fragments of $\mathcal{ALC}$ now.

To begin with, we will focus on the sublogic $\mathcal{EL}$ [1] of $\mathcal{ALC}$ that does not allow for negation and universal quantification. Formally, $\mathcal{EL}$ concepts are constructed by the grammar rule $C ::= A \mid \top \mid C \sqcap C \mid \exists r.C$, where $A \in N_C$ and $r \in N_R$. A *statistical $\mathcal{EL}$ KB* is a statistical $\mathcal{ALC}$ KB where conditionals are restricted to $\mathcal{EL}$ concepts. Notice that, due to the upper bounds in conditionals, statistical $\mathcal{EL}$ KBs are capable of expressing some weak variants of negations. For instance, a statement $(C \mid \top)[\ell, u]$ with $u < 1$ restricts every model $\mathcal{I} = (\Delta^{\mathcal{I}}, \cdot^{\mathcal{I}})$ to contain at least one element $\delta \in \Delta^{\mathcal{I}} \setminus C^{\mathcal{I}}$. Thus, contrary to classical $\mathcal{EL}$, statistical $\mathcal{EL}$ KBs may be inconsistent.

Example 14. Consider the KB $\mathcal{K}_1 = (\emptyset, \mathcal{C}_1)$, where

$$\mathcal{C}_1 = \{(A \mid \top)[0, 0.2], \ (A \mid \top)[0.3, 1]\}.$$

Since $\top^{\mathcal{I}} = \Delta^{\mathcal{I}} \neq \emptyset$, every model $\mathcal{I} = (\Delta^{\mathcal{I}}, \cdot^{\mathcal{I}})$ of $\mathcal{K}_1$ must satisfy

$$[A]^{\mathcal{I}} \leq 0.2[\top]^{\mathcal{I}} < 0.3[\top]^{\mathcal{I}} \leq [A]^{\mathcal{I}},$$

which is clearly a contradiction. Thus, $\mathcal{K}_1$ is inconsistent.

More interestingly, though, it is possible to simulate valuations over a finite set of propositional formulas wit the help of conditional statements. Thus, the satisfiability problem is at least NP-hard even for Statistical $\mathcal{EL}$.

Theorem 15. *The satisfiability problem for Statistical $\mathcal{EL}$ is NP-hard.*

On the other hand, consistency can be decided in non-deterministic exponential time, through a reduction to integer programming. Before describing the reduction in detail, we introduce a few simplifications.

Recall from Proposition 4 that a conditionals of the form $(D \mid C)[1, 1]$ is equivalent to the classical GCI $C \sqsubseteq D$. Thus, in the following we will often express statistical $\mathcal{EL}$ KBs as pairs $\mathcal{K} = (\mathcal{T}, \mathcal{C})$, where $\mathcal{T}$ is a classical TBox (i.e., a finite set of GCIs), and $\mathcal{C}$ is a set of conditionals. A statistical $\mathcal{EL}$ KB $\mathcal{K} = (\mathcal{T}, \mathcal{C})$ is said to be in *normal form* if all the GCIs in $\mathcal{T}$ are of the form

$$A_1 \sqcap A_2 \sqsubseteq B, \qquad A \sqsubseteq \exists r.B, \qquad \exists r.A \sqsubseteq B$$

and all its conditionals are of the form

$$(A \mid B)[\ell, u]$$

where $A, B \in N_C \cup \{\top\}$, and $r \in N_R$. Informally, a KB is in normal form if at most one constructor is used in any GCI, and all conditionals are atomic (i.e., between concept names). Every KB can be transformed to an equivalent one (w.r.t. the original signature) in linear time using the normalization rules from [1], and introducing new concept names for complex concepts appearing in conditionals. More precisely, we replace any conditional of the form $(C \mid D)[\ell, u]$ with the statement $(A \mid B)[\ell, u]$, where A, B are two fresh concept names, and extend the TBox with the axioms $A \equiv C$, and $B \equiv D$.

The main idea behind our consistency algorithm is to partition the finite domain of a model into the different types that they define, and use integer programming to verify that all the logical and conditional constraints are satisfied. Let $N_C(\mathcal{K})$ denote the set of all concept names appearing in the KB $\mathcal{K}$. We call any subset $\theta \subseteq N_C(\mathcal{K})$ a *type* for $\mathcal{K}$. Intuitively, such a type θ represents all the elements of the domain that are interpreted to belong to all concept names $A \in \theta$ and no concept name $A \notin \theta$. We denote as $\Theta(\mathcal{K})$ the set of all types of $\mathcal{K}$. To simplify the presentation, in the following we treat $\top$ as a concept name that belongs to all types.

Given a statistical $\mathcal{EL}$ KB $\mathcal{K} = (\mathcal{T}, \mathcal{C})$ in normal form, we consider an integer variable x_θ for every type $\theta \in \Theta(\mathcal{K})$. These variables will express the number of domain elements that belong to the corresponding type. In addition, $x_\top$ will be used to represent the total size of the domain. We build a system of linear inequalities over these variables as follows. First, we require that all variables have a value at least 0, and that the sizes of all types add exactly the size of the domain.

$$\sum_{\theta \in \Theta(\mathcal{K})} x_\theta = x_\top \tag{2}$$

$$0 \leq x_\theta \qquad \text{for all } \theta \in \Theta(\mathcal{K}) \tag{3}$$

Then, we ensure that all the conditional statements from the KB are satisfied by adding, for each statement $(A \mid B)[\ell, u] \in \mathcal{C}$ the constraint

$$\ell \cdot \sum_{B \in \theta} x_\theta \leq \sum_{A, B \in \theta} x_\theta \leq u \cdot \sum_{B \in \theta} x_\theta, \tag{4}$$

Finally, we must ensure that the types satisfy all the logical constraints introduced by the TBox. The GCI $A_1 \sqcap A_2 \sqsubseteq B$ states that every element that belongs to both A_1 and A_2 must also belong to B. This means that types containing A_1, A_2 but excluding B should not be populated. We thus introduce the inequality

$$x_\theta = 0 \qquad \text{if } A_1 \sqcap A_2 \sqsubseteq B \in \mathcal{T}, A_1, A_2 \in \theta, \text{and } B \notin \theta \tag{5}$$

Dealing with existential restrictions requires checking different alternatives, which we solve by creating different linear programs. The GCI $A \sqsubseteq \exists r.B$ implies that, whenever there exists an element in A, there must also exist at least one element in B. Thus, to satisfy this axiom, either A should be empty (i.e., $\sum_{A \in \theta} x_\theta = 0$), or $\sum_{B \in \theta} x_\theta \geq 1$. Hence, for every existential restriction of the form $A \sqsubseteq \exists r.B$, we define the set

$$\mathcal{E}_{A,B} := \{\sum_{A \in \theta} x_\theta = 0, \sum_{B \in \theta} x_\theta \geq 1\}$$

To deal with GCIs of the form $\exists r.A \sqsubseteq B$, we follow a similar approach, together with the ideas of the completion algorithm for classical $\mathcal{EL}$. For every pair of existential restrictions $A \sqsubseteq \exists r.B, \exists r.C \sqsubseteq D$, we define the set

$$\mathcal{F}_{A,B,C,D} := \{ \sum_{A \in \theta, D \notin \theta} x_\theta = 0, \quad \sum_{B \in \theta, C \notin \theta} x_\theta \geq 1 \}$$

Intuitively, $\sum_{A \in \theta, D \notin \theta} x_\theta \geq 1$ whenever there exists an element that belongs to A but not to D. If this is the case, and the GCIs $A \sqsubseteq \exists r.B, \exists r.C \sqsubseteq D$ belong to the TBox $\mathcal{T}$, then there must exist some element that belongs to B but not to C.

We call the hitting sets of

$$\{\mathcal{E}_{A,B} \mid A \sqsubseteq \exists r.B \in \mathcal{T}\} \cup \{\mathcal{F}_{A,B,C,D} \mid A \sqsubseteq \exists r.B, \exists r.C \sqsubseteq D \in \mathcal{T}\}$$

choices for $\mathcal{T}$. A *program for* $\mathcal{K}$ is an integer program containing all the inequalities (2)–(5) and a choice for $\mathcal{T}$. Then we get the following result.

Lemma 16. $\mathcal{K}$ *is consistent iff there exists a program for* $\mathcal{K}$ *that is satisfiable.*

Proof. The "only if" direction is straight-forward since the inequalities are sound w.r.t. the semantics of statistical KBs. We focus on the "if" direction only.

Given a solution of the integer program, we construct an interpretation $\mathcal{I} = (\Delta, \cdot^{\mathcal{I}})$ as follows. We create a domain Δ with $x_\top$ elements, and partition it such that for every type $\theta \in \Theta(\mathcal{K})$, there is a class $[[\theta]]$ containing exactly x_θ elements. For every non-empty class, select a representative element $\delta_\theta \in [[\theta]]$.

The interpretation function $\cdot^{\mathcal{I}}$ maps every concept name A to the set

$$A^{\mathcal{I}} := \bigcup_{A \in \theta} [[\theta]].$$

Given a non-empty class $[[\theta]]$ such that $A \in \theta$ and $A \sqsubseteq \exists r.B \in \mathcal{T}$, let τ be a type such that $B \in \tau$, $x_\tau > 0$, and for every $\exists r.C \sqsubseteq D \in \mathcal{T}$, if $D \notin \theta$, then $C \notin \tau$. Notice that such a τ must exist because the solution must satisfy at least one restriction in each $\mathcal{F}_{A,B,C,D}$. We define $r_{A,B}^\theta := \theta \times \{\delta_\tau\}$ and set

$$r^{\mathcal{I}} := \bigcup_{A \in \theta, A \sqsubseteq \exists r.B \in \mathcal{T}} r_{A,B}^\theta.$$

It remains to be shown that $\mathcal{I}$ is a model of $\mathcal{K}$.

Notice that for two concept names A, B, it holds that $(A \sqcap B)^{\mathcal{I}} = \bigcup_{A,B \in \theta} [[\theta]]$ and hence $[A \sqcap B]^{\mathcal{I}}| = \sum_{A,B \in \theta} x_\theta$. Given a conditional statement $(A \mid B)[\ell, u] \in \mathcal{C}$, since the solution must satisfy the inequality (4), it holds that

$$\ell \cdot [B]^{\mathcal{I}} \leq [A \sqcap B]^{\mathcal{I}} \leq u \cdot [B]^{\mathcal{I}}.$$

For a GCI $A_1 \sqcap A_2 \sqsubseteq B \in \mathcal{T}$, by the inequality (5) it follows that for every type θ containing both A_1, A_2, but not B, $[[\theta]] = \emptyset$. Hence $A_1^{\mathcal{I}} \cap A_2^{\mathcal{I}} \subseteq B^{\mathcal{I}}$. For every $A \sqsubseteq \exists r.B \in \mathcal{T}$, and every $\gamma \in \Delta$, if $\gamma \in A^{\mathcal{I}}$ then by construction there is an element γ' such that $(\gamma, \gamma') \in r^{\mathcal{I}}$.

Finally, if $(\gamma, \gamma') \in r^{\mathcal{I}}$, then by construction there exists a type θ and an axiom $A \sqsubseteq \exists r.B \in \mathcal{T}$ such that $\gamma \in [[\theta]]$ and $\gamma' = \delta_\tau$. Then, for every GCI $\exists r.C \sqsubseteq D \in \mathcal{T}$, $\gamma' \in C^{\mathcal{I}}$ implies $C \in \tau$ and hence $D \in \theta$ which means that $\gamma \in D^{\mathcal{I}}$.   □

Notice that the construction produces exponentially many integer programs, each of which uses exponentially many variables, measured on the size of the KB. Since satisfiability of integer linear programs is decidable in non-deterministic polynomial time on the size of the program, we obtain a non-deterministic exponential time upper bound for deciding consistency of statistical $\mathcal{EL}$ KBs.

Theorem 17. *Consistency of statistical $\mathcal{EL}$ KBs is in* NEXPTIME.

5   Reasoning with Open Minded KBs

In order to regain tractability, we now further restrict statistical $\mathcal{EL}$ KBs by disallowing upper bounds in the conditional statements. We call such knowledge bases open minded.

Definition 18 (Open Minded KBs). *A statistical $\mathcal{EL}$ KB $\mathcal{K} = (\mathcal{T}, \mathcal{C})$ is open minded iff all the conditional statements $(C \mid D)[\ell, u] \in \mathcal{C}$ are such that $u = 1$.*

For the scope of this section, we consider only open minded KBs. The first obvious consequence of restricting to this class of KBs is that negations cannot be simulated. In fact, every open minded KB is consistent and, as in classical $\mathcal{EL}$, can be satisfied in a simple universal model.

Theorem 19. *Every open minded KB is consistent.*

Proof. Consider the interpretation $\mathcal{I} = (\{\delta\}, \cdot^{\mathcal{I}})$ where the interpretation function maps every concept name A to $A^{\mathcal{I}} := \{\delta\}$ and every role name r to $r^{\mathcal{I}} := \{(\delta, \delta)\}$. It is easy to see that this interpretation is such that $C^{\mathcal{I}} = \{\delta\}$ holds for every $\mathcal{EL}$ concept C. Hence, $\mathcal{I}$ satisfies all $\mathcal{EL}$ GCIs and in addition $[C \sqcap D]^{\mathcal{I}} = [C]^{\mathcal{I}} = 1$ which implies that all conditionals are also satisfied.   □

Recall that, intuitively, conditionals specify that a proportion of the population satisfies some given properties. One interesting special case of p-entailment is the question how likely it is to observe an individual that belongs to a given concept.

Definition 20. *Let $\mathcal{K}$ be an open minded KB, C a concept, and $m \in [0, 1]$. C is m-necessary in $\mathcal{K}$ if $\mathcal{K}$ p-entails $(C \mid \top)[m, 1]$. The problem of m-necessity consists in deciding whether C is m-necessary in $\mathcal{K}$.*

We show that this problem can be solved in polynomial time. As in the previous section, we assume that the KB is in normal form and additionally, that all conditional statements $(A \mid B)[\ell, 1] \in \mathcal{C}$ are such that $\ell < 1$. This latter assumption is made w.l.o.g. since the conditional statement $(A \mid B)[1, 1]$ can be equivalently replaced by the GCI $B \sqsubseteq A$ (see Proposition 4). Moreover, checking m-necessity of a complex concept C w.r.t. the KB $(\mathcal{T}, \mathcal{C})$ is equivalent to deciding m-necessity of a new concept name A w.r.t. the KB $(\mathcal{T} \cup \{A \equiv C\}, \mathcal{C})$. Thus, in the following we consider w.l.o.g. only the problem of deciding m-necessity of a concept name w.r.t. to a KB in normal form.

Our algorithm extends the completion algorithm for classification of $\mathcal{EL}$ TBoxes to in addition keep track of the lower bounds of necessity for all relevant concept names. The algorithm keeps as data structure a set $\mathcal{S}$ of tuples of the form (A, B) and (A, r, B) for $A, B \in N_C \cup \{\top\}$. These intuitively express that the TBox $\mathcal{T}$ entails the subsumptions $A \sqsubseteq B$ and $A \sqsubseteq \exists r.B$, respectively. Additionally, we keep a function $\mathcal{L}$ that maps every element $A \in N_C \cup \{\top\}$ to a number $\mathcal{L}(A) \in [0,1]$. Intuitively, $\mathcal{L}(A) = n$ expresses that $\mathcal{K}$ p-entails $(A \mid \top)[n, 1]$.

The algorithm initializes the structures $\mathcal{S}$ and $\mathcal{L}$ as

$$\mathcal{S} := \{(A, A), (A, \top) \mid A \in N_C(\mathcal{K}) \cup \{\top\}\}$$

$$\mathcal{L}(A) := \begin{cases} 0 & \text{if } A \in N_C(\mathcal{K}) \\ 1 & \text{if } A = \top. \end{cases}$$

These structures are then updated using the rules from Table 1. In each case, a rule is only applied if its execution extends the available knowledge; that is, if either $\mathcal{S}$ is extended to include one more tuple, or a lower bound in $\mathcal{L}$ is increased. In the latter case, only the larger value is kept through the function $\mathcal{L}$.

Table 1. Rules for deciding m-necessity

$\mathbf{C_1}$	if $\{(X, A_1), (X, A_2)\} \subseteq \mathcal{S}$	and $A_1 \sqcap A_2 \sqsubseteq B \in \mathcal{T}$	then add (X, B) to $\mathcal{S}$
$\mathbf{C_2}$	if $(X, A) \in \mathcal{S}$	and $A \sqsubseteq \exists r.B \in \mathcal{T}$	then add (A, r, B) to $\mathcal{S}$
$\mathbf{C_3}$	if $\{(X, r, Y), (Y, A)\} \subseteq \mathcal{S}$	and $\exists r.A \sqsubseteq B \in \mathcal{T}$	then add (X, B) to $\mathcal{S}$
$\mathbf{L_1}$	if	$(A \mid B)[\ell, 1] \in \mathcal{C}$	then $\mathcal{L}(A) \leftarrow \ell \cdot \mathcal{L}(B)$
$\mathbf{L_2}$	if	$A_1 \sqcap A_2 \sqsubseteq B \in \mathcal{T}$	then $\mathcal{L}(B) \leftarrow \mathcal{L}(A_1) + \mathcal{L}(A_2) - 1$
$\mathbf{L_3}$	if $(B, A) \in \mathcal{S}$		then $\mathcal{L}(A) \leftarrow \mathcal{L}(B)$

The first three rules in Table 1 are the standard completion rules for classical $\mathcal{EL}$. The remaining rules update the lower bounds for the likelihood of all relevant concept names, taking into account their logical relationship, as explained next.

Rule $\mathbf{L_1}$ applies the obvious inference associated to conditional statements: from all the individuals that belong to B, $(A \mid B)[\ell, 1]$ states that at least $100\ell\%$ belong also to A. Thus, assuming that $\mathcal{L}(B)$ is the lowest proportion of elements in B possible, the proportion of elements in A must be at least $\ell \cdot \mathcal{L}(B)$. $\mathbf{L_3}$ expresses that if every element of B must also belong to A, then there must be at least as many elements in A as there are in B. Finally, $\mathbf{L_2}$ deals with the fact that two concepts that are proportionally large must necessarily overlap.

For example, if 60% of all individuals belong to A and 50% belong to B, then at least 10% must belong to both A and B; otherwise, together they would cover more than the whole domain.

The algorithm executes all the rules until *saturation*; that is, until no rule is applicable. Once it is saturated, we can decide m-necessity from the function $\mathcal{L}$ as follows: A is m-necessary iff $m \leq \mathcal{L}(A)$. Before showing the correctness of this algorithm, we show an important property.

Notice that the likelihood information from $\mathcal{L}$ is never transferred through roles. The reason for this is that an existential restriction $\exists r.B$ only guarantee the existence of one element belonging to the concept B. Proportionally, the number of elements that belong to B tends to 0.

Example 21. Consider the KB $(\{\top \sqsubseteq \exists r.A\}, \emptyset)$. For any $n \in \mathbb{N}$, construct the interpretation $\mathcal{I}_n := (\{0, \ldots, n\}, \cdot^{\mathcal{I}_n})$, where $A^{\mathcal{I}_n} = \{0\}$ and $r^{\mathcal{I}_n} = \{(k, 0) \mid 0 \leq k \leq n\}$. It is easy to see that $\mathcal{I}_n$ is a model of the KB and $[A]^{\mathcal{I}_n}/[\top]^{\mathcal{I}_n} < 1/n$. Thus, the best lower bound for m-necessity of A is 0, as correctly given by the algorithm.

Theorem 22 (correctness). *Let $\mathcal{L}$ be the function obtained by the application of the rules until saturation and $A_0 \in N_C$. Then A_0 is m-necessary iff $m \leq \mathcal{L}(A)$.*

Proof (sketch). It is easy to see that all the rules are sound, which proves the "if" direction. For the converse direction, we consider a finite domain Δ and an interpretation $\cdot^{\mathcal{I}}$ of the concept names such that $[A]^{\mathcal{I}}/|\Delta| = \mathcal{L}(A)$ and the post-conditions of the rules $\mathbf{L}_1$–$\mathbf{L}_3$ are satisfied. Such interpretation can be obtained recursively by considering the last rule application that updated $\mathcal{L}(A)$. Assume w.l.o.g. that the domain is large enough so that $c/|\Delta| < m - \mathcal{L}(A_0)$, where c is the number of concept names appearing in $\mathcal{K}$. It is easy to see that this interpretation satisfies all conditional statements and the GCIs $A_1 \sqcap A_2 \sqsubseteq B \in \mathcal{T}$. For every concept name A, create a new domain element δ_A and extend the interpretation $\mathcal{I}$ such that $\delta_A \in B$ iff $(A, B) \in \mathcal{S}$. Given a role name r, we define $r^{\mathcal{I}} := \{(\gamma, \delta_B) \mid A \sqsubseteq \exists r.B, \gamma \in A^{\mathcal{I}}\}$. Then, this interpretation satisfies the KB $\mathcal{K}$, and $[A_0]^{\mathcal{I}}/|\Delta| \leq \mathcal{L}(A_0) + c/|\Delta| < m$. $\qquad\square$

Thus, the algorithm can correctly decide m-necessity of a given concept name. It remains only to be shown that the process terminates after polynomially many rule applications. To guarantee this, we impose an ordering in the rule applications. First, we apply all the classical rules $\mathbf{C}_1$–$\mathbf{C}_3$, and only when no such rules are applicable, we update the function $\mathcal{L}$ through the rules $\mathbf{L}_1$–$\mathbf{L}_3$. In this case, the rule that will update to the largest possible value is applied first. It is known that only polynomially many classical rules (on the size of $\mathcal{T}$) can be applied [1]. Deciding which bound rule to apply next requires polynomial time on the number of concept names in $\mathcal{K}$. Moreover, since the largest update is applied first, the value of $\mathcal{L}(A)$ is changed at most once for every concept name A. Hence, only linearly many rules are applied. Overall, this means that the algorithm terminates after polynomially many rule applications, which yields the following result.

Theorem 23. *Deciding m-necessity is in* P.

6   Related Work

Over the years, various probabilistic extensions of description logics have been investigated, see, for instance, [3,7,8,10,12,15,18]. The one that is closest to our approach is the type 1 extension of $\mathcal{ALC}$ proposed in the appendix of [11]. Briefly, [11] introduces probabilistic constraints of the form $P(C \mid D) \leq p$, $P(C \mid D) = p$, $P(C \mid D) \geq p$ for $\mathcal{ALC}$ concepts C, D. These correspond to the conditionals $(C \mid D)[0,p]$, $(C \mid D)[p,p]$, $(C \mid D)[p,1]$, respectively. Conversely, each conditional can be rewritten as such a probabilistic constraint. However, there is a subtle but fundamental difference in the semantics. While the definition in [11] allows for probability distributions over arbitrary domains, we do not consider uncertainty over the domain. This comes down to allowing only finite domains and only the uniform distribution over this domain; that is, our approach further restricts the class of models of a KB. One fundamental difference between the two approaches is that Proposition 4 does not hold in [11]: the reason is that the conditional $(C \mid D)[1,1]$ can be satisfied by an interpretation $\mathcal{I}$ that contains an element $x \in (C \sqcap \neg D)^{\mathcal{I}}$, where x has probability 0.

This difference is the main reason why the EXPTIME algorithm proposed by Lutz and Schröder cannot be transferred to our setting. It does not suffice to consider the satisfiable types independently, but other implicit subsumption relations may depend on the conditionals only.

Example 24. Consider the statistical $\mathcal{EL}$ KB $\mathcal{K} = (\mathcal{T}, \mathcal{C})$ with

$$\mathcal{T} := \{\top \sqsubseteq \exists r.A, \quad \exists r.B \sqsubseteq C\}$$
$$\mathcal{C} := \{(B \mid \top)[0.5,1], \quad (A \mid B)[0.5,1], \quad (A \mid \top)[0,0.25]\}$$

From $\mathcal{C}$ it follows that every element of A must also belong to B, and hence every domain element must be an element of C. However, $\neg C$ defines a satisfiable type (w.r.t. $\mathcal{T}$) which will be interpreted as non-empty in the model generated by the approach in [11].

7   Conclusions

We have introduced Statistical $\mathcal{ALC}$, a new probabilistic extension of the description logic $\mathcal{ALC}$ for statistical reasoning. We analyzed the basic properties of this logic and introduced some reasoning problems that we are interested in. As a first step towards effective reasoning in Statistical $\mathcal{ALC}$, we focused on $\mathcal{EL}$, a well-known sublogic of $\mathcal{ALC}$ that, in its classical form, allows for polynomial-time reasoning. We showed that upper bounds in conditional constraints make the satisfiability problem in statistical $\mathcal{EL}$ NP-hard and gave an NEXPTIME algorithm to decide satisfiability. We showed that tractability can be regained by disallowing strict upper bounds in the conditional statements.

We are going to provide more algorithms and a more complete picture of the complexity of reasoning for Statistical $\mathcal{ALC}$ and its fragments in future work. A combination of integer programming and the inclusion-exclusion principle may be fruitful to design first algorithms for reasoning in full Statistical $\mathcal{ALC}$.

References

1. Baader, F., Brandt, S., Lutz, C.: Pushing the $\mathcal{EL}$ envelope. In: Kaelbling, L.P., Saffiotti, A. (eds.) Proceedings of IJCAI 2005, pp. 364–369. Morgan-Kaufmann (2005)
2. Beierle, C., Kern-Isberner, G., Finthammer, M., Potyka, N.: Extending and completing probabilistic knowledge and beliefs without bias. KI-Künstliche Intelligenz 29(3), 255–262 (2015)
3. Ceylan, İ.İ., Peñaloza, R.: The bayesian ontology language $\mathcal{BEL}$. J. Autom. Reasoning 58(1), 67–95 (2017)
4. Grove, A.J., Halpern, J.Y., Koller, D.: Random worlds and maximum entropy. In: Proceedings of the Seventh Annual IEEE Symposium on Logic in Computer Science, 1992, LICS 1992, pp. 22–33. IEEE (1992)
5. Halpern, J.Y.: An analysis of first-order logics of probability. Artif. Intell. 46(3), 311–350 (1990)
6. Hansen, P., Jaumard, B.: Probabilistic satisfiability. In: Kohlas, J., Moral, S. (eds.) Handbook of Defeasible Reasoning and Uncertainty Management Systems, vol. 5, pp. 321–367. Springer, Netherlands (2000)
7. Klinov, P., Parsia, B.: Pronto: a practical probabilistic description logic reasoner. In: Bobillo, F., Costa, P.C.G., d'Amato, C., Fanizzi, N., Laskey, K.B., Laskey, K.J., Lukasiewicz, T., Nickles, M., Pool, M. (eds.) UniDL/URSW 2008-2010. LNCS, vol. 7123, pp. 59–79. Springer, Heidelberg (2013). doi:10.1007/978-3-642-35975-0_4
8. Koller, D., Levy, A., Pfeffer, A.: P-classic: a tractable probablistic description logic. AAAI/IAAI **1997**, 390–397 (1997)
9. Lukasiewicz, T.: Probabilistic logic programming with conditional constraints. ACM Trans. Comput. Logic 2(3), 289–339 (2001)
10. Lukasiewicz, T., Straccia, U.: Managing uncertainty and vagueness in description logics for the semantic web. JWS 6(4), 291–308 (2008)
11. Lutz, C., Schröder, L.: Probabilistic description logics for subjective uncertainty. In: Proceedings of KR 2010. AAAI Press (2010)
12. Niepert, M., Noessner, J., Stuckenschmidt, H.: Log-linear description logics. In: IJCAI, pp. 2153–2158 (2011)
13. Nilsson, N.J.: Probabilistic logic. Artif. Intell. **28**, 71–88 (1986)
14. Paris, J.B.: The Uncertain Reasoner's Companion - A Mathematical Perspective. Cambridge University Press, Cambridge (1994)
15. Peñaloza, R., Potyka, N.: Probabilistic reasoning in the description logic $\mathcal{ALCP}$ with the principle of maximum entropy. In: Schockaert, S., Senellart, P. (eds.) SUM 2016. LNCS, vol. 9858, pp. 246–259. Springer, Cham (2016). doi:10.1007/978-3-319-45856-4_17
16. Peñaloza, R., Potyka, N.: Towards statistical reasoning in description logics over finite domains (full version). CoRR abs/1706.03207 (2017). http://arxiv.org/abs/1706.03207
17. Potyka, N., Thimm, M.: Probabilistic reasoning with inconsistent beliefs using inconsistency measures. In: IJCAI, pp. 3156–3163 (2015)
18. Riguzzi, F., Bellodi, E., Lamma, E., Zese, R.: Probabilistic description logics under the distribution semantics. Semant. Web 6(5), 477–501 (2015)

Bankruptcy Scenario Query: B-SQ

Carlos Molina[1]([⊠]), Belén Prados-Suárez[2], and Antonio Cortes-Romero[3]

[1] Department of Computer Sciences, University of Jaen, Jaen, Spain
carlosmo@ujaen.es
[2] Department of Software Engineering, University of Granada, Granada, Spain
belenps@ugr.es
[3] Department of Accounting and Finance, University of Granada, Granada, Spain
amcortes@ugr.es

Abstract. There has been increasing interest in risk scoring and bankruptcy prediction in recent years. Most of the current proposals analyse a set of parameters to classify companies as either *active* or *default*. What banks really need, however, is to be able to predict the *probability* of bankruptcy occurring in the future. Current approaches do not enable a deeper analysis to estimate the direction of a company as the parameters under study evolve. This article proposes a system for the *Bankruptcy Scenario Query* (B-SQ) which is based on association rules to allow users to conduct *"What if...?"* queries, and obtain as a response what usually happens under similar scenarios with the corresponding probability of it occurring.

1 Introduction

The development of accurate bankruptcy prediction models has received a major boost in recent years [1,4,16,22] thanks to the introduction of the Basel II guidelines [3] which require banks to implement credit risk prediction models to properly adjust their risk assumptions to the current economic situation.

Current approaches analyse a set of variables to classify companies as either *active* or *default*. Most of these approaches can be divided into two categories: statistical methods [14,20], with a solid mathematical basis and efficient calculation, but mainly suitable for simple classification cases with single relations between the variables; and artificial intelligence and data mining proposals [10,15,19] that can gather the complexity of greater problems but find it difficult to work with large data sets and unbalanced problems. This second group includes *tree-based* methods [5,22] which are interpretable but sensitive to noise and over-fitting; *genetic algorithms* [4,6] which are accurate but can only work on small problems and this makes them inconsistent between executions due to the pre-selection of the required variables; *support vector machine* approaches [16,21] which may encounter difficulties in unbalanced problems due to their search for an overall solution; and *neural network* proposals [1,8] which are suitable for large amounts of data and unbalanced problems but are too complex for the user to understand.

© Springer International Publishing AG 2017
S. Moral et al. (Eds.): SUM 2017, LNAI 10564, pp. 295–306, 2017.
DOI: 10.1007/978-3-319-67582-4_21

Although these approaches are useful for classification purposes, most have three disadvantages. Firstly, they return a binary response (such as *active* or *default*) for the current situation but they do not give *the probability* of the new case being in one of these classes in the future, and this is what banks really need to know [7]. Secondly, companies are studied from a static perspective, without taking into account the fact that business evolve and that this *evolution* conditions their future. Tools are therefore necessary to study how changing conditions (variables) can influence or modify this future [9]. A model based on the use of scenarios would enable this type of analysis to answer *"What if...?"* queries. Finally, although these methods are applied to support decision-making, they mainly focus on improving classification accuracy and ignore *understandability*, which is a growing concern [12]. In more understandable approaches, the results must follow long paths through decision trees or look through very large sets of rules in order to understand the line of reasoning followed by the system. It is for this reason that in this article we examine the problem of designing an *understandable* approach to analyze corporate *evolution* and predict the *probability* of a business remaining *active* according to various different possible situations or *scenarios*.

For this purpose, we present the *Bankruptcy Scenario Query* (B-SQ), which is an association rule-based system directed towards supporting financial decision-making based on scenarios. The main idea consists in creating a set of association rules and building a knowledge base on this for an inference system. This system enables the user to make a query about a given scenario with queries of the type *"What if..?"*, and then returns the facts or situations that usually occur under similar conditions.

Section 2 describes the structure of the B-SQ approach; Sect. 3 presents a brief to illustrate the system; and finally, Sect. 4 outlines our conclusions.

2   Scenario Query System

The *Bankruptcy Scenario Query* system that we propose in this article has three modules and these are shown in Fig. 1 and described in the following subsections.

2.1   Knowledge Base

Our proposal is based on a fuzzy multidimensional model which is described in detail in [13]. By way of summary, this model organizes the data into dimensions (e.g. "time") which can then be analyzed through a kinship relation according to different detail levels (day, week, month, etc.). It also contains the measures or variables to be analysed (e.g. sales or profits). In this section, we describe how to build this model by using, for example, the data obtained from the database built by the company Axesor[1] with 88534 companies from three sectors using the National Classification of Economic Activities (CNAE). We will now explain

---

[1] URL: https://www.axesor.com.

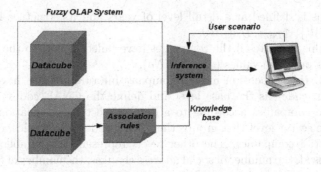

Fig. 1. B-SQ system structure

how to apply an association rule extraction algorithm on this in order to obtain the knowledge base which will be used as input for the inference system.

$C_{Financial}$ **Datacube construction.** In order to build the datacube, we differentiate between failed and successful companies (from the service, commercial and industrial sectors) according to Spanish Law. We have considered three economic-financial variables: return on assets, working capital, and indebtedness cost in the period 2014–2016. The structure of the datacube is shown in Fig. 2. The following section describes the dimensions used for the analysis.

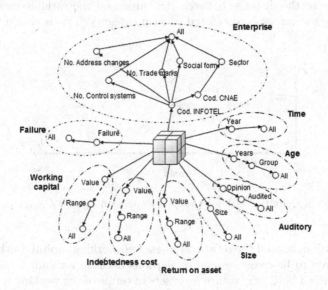

Fig. 2. Business datacube

Dimensions. We have defined five dimensions using the minimum and maximum operators as t-norm and t-conorm when calculating the extended kinship relationship as follows:

- *Time*: this is defined at a detail level of years and its structure is: $Time = (\{Year, All\}, \leq_{Time}, Year, All)$.
- *Failure*: this indicates if the companies have failed or not, so the basic level (*Fail*) has only two values (*Yes*) or (*No*).
- *Company*: this dimension models company information. We have used the INFOTEL code[2] as the base level and define the CNAE codes[3] on this to group the companies according to a detailed sector classification. We then define the sector level that groups the CNAE codes into service, commercial or industrial companies. The other levels represent the number of control systems used, the number of social address changes, the number of trademarks obtained by the company and the company type.
- *Age*: the base level of this dimension is the number of years that a company has been in operation. On this level, we define another to classify the companies as *very young*, *young*, *mature* or *very mature*. This kind of concept is normally defined using crisp intervals, which is not how they are normally used. Fuzzy logic enables these concepts to be characterized more intuitively, as illustrated in Fig. 3.
- *Return on assets*, *Cost of debt* and *Working capital*: for these three dimensions defined on the economic-financial variables, we define the base level using the values observed in the data set. Since numeric values (e.g. a 6.51 return on assets) are less intuitive, we have defined the next level, *Range*, by grouping the values into five categories (*very low*, *low*, *average*, *high* and *very high*) according to the distance between the value and the variable mean and by fuzzifying the intervals associated to each category. This is shown in Fig. 4.

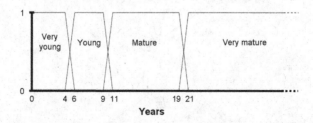

Fig. 3. Definition of level *Group* in dimension *Age* for $C_{Financial}$

Measures. We have used the return on assets and working capital. Both variables are considered to be measures and dimensions because we want to analyse the relation between both (e.g. return on assets in terms of the working capital, and vice versa). Since all the data have been obtained from a reliable source, we assign the maximum confidence value to each fact.

---

[2] Axesor assigns a code to each company in the database.
[3] URL: http://www.cnae.com.es/.

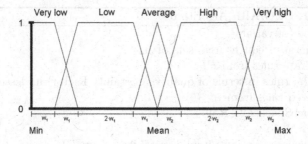

Fig. 4. Definition of ranges over the economic-financial variables for $C_{Financial}$

Association Rule Extraction: COGIARE. In this system, we use the COGIARE algorithm [17] shown in Algorithm 1. This method is based on the extraction of association rules on multidimensional models for classification purposes. The approach uses hierarchies on the dimensions to reduce rule complexity by considering a trade-off between the precision and understandability of the rule set.

It applies an iterative process (Fig. 5): it first builds a set of rules with a single antecedent; it then generates rules with two antecedents and compares the quality of the rules. If better results are obtained, then the process continues to rules with three items in the antecedent. The iterative process stops when the new rule set is not better than the previous one. More details can be found in [17].

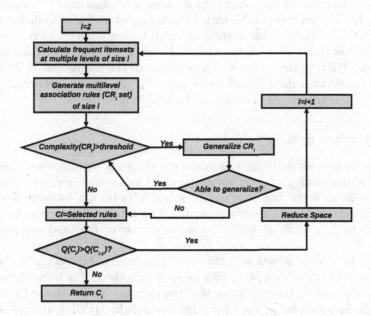

Fig. 5. COGIARE algorithm

Algorithm 1. COGIARE Algorithm

N = Number of variables
FI_2 = Generate frequent item sets of size 2
R_2 = Generate rules from FI_2
R' = Sort R_2 rules in terms of quality (Certainty Factor) in descending order
R_a = Selected rules from R'
FI_2 = reduceSpace(FI_2,R_a)
for $i = 3$ to N **do**
$\quad FI_i$ = Generate frequent item sets of size i from FI_{i-1}
$\quad R_i$ = Generate rules from FI_i
$\quad R = \cup_{j=2}^{i} R_i$
$\quad R'$ = Sort R rules in terms of quality in descending order
$\quad R_c$ = Selected rule from R'
$\quad$ **if** $Quality(R_c) > Quality(R_a)$ **then**
$\quad\quad R_a = R_c$
$\quad\quad FI_i$=reduceSpace(FI_i,R_a)
$\quad$ **else**
$\quad\quad$ **return** R_a
$\quad$ **end if**
end for
return R_a

The function *reduceSpace* that appears in the algorithm reduces the number of itemsets to generate on each iteration. This is implemented by ignoring any itemset that has generated rules with a quality equal to 1 as a candidate for the next iteration. This is an application of a well-known property of rule systems: if rule R1 has the highest quality (1 in the case of CF) and R2 is another rule built using R1 with the addition of an element to the antecedent, if R2 is fired then R1 is also fired since R1 subsumes R2 so R2 cannot be poorer quality than R1. Rule R2 is therefore redundant in the knowledge base.

2.2   Inference System

There is one problem of using association rules in an inference system: the propagation of the quality measure values (support and confidence) when applying the rules [2]. However, this issue can be solved with the CF measure, first used in MYCIN [18] to enable inexact reasoning. The authors also proposed an inference system in order to deal with the certainty factor (CF) and propagate the values.

Since the rules obtained with the algorithm described in Sect. 2.1 have been calculated with this factor, they can be used directly in this inference system. In this article, we therefore propose that the rules obtained with the COGIARE process be incorporated as the knowledge base for the MYCIN inference system in an interactive interface. With this approach, the rules are hidden from the user so there is no need to interpret them. The user simply has to establish the known values that define the scenario of interest (i.e. the elements in the datacube dimensions) and the system will apply the obtained rules to infer other values that are related to these.

2.3 User Interaction

We propose an interactive process to answer the user's scenario queries:

1. The user chooses the values that define the scenario for one or several datacube dimensions.
2. The inference process is applied and the elements related to those selected are shown to the user.
3. The user may interact with the system by adding new values or deleting some of the previously selected values to establish a new scenario.
4. After each change, the system again applies the inference process and shows the new results to the user.

This scheme enables the user to "play" with conditions and identify the company's future evolution as parameters or overall conditions change, and to redirect the business in the desired direction.

3 Example

There are two parts to this example: firstly, we show the operation of the OLAM system; and secondly, the scenario query is exemplified.

3.1 OLAM Process

In order to extract the rules, we use the COGIARE algorithm on the $C_{Financial}$ datacube. The following main parameters are used:

- *Support*: 0.01.
- *CF*: 0.4 (minimum value of CF to consider a rule).
- *Abstraction function*: we use *Generality* as the abstraction function defined as the number of elements grouped by an item compared to the total number of elements at the most detailed dimension level (the base level) (see [11] for more details).
- *Complexity due to the number of rules*: in this case, we use the function N, defined as the relation between the number of rules and the number of possible items in the datacube (see [17] for details).

From this process we obtain 764 association rules. These rules will be used in the following step to enable the user to request possible scenarios.

Quality: In order to measure the quality of the rule set, we have applied a 10-fold cross validation with the same parameters and the results of this was an average precision of 89.2%.

3.2   *What-if...* Process

Figure 6 shows the initial system screen. Let us suppose we are analysing the default probability for a *Young* Company with *Company Type Limited* (in Spanish: *Sociedad Anonima*), where the *Cost of debt* is considered to be *High*, and the *Return on assets* is considered *Low*. The system then applies the inference process and responds (Fig. 7) that the *default* probability is very high (value 0.986). The system explains the result by showing the applied rules. This example has the following rules:

- *Return on assets* is *Low* AND *Cost of debt* is *High* AND *The company* is *Young* THEN *Default* is *Yes* with *Support* 0.015 and *CF*: 0.9664
- *Company Type* is *Limited* and *Return on assets* is *Low* AND *Cost of debt* is *High* THEN *Default* is *Yes* with Support 0.045 and CF: 0.584

By looking at the rules, the user can see that the combination of values for *indebtedness cost* and *return on assets* is presented in both so these factors are important. If we want to know what effect the age of the company has, we can change the value to *Mature* instead of *Young*. In this case, the probability

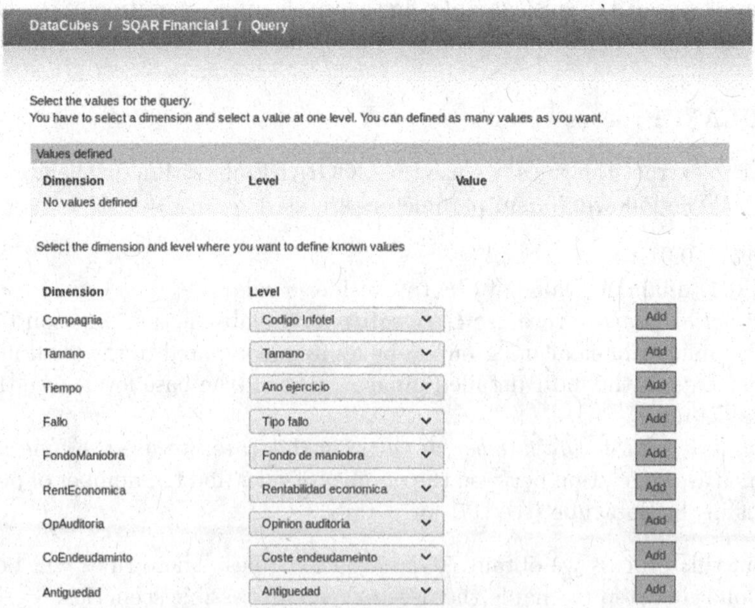

Fig. 6. Initial screen

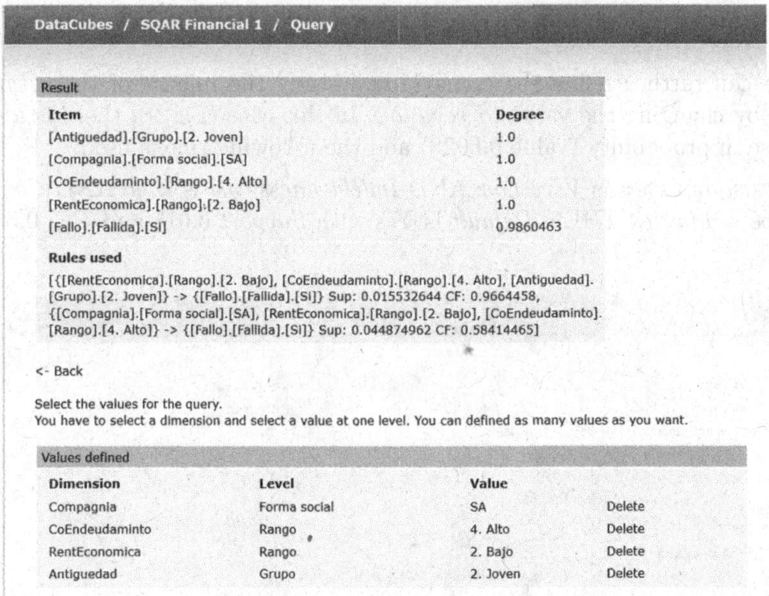

DataCubes / SQAR Financial 1 / Query

Result

Item	Degree
[Antiguedad].[Grupo].[2. Joven]	1.0
[Compagnia].[Forma social].[SA]	1.0
[CoEndeudaminto].[Rango].[4. Alto]	1.0
[RentEconomica].[Rango].[2. Bajo]	1.0
[Fallo].[Fallida].[Si]	0.9860463

Rules used

[{[RentEconomica].[Rango].[2. Bajo], [CoEndeudaminto].[Rango].[4. Alto], [Antiguedad].
[Grupo].[2. Joven]} -> {[Fallo].[Fallida].[Si]} Sup: 0.015532644 CF: 0.9664458,
{[Compagnia].[Forma social].[SA], [RentEconomica].[Rango].[2. Bajo], [CoEndeudaminto].
[Rango].[4. Alto]} -> {[Fallo].[Fallida].[Si]} Sup: 0.044874962 CF: 0.58414465]

<- Back

Select the values for the query.
You have to select a dimension and select a value at one level. You can defined as many values as you want.

Values defined

Dimension	Level	Value	
Compagnia	Forma social	SA	Delete
CoEndeudaminto	Rango	4. Alto	Delete
RentEconomica	Rango	2. Bajo	Delete
Antiguedad	Grupo	2. Joven	Delete

Fig. 7. Initial scenario

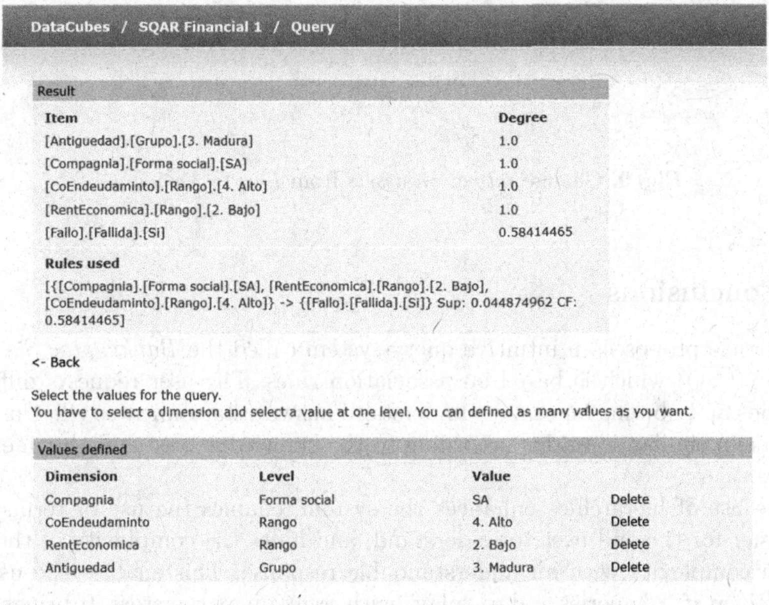

DataCubes / SQAR Financial 1 / Query

Result

Item	Degree
[Antiguedad].[Grupo].[3. Madura]	1.0
[Compagnia].[Forma social].[SA]	1.0
[CoEndeudaminto].[Rango].[4. Alto]	1.0
[RentEconomica].[Rango].[2. Bajo]	1.0
[Fallo].[Fallida].[Si]	0.58414465

Rules used

[{[Compagnia].[Forma social].[SA], [RentEconomica].[Rango].[2. Bajo],
[CoEndeudaminto].[Rango].[4. Alto]} -> {[Fallo].[Fallida].[Si]} Sup: 0.044874962 CF:
0.58414465]

<- Back

Select the values for the query.
You have to select a dimension and select a value at one level. You can defined as many values as you want.

Values defined

Dimension	Level	Value	
Compagnia	Forma social	SA	Delete
CoEndeudaminto	Rango	4. Alto	Delete
RentEconomica	Rango	2. Bajo	Delete
Antiguedad	Grupo	3. Madura	Delete

Fig. 8. Change *Age* to *mature*

decreases to 0.584, only the second rule is applied and no rule that considers age is used. Another important factor, therefore, is the time the company has been active (Fig. 8).

We can further refine the scenario and study the impact of the *return on assets* by changing the value to *very low*. In this case (Fig. 9), there is a slight increase in probability (value 0.6028) and the following rule is used:

– *Return on assets* is *Very Low* AND *Indebtedness cost* is *High* AND *Company Type* is *Limited* THEN *Default* is *Yes* with *Support* 0.011 and *CF*: 0.603

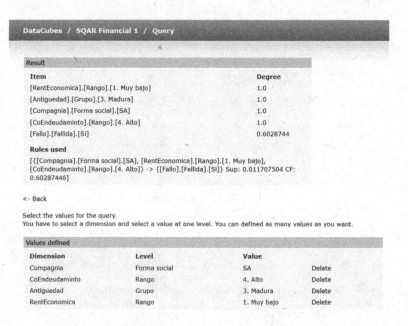

Fig. 9. Change *return on assets* from *Low* to *Very low*

4   Conclusions

This article proposes an intuitive query system called the *Bankruptcy Scenario Query* (B-SQ) which is based on association rules. The user requests different scenarios and obtains an easy-to-understand answer that indicates what usually happens in similar scenarios according to the knowledge base and the inference process.

The use of hierarchies enhances the system, enables the use of terms that are easier for the end user to understand, and hides the complexity of the rule system complexity with an understandable response. This enables the user to pose *"What if..?"* queries and to "play" with company parameters. It brings rule-based decision support systems closer to non-expert users to enable them to foresee the evolution of the company in order to guide it in the desired direction.

Acknowledgements. This research is partially funded by Spanish Ministry of Economy and Competitiveness and the European Regional Development Fund (FEDER) under project TIN2014-58227-P *Descripción lingüística de información visual mediante técnicas de minería de datos y computación flexible.* and project TIC1582 *Mejora de la Accesibilidad a la información mediante el uso de contextos e interprtaciones adaptadas al usuario* of the Junta de Andalucia (Spain).

References

1. Azayite, F.Z., Achchab, S.: Hybrid discriminant neural networks for bankruptcy prediction and risk scoring. Procedia Comput. Sci. **83**, 670–674 (2016)
2. Balcázar, J.L.: Redundancy, deduction schemes, and minimum-size bases for association rules. Log. Methods Comput. Sci. **6**(2), 1–33 (2010)
3. Basel Committee on Banking Supervision2: Basel II: International Convergence of Capital Measurement and Capital Standards: A Revised Framework - Comprehensive Version (2006)
4. Fallahpour, S., Lakvan, E.N., Zadeh, M.H.: Using an ensemble classifier based on sequential floating forward selection for financial distress prediction problem. J. Retail. Consum. Serv. **34**, 159–167 (2017)
5. Gepp, A., Kumar, K.: Predicting financial distress: a comparison of survival analysis and decision tree techniques. Procedia Comput. Sci. **54**, 396–404 (2015)
6. Gordini, N.: A genetic algorithm approach for SMEs bankruptcy prediction: empirical evidence from Italy. Expert Syst. Appl. **41**(14), 6433–6445 (2014)
7. Huang, Y.: Prediction of contractor default probability using structural models of credit risk: an empirical investigation. Constr. Manage. Econ. **27**(6), 581–596 (2009)
8. Iturriaga, F.J.L., Sanz, I.P.: Bankruptcy visualization and prediction using neural networks: a study of U.S. commercial banks. Expert Syst. Appl. **42**(6), 2857–2869 (2015)
9. du Jardin, P.: Dynamics of firm financial evolution and bankruptcy prediction. Expert Syst. Appl. **75**, 25–43 (2017)
10. Lin, W.Y., Hu, Y.H., Tsai, C.F.: Machine learning in financial crisis prediction: a survey. IEEE Trans. Syst. Man Cybern. (C) **42**(4), 421–436 (2012)
11. Marín, N., Molina, C., Serrano, J.M., Vila, A.: A complexity guided algorithm for association rule extraction on fuzzy datacubes. IEEE Tras. Fuzzy Syst. **16**, 693–714 (2008)
12. Martin-Barragan, B., Lillo, R., Romo, J.: Interpretable support vector machines for functional data. Europ. J. Oper. Res. **232**(1), 146–155 (2014)
13. Molina, C., Rodriguez-Ariza, L., Sanchez, D., Vila, A.: A new fuzzy multidimensional model. IEEE Trans. Fuzzy Syst. **14**(6), 897–912 (2006)
14. Ng, S.T., Wong, J.M., Zhang, J.: Applying z-score model to distinguish insolvent construction companies in China. Habitat Int. **35**(4), 599–607 (2011)
15. Olson, D.L., Delen, D., Meng, Y.: Comparative analysis of data mining methods for bankruptcy prediction. Decis. Support Syst. **52**(2), 464–473 (2012)
16. Pal, R., Kupka, K., Aneja, A.P., Militky, J.: Business health characterization: a hybrid regression and support vector machine analysis. Expert Syst. Appl. **49**, 48–59 (2016)
17. de Reyes, M.A.P., Molina, C., Yáñez, M.C.P.: Interpretable associations over datacubes: application to hospital managerial decision making. In: Conference Proceedings of MIE 2014 (2014)

18. Shortliffe, E., Buchanan, B.: A model of inexact reasoning in medicine. Math. Biosci. **23**(3–4), 351–379 (1975)
19. Tsai, C.F., Hsu, Y.F., Yen, D.C.: A comparative study of classifier ensembles for bankruptcy prediction. Appl. Soft Comput. **24**, 977–984 (2014)
20. Tserng, H.P., Liao, H.H., Jaselskis, E.J., Tsai, L.K., Chen, P.C.: Predicting construction contractor default with barrier option model. J. Constr. Eng. Manage. **138**(5), 621–630 (2012)
21. Yeh, C.C., Chi, D.J., Hsu, M.F.: A hybrid approach of DEA, rough set and support vector machines for business failure prediction. Expert Syst. Appl. **37**(2), 1535–1541 (2010)
22. Ziba, M., Tomczak, S.K., Tomczak, J.M.: Ensemble boosted trees with synthetic features generation in application to bankruptcy prediction. Expert Syst. Appl. **58**, 93–101 (2016)

A View of f-indexes of Inclusion Under Different Axiomatic Definitions of Fuzzy Inclusion

Nicolás Madrid$^{(\boxtimes)}$ and Manuel Ojeda-Aciego

Departamento de Matemática Aplicada, Universidad de Málaga, Málaga, Spain
nmadrid@ctima.uma.es

Abstract. In this paper we analyze the novel constructive definition of f-index of inclusion with respect to four of the most common axiomatic definitions of inclusion measure, namely Sinha-Dougherty, Kitainik, Young and Fan-Xie-Pei. There exist an important difference between the f-index and these axiomatic definitions of inclusion measure: the f-index represents the inclusion in terms of a *mapping* in unit interval, whereas the inclusion measure represents such an inclusion as a *value* in the unit interval.

1 Introduction

Extending crisp operations and relations to the fuzzy case has taken the attention of researchers since Zadeh introduced the notion of fuzzy sets [21]. However, there is not consensus about how to extend some of them and, due to intrinsic features of fuzzy sets, it looks that all those different ways are acceptable; the choice depends on the task or the context where fuzzy sets are defined. One example of this fact is the fuzzy extension of the relation of inclusion, for which there are two different kind of approaches, the *constructive* ones (which provide a formula to represent the inclusion relation) and the *axiomatic* ones (which present some basic properties that must be satisfied by any inclusion measure). In the former case, we can distinguish those based on fuzzy implications [1,12], probability [17] and overlapping [7,8,14]. In the latter class, we can distinguish also between other two subclasses, those allowing a non-null degree of inclusion of some fuzzy sets into the empty set [13,18] and those that not, which are also related to entropy measures and overlapping [11,20]. In the literature, one can find many theoretical and practical studies on such families of axiomatic definitions [4–6,9,10,12,19].

Most of the generalizations in the literature about fuzzy inclusion have a common feature; namely, they are relations that assign a value in the unit interval to each pair of fuzzy sets, A and B, that determines the degree of inclusion of A in B. One exception is [16], where the notion of inclusion is represented by assigning to each pair of fuzzy sets a mapping between the unit interval. Despite this differential feature, in this paper we analize the f-index of inclusion under the view of the four most common axiomatic definitions of measure of inclusion namely, Sinha-Dougherty [18], Kitainik [13], Young [20] and Fan-Xie-Pei [11].

© Springer International Publishing AG 2017
S. Moral et al. (Eds.): SUM 2017, LNAI 10564, pp. 307–318, 2017.
DOI: 10.1007/978-3-319-67582-4_22

Actually, we show that the f-index of inclusion satisfies, somehow, the axioms of Sinha-Dougherty with the exception of the relationship with the complementary, which needs to be rewritten in terms of Galois connections.

The structure of this paper is the following. In Sect. 2 we present the preliminaries which includes the four respective axiomatic definitions of inclusion measures and the notion of the f-index of inclusion. Subsequently, in Sect. 3, we check each axiom in the context of f-indexes of inclusion, by showing which axioms hold and under which circumstances. Finally, in Sect. 4 we present conclusions and prospects for future work.

2   Preliminaries

A fuzzy set A is a pair $(\mathcal{U}, \mu_A)$ where $\mathcal{U}$ is a non empty set (called the universe of A) and μ_A is a mapping from $\mathcal{U}$ to $[0, 1]$ (called membership function of A). In general, the universe is a fixed set for all the fuzzy sets considered and therefore, each fuzzy sets is determined by its membership function. Hence, for the sake of clarity, we identify fuzzy sets with membership functions (i.e., $A(u) = \mu_A(u)$).

On the set of fuzzy sets defined on the universe $\mathcal{U}$, denoted $\mathcal{F}(\mathcal{U})$, we can extend the usual crisp operations of union, intersection and complement as follows. Given two fuzzy sets A and B, we define

- (union) $A \cup B(u) = \max\{A(u), B(u)\}$
- (intersection) $A \cap B(u) = \min\{A(u), B(u)\}$
- (complement) $A^c(u) = n(A(u))$

where $n \colon [0, 1] \to [0, 1]$ is a negation operator; i.e., n is a decreasing mapping such that $n(0) = 1$ and $n(1) = 0$. In a considerable number of papers, the negation considered is involutive (*strong negation* in fuzzy settings) which adds to n the condition $n(n(x)) = x$ for all $x \in [0, 1]$. This condition is crucial to have in fuzzy sets the equality $(A^c)^c = A$. In this paper we assume that *the complement of a fuzzy set is always defined in terms of an involutive negation*.

An implication $I \colon [0, 1] \times [0, 1]$ is any mapping decreasing in its first component, increasing in the second component and such that $I(0, 0) = I(0, 1) = I(1, 1) = 1$ and $I(0, 1) = 0$.

Any transformation in the universe $T \colon \mathcal{U} \to \mathcal{U}$ can be extended to $\mathcal{F}(\mathcal{U})$ by defining for each $A \in \mathcal{F}(\mathcal{U})$ the fuzzy set $T(A)(x) = A(T(x))$.

In the rest of this section we deal with different approaches to the notion of inclusion between fuzzy sets which can be found in the literature. We recall below some of them which will be alter considered in the framework of our f-indexes of inclusion in Sect. 3.

2.1   Sinha-Dougherty Axioms

One of the most common measures of inclusion was originally proposed by Sinha-Dougherty in [18].

Definition 1. *A mapping* $\mathcal{I}\colon \mathcal{F}(\mathcal{U}) \times \mathcal{F}(\mathcal{U}) \to [0,1]$ *is called an* SD-inclusion relation *if it satisfies the following axioms for all fuzzy sets A, B and C:*

(SD1) $\mathcal{I}(A, B) = 1$ if and only if $A(u) \leq B(u)$ for all $u \in \mathcal{U}$.

(SD2) $\mathcal{I}(A, B) = 0$ if and only if there exists $u \in \mathcal{U}$ such tat $A(u) = 1$ and
  $B(u) = 0$.

(SD3) If $B(u) \leq C(u)$ for all $u \in \mathcal{U}$ then $\mathcal{I}(A, B) \leq \mathcal{I}(A, C)$.

(SD4) If $B(u) \leq C(u)$ for all $u \in \mathcal{U}$ then $\mathcal{I}(C, A) \leq \mathcal{I}(B, A)$.

(SD5) If $T\colon \mathcal{U} \to \mathcal{U}$ is a bijective transformation on the universe, then $\mathcal{I}(A, B) =$
  $\mathcal{I}(T(A), T(B))$.

(SD6) $\mathcal{I}(A, B) = \mathcal{I}(B^c, A^c)$.

(SD7) $\mathcal{I}(A \cup B, C) = \min\{\mathcal{I}(A, C), \mathcal{I}(B, C)\}$.

(SD8) $\mathcal{I}(A, B \cap C) = \min\{\mathcal{I}(A, B), \mathcal{I}(A, C)\}$.

Sinha and dougherty included also that $\mathcal{I}(A, B \cup C) \geq \max\{\mathcal{I}(A, B), \mathcal{I}(A, C)\}$ for all fuzzy sets A, B and C. However, we do not consider it here since it is a direct consequence of Axiom (SD3) [2,6].

2.2 Kitainik Axioms

In 1987, Kitainik [13] proposed an axiomatic definition of fuzzy subsethood which captures the essential of inclusion measures based on implications. One of the main differences with respect to the axiomatic definition of fuzzy inclusion is the independence with respect to Zadeh's definition, i.e. $(SD1)$.

Definition 2. *A mapping $\mathcal{I}\colon \mathcal{F}(\mathcal{U}) \times \mathcal{F}(\mathcal{U}) \to [0, 1]$ is called* K-inclusion relation *if it satisfies the following axioms for all fuzzy sets A, B and C:*

(K1) $\mathcal{I}(A, B) = \mathcal{I}(B^c, A^c)$.

(K2) $\mathcal{I}(A, B \cap C) = \min\{\mathcal{I}(A, B), \mathcal{I}(A, C)\}$.

(K3) If $T\colon \mathcal{U} \to \mathcal{U}$ is a bijective transformation on the universe, then $\mathcal{I}(A, B) =$
  $\mathcal{I}(T(A), T(B))$.

(K4) If A and B are crisp then $\mathcal{I}(A, B) = 1$ if and only if $A \subseteq B$.

(K5) If A and B are crisp then $\mathcal{I}(A, B) = 0$ if and only if $A \not\subseteq B$.

It is not difficult to check that every Kitainik inclusion measure is also a Sinha-Dougherty measure.

Moreover, Fodor and Yager showed [12], by using a representation result already published by Kitianik, that for every K-measure of inclusion $\mathcal{I}$ there exists an implication I such that, for all fuzzy sets A and B, it holds

$$\mathcal{I}(A, B) = \inf\{I(A(u), B(u)) \mid u \in \mathcal{U}\}.$$

2.3   Young Axioms

The axioms proposed by Young [20] for a measure of inclusion are based on measures of entropy [15]. Specifically, the following relationship between both measures is proposed: if $\mathcal{I}$ is a measure of inclusion, then $\mathcal{E}(A) = \mathcal{I}(A \cup A^c, A \cap A^c)$ defines a measure of entropy. Based on such an idea, the following axiomatic definition was given:

Definition 3. *A mapping* $\mathcal{I} \colon \mathcal{F}(\mathcal{U}) \times \mathcal{F}(\mathcal{U}) \to [0,1]$ *is called* Y-inclusion relation *if it satisfies the following axioms for all fuzzy sets* A, B *and* C:

(Y1) $\mathcal{I}(A, B) = 1$ *if and only if* $A(u) \leq B(u)$ *for all* $u \in \mathcal{U}$.
(Y2) *If* $A(u) \geq 0.5$ *for all* $u \in \mathcal{U}$, *then* $\mathcal{I}(A, A^c) = 0$ *if and only if* $A = \mathcal{U}$; *i.e.,* $A(u) = 1$ *for all* $u \in \mathcal{U}$.
(Y3) *If* $A(u) \leq B(u) \leq C(u)$ *for all* $u \in \mathcal{U}$ *then* $\mathcal{I}(C, A) \leq \mathcal{I}(B, A)$ *for all fuzzy set* $A \in \mathcal{F}(\mathcal{U})$.
(Y4) *If* $B(u) \leq C(u)$ *for all* $u \in \mathcal{U}$ *then* $\mathcal{I}(A, B) \leq \mathcal{I}(A, C)$ *for all fuzzy set* $A \in \mathcal{F}(\mathcal{U})$.

In the original definition [20] axioms (Y3) and (Y4) are written jointly as one axiom.

2.4   Fan-Xie-Pei Axioms

The definition of Young was analyzed and modified slightly by Fan, Xie and Pie [11]. Firstly they criticize the axiom (Y4) and propose to change it by

(FX4) If $A(u) \leq B(u) \leq C(u)$ for all $u \in \mathcal{U}$ then $\mathcal{I}(A, B) \leq \mathcal{I}(A, C)$ for all fuzzy set $A \in \mathcal{F}(\mathcal{U})$.

Subsequently, they propose two other different definitions of measure of inclusion, called weak and strong respectively, by modifying the axioms in Young's definition.

Definition 4. *A mapping* $\mathcal{I} \colon \mathcal{F}(\mathcal{U}) \times \mathcal{F}(\mathcal{U}) \to [0,1]$ *is said to be a* strong FX-inclusion relation *if it satisfies the following axioms for all fuzzy sets* A, B *and* C:

(sFX1) $\mathcal{I}(A, B) = 1$ *if and only if* $A(u) \leq B(u)$ *for all* $u \in \mathcal{U}$.
(sFX2) *If* $A \neq \varnothing$ *and* $A \cap B = \varnothing$ *then,* $\mathcal{I}(A, B) = 0$.
(sFX3) *If* $A(u) \leq B(u) \leq C(u)$ *for all* $u \in \mathcal{U}$ *then* $\mathcal{I}(C, A) \leq \mathcal{I}(B, A)$ *and* $\mathcal{I}(A, B) \leq \mathcal{I}(A, C)$ *for all fuzzy set* $A \in \mathcal{F}(\mathcal{U})$.

Definition 5. *A mapping* $\mathcal{I} \colon \mathcal{F}(\mathcal{U}) \times \mathcal{F}(\mathcal{U}) \to [0,1]$ *is said to be* weak FX-inclusion relation *if it satisfies the following axioms for all fuzzy sets* A, B *and* C:

(wFX1) $\mathcal{I}(\varnothing, \varnothing) = \mathcal{I}(\varnothing, \mathcal{U}) = \mathcal{I}(\mathcal{U}, \mathcal{U}) = 1$; *where* $\mathcal{U}(u) = 1$ *for all* $u \in \mathcal{U}$.
(wFX2) $\mathcal{I}(A, \varnothing) = 0$
(wFX3) *If* $A(u) \leq B(u) \leq C(u)$ *for all* $u \in \mathcal{U}$ *then* $\mathcal{I}(C, A) \leq \mathcal{I}(B, A)$ *and* $\mathcal{I}(A, B) \leq \mathcal{I}(A, C)$ *for all fuzzy set* $A \in \mathcal{F}(\mathcal{U})$.

In the original paper of Fan, Xie and Pie [11] the reader can find relationships between these measures and fuzzy implications.

2.5 f-indexes of Inclusion

Although most of the approaches for extending the relation of inclusion between fuzzy sets consider mappings of the type $\mathcal{I}\colon \mathcal{F}(\mathcal{U}) \times \mathcal{F}(\mathcal{U}) \to [0,1]$, as done by Sinha and Dougherty, there are also other approaches that differ from such idea. One example is [16] which assigns to each pair of fuzzy sets a mapping $f\colon [0,1] \to [0,1]$ to represent their degree of inclusion. This approach is based on the definition of f-inclusion given at following.

Definition 6. *Let A and B be two fuzzy sets and let $f\colon [0,1] \to [0,1]$ be a mapping such that $f(x) \leq x$ for all $x \in [0,1]$. We say that A is f-included in B if the inequality $f(A(u)) \leq B(u)$ holds for all $u \in \mathcal{U}$.*

Note that the f-inclusion is a crisp relationship in the sense that A is either f-included in B or not. Thus, it somehow reminds the original definition of fuzzy inclusion given by Zadeh. The fuzziness in the definition above is that each f should be considered a degree of inclusion. Let us try to clarify this point. For the sake of presentation, let us define Ω as the set of functions $f\colon [0,1] \to [0,1]$ such that $f(x) \leq x$ for all $x \in [0,1]$; i.e., Ω is our set of inclusion degrees. Note that Ω has the structure of a complete lattice, where id (i.e. $id(x) = x$ for all $x \in [0,1]$) and 0 (i.e. $0(x) = 0$ for all $x \in [0,1]$) are the top and least element, respectively.

Given two functions $f, g \in \Omega$ such that $f \leq g$ then, $A \subseteq_f B$ implies $A \subseteq_g B$ (see Proposition 7 in Sect. 3). As a result, the greater the mapping f, the stronger the restriction imposed by the f-inclusion. So, each mapping f in Ω represents a degree of inclusion between fuzzy sets according to the strength of the restriction imposed by the respective f-inclusion relation. In such a way, the mapping 0 represents the null degree of inclusion (actually all pairs of fuzzy sets A and B satisfy $A \subseteq_0 B$) whereas id represents the highest degree (actually if a pair of fuzzy sets A and B satisfy $A \subseteq_{id} B$ then $A \subseteq_f B$ for all $f \in \Omega$). The reader is also referred to [16] for deeper motivational aspects of this set of f-indexes of inclusions.

In order to assign a convenient f-index of inclusion to a pair of fuzzy sets, it can be proved that given two fuzzy sets A and B, the following set

$$\{f \in \Omega \mid A \subseteq_f B\}$$

is closed under suprema. Therefore, its greatest element (denoted hereafter by f_{AB}) seems to be the most appropriated f-index of inclusion for the relation $A \subseteq B$. Moreover, such a mapping is determined by the following theorem.

Theorem 1 ([16]). *Let A and B be two fuzzy sets. Then, the greatest element of $\{f \in \Omega \mid A \subseteq_f B\}$ is*

$$f_{AB}(x) = \min\{x, \inf_{u \in \mathcal{U}}\{B(u) \mid x \leq A(u)\}\} \tag{1}$$

Now, in order to provide evidences about why f_{AB} is an appropriated f-index for the relation $A \subseteq B$, in the next section we show that almost all the axioms given by Sinha and Dougherty hold for such an index under a convenient and natural rewriting.

3   Checking Axioms for the f-index of Inclusion

In the previous section, we have recalled the definition of the f-index of inclusion together with several axiomatic definitions of inclusion measure. To begin with, it is worth to note that the usual notion of an inclusion measure is that of a mapping $\mathcal{F}(\mathcal{U}) \times \mathcal{F}(\mathcal{U}) \to [0,1]$, whereas the f-index is a mapping $\mathcal{F}(\mathcal{U}) \times \mathcal{F}(\mathcal{U}) \to \Omega$, where Ω is the set of functions in the unit interval which are smaller than the identity. Fortunately, thanks to the lattice structure of Ω, the translation of all the axioms is straightforward.

It is also remarkable that many of the axioms in different axiomatic systems are identical or very related. For that reason and for the sake of the presentation, these axioms are grouped together in this section under a common feature, and the relationship with the f-index of inclusion is analyzed jointly.

3.1   Relationship with Zadeh's Definition

The original definition of fuzzy inclusion introduced by Zadeh [21] states that, for any two fuzzy sets $A, B \in \mathcal{F}(\mathcal{U})$,

$$A \subseteq B \text{ if and only if } A(u) \leq B(u) \text{ for all } u \in \mathcal{U}.$$

Note that axioms (SD1), (Y1) and (sFX1) are almost identical to that definition. Actually, those axioms can be rewritten as *"the degree of inclusion of A in B is 1 if and only if A is contained in B in Zadeh's sense"*. The following result, already proved in [16, Corollary 2], shows that the f-index of inclusion satisfies exactly this condition.

Proposition 1. *Let A and B be two fuzzy sets. Then, $f_{AB} = id$ if and only if $A(u) \leq B(u)$ for all $u \in \mathcal{U}$.*

It is convenient to mention that axioms (K4) and (wFX1) are weaker assumptions than Zadeh's inclusion, and therefore, they are also satisfied by f-indexes.

3.2   The Case of Null Inclusion

Different axiomatic systems treat differently the particular case of the null inclusion, some of them introduce a characterization (if and only if) of the situations in which the degree of inclusion is zero, whereas others simply state a condition (if ... then) that it should satisfy. The axioms related to null inclusion are (SD2), (K5), (Y2), (sFX2) and (wFX2).

It is worth to mention here that, these axioms are slightly controversial. In fact, Sinha and Dougherty stated in [18]:

"[...] Axiom 2 may seem unnatural to many readers. In particular, this axiom causes 'problems' if we wish to model the entropy of a set via Kosko's method [...]" and also

" [...] one may want to add another requirement that $\mathcal{I}(A, \varnothing) = 0$. This is not consistent with Axiom 2. "

The importance of not assuming $\mathcal{I}(A, \varnothing) = 0$ (note that this equality does not hold in our approach either) is that

"$\mathcal{I}(A, \varnothing)$ denotes the degree to which A can be classified as the empty set".

In other words, (SD2) is contradictory with axioms (Y2), (sFX2) and (wFX2). Sinha and Dougherty also propose in [18] the following weak version of Axiom (SD2): for every pair of fuzzy sets A and B

$(SD2^*)$ $(\exists u \in \mathcal{U}$ such that $A(u) = 1$ and $B(u) = 0)$ implies $\mathcal{I}(A, B) = 0$.

This weaker version of the axiom $(SD2)$ holds in the context of f-indexes.

Proposition 2 ([16]). *Let A and B be two fuzzy sets. If there exists $u \in \mathcal{U}$ such that $A(u) = 1$ and $B(u) = 0$, then $f_{AB} = 0$.*

Note that the previous result implies, in general, that the f-index of inclusion does not satisfy either axiom (Y2) or (sFX2) or (wFX2), as the following example shows.

Example 1. Consider $\mathcal{U} = \{a, b\}$ and let A be the fuzzy set defined by $A(a) = 0.6$ and $A(b) = 0.8$. Then, by Eq. (1), we have that:

$$f_{A\varnothing}(x) = \begin{cases} 0 & \text{if } x \leq 0.8 \\ x & \text{otherwise} \end{cases}$$

which is obviously different from the function 0. This fact contradicts (sFX2) and (wFX2), since for any measure of inclusion $\mathcal{I}$ holding such axioms we have $\mathcal{I}(A, \varnothing) = 0$. Moreover, by Eq. (1) again, we have

$$f_{A,A^c}(x) = \begin{cases} 0.2 & \text{if } 0.2 \leq x \leq 0.8 \\ x & \text{otherwise} \end{cases}$$

Note that $f_{A,A^c} \neq 0$ and that for any measure of inclusion $\mathcal{I}$ holding (Y2), we have $\mathcal{I}(A, A^c) = 0$.

We study now, more in depth, the relationship of f-indexes with the axioms (SD2) and (K4). For this, let us recall the following characterization for the index $f = 0$.

Proposition 3 ([16]). *Let A and B be two fuzzy sets. $f_{AB} = 0$ if and only if there exists a sequence $\{u_n\}_{n\in\mathbb{N}} \subseteq \mathcal{U}$ such that $A(u_n) = 1$ and $\lim B(u_n) = 0$.*

As a result, we have that axiom (K4) holds for the f-index of inclusion.

Corollary 1. *Let A and B be two crisp sets then, $f_{AB} = 0$ if and only if $A \subseteq B$.*

Finally, note that Proposition 3 is very close but not equal to axiom $(SD2)$. The difference is that the for the f-index to be 0 there should not exist an element fully in A that is not fully in B but, instead, it is sufficient that for each $\varepsilon < 0$ there exists an element fully in A which is in B in degree smaller than ε. Obviously, if the underlying universe $\mathcal{U}$ is finite, we get exactly axiom (SD2).

Corollary 2. *Let A and B be two fuzzy sets on a finite universe $\mathcal{U}$. $f_{AB} = 0$ if and only if there exists $u \in \mathcal{U}$ such that $A(u) = 1$ and $B(u) = 0$.*

3.3 About Monotonicity

The axioms related to the monotonicity of measures of inclusion are (SD3), (SD4), (Y3),(Y4), (FX4), (sFX3) and (wFX3).

We show that (SD3) and (SD4) hold for the f-indexes of inclusion as a consequence of the following result which establishes some monotonic properties for the f-index

Proposition 4 ([16]). *Let A, B, C, and D be four fuzzy sets such that $A(u) \leq B(u)$ and $C(u) \leq D(u)$ for all $u \in \mathcal{U}$ then, $B \subseteq_f C$ implies $A \subseteq_f D$.*

As a direct consequence, the axioms (SD3) and (SD4) hold in the context of f-indexes as well.

Corollary 3. *Let A, B and C be three fuzzy sets:*

- *if $B(u) \leq C(u)$ for all $u \in \mathcal{U}$ then, $f_{AB} \leq f_{AC}$;*
- *if $B(u) \leq C(u)$ for all $u \in \mathcal{U}$ then, $f_{CA} \leq f_{BA}$.*

The rest of axioms, namely (Y3),(Y4), (FX4), (sFX3) and (wFX3) also hold for the f-indexes since are weaker forms of (SD3) and (SD4).

3.4 Transformation Invariance

The only two axioms related to transformations on the universe $\mathcal{U}$ are (SD5) and (K3), and are identical. Let us recall that the axiom (SD5) states that for any fuzzy inclusion $\mathcal{I}$, if $T : \mathcal{U} \to \mathcal{U}$ is a transformation (i.e. a one-to-one mapping) on the universe, then

$$\mathcal{I}(A, B) = \mathcal{I}(T(A), T(B))$$

for all fuzzy sets A and B. This axiom comes from the crisp environment, where the inclusion relationship is not modified if it is applied any kind of transformation; as reflexion, translations, etc. Let us see that it is also satisfied in the context of f-indexes.

Proposition 5. *Let A and B be two fuzzy sets and let $T : \mathcal{U} \to \mathcal{U}$ be a transformation on $\mathcal{U}$, then $f_{AB} = f_{T(A)T(B)}$.*

Proof. Since f_{AB} and $f_{T(A)T(B)}$ are, by definition, the suprema of the sets $\{f \in \Omega \mid A \subseteq_f B\}$ and $\{f \in \Omega \mid T(A) \subseteq_f T(B)\}$, respectively, we prove the result by showing that both sets are the same. Consider $f \in \Omega$ such that $A \subseteq_f B$; then, for all $u \in \mathcal{U}$ we have $f(A(u)) \leq B(u)$ which, by the bijectivity of T, is equivalent to say that for all $u \in \mathcal{U}$ we have $f(A(T(u))) \leq B(T(u))$, which is equivalent to $T(A) \subseteq_f T(B)$.

3.5   Relationship with the Complement

The complement appears in axioms (Y2), (SD6) and (K1). The case of (Y2) is more related with the null degree of inclusion than with the relation between inclusion and complement, and for such a reason, it was studied in Sect. 3.2; the other two axioms (SD6) and (K1) are identical, and state that that for any fuzzy inclusion $\mathcal{I}$ and every pair of fuzzy sets A and B, the equality $\mathcal{I}(A,B) = \mathcal{I}(B^c, A^c)$ holds.

In general, neither the equality $f_{AB} = f_{B^c A^c}$ nor the relation $A \subseteq_f B$ implies $B^c \subseteq_f A^c$ holds for $f \in \Omega$. However, it is possible to establish some relationships between both f-indexes via adjoint pairs. Let us recall that two mappings $f, g \colon [0,1] \to [0,1]$ form an adjoint pair if

$$f(x) \leq y \iff x \leq g(y) \qquad \text{for all } x \in [0,1] \tag{2}$$

The first result connects the f-indexes of inclusion of A in B with those related to the inclusion of B in A via the negation n used to define the complement and adjoint pairs.

Proposition 6. *Let A and B be two fuzzy sets and let (f, g) be an adjoint pair. Then $A \subseteq_f B$ if and only if $B^c \subseteq_{nogon} A^c$.*

Proof. Let us begin by proving that $f \in \Omega$ if and only if $n \circ g \circ n \in \Omega$.

On the one hand, since (f, g) forms an adjoint pair, both mappings f and g are monotonic. This is a straightforward consequence of the definition: by Eq. (2) and from $f(x) \leq f(x)$ we get $x \leq g \circ f(x)$ for all $x \in [0,1]$. Now, monotonicity comes from adjointness, since

$$x_1 \leq x_2 \leq g \circ f(x_2) \iff f(x_1) \leq f(x_2) \qquad \text{for all } x_1, x_2 \in [0,1]$$

The monotonicity of g is proved similarly.

On the other hand, let us assume $f \in \Omega$, that is, $f(x) \leq x$ for all $x \in [0,1]$. Then by the adjoint property we have the following chain of equivalences for all $x \in [0,1]$.

$$f(x) \leq x \iff f(n(x)) \leq n(x) \iff n(x) \leq g(n(x)) \iff x \geq n(g(n(x)))$$

So, $f \in \Omega$ if and only if $n \circ g \circ n \in \Omega$.

Let us assume now that $A \subseteq_f B$. Then, for any $u \in \mathcal{U}$ we have:

$$f(A(u)) \leq B(u) \iff A(u) \leq g(B(u)) \iff n(A(u)) \geq n(g(B(u))).$$

Finally, by using that $n \circ n = id$, we have that

$$f(A(u)) \leq B(u) \iff n(A(u)) \geq n(g(n(n(B(u))))),$$

or equivalently, $B^c \subseteq_{nogon} A^c$.

The following theorem shows that the f-index of A included in B and B^c included in A^c are related by adjointness in the case of a finite universe.

Theorem 2. *Let A and B be two fuzzy sets on a finite universe $\mathcal{U}$. Then, $(f_{AB}, n \circ f_{B^c A^c} \circ n)$ forms an adjoint pair.*

Proof. Let us begin by noticing that f_{AB} is always the left adjoint of an isotone Galois connection. This is equivalent to prove that $f_{AB} (\sup_{i \in \mathbb{I}} x_i) = \sup_{i \in \mathbb{I}} f_{AB}(x_i)$ and that equality comes from the structure of f_{AB}

$$f_{AB}(x) = \min\{x, \inf_{u \in \mathcal{U}} \{B(u) \mid x \leq A(u)\}\}$$

given by Eq. (1) and the fact that the universe is finite.

To prove now that $(f_{AB}, n \circ f_{B^c A^c} \circ n)$ forms an adjoint pair we use a result that states that if (f_{AB}, g) is an isotone Galois connection, then $g(y) = \sup\{x \in [0,1] \mid f_{AB}(x) \leq y\}$. So let us check that $g(y) = n \circ f_{B^c A^c} \circ n(y)$ for all $y \in [0,1]$:

$$g(y) = \sup\{x \in [0,1] \mid f_{AB}(x) \leq y\}$$

(By definition of f_{AB})
$$= \sup\{x \in [0,1] \mid \min\{x, \min_{u \in \mathcal{U}}\{B(u) \mid x \leq A(u)\} \leq y\}$$

(By associativity of min)
$$= \sup_{u \in \mathcal{U}}\{x \in [0,1] \mid x \leq y, B(u) \leq y, x \leq A(u)\}$$

(By n involutive)
$$= \sup_{u \in \mathcal{U}}\{n(n(x)) \in [0,1] \mid x \leq n(n(y)), B(u) \leq n(n(y)), n(n(x)) \leq A(u)\}$$

(By n decreasing and involutive)
$$= n \left(\inf_{u \in \mathcal{U}} \{n(x) \in [0,1] \mid n(y) \leq n(x), n(y) \leq n\left(B(u)\right), n(A(u)) \leq n(x)\} \right)$$

(By associativity of min)
$$= n \left(\inf\{n(x) \in [0,1] \mid \min\{n(y), \min_{u \in \mathcal{U}}\{n(A(u)) \mid n(y) \leq n(B(u))\} \leq n(x)\} \right)$$

$$= n \left(\min\{n(y), \min_{u \in \mathcal{U}}\{n(A(u)) \mid n(y) \leq n(B(u))\} \right) = n(f_{B^c A^c}(n(y)))$$

3.6 Relationship with Union and Intersection

The axioms related with union and intersection are (SD7), (SD8) and (K8). Note that (SD8) coincides with (K8). They state that for any fuzzy inclusion $\mathcal{I}$ and three fuzzy sets A, B and C we have the following equalities:

- $\mathcal{I}(A \cup B, C) = \min\{\mathcal{I}(A, C), \mathcal{I}(B, C)\}$
- $\mathcal{I}(A, B \cap C) = \min\{\mathcal{I}(A, B), \mathcal{I}(A, C)\}$

Once again, let us recall a result concerning ordering between f-indexes.

Proposition 7 ([16]). *Let A and B be two fuzzy sets and let $f, g \in \Omega$ such that $f \geq g$. Then, $A \subseteq_f B$ implies $A \subseteq_g B$.*

As a consequence of Propositions 1 and 7, we obtain:

Corollary 4. *Let A, B and C be three fuzzy sets and let $f, g \in \Omega$.*

- *If $A \subseteq_f C$ and $B \subseteq_g C$, then $A \cup B \subseteq_{f \wedge g} C$.*
- *If $A \subseteq_f B$ and $A \subseteq_g C$, then $A \subseteq_{f \wedge g} B \cap C$.*

But we can go further and prove the following theorem.

Theorem 3. *Let A, B and C be three fuzzy sets then,*

$$f_{A \cup B, C} = \min\{f_{AC}, f_{BC}\} \quad and \quad f_{A, B \cap C} = \min\{f_{AB}, f_{AC}\}.$$

Proof. Let us prove the first equality $f_{A \cup B, C} = \min\{f_{AC}, f_{BC}\}$. By Theorem 1 and definition of $A \cup B$, we know that

$$f_{A \cup B, C} = \min\{x, \inf_{u \in \mathcal{U}} \{C(u) \mid x \leq \max\{A(u), B(u)\}\}\}$$

and by properties of infimum and maximum we have:

$$f_{A \cup B, C} = \min\{x, \inf_{u \in \mathcal{U}} \{C(u) \mid x \leq A(u)\}, \inf_{u \in \mathcal{U}} \{C(u) \mid x \leq B(u)\}\}$$

which is equivalent to say $f_{A \cup B, C} = \min\{f_{AC}, f_{BC}\}$.

The second equality, i.e., $f_{A \cup B, C} = \min\{f_{AC}, f_{BC}\}$, is proved similarly. By Theorem 1 and definition of $B \cap C$, we have that

$$f_{A, B \cap C} = \min\{x, \inf\{\min_{u \in \mathcal{U}}\{B(u), C(u)\} \mid x \leq A(u)\}\}$$

and by properties of infimum and maximum we have:

$$f_{A \cup B, C} = \min\{x, \inf_{u \in \mathcal{U}} \{B(u) \mid x \leq A(u)\}, \inf_{u \in \mathcal{U}} \{C(u) \mid x \leq A(u)\}\}$$

which is equivalent to say $f_{A, B \cap C} = \min\{f_{AB}, f_{AC}\}$.

4 Conclusions and Future Work

We have studied the relationships between the f-index of inclusion presented in [16] and some axiomatic inclusion measures used commonly in the literature, namely, Sinha-Dougherty [18], Kitainik [13], Young [20] and Fan-Xie-Pei [11]. Despite the f-index cannot be considered, by definition, any of those inclusion measures, we show that it is very close to the Sinha-Dougherty axioms. Actually we show that for a finite universe all the axioms of Sinha-Dougherty (and therefore also those of Kitainik) hold except the one related to the complement (SD6). With respect to the complements, there is a natural relationship between the f-index of A in B and the one of B^c in A^c by means of Galois connections.

As future work it would be interesting to continue the motivation of the f-index of inclusion as a convenient representation of the relationship $A \subseteq B$. Moreover, it would be interesting also to establish relationships with the n-weak contradiction [3] and to define an f-index of similarity.

References

1. Bandler, W., Kohout, L.: Fuzzy power sets and fuzzy implication operators. Fuzzy Sets Syst. **4**(1), 13–30 (1980)
2. Burillo, P., Frago, N., Fuentes, R.: Inclusion grade and fuzzy implication operators. Fuzzy Sets Syst. **114**(3), 417–429 (2000)
3. Bustince, H., Madrid, N., Ojeda-Aciego, M.: The notion of weak-contradiction: definition and measures. IEEE Trans. Fuzzy Syst. **23**(4), 1057–1069 (2015)
4. Bustince, H., Mohedano, V., Barrenechea, E., Pagola, M.: Definition and construction of fuzzy DI-subsethood measures. Inf. Sci. **176**(21), 3190–3231 (2006)
5. Bustince, H., Pagola, M., Barrenechea, E.: Construction of fuzzy indices from fuzzy DI-subsethood measures: application to the global comparison of images. Inf. Sci. **177**(3), 906–929 (2007)
6. Cornelis, C., Van der Donck, C., Kerre, E.: Sinha-Dougherty approach to the fuzzification of set inclusion revisited. Fuzzy Sets Syst. **134**(2), 283–295 (2003)
7. De Baets, B., Meyer, H., Naessens, H.: A class of rational cardinality-based similarity measures. J. Comput. Appl. Math. **132**(1), 51–69 (2001)
8. De Baets, B., Meyer, H., Naessens, H.: On rational cardinality-based inclusion measures. Fuzzy Sets Syst. **128**(2), 169–183 (2002)
9. Deng, G., Jiang, Y., Fu, J.: Monotonic similarity measures between fuzzy sets and their relationship with entropy and inclusion measure. Fuzzy Sets Syst. **287**, 97–118 (2016)
10. Esmi, E.L., Sussner, P.: A fuzzy associative memory based on Kosko's subsethood measure. In: The 2010 International Joint Conference on Neural Networks (IJCNN), pp. 1–8 (2010)
11. Fan, J., Xie, W., Pei, J.: Subsethood measure: new definitions. Fuzzy Sets Syst. **106**(2), 201–209 (1999)
12. Fodor, J., Yager, R.R.: Fuzzy set-theoretic operators and quantifiers. In: Dubois, D., Prade, H. (eds.) Fundamentals of Fuzzy Sets. The Handbooks of Fuzzy Sets Series, vol. 7, pp. 125–193. Springer, Boston (2000). doi:10.1007/978-1-4615-4429-6_3
13. Kitainik, L.M.: Fuzzy inclusions and fuzzy dichotomous decision procedures. In: Kacprzyk, J., Orlovski, S.A. (eds.) Optimization Models Using Fuzzy Sets and Possibility Theory. Theory and Decision Library. Series B: Mathematical and Statistical Methods, vol. 4, pp. 154–170. Springer, Dordrecht (1987). doi:10.1007/978-94-009-3869-4_11
14. Kosko, B.: Fuzziness vs. probability. Int. J. Gen. Syst. **17**(2–3), 211–240 (1990)
15. Luca, A.D., Termini, S.: A definition of a nonprobabilistic entropy in the setting of fuzzy sets theory. Inf. Control **20**(4), 301–312 (1972)
16. Madrid, N., Ojeda-Aciego, M., Perfilieva, I.: f-inclusion indexes between fuzzy sets. In: Proceedings of IFSA-EUSFLAT (2015)
17. Scozzafava, R., Vantaggi, B.: Fuzzy inclusion and similarity through coherent conditional probability. Fuzzy Sets Syst. **160**(3), 292–305 (2009)
18. Sinha, D., Dougherty, E.R.: Fuzzification of set inclusion: theory and applications. Fuzzy Sets Syst. **55**(1), 15–42 (1993)
19. Tsiporkova-Hristoskova, E., De Baets, B., Kerre, E.: A fuzzy inclusion based approach to upper inverse images under fuzzy multivalued mappings. Fuzzy Sets Syst. **85**(1), 93–108 (1997)
20. Young, V.R.: Fuzzy subsethood. Fuzzy Sets Syst. **77**(3), 371–384 (1996)
21. Zadeh, L.: Fuzzy sets. Inf. Control **8**(3), 338–353 (1965)

An Integer 0-1 Linear Programming Approach for Computing Inconsistency Degree in Product-Based Possibilistic DL-Lite

Salem Benferhat[1], Khaoula Boutouhami[1,2], Faiza Khellaf[2], and Farid Nouioua[3(✉)]

[1] CRIL - CNRS UMR 8188, University of Artois, 62307 Lens, France
[2] RIIMA, University of Sciences and Technology Houari Boumediene, Bab Ezzouar, Algeria
[3] Aix-Marseille University, CNRS, ENSAM, University of Toulon, LSIS UMR 7296, Marseille, France
farid.nouioua@univ-amu.fr

Abstract. This paper considers the problem of computing inconsistency degree of uncertain knowledge bases expressed in product-based possibilistic DL-Lite, which is an extension of DL-Lite to deal with uncertainty in the product-based possibility theory framework. Indeed, computing the inconsistency degree is at the heart of any query answering process in such knowledge bases. Unlike previous work where uncertainty is only considered at the ABox level, in the present work both ABox and TBox may be uncertain. We discuss the new form of conflicts and how to obtain them by a generalized negative closure procedure. Then, we model the inconsistency degree computation as an integer 0-1 linear programming problem and we show the efficiency of this choice by a comparison with two other solutions, using the weighted Max-SAT and the approximate greedy algorithm for the weighted set cover problem, respectively.

Keywords: DL-Lite · Product-based possibility theory · Integer 0-1 linear programming · Inconsistency degree · Instance checking

1 Introduction

Ontologies play an important role in the success of the semantic web as they provide shared vocabularies for different resources and applications. Among the representation languages for ontologies, description logics (DLs) [3,16] are proven to be a successful formalism for representing and reasoning about knowledge thanks to their clear semantics and formal properties. Besides, despite its syntactical restrictions, the DL-Lite family [1,2,18] enjoys good computational properties while still offering interesting capabilities in representing terminological knowledge. That is why many works have been recently dedicated to this family and this paper is a contribution to this general research line.

In many practical applications, DL knowledge bases (KBs) may be inconsistent and/or uncertain. Handling inconsistency in DL is an active research topic

© Springer International Publishing AG 2017
S. Moral et al. (Eds.): SUM 2017, LNAI 10564, pp. 319–333, 2017.
DOI: 10.1007/978-3-319-67582-4_23

which led to several works on the notions of repairs (finding maximal consistent subontologies), axiom pinpointing (finding minimal inconsistent subontologies), and inconsistency-tolerant query answering (see [7, 10, 12]).

The problem of handling uncertainty in DLs has increasingly received attention during the last years. Many approaches have been proposed to extend DLs. Some of these extensions have been done in a probability theory framework (e.g., [11, 15]). Other extensions of description logics use fuzzy set theory (e.g., [8, 17]). Besides, the extensions based on a possibility theory framework have also received a lot of attention. These different extensions are adapted to different situations according to the nature of uncertainty present in data.

Possibility theory offers two major incomparable definitions [9]: min-based and product-based possibility theories. At the semantic level, these two theories share several basic definitions, including the concepts of possibility distributions and the necessity and possibility measures. However, they differ in the way they define conditioning and also in the way possibility degrees are defined over interpretations. Min-based possibility theory is appropriate when the uncertainty scale only encodes a plausibility ordering (a total pre-order) over interpretations. In min-based theory, the certainty degree of an interpretation only depends on the maximal certainty degree over the formulas falsified by this interpretation. Product-based possibility theory is appropriate when uncertainty degrees represent degrees of surprise (in the sense of Spohn ordinal conditional functions) or reflect the result of transforming a probability distribution into a possibility distribution. In these cases, the certainty degrees of all formulas falsified by an interpretation contribute to the determination of its possibility degree.

Most of existing works on possibilistic DL are based on the min-based possibility theory. For example, [13, 14] study min-based possibilistic extensions of general DLs and [4] focuses on min-based possibilistic DL-Lite and shows that it is done without extra computational cost. The product-based extension of DL-Lite has been recently considered in [5, 6]. However, these last works suppose that only ABox assertions may be uncertain while TBox axioms are considered stable and should not be questioned in the presence of inconsistencies. The contribution of this paper is to propose a product-based extension of DL-Lite where both the TBox and the ABox are uncertain. Moreover, the paper studies a new approach based on integer 0-1 linear programming to compute the inconsistency degree and to answer instance checking queries on product-based possibilistic DL-Lite KBs. To achieve our objective, the following problems will be considered: (1) defining the new notion of supported conflict which contains the relevant elements of a minimal conflict in the new setting; (2) extending the standard rules used to compute the negative closure of a DL-Lite KB [2] in order to compute the supported conflicts and (3) using integer 0-1 linear programming for computing the inconsistency degree of a KB and answering instance checking queries.

Section 2 introduces the syntax and the semantics of product-based possibilistic DL-Lite with uncertain ABox and TBox. In Sect. 3, we introduce the notion of supported conflict and we explain how to obtain them by extending the standard rules of negative closure. We also present in this section a new modeling of the inconsistency degree computation problem by an integer 0-1 linear program. Section 4 compares our approach with two other algorithms proposed in previous works. Finally, Sect. 5 concludes the paper.

2 Extended Framework: Product-Based Possibilistic DL-Lite with Uncertain ABox and TBox

Let us first briefly recall that DL-Lite is a family of description logics that aims to capture some of the most popular conceptual modeling formalisms. The syntax of the DL-Lite$_{core}$ language is defined as follows:

$$B \to A|\exists R \quad C \to B|\neg B$$
$$R \to P|P^- \quad E \to R|\neg R \tag{1}$$

where A denotes an atomic concept, P an atomic role, P^- the inverse of the atomic role P. B (resp. C) are called basic (resp. complex) concepts and R (resp. E) are called basic (resp. complex) roles.

The DL-Lite$_R$ language extends DL-Lite$_{core}$ with the ability of specifying in the TBox inclusion axioms between roles of the form: $R \sqsubseteq E$ while the DL-Lite$_F$ language extends DL-Lite$_{core}$ with the ability of specifying functionality on roles or on their inverses using axioms of the form: $(funct\ R)$.

A DL KB $\mathcal{K} = \langle \mathcal{T}, \mathcal{A} \rangle$ consists of a finite set $\mathcal{T}$ (called TBox) of inclusion axioms of the form: $B \sqsubseteq C$ and a finite set $\mathcal{A}$ (called ABox) of membership assertions of the form: $A(a)$ or $P(a,b)$ where a, b are individual names.

2.1 Syntax and Semantics of Product-Based Possibilistic DL-Lite

A product-based possibilistic DL-Lite KB (Pb-π-DL-Lite KB) $\mathcal{K} = \langle \mathcal{T}, \mathcal{A} \rangle$ consists of a product-based possibilistic TBox $\mathcal{T}$ and a product-based possibilistic ABox $\mathcal{A}$. A product-based possibilistic TBox (resp. ABox) is a finite set of possibilistic axioms of the form $\langle \phi_i, \alpha_i \rangle$, where ϕ_i is a DL-Lite TBox axiom (resp. a DL-Lite ABox assertion) and $\alpha_i \in [0,1]$ represents the certainty degree.

In the rest of the paper, for a set of weighted TBox axioms $\mathcal{T}'$ (resp. weighted ABox assertions $\mathcal{A}'$), we denote by $\mathcal{T}'_u$ (resp. $\mathcal{A}'_u$) the set of the corresponding unweighted axioms (resp. unweighted assertions).

Example 1. *Consider the following Pb-π-DL-Lite KB $\mathcal{K} = \langle \mathcal{T}, \mathcal{A} \rangle$ where:*

$\mathcal{T} = \{$ $\langle \exists teachesTo \sqsubseteq Professor, 0.79 \rangle,$ $\langle \exists hasTutor \sqsubseteq Student, 0.77 \rangle,$

$\langle \exists teachesTo^- \sqsubseteq Student, 0.75 \rangle,$ $\langle \exists hasTutor^- \sqsubseteq Professor, 0.9 \rangle,$

$\langle Professor \sqsubseteq \neg Student, 1.0 \rangle \}.$

$\mathcal{A} = \{$ $\langle Professor(b), 0.95 \rangle,$ $\langle Student(a), 0.9 \rangle,$

$\langle teachesTo(a,b), 0.87 \rangle,$ $\langle hasTutor(b,a), 0.65 \rangle \}.$

Student and Professor are two concepts while teachesTo and hasTutor are roles. For instance the assertional fact $\langle Professor(b), 0.95 \rangle$ states that the individual b is a professor with the certainty degree of 0.95.

322   S. Benferhat et al.

The semantics of a Pb-π-DL-Lite KB is given as usual by the concept of a possibility distribution $\pi_\mathcal{K}$ over the set of all interpretations Ω of the underlying DL-Lite language: $\pi_\mathcal{K} : \Omega \rightarrow [0,1]$. A possibility distribution π is said to be normalized if there exists at least one totally possible interpretation. Namely, $\exists I \in \Omega$ such that $\pi(I) = 1$. Otherwise, we say that π is sub-normalized. In this case, the expression: $h(\pi) = \max_{I \in \Omega} \pi(I)$ is called the consistency degree of π. If an interpretation I is a model of each axiom of $\mathcal{T}$ and each assertion of $\mathcal{A}$ then its possibility degree is equal to 1. This reflects the fact that I is fully compatible with $\langle \mathcal{T}, \mathcal{A} \rangle$. It also obviously means that $\langle \mathcal{T}, \mathcal{A} \rangle$ is consistent. More generally, if an interpretation I falsifies some assertions of the ABox or axiom of the TBOX or both, then its possibility degree is inversely proportional to the product of the weights of the assertions and axioms that it falsifies. More formally:

Definition 1. *For all $I \in \Omega$,*

$$\pi_K(I) = \begin{cases} 1 & if \ \forall \langle \phi_i, \alpha_i \rangle \in \mathcal{K}, I \models \phi_i \\ \prod_{\langle \phi_i, \alpha_i \rangle \in \mathcal{K}, I \nvDash \phi_i} (1 - \alpha_i) & otherwise \end{cases} \tag{2}$$

s.t. $\models$ is the satisfaction relation between DL-Lite interpretations and formulas.

The possibility distribution defined by Eq. 2 represents the least specific possibility distribution satisfying $\mathcal{K}$.

Example 1 (Cont). Consider again Example 1. Here are the possibility degrees, obtained by Definition 1, for three interpretations over the domain $\triangle = \{a, b\}$.

I	$\cdot^I$	$\pi_\mathcal{K}(I)$
I_1	$Professor^I = \{a\}, Student^I = \{b\}, hasTutor^I = \{(a,b)\}$ $teachesTo^I = \{(b,a)\}$	0.001
I_2	$Professor^I = \{a\}, Student^I = \{b\}, hasTutor^I = \{(b,a)\}$ $teachesTo^I = \{(a,b)\}$	0.007
I_3	$Professor^I = \{b,a\}, Student^I = \{a,b\}, hasTutor^I = \{(a,b)\}$ $teachesTo^I = \{(b,a)\}$	0

For instance, the interpretation I_1 falsifies the assertions $\langle teachesTo(a,b), 0.87 \rangle$ and $\langle hasTutor(b,a), 0.65 \rangle$. Thus, its possibility degree is $\pi_\mathcal{K}(I_1) = (1 - 0.87) \times (1 - 0.65) = 0.001$. Note that $\pi_\mathcal{K}(I_3) = 0$ since it falsifies the certain axiom $\langle Professor \sqsubseteq \neg Student, 1.0 \rangle$. None of the three interpretations is a model of $\mathcal{K}$.

2.2 Inconsistency Degree

Inconsistency in DL-Lite (as well as in most logical languages) corresponds to the absence of an interpretation satisfying all formulas of the KB. A Pb-π-DL-Lite KB $\mathcal{K}$ is said to be fully consistent if there exists an interpretation I such that $\pi_\mathcal{K}(I) = 1$. Otherwise, $\mathcal{K}$ is said to be somewhat inconsistent. More formally:

Definition 2 *Let Ω be the set of all possible interpretations. Let $\mathcal{K}$ be a Pb-π-DL-Lite KB and $\pi_{\mathcal{K}}$ be the possibility distribution induced by $\mathcal{K}$ and obtained by Definition 1. The inconsistency degree of $\mathcal{K}$, denoted by $Inc(\mathcal{K})$, is semantically defined as follows:*

$$Inc(\mathcal{K}) = 1 - \max_{I \in \Omega}(\pi_{\mathcal{K}}(I)). \tag{3}$$

The inconsistency degree $Inc(\mathcal{K})$ is the dual of the normalized degree. The duality between $Inc(\mathcal{K})$ and $h(\pi)$ is expressed by: $Inc(\mathcal{K}) = 1 - h(\pi)$. Namely, the inconsistency degree of a DL-Lite KB is given by the inverse of the normalisation degree of the possibility distribution π associated with $\mathcal{K}$.

Example 1 (Cont). *One can show that the inconsistency degree of the KB $\mathcal{K}$, presented in Example 1 is: $Inc(\mathcal{K}) = 1 - \max_{I \in \Omega}(\pi_{\mathcal{K}}(I)) = 0.983$.*

The problem of standard query answering is closely related to the ontology-based data access problem which takes as input a set of assertions, an ontology and a conjunctive query q and aims to find all answers to q over the set of data. As pointed out in [6], the inconsistency degree plays a central role in answering such queries in Pb-π-DL-Lite. We focus here on a very basic case in query answering which is instance checking. The instance checking problem, in standard DL-Lite consists in deciding, given an individual a (resp. a pair of individuals (a,b)) a concept B (resp. a role R) and a DL-Lite KB $\mathcal{K} = \langle \mathcal{T}, \mathcal{A} \rangle$, whether $B(a)$ (resp. $R(a,b)$) follows from $\langle \mathcal{T}, \mathcal{A} \rangle$. For an instance checking query q, let us recall that its necessity degree denoted by $N_{\mathcal{K}}(q)$ is given by:

$$N_{\mathcal{K}}(q) = 1 - \max_{I \in \Omega, I \not\models q}(\pi_{\mathcal{K}}(I)) \tag{4}$$

and the satisfaction of the query q by the KB $\mathcal{K}$ is defined by:

$$\mathcal{K} \models_{\pi} q \text{ if and only if } N_{\mathcal{K}}(q) > Inc(\mathcal{K}) \tag{5}$$

The following result shown in [6] for Pb-π-DL-Lite KBs with certain TBox continues to hold in our case where the TBox may be uncertain. This result states that answering an instance checking query q from a Pb-π-DL-Lite KB $\mathcal{K}$ comes down to compare the inconsistency degrees of two Pb-π-DL-Lite KBs: $\mathcal{K}$ itself and a new KB $\mathcal{K}_1$ which results from $\mathcal{K}$ by adding to it the fact that q is surely false. Notice that we limit ourselves here to answer only unweighted queries, i.e., we determine if a query q is a plausible conclusion of a KB without determining to what extent q follows from this KB. This issue is left for future work.

Proposition 1. *Let $\mathcal{K} = \langle \mathcal{T}, \mathcal{A} \rangle$ be a Pb-π-DL-Lite KB, $\pi_{\mathcal{K}}$ be the possibility distribution associated to $\mathcal{K}$ and $N_{\mathcal{K}}$ the corresponding necessity distribution. Let B be a concept (resp. R be a role) and a, b be two individuals. Let D_B (resp. D_R) be an atomic concept (resp. an atomic role) not appearing in $\mathcal{T}$. Then:*

1. $N_{\mathcal{K}}(B(a)) = Inc(\mathcal{K}_1)$ *(resp. $N_{\mathcal{K}}(R(a,b)) = Inc(\mathcal{K}_1)$) where $\mathcal{K}_1 = \langle \mathcal{T}_1, \mathcal{A}_1 \rangle$ with $\mathcal{T}_1 = \mathcal{T} \cup \{(D_B \sqsubseteq \neg B, 1)\}$ (resp. $\mathcal{T}_1 = \mathcal{T} \cup \{(D_R \sqsubseteq \neg R, 1)\}$) and $\mathcal{A}_1 = \mathcal{A} \cup \{(D_B(a), 1)\}$ (resp. $\mathcal{A}_1 = \mathcal{A} \cup \{(D_R(a,b), 1)\}$).*

2. $\mathcal{K} \models_\pi B(a)$ *(resp.* $\mathcal{K} \models_\pi R(a,b)$*) if* $Inc(\mathcal{K}_1) > Inc(\mathcal{K})$.

Example 1 (Cont). *Take again the KB of Example 1. Consider the query:* $q \leftarrow Professor(a)$. *We introduce a new concept* D_{Prof} *and we construct the new knowledge base* $\mathcal{K}_1 = \langle \mathcal{T}_1, \mathcal{A}_1 \rangle$ *with* $\mathcal{T}_1 = \mathcal{T} \cup \{\langle D_{Prof} \sqsubseteq \neg Professor, 1.0 \rangle\}$ *and* $\mathcal{A}_1 = \mathcal{A} \cup \{\langle D_{Prof}(a), 1.0 \rangle\}$.

We have already shown that $Inc(\mathcal{K}) = 0.983$. *Now, one can check, by using Definition 2, that the inconsistency degree of the augmented base* $\mathcal{K}_1$ *is* $Inc(\mathcal{K}_1) = Inc(\mathcal{K}) = 0.983$. *Hence, q is not a consequence of* $\mathcal{K}$.

3   Computing the Inconsistency Degree

3.1   Supported Conflicts and Pb-based Negative Closure

In a Pb-π DL-Lite setting, computing the inconsistency degree of a KB comes down to determine a subset of the axioms and assertions of the KB such that: (i) removing this subset allows one to restore the consistency of the KB and (ii) the possibility degree of the corresponding interpretation that falsifies exactly the removed formulas (in the sense of Eq. 3) is maximal. In [5,6], the TBox axioms are considered certain, i.e., every TBox axiom has a certainty degree equal to 1. Hence, breaking the conflicts of a KB may only be done by removing ABox assertions. Consequently, to compute the inconsistency degree, one only needs to compute the binary assertional conflicts that can be obtained by simply applying the standard negated closure defined for standard DL-Lite KBs [2]. In the present paper, both ABox assertions and TBox axioms may be uncertain. So, in addition to removing ABox assertions, breaking the conflicts of a KB may also be done by removing TBox axioms that are not fully certain. Notice that every negative axiom $C \sqsubseteq \neg D$ of $Cln(\mathcal{T})$ may be derived by applying at least one sequence of standard rules which involve a set of TBox axioms containing exactly one negative axiom and a subset (possibly empty if $C \sqsubseteq \neg D \in \mathcal{T}$) of positive axioms. Therefore, for every binary conflict based on a negative axiom $C \sqsubseteq \neg D$ of $Cln(\mathcal{T})$, we need to explicit all the different supports in $\mathcal{T}$ of this negative axiom, i.e., the subsets of $\mathcal{T}$ that allow us to derive this negative axiom. This leads to the introduction of the notion of *supported conflict* defined as follows:

Definition 3. *Let* $\mathcal{K} = \langle \mathcal{T}, \mathcal{A} \rangle$ *be a Pb-π-DL-Lite KB. Let* $\langle C(\overrightarrow{a}), \alpha_1 \rangle^1$ *and* $\langle D(\overrightarrow{a}), \alpha_2 \rangle$ *be two ABox assertions such that each of C and D is either a concept or a role. Let* $\mathcal{T}' \subseteq \mathcal{T}$ *be a subset of TBox axioms containing exactly one negative axiom.* $\mathcal{C} = \{\langle C(\overrightarrow{a}), \alpha_1 \rangle, \langle D(\overrightarrow{a}), \alpha_2 \rangle\} \cup \mathcal{T}'$ *is a supported conflict in* $\mathcal{K}$ *if and only if: (1)* $\langle \mathcal{T}'_u, \{C(\overrightarrow{a}), D(\overrightarrow{a})\} \rangle$ *is inconsistent and (2) for every subsets* $\mathcal{T}''_u$ *of* $\mathcal{T}'_u$ *and* $\mathcal{A}'_u$ *of* $\{C(\overrightarrow{a}), D(\overrightarrow{a})\}$ *such that* $\mathcal{T}''_u \subset \mathcal{T}'_u$ *or* $\mathcal{A}'_u \subset \{C(\overrightarrow{a}), D(\overrightarrow{a})\}$, $\langle \mathcal{T}''_u, \mathcal{A}'_u \rangle$ *is consistent. We call* $\mathcal{T}'$ *the support of* $\mathcal{C}$ *and we write:* $\mathcal{T}' = support(\mathcal{C})$.

---

[1] When we write $C(\overrightarrow{a})$, $\overrightarrow{a}$ is an individual if C is a concept and a pair of individuals if C is a role.

Notice that supported conflicts are minimal conflicts and since inconsistency is due to negative axioms, a supported conflict contains only one negative axiom. In standard DL-Lite, conflicts are obtained by computing the negative closure $Cln(\mathcal{T})$ of the Tbox $\mathcal{T}$ containing all the negative axioms that follow from $\mathcal{T}$ [2]. In our setting, the Pb-based negative closure of $\mathcal{T}$, denoted $Pb\_Cln(\mathcal{T})$, consists of a set of pairs of the form $[Axiom, Support]$ where: $Axiom$ is an unweighted negative axiom such that $Axiom \in Cln(\mathcal{T}_u)$ and $Support \subseteq \mathcal{T}$ is a minimal (w.r.t. set inclusion) subset of (weighted) axioms allowing one to derive $Axiom$. More precisely, the set of adapted rules used to generate the product based possibilistic negative closure $Pb\_Cln(\mathcal{T})$ of the product based possibilistic TBox $\mathcal{T}$ are given as follows:

1. For every negative axiom $\langle \Phi, \alpha \rangle$ in $\mathcal{T}$, add $[\Phi, \{\langle \Phi, \alpha \rangle\}]$ to $Pb\_Cln(\mathcal{T})$.
2. If $\langle B_1 \sqsubseteq B_2, \alpha \rangle$ is in $\mathcal{T}$ then,
   for every $[B_2 \sqsubseteq \neg B_3, Support]$ or $[B_3 \sqsubseteq \neg B_2, Support]$ in $Pb\_Cln(\mathcal{T})$, add $[B_1 \sqsubseteq \neg B_3, Support \cup \{\langle B_1 \sqsubseteq B_2, \alpha \rangle\}]$ to $Pb\_Cln(\mathcal{T})$.
3. If $\langle R_1 \sqsubseteq R_2, \alpha \rangle$ is in $\mathcal{T}$ then,
   for every $[\exists R_2 \sqsubseteq \neg B, Support]$ or $[B \sqsubseteq \neg \exists R_2, Support]$ in $Pb\_Cln(\mathcal{T})$, add $[\exists R_1 \sqsubseteq \neg B, Support \cup \{\langle R_1 \sqsubseteq R_2, \alpha \rangle\}]$ to $Pb\_Cln(\mathcal{T})$.
4. If $\langle R_1 \sqsubseteq R_2, \alpha \rangle$ is in $\mathcal{T}$ then,
   for every $[\exists R_2^- \sqsubseteq \neg B, Support]$ or $[B \sqsubseteq \neg \exists R_2^-, Support]$ in $Pb\_Cln(\mathcal{T})$, add $[\exists R_1^- \sqsubseteq \neg B, Support \cup \{\langle R_1 \sqsubseteq R_2, \alpha \rangle\}]$ to $Pb\_Cln(\mathcal{T})$.
5. If $\langle R_1 \sqsubseteq R_2, \alpha \rangle$ is in $\mathcal{T}$ then,
   for every $[R_2 \sqsubseteq \neg R_3, Support]$ or $[R_3 \sqsubseteq \neg R_2, Support]$ in $Pb\_Cln(\mathcal{T})$, add $[R_1 \sqsubseteq \neg R_3, Support \cup \{\langle R_1 \sqsubseteq R_2, \alpha \rangle\}]$ to $Pb\_Cln(\mathcal{T})$.
6. For every functionality axiom $\langle funct\ R, \alpha \rangle$ in $\mathcal{T}$, add $[funct\ R, \{\langle funct\ R, \alpha \rangle\}]$ to $Pb\_Cln(\mathcal{T})$.
7. (a) For DL-Lite$_R$: If one of the expressions $[\exists R \sqsubseteq \neg \exists R, Support]$, $[\exists R^- \sqsubseteq \neg \exists R^-, Support]$ or $[R \sqsubseteq \neg R, Support]$ is in $Pb\_Cln(\mathcal{T})$, then add all these three expressions to $Pb\_Cln(\mathcal{T})$.
   (b) For DL-Lite$_F$: if one of the expressions $[\exists R \sqsubseteq \neg \exists R, Support]$ or $[\exists R^- \sqsubseteq \neg \exists R^-, Support]$ is in $Pb\_Cln(\mathcal{T})$, then add these two expressions to $Pb\_Cln(\mathcal{T})$.

Then, an expression $[Axiom, Support]$ may give size to a conflict: $\{\langle C(\overrightarrow{a}), \alpha \rangle, \langle D(\overrightarrow{a}), \beta \rangle\} \cup Support$ where $\langle C(\overrightarrow{a}), \alpha \rangle, \langle D(\overrightarrow{a}), \beta \rangle \in \mathcal{A}$ and $C(\overrightarrow{a}), D(\overrightarrow{a})$ are extracted from $\mathcal{A}_u$ based on $Axiom$ as in the standard case (see e.g. [2,4]).

Example 1 (Cont). *Consider again our KB* $\mathcal{K}$, $Pb\_Cln(\mathcal{T})$ *is the following:*

$Pb\_Cln(\mathcal{T}) = \{ \; [Professor \sqsubseteq \neg Student, \{\langle Professor \sqsubseteq \neg Student, 1.0\rangle\}]$

$\quad [\exists TeachesTo \sqsubseteq \neg Student, \{\langle \exists TeachesTo \sqsubseteq Professor, 0.89\rangle,$
$\quad \langle Professor \sqsubseteq \neg Student, 1.0\rangle\}]$

$\quad [\exists HasTutor^- \sqsubseteq \neg Student, \{\langle \exists HasTutor^- \sqsubseteq Professor, 0.9\rangle,$
$\quad \langle Professor \sqsubseteq \neg Student, 1.0\rangle\}]$

$\quad [\exists TeachesTo^- \sqsubseteq \neg Professor, \{\langle \exists teachesTo^- \sqsubseteq Student, 0.85\rangle,$
$\quad \langle Professor \sqsubseteq \neg Student, 1.0\rangle\}]$

$\quad [\exists HasTutor \sqsubseteq \neg Professor, \{\langle \exists HasTutor \sqsubseteq Student, 0.95\rangle$
$\quad \langle Professor \sqsubseteq \neg Student, 1.0\rangle\}]\}.$

The set of supported conflicts is $\zeta = \{\mathcal{C}_1, \mathcal{C}_2, \mathcal{C}_3, \mathcal{C}_4\}$ *such that:*

$\mathcal{C}_1 = \{\langle Professor(b), 0.95\rangle, \langle teachesTo(a,b), 0.98\rangle, \langle \exists teachesTo^- \sqsubseteq Student, 0.77\rangle,$
$\quad \langle Professor \sqsubseteq \neg Student, 1.0\rangle\}$

$\mathcal{C}_2 = \{\langle Professor(b), 0.95\rangle, \langle HasTutor(b,a), 0.65\rangle, \langle \exists HasTutor \sqsubseteq Student, 0.75\rangle,$
$\quad \langle Professor \sqsubseteq \neg Student, 1.0\rangle\}$

$\mathcal{C}_3 = \{\langle Student(a), 0.9\rangle, \langle teachesTo(a,b), 0.98\rangle, \langle \exists TeachesTo \sqsubseteq Professor, 0.79\rangle,$
$\quad \langle Professor \sqsubseteq \neg Student, 1.0\rangle\}$

$\mathcal{C}_4 = \{\langle Student(a), 0.9\rangle, \langle HasTutor(b,a), 0.65\rangle, \langle \exists HasTutor^- \sqsubseteq Professor, 0.9\rangle,$
$\quad \langle Professor \sqsubseteq \neg Student, 1.0\rangle\}$

A subset to remove in order to break all the conflicts while ensuring a maximal possibility degree of the corresponding interpretation is: $\{\langle HasTutor(b,a), 0.65\rangle,$ $\langle \exists teachesTo^- \sqsubseteq Student, 0.77\rangle, \langle \exists TeachesTo \sqsubseteq Professor, 0.79\rangle\}$. *The inconsistency degree of* $\mathcal{K}$ *is then:* $Inc(\mathcal{K}) = 1 - (1 - 0.65)(1 - 0.77)(1 - 0.79) = 0.983.$

3.2 Inconsistency Degree Computation as an Integer 0-1 Linear Program

As explained above, the computation of the inconsistency degree of a Pb-π-DL-Lite KB is done by determining a subset of assertions and axioms such that: (i) removing this subset from the KB breaks all the supported conflicts present in it and hence restores its consistency and (ii) this subset ensures a maximal value of the possibility degree of the corresponding interpretation (which falsifies exactly the removed formulas). In this subsection, we provide an encoding of the problem of computing the inconsistency degree of a Pb-π-DL-Lite KB as an integer linear program (ILP). Notice that this approach is not specific to the DL-Lite setting and can be used in any possibilistic KB by considering minimal inconsistent subsets instead of supported conflicts. Integer linear programming is like linear programming, with the additional constraint that all variables must take integer values.

Definition 4 *Let $\mathcal{K}$ be a Pb-π-DL-Lite KB, $\zeta = \{\mathcal{C}_1, \ldots, \mathcal{C}_m\}$ be the set of all supported conflicts present in $\mathcal{K}$, $\Gamma = \{\langle \Phi_1, \alpha_1 \rangle, \ldots, \langle \Phi_n, \alpha_n \rangle\}$ be all the non fully certain weighted formulas (axioms and assertions) involved in ζ. Let F be a scale changing function defined by[2]: $F(y) = -(ln(1-y))$. If there is no supported conflict $\mathcal{C} \in \zeta$ such that $\Gamma \cap \mathcal{C} = \emptyset$, then the computation of the inconsistency degree of $\mathcal{K}$ is modeled by the integer linear program $LP_{\mathcal{K}}$ defined by:*

- *Each formula $\langle \Phi_i, \alpha_i \rangle \in \Gamma$ is represented by a binary variable x_i such that in the solution, $x_i = 1$ if $\langle \Phi_i, \alpha_i \rangle$ should be removed and $x_i = 0$ otherwise.*
- *The aim is to minimize the following objective function, expressing the total weight of removed formulas:*

$$\sum_{i=1}^{n} F(\alpha_i).x_i$$

- *Each supported conflict $\mathcal{C}_j$ gives rise to a constraint C_j which says that this supported conflict should be broken by removing at least one of the formulas it contains, i.e., at least one formula $\langle \Phi_i, \alpha_i \rangle \in \mathcal{C}_j$ is such that its representative variable x_i equals 1:*

$$C_j : \sum_{\langle \Phi_i, \alpha_i \rangle \in \mathcal{C}_j} x_i \geq 1$$

Remark 1 The construction of $LP_{\mathcal{K}}$ is conditioned in Definition 4 by the fact that there is no supported conflict $\mathcal{C} \in \zeta$ such that $\Gamma \cap \mathcal{C} = \emptyset$ because in this case $Inc(\mathcal{K})$ can be computed without constructing $LP_{\mathcal{K}}$. Indeed, in this case the supported conflict $\mathcal{C}$ contains only certain formulas, i.e., $\mathcal{C}$ cannot be broken since certain formulas can't be removed. So, every interpretation falsifies at least a formula from $\mathcal{C}$. This simply means that $Inc(\mathcal{K}) = 1$.

Let us show how the ILP is obtained from our running example.

Example 1 (Cont). The ILP obtained by Definition 4 for our KB $\mathcal{K}$ is as follows:

Minimize: $F(0.8).x_1 + F(0.7).x_2 + F(0.85).x_3 + F(0.5).x_4 +$
              $F(0.95).x_5 + F(0.9).x_6 + F(0.89).x_7 + F(0.9).x_8$

subject to: $C_1 : x_1 + x_2 + x_3 \geq 1$
               $C_2 : x_1 + x_4 + x_5 \geq 1$
               $C_3 : x_6 + x_2 + x_7 \geq 1$
               $C_4 : x_6 + x_4 + x_8 \geq 1$
               $x_i \in \{0,1\} \ (1 \leq i \leq 9)$

---

[2] Here, ln denotes the natural logarithm function.

where: x_1 corresponds to $\langle Professor(b), 0.8 \rangle$, x_2 to $\langle t\acute{e}achesTo(a, b), 0.7 \rangle$, x_3 to $\langle \exists teachesTo^- \sqsubseteq Student, 0.85 \rangle$, x_4 to $\langle HasTutor(b, a), 0.5 \rangle$, x_5 to $\langle \exists HasTutor \sqsubseteq Student, 0.95 \rangle$, x_6 to $\langle Student(a), 0.9 \rangle$, x_7 to $\langle \exists TeachesTo \sqsubseteq Professor, 0.89 \rangle$ and x_8 to $\langle \exists HasTutor^- \sqsubseteq Professor, 0.9 \rangle$.

In this work, we have used the commercial linear programming tool Cplex (IBM) which is an optimization software developed and sold by ILOG, Inc.

The solution of our ILP consists in putting x_3, x_4 and x_7 to 1 and all the other variables to 0. We obtain $Inc(K) = F^{-1}(F(\alpha_{x_3}) + F(\alpha_{x_4}) + F(\alpha_{x_7})) = 0.983$.

Now, the following proposition states the soundness and completeness of the previous encoding.

Proposition 2. *Let $\mathcal{K}$ be a Pb-π-DL-Lite KB and LP_K be the corresponding integer linear program given by Definition 4. It holds that $Inc(\mathcal{K}) = \lambda$ if and only if $F(\lambda)$ is the value of the objective function of an optimal solution of $LP_{\mathcal{K}}$.*

Based on Definition 4 and Proposition 2, $Inc(\mathcal{K})$ is computed as follows: Given a Pb-π-DL-Lite KB $\mathcal{K}$ as input, to compute $Inc(\mathcal{K})$. Firstly, the set ζ of supported conflicts present in this KB is computed. Then, this set of supported conflicts is transformed into an ILP by using Definition 4. Lastly, the inconsistency degree is computed by a call to a linear programming solver.

4   Experimental Results

The aim of this section is to evaluate the performance of our algorithm based on integer linear programming and using the Cplex (IBM) tool[3] for computing the inconsistency degrees of Pb-π-DL-Lite KBs. For that purpose, we compare this algorithm with two other algorithms proposed in previous work, namely:

- the first algorithm is an exact one which generalizes the encoding used in [6] to represent the set of supported conflicts as a weighted-Max-SAT base;
- the second algorithm is approximate. It is based on an encoding of our problem as a weighted set cover problem (W-SCP) in a similar way as that used in [5] for the particular case where only the ABox may be uncertain. The well-known polynomial greedy algorithm is adapted to our general case and can be used to compute an approximate value of the inconsistency degree.

We have conducted two experiments. The first one aims at assessing the quality of the results obtained by the approximate greedy algorithm for W-SCP in computing the inconsistency degree and answering instance checking queries.

- we have randomly generated 400 Pb-π-DL-Lite KBs. For each KB $\mathcal{K}$, we computed the gap between the exact value of $Inc(\mathcal{K})$ and its approximate value found by the greedy algorithm. The average error of the greedy algorithm equals to **0.0023** which confirms the very good quality of the approximate inconsistency degrees computed by this algorithm;

---

[3] ILOG. Cplex 10.0.: http://www.ilog.com/products/cplex/.

- we have randomly generated 400 pairs $(\mathcal{K}, query)$, where $\mathcal{K}$ is a Pb-π-DL-Lite KB and $query$ is an instance checking query. For each pairs $(\mathcal{K}, query)$ the augmented KB $\mathcal{K}_1$ is constructed as in Proposition 1. Then the answer is obtained by comparing the approximate values of $Inc(\mathcal{K})$ and $Inc(\mathcal{K}_1)$ obtained by using the greedy algorithm of W-SCP. The percentage of cases where the found answer coincides with the exact one is **81,26%**.

In the second experiment, we have compared the three algorithms in terms of execution time necessary to compute the inconsistency degree. More precisely:

- we generated 588 random instances of Pb-π-DL-Lite KBs. We found that the number of conflicts in the generated KBs varies between 1 and 2365;
- for each KB $\mathcal{K}$, we have considered the time needed to compute $Inc(\mathcal{K})$ using Clpex, W-Max-SAT and the greedy approximate algorithm.

The obtained results are given in Fig. 1. The X-axis represents the number of conflicts in the input KBs. For a given x, the corresponding y in the Y-axis is the execution time in seconds.

The experimental results are very encouraging: We can observe that the approximate greedy algorithm for W-SCP is faster than the exact algorithm using W-Max-SAT which is not surprising. We observe that Cplex also performs better than W-Max-SAT. Finally, it is clear that Cplex and the approximate greedy algorithm are very competitive and their performance are very close: The average execution time for the whole set of generated KBs is **0.119 s** for

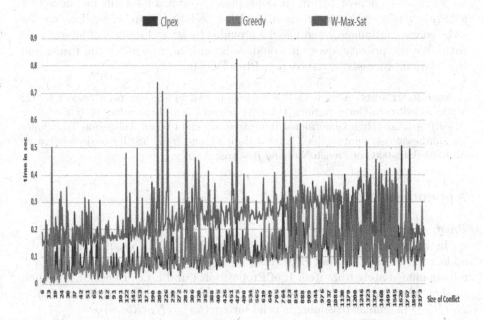

Fig. 1. Comparison between Clpex, W-Max-Sat and Greedy algorithm

Cplex and **0.117** for the approximate greedy algorithm. This is very interesting since it shows that, in addition to the fact that Cplex gives exact values, it achieves in practice a performance comparable to a polynomial algorithm even if its theoretical complexity in the worst case remains exponential. This makes Cplex a very good choice for computing inconsistency degrees of Pb-π-DL-Lite KBs and hence to query-answering in such KBs.

5   Conclusions and Future Work

This paper proposes an approach to represent uncertain lightweight ontologies encoded in Pb-π-DL-Lite, an extension of DL-Lite to represent uncertainty in product-based possibilistic theory. In particular, unlike previous work where only ABox assertions may be uncertain, this paper studies the case where both ABox and TBox may be uncertain. After the presentation of the syntax and semantics of the proposed language, the paper focuses on the notion of inconsistency degree since it is at the heart of the query answering process in Pb-π-DL-Lite setting. We propose the modeling of inconsistency degree computation as an integer linear program and the use of Cplex solver as a practical tool for computing inconsistency degree. To validate this choice, we show experimentally that in terms of execution time, Cplex outperforms W-Max-Sat solver and is very close to the polynomial greedy algorithm of weighted set cover problem which only gives approximate solutions. These results show clearly that Cplex is a good candidate solver for query answering in Pb-π-DL-Lite.

We plan in a near future to implement a complete system for answering arbitrary conjunctive queries on Pb-π-DL-Lite KBs. Then, since data on the web evolves continuously and may be provided by several sources of information with different priority levels, it would be interesting to explore the fusion and the revision processes in the context Pb-π DL-Lite.

Acknowledgments. This work has been supported by the european project H2020 Marie Sklodowska-Curie Actions (MSCA) research and Innovation Staff Exchange (RISE): AniAge (High Dimensional Heterogeneous Data based Animation Techniques for Southeast Asian Intangible Cultural Heritage and from ASPIQ project reference ANR-12-BS02-0003 of French National Research Agency.

Appendix: Proofs

Proof. of Proposition 1.

In Proposition 1, $\mathcal{K}_1$ represents a Pb-π-DL-Lite KB obtained from $\mathcal{K}$ by adding the assumption that $B(a)$ (resp. $R(a,b)$) is surely false. Note that item 2. follows immediately from item 1. of Proposition 1 and Eq. 5. Hence, it is enough to show that item 1. of Proposition 1 holds. So, let us show that $Inc(\mathcal{K}_1) = N_{\mathcal{K}}(B(a))$ (a similar reasoning is valid for $Inc(\mathcal{K}_1) = N_{\mathcal{K}}(R(a,b))$).

Let I be a DL-lite interpretation. Since $\mathcal{K}_1$ is composed of $\mathcal{K} \cup \{(D \sqsubseteq \neg B, 1), (D(a), 1)\}$, we have two cases:

- If $I \models D(a)$ and $I \models D \sqsubseteq \neg B$ (Hence $I \nvDash B(a)$) then $\pi_{\mathcal{K}_1}(I) = \pi_{\mathcal{K}}(I)$.
- If $I \nvDash D(a)$ or $I \nvDash D \sqsubseteq \neg B$ then $\pi_{\mathcal{K}_1}(I) = 0$.

Hence, we have: $\max_{I \in \Omega}(\pi_{\mathcal{K}_1}(I)) = \max_{I \in \Omega, I \nvDash B(a)}(\pi_{\mathcal{K}}(I))$. It follows from Eqs. 3 and 4 that $Inc(\mathcal{K}_1) = 1 - \max_{I \in \Omega, I \nvDash B(a)}(\pi_{\mathcal{K}}(I)) = N_{\mathcal{K}}(B(a))$ ∎

Proof. of Proposition 2.

Let K be a Pb-π-DL-Lite KB, $LP_{\mathcal{K}}$ be the corresponding integer linear program defined as in Definition 4. We consider that the formulas (axioms and assertions) of $\mathcal{K}$ are indexed from 1 to $n = |\mathcal{T} \cup \mathcal{A}|$. The formula $\langle \Phi_i, \alpha_i \rangle$ in $\mathcal{K}$ corresponds to the variable of index i in $LP_{\mathcal{K}}$.

⇒) Suppose that $Inc(\mathcal{K}) = \lambda$. Then by Definition 3, $\lambda = 1 - \max_{I \in \Omega}(\pi_{\mathcal{K}}(I))$. Let I_1 be an interpretation such that $\pi_{\mathcal{K}}(I_1) = \max_{I \in \Omega}(\pi_{\mathcal{K}}(I))$.

Let J be the set of indices of formulas falsified by I_1, i.e., $J = \{i \mid I_1 \nvDash \langle \Phi_i, \alpha_i \rangle\}$. By Definition 2, $\pi_{\mathcal{K}}(I_1) = \prod_{i \in J}(1 - \alpha_i)$ and hence $\lambda = 1 - \pi_{\mathcal{K}}(I_1)$.

Let $Y = (y_1, \ldots, y_n)$ be a binary vector such that $y_i = 1$ if $i \in J$ and $y_i = 0$ otherwise. Let us show that Y is an optimal solution of $LP_{\mathcal{K}}$ whose objective function is $F(\lambda)$.

Since for all $i \notin J$, $I_1 \models \langle \Phi_i, \alpha_i \rangle$, it is clear that the set $S = \{\langle \Phi_i, \alpha_i \rangle \in \mathcal{T} \cup \mathcal{A} \mid i \notin J\}$ is consistent and hence does not contain any supported conflict. This means that for every supported conflict $\mathcal{C}_j$ there is a formula $\langle \Phi_i, \alpha_i \rangle \in \mathcal{C}_j$ such that $i \in J$, i.e., $y_i = 1$. It follows that the corresponding constraint C_j is satisfied by Y.

The objective function of Y is: $\sum_{i=1}^n F(\alpha_i).y_i = \sum_{i \in J} F(\alpha_i) = \sum_{i \in J} -ln(1 - \alpha_i) = -ln(\prod_{i \in J}(1 - \alpha_i)) = F(1 - \prod_{i \in J}(1 - \alpha_i)) = F(1 - \pi_{\mathcal{K}}(I_1)) = F(\lambda)$.

Now, to show that Y is an optimal solution, suppose for the sake of contradiction that there is a solution $Z = (z_1, \ldots, z_n)$ of $LP_{\mathcal{K}}$ such that $\sum_{i=1}^n F(\alpha_i).z_i < \sum_{i=1}^n F(\alpha_i).y_i$. Let J' be the set of indices of variables put to 1 in Z: $J' = \{i \mid z_i = 1\}$ and let I_2 be an interpretation that satisfies all the formulas $\langle \Phi_i, \alpha_i \rangle$ where $i \notin J'$. Such an interpretation exists because the set of formulas $H = \{\langle \Phi_i, \alpha_i \rangle \mid i \notin J'\}$ is consistent. Indeed if we suppose that this is not the case, it follows that there is a supported conflict $\mathcal{C}_j \subseteq H$ i.e., for all $\langle \Phi_i, \alpha_i \rangle \in \mathcal{C}_j$ we have $z_i = 0$. This means that the constraint C_j is not satisfied by Z which contradicts the fact that Z is a solution of $LP_{\mathcal{K}}$. The possibility degree of I_2 is given by: $\pi_{\mathcal{K}}(I_2) = \prod_{i \in J'}(1 - \alpha_i)$. Now, it holds that:

$\sum_{i=1}^n F(\alpha_i).z_i < \sum_{i=1}^n F(\alpha_i).y_i \Leftrightarrow \sum_{i \in J'} F(\alpha_i) < \sum_{i \in J} F(\alpha_i)$
$\Leftrightarrow \sum_{i \in J'} -ln(1 - \alpha_i) < \sum_{i \in J} -ln(1 - \alpha_i)$
$\Leftrightarrow -ln(\prod_{i \in J'}(1 - \alpha_i)) < -ln(\prod_{i \in J}(1 - \alpha_i))$
$\Leftrightarrow ln(\prod_{i \in J'}(1 - \alpha_i)) > ln(\prod_{i \in J}(1 - \alpha_i))$
$\Leftrightarrow \prod_{i \in J'}(1 - \alpha_i) > \prod_{i \in J}(1 - \alpha_i) \Leftrightarrow \pi_{\mathcal{K}}(I_2) > \pi_{\mathcal{K}}(I_1)$.

This contradicts the fact that λ is the inconsistency degree of $\mathcal{K}$.

⇐) Let $Y = (y_1, \ldots, y_n)$ be an optimal solution of $LP_{\mathcal{K}}$ whose objective function value is $F(\lambda) = \sum_{i=1}^n F(\alpha_i).y_i$. Let $J = \{i \mid y_i = 1\}$. It is easy to check that: $\lambda = 1 - \prod_{i \in J}(1 - \alpha_i)$.

Let I_1 be an interpretation that satisfies all the formulas $\langle \Phi_i, \alpha_i \rangle$ where $i \notin J$. Such an interpretation exists because the set of formulas $H = \{\langle \Phi_i, \alpha_i \rangle \mid i \notin J\}$ is consistent. Indeed if we suppose that this is not the case, it follows that there is a supported conflict $\mathcal{C}_j \subseteq H$ i.e., for all $\langle \Phi_i, \alpha_i \rangle \in \mathcal{C}_j$ we have $y_i = 0$. This means that the constraint C_j is not satisfied by Y which contradicts the fact that Y is a solution of $LP_\mathcal{K}$. The possibility degree of I_1 is given by $\pi_\mathcal{K}(I_1) = \prod_{i \in J}(1 - \alpha_i)$ and hence, $\lambda = 1 - \pi_\mathcal{K}(I_1)$. To show that λ is the inconsistency degree of $\mathcal{K}$ it suffices to show that $\pi_\mathcal{K}(I_1) = \max_{I \in \Omega}(\pi_\mathcal{K}(I))$.

Suppose for the sake of contradiction that there is an interpretation I_2 such that $\pi_\mathcal{K}(I_2) > \pi_\mathcal{K}(I_1)$. Let J' be the set of indices of formulas falsified by I_2: $J' = \{i \mid I_2 \not\models \langle \Phi_i, \alpha_i \rangle\}$ and let $Z = (z_1, \ldots, z_n)$ be a binary vector such that $z_i = 1$ if $i \in J'$ and $z_i = 0$ otherwise.

Let us show that Z satisfies all the constraints of $LP_\mathcal{K}$. Since for all $i \notin J'$, $I_2 \models \langle \Phi_i, \alpha_i \rangle$, it is clear that the set $S = \{\langle \Phi_i, \alpha_i \rangle \in \mathcal{T} \cup \mathcal{A} \mid i \notin J'\}$ is consistent, i.e., does not contain any supported conflict. This means that for every supported conflict $\mathcal{C}_j$ there is a formula $\langle \Phi_i, \alpha_i \rangle \in \mathcal{C}_j$ such that $i \in J$, i.e., $z_i = 1$. It follows that the corresponding constraint C_j is satisfied by Y. Now, it holds that:

$$\pi_\mathcal{K}(I_2) > \pi_\mathcal{K}(I_1) \Leftrightarrow \prod_{i \in J'}(1 - \alpha_i) > \prod_{i \in J}(1 - \alpha_i)$$
$$\Leftrightarrow ln(\prod_{i \in J'}(1-\alpha_i)) > ln(\prod_{i \in J}(1-\alpha_i)) \Leftrightarrow \sum_{i \in J'} ln(1-\alpha_i) > \sum_{i \in J} ln(1-\alpha_i)$$
$$\Leftrightarrow - \sum_{i \in J'} ln(1 - \alpha_i) < - \sum_{i \in J} ln(1 - \alpha_i)$$
$$\Leftrightarrow \sum_{i \in J'} -ln(1 - \alpha_i) < \sum_{i \in J} -ln(1 - \alpha_i)$$
$$\Leftrightarrow \sum_{i \in J'} F(\alpha_i) < \sum_{i \in J} F(\alpha_i) \Leftrightarrow \sum_{i=1}^n F(\alpha_i).z_i < \sum_{i=1}^n F(\alpha_i).y_i.$$

But this contradicts the fact that Y is an optimal solution of $LP_\mathcal{K}$ ∎

References

1. Artale, A., Ryzhikov, V., Kontchakov, R.: DL-Lite with attributes and datatypes. In: 20th European Conference on Artificial Intelligence (ECAI-2012), pp. 61–66 (2012)
2. Artale, A., Calvanese, D., Kontchakov, R., Zakharyaschev, M.: The DL-lite family and relations. J. Artif. Intell. Res. **36**, 1–69 (2009)
3. Baader, E., Calvanese, D., McGuinness, D.L., Nardi, D., Patel-Schneider, P.F. (eds.): The Description Logic Handbook. Theory, Implementation, and Applications. Cambridge University Press, New York (2003)
4. Benferhat, S., Bouraoui, Z.: Min-based possibilistic DL-Lite. J. Logic Comput. **27**(1), 261–297 (2017). First published online April 12 (2015)
5. Benferhat, S., Boutouhami, K., Khellaf, F., Nouioua, F.: Algorithms for quantitative-based possibilistic lightweight ontologies. In: Fujita, H., Ali, M., Selamat, A., Sasaki, J., Kurematsu, M. (eds.) IEA/AIE 2016. LNCS, vol. 9799, pp. 364–372. Springer, Cham (2016). doi:10.1007/978-3-319-42007-3_31
6. Benferhat, S., Boutouhami, K., Khellaf, F., Nouioua, F.: Representing lightweight ontologies in a product-based possibility theory framework. In: Ferraro, M.B., Giordani, P., Vantaggi, B., Gagolewski, M., Gil, M.Á., Grzegorzewski, P., Hryniewicz, O. (eds.) Soft Methods for Data Science. AISC, vol. 456, pp. 45–52. Springer, Cham (2017). doi:10.1007/978-3-319-42972-4_6
7. Bienvenu, M., Rosati, R.: Tractable approximations of consistent query answering for robust ontology-based data access. In: 23rd International Joint Conference on Artificial Intelligence, pp. 775–781 (2013)

8. Bobillo, F., Delgado, M., Gómez-Romero, J.: Reasoning in fuzzy OWL 2 with DeLorean. In: Bobillo, F., Costa, P.C.G., d'Amato, C., Fanizzi, N., Laskey, K.B., Laskey, K.J., Lukasiewicz, T., Nickles, M., Pool, M. (eds.) UniDL/URSW 2008-2010. LNCS, vol. 7123, pp. 119–138. Springer, Heidelberg (2013). doi:10.1007/978-3-642-35975-0_7

9. Dubois, D., Prade, H.: Possibility Theory. Plenum Press, New York (1988)

10. Eiter, T., Lukasiewicz, T., Predoiu, L.: Generalized consistent query answering under existential rules. In: 15th International Conference on Principles of Knowledge Representation and Reasoning, pp. 359–368 (2016)

11. Heinsohn, J.: Probabilistic description logics. In: 10th International Conference on Uncertainty in Artificial Intelligence, pp. 311–318 (1994)

12. Kalyanpur, A., Parsia, B., Horridge, M., Sirin, E.: Finding all justifications of OWL DL entailments. In: Aberer, K., Choi, K.-S., Noy, N., Allemang, D., Lee, K.-I., Nixon, L., Golbeck, J., Mika, P., Maynard, D., Mizoguchi, R., Schreiber, G., Cudré-Mauroux, P. (eds.) ASWC/ISWC -2007. LNCS, vol. 4825, pp. 267–280. Springer, Heidelberg (2007). doi:10.1007/978-3-540-76298-0_20

13. Qi, G., Ji, Q., Pan, J.Z., Du, J.: Extending description logics with uncertainty reasoning in possibilistic logic. Int. J. Intell. Syst. **26**(4), 353–381 (2011)

14. Qi, G., Ji, Q., Pan, J.Z., Du, J.: PossDL — a possibilistic DL reasoner for uncertainty reasoning and inconsistency handling. In: Aroyo, L., Antoniou, G., Hyvönen, E., ten Teije, A., Stuckenschmidt, H., Cabral, L., Tudorache, T. (eds.) ESWC 2010. LNCS, vol. 6089, pp. 416–420. Springer, Heidelberg (2010). doi:10.1007/978-3-642-13489-0_35

15. Ramachandran, R., Qi, G., Wang, K., Wang, J., Thornton, J.: Probabilistic reasoning in DL-Lite. In: Anthony, P., Ishizuka, M., Lukose, D. (eds.) PRICAI 2012. LNCS, vol. 7458, pp. 480–491. Springer, Heidelberg (2012). doi:10.1007/978-3-642-32695-0_43

16. Rosati, R.: On the complexity of dealing with inconsistency in description logic ontologies. In: International Joint Conference on Artificial Intelligence, pp. 1057–1062 (2011)

17. Straccia, U.: Reasoning within fuzzy description logics. J. Artif. Intell. Res. **14**, 137–166 (2001)

18. Wang, Z., Wang, K., Zhuang, Z., Qi, G.: Instance-driven ontology evolution in DL-Lite. In: 29th AAAI Conference, pp. 1656–1662 (2015)

A Semantic Characterization for ASP Base Revision

Laurent Garcia[1(✉)], Claire Lefèvre[1], Odile Papini[2], Igor Stéphan[1],
and Éric Würbel[2]

[1] LERIA Université d'Angers, Angers, France
{garcia,claire.lefevre,stephan}@info.univ-angers.fr
[2] LSIS-CNRS UMR 7296, Aix-Marseille Université, Marseille, France
{papini,wurbel}@univ-amu.fr

Abstract. The paper deals with base revision for Answer Set Programming (ASP). Base revision in classical logic is done by the removal of formulas. Exploiting the non-monotonicity of ASP allows one to propose other revision strategies, namely addition strategy or removal and/or addition strategy. These strategies allow one to define families of rule-based revision operators. The paper presents a semantic characterization of these families of revision operators in terms of answer sets. This characterization allows one to equivalently consider the evolution of syntactic logic programs and the evolution of their semantic content.

Keywords: Answer set programming · Base revision · Belief revision · Belief change · Non-monotonic reasoning

1 Introduction

Answer Set Programming (ASP) is an efficient unified formalism for both knowledge representation and reasoning in Artificial Intelligence (AI). It has its roots in non-monotonic reasoning and logic programming and gave rise to intensive research since Gelfond & Lifschitz's seminal paper [12]. ASP has an elegant and conceptually simple theoretical foundation and has been proved useful for solving a wide range of problems in various domains [26]. Beyond its ability to formalize various problems from AI and to encode combinatorial problems [3,22], ASP provides also an interesting way to practically solve such problems since some efficient solvers are available [11,21]. But in most domains, information is subject to change, and so ASP logic programs are subject to change by the addition and/or withdrawal of rules.

Belief change in a classical logic setting, in particular belief revision, has been extensively studied for decades. It applies to situations where an agent faces incomplete or uncertain information and where new and more reliable information may be contradictory with its initial beliefs. Belief revision consists

---

This work was supported by the project ASPIQ (ANR-12-BS02-0003).

in modifying the initial agent's beliefs while taking into account new information and ensuring the consistency of the result. Belief revision relies on three main principles: (*i*) *Success*: Change must succeed, new information has to be accepted, (*ii*) *Consistency*: The result of the revision operation must be a consistent set of beliefs, and (*iii*) *Minimal change*: The initial beliefs have to be changed as little as possible. Two main frameworks became standards according to the nature of the involved representation of beliefs: AGM paradigm [1] for belief set revision, rephrased by Katsuno and Mendelzon [18] for model-based revision and Hansson's approach [14] for formula-based revision (or base revision). Several concrete base revision operators have been proposed. Most approaches focus on the construction of consistent subbases maximal w.r.t. several criteria [5,20]. From a dual point of view, others stem from the minimal withdrawal of formulas in order to restore consistency with new information like Kernel revision [13] or like Removed Sets Revision (RSR) [4,23,30] that focuses on subsets of formulas minimal w.r.t. cardinality to remove. All these approaches require selection functions that encode the revision strategies for selecting among subbases or among subsets of formulas to remove.

This paper aims at studying base revision when beliefs are represented by ASP logic programs. Due to the non-monotonic nature of logic programs under answer set semantics, the problem of change in ASP is different and more difficult than in classical logic.

The first approaches dealing with logic programs in a dynamic setting focused on the problem of logic program update [2,10,25,31]. The first work bridging ASP in a dynamic setting and belief change has been proposed by Delgrande et al. [7–9]. Their approach uses a semantic of logic programs in terms of SE-models [29]. Model-based revision and merging stemming from a distance between interpretations have been extended to logic programs. However, they noted that this approach has the drawback that arbitrary sets of SE-models may not necessarily be expressed via a logic program. Recently this drawback has been avoided for classes of logic programs satisfying an AGM-compliance condition on SE-models and a new postulate [6]. Model-based update has been addressed in the same spirit by Slota and Leite [27,28].

From a syntactic point of view, formula-based belief merging and revision have been extented to ASP. The "removed sets" approach for fusion and revision (RSF) [15] and (RSR) [4] respectively, proposed in propositional logic have been extended to ASP [16,17]. The "remainder sets" approach for screened consolidation in a classical setting has been extended to ASP [19]. The strategy of these two approaches stems from the removal of some rules in order to restore consistency. More recently, a new approach for extending belief base revision to ASP has been proposed with additional strategies stemming from the addition and the addition and/or removal of some rules [33].

This paper focuses on three different families of ASP base revision operators. It first reviews the RSR family, it then introduces the notions of "added Sets" and "modified Sets" and proposes the Added Set Revision (ASR) and the Modified Set Revision (MSR). Note that these families of operators differ from the ones provided in [33] since the minimality criterion for the removed, added or modified sets is cardinality and not set inclusion. For each family of ASP base revision operators the paper provides a semantic counter-part that characterizes the operators in terms of answer sets.

The main contribution of the paper is the characterization of ASP base revision operators which also covers the family of SLP operators proposed in [33]. This is an important result since it provides a new semantic characterization of logic program revision in terms of answer sets and allows one to change the focus from the evolution of a syntactic logic program to the evolution of its semantic content.

The paper is organized as follows. Section 2 gives a refresher on ASP and on belief base revision. Section 3 recalls RSR revision and provides a semantic characterization of removed sets. Section 4 introduces the notions of added set and ASR revision, it then gives a semantic characterization of added sets. Section 5 introduces the notions of modified set and MSR revision, it then provides a semantic characterization of modified sets. Finally Sect. 6 concludes the paper.

Due to space limitations, the complete proofs of the theorems are not included, they can be found on http://aspiq.lsis.org/aspiq/pdf/proofs-2017.pdf.

2   Preliminaries

In this paper we only consider normal logic programs. Let $\mathcal{A}$ be a set of propositional atoms, a logic program is a finite set of rules of the form:

$(c \leftarrow a_1, \ldots, a_n, not\ b_1, \ldots, not\ b_m.)$ $n \geq 0, m \geq 0$ where $c, a_1, \ldots, a_n, b_1, \ldots b_m \in \mathcal{A}$. The set of all logic programs is denoted by $\mathcal{P}$. The symbol "not" represents default negation and such a program may be seen as a sub-case of the default theory of Reiter [24]. A negation-free program is a definite program. For each rule r, let $head(r) = c$, $body^+(r) = \{a_1, \ldots, a_n\}$ and $body^-(r) = \{b_1, \ldots, b_m\}$. If $body^+(r) = \emptyset$ and $body^-(r) = \emptyset$ then the rule is simply written $(c.)$ and is called a fact. For a set of rules R, $Head(R) = \{head(r) \mid r \in R\}$.

Let X be a set of atoms. A rule r is *applicable in* X if $body^+(r) \subseteq X$. $App(P, X)$ denotes the set of applicable rules of P in X. The least Herbrand model of a definite program P, denoted $Cn(P)$, is the smallest set of atoms closed under P and can be computed as the least fix-point of the following consequence operator: $T_P : 2^{\mathcal{A}} \to 2^{\mathcal{A}}$ such that $T_P(X) = Head(App(P, X))$.

The Gelfond-Lifschitz reduct of a program P by a set of atoms X [12] is the program $P^X = \{head(r) \leftarrow body^+(r) \mid r \in P, body^-(r) \cap X = \emptyset\}$. Since it has no default negation, such a program is definite and then it has a unique minimal Herbrand model. By definition, an answer set (or stable model) of P is a set of atoms $X \subseteq \mathcal{A}$ such that $X = Cn(P^X)$. The set of answer sets of a logic program P is denoted by $AS(P)$ and if $AS(P) \neq \emptyset$ the program is said *consistent* otherwise it is said *inconsistent*.

$GR(P, X) = \{r \in P \mid body^+(r) \subseteq X$ and $body^-(r) \cap X = \emptyset\}$ denotes the set of the generating rules of a logic program P w.r.t. a set of atoms X. A set of rules $R \subseteq P$ is *grounded* if there exists some enumeration $\langle r_i \rangle_{i=1}^n$ of the rules of R such that $\forall i > 0, body^+(r_i) \subseteq \{head(r_j) \mid j < i\}$. With those definitions the following result holds: $X \in AS(P)$ if and only if $X = Cn(GR(P, X)^+)$ if and only if $X = Head(GR(P, X))$ and $GR(P, X)$ is grounded.

A *constraint* is a rule without head ($\leftarrow a_1, \ldots, a_n, not\ b_1, \ldots, not\ b_m$.) that should be read as ($h \leftarrow a_1, \ldots, a_n, not\ b_1, \ldots, not\ b_m, not\ h.$) where h is a new atom symbol appearing nowhere else in the program.

We now consider a rule ($c \leftarrow a_1, \ldots, a_n, not\ b_1, \ldots, not\ b_m$.) as a classical implication ($a_1 \wedge \cdots \wedge a_n \wedge \neg b_1 \wedge \cdots \wedge \neg b_m \rightarrow c$) which is equivalent to ($\neg a_1 \vee \cdots \vee \neg a_n \vee b_1 \vee \cdots \vee b_m \vee c$). Hence, an *interpretation* of P is a set of atoms $m \subseteq \mathcal{A}$. An interpretation m *satisfies* a rule r if $body^+(r) \not\subseteq m$ or $body^-(r) \cap m \neq \emptyset$ or $head(r) \in m$. An interpretation m is a *model* of a program P if m satisfies all rules from P. $Mod(P)$ denotes the set of all the models of a logic program P. Conversely, an interpretation m *falsifies* a rule r if m does not satisfy r: $body^+(r) \subseteq m$ and $body^-(r) \cap m = \emptyset$ and $head(r) \notin m$. $Fal(P, m) = \{r \in P : body^+(r) \subseteq m, body^-(r) \cap m = \emptyset, head(r) \notin m\}$ denotes the set of the rules of a logic program P that are falsified w.r.t. an interpretation m.

The set $m \backslash Cn(GR(P, m)^+)$ denotes the atoms of an interpretation m not deduced from this interpretation for a logic program P and $Nded(m, P) = fact(m \backslash Cn(GR(P, m)^+))$ denotes the previous set of atoms considered as a set of facts.

We review some notions and notations useful in subsequent sections.

A preorder on a set A is a reflexive and transitive binary relation. A total preorder, denoted by $\leq$, is a preorder such that $\forall x, y \in A$ either $x \leq y$ or $y \leq x$ holds. Equivalence is defined by $x \simeq y$ if and only if $x \leq y$ and $y \leq x$. The corresponding strict total preorder, denoted by $<$, is the relation defined by $x < y$ if and only if $x \leq y$ holds but $x \simeq y$ does not hold. Let M be a subset of A, the set of minimal elements of M with respect to $\leq$, denoted by $Min(M, \leq)$, is defined as: $Min(M, \leq) = \{x \in M, \nexists y \in M : y < x\}$.

Let X and Y be two sets, $|X|$ (resp. $|Y|$) denotes the cardinality of X (resp. of Y) and $X \leq Y$ if $|X| \leq |Y|$. $X \leq Y$ means that X is preferred to Y.

Let A be a finite set, a selection function denoted by f is a function from $2^A \backslash \emptyset$ to A which for any set $X \in 2^A$ returns an element $f(X)$ such that $f(X) \in X$.

3   ASP Base Revision by Removal

Let P and Q be logic programs, revising P by Q is providing a new consistent logic program containing Q and differing as little as possible from $P \cup Q$. This section is dedicated to ASP base revision by removal. This revision strategy stems from the suppression of rules of P when $P \cup Q$ is inconsistent. This strategy is a direct application of the one used for revising belief bases in a classical setting, however it differs from it due to the non-monotonicity of logic programs. For instance, in ASP, P and Q can be inconsistent while $P \cup Q$ is consistent. This is the reason why, we allow both P and Q to be inconsistent (note that, in a classical setting, both P and Q must be consistent, only $P \cup Q$ may be inconsistent). However, this means that revision is not always possible (for example, when Q does not admit any classical model since the revision strategy only removes rules from P). Throughout the paper Q is the program used to revise the program P.

3.1   Rule-Based Revision by Removal

We review Removed Sets Revision (RSR) extended to ASP [17]. This strategy focuses on the minimal number of rules needed to remove in order to restore the consistency. We first review the notion of *potential removed set*.

Definition 1 (potential removed set *[17]*). *Let P and Q be two logic programs, let X be a set of rules. A potential removed set X is such that: (i) $X \subseteq P$. (ii) $(P \setminus X) \cup Q$ is consistent. (iii) For each $X' \subset X$, $(P \setminus X') \cup Q$ is inconsistent.*

$\mathcal{PR}(P, Q)$ denotes the set of potential removed sets for P and Q. According to the definition, if $P \cup Q$ is consistent then $\mathcal{PR}(P, Q) = \{\emptyset\}$. Since potential removed sets are built by removing only rules from P in order to restore consistency of $P \cup Q$, it may be possible that the set of potential removed sets for an inconsistent set Q is empty.

Example 1. Let P and Q be two logic programs such that $P = \{r_1 : a \leftarrow b., r_2 : a., r_3 : b., r_4 : c \leftarrow not\, a., r_5 : d., r_6 : d \leftarrow not\, b.\}$ and $Q = \{\leftarrow a, b., \leftarrow not\, c, d.\}$. These two logic programs are consistent since $AS(P) = \{\{a, b, d\}\}$ and $AS(Q) = \{\emptyset\}$ but $P \cup Q$ is inconsistent. $\mathcal{PR}(P, Q) = \{\{r_3, r_5, r_6\}, \{r_2, r_3\}, \{r_1, r_2\}\}$.

For RSR, the minimality criterion is cardinality, we review the notion of *removed set* by selecting the potential removed sets minimal w.r.t. cardinality.

Definition 2 (removed set *[17]*). *Let P and Q be two logic programs, let X be a set of rules. A removed set X is such that: (i) X is a potential removed set. (ii) There is no potential removed set Y such that $Y < X$.*

$\mathcal{R}(P, Q)$ denotes the set of removed sets for P and Q. According to the definition $\mathcal{R}(P, Q) = Min(\mathcal{PR}(P, Q), \leq)$ and if $P \cup Q$ is consistent then $\mathcal{R}(P, Q) = \{\emptyset\}$.

Example 2 (Example 1 continued). $\mathcal{R}(P, Q) = \{\{r_2, r_3\}, \{r_1, r_2\}\}$.

We now review the *Removed Set Revision* (RSR) family of operators.

Definition 3 (RSR operators *[17]*). *Let P and Q be two logic programs, $\mathcal{R}(P, Q)$ the set of removed sets and f a selection function. The revision operator denoted by $\star_{RSR(f)}$ is a function from $\mathcal{P} \times \mathcal{P}$ to $\mathcal{P}$ such that $P \star_{RSR(f)} Q = (P \setminus f(\mathcal{R}(P, Q))) \cup Q$.*

Note that if $\mathcal{R}(P, Q) = \emptyset$, $f(\mathcal{R}(P, Q))$ is not defined. This means that the program P cannot be revised by Q.

3.2   Semantic Characterization of ASP Base Revision by Removal

We now present the semantical counterparts of the potential removed set and removed set notions. We first introduce the notion of *canonical removed set*. Intuitively, given P and Q two logic programs, a *canonical removed set* is a set of rules of P falsified by a model of Q.

Definition 4 (canonical removed set). *Let P and Q be two logic programs and m a model of Q. A* canonical removed set X *is such that:* (i) $X = Fal(P, m)$. (ii) $m \in AS((P \setminus X) \cup Q)$.

$CR(P, Q, m) = \{X \mid X = Fal(P, m) \text{ and } m \in AS((P \setminus X) \cup Q)\}$ denotes the set of all canonical removed sets for m and $CR(P, Q) = \bigcup_{m \in Mod(Q)} CR(P, Q, m)$ denotes the union set of all canonical removed sets for the models of a program Q w.r.t. a program P. Note that for one given interpretation m, there is zero or one canonical removed set and if Q has no model then $CR(P, Q) = \emptyset$.

Example 3 (Example 1 continued). $P = \{r_1 : a \leftarrow b., r_2 : a., r_3 : b., r_4 : c \leftarrow not\, a., r_5 : d., r_6 : d \leftarrow not\, b.\}$ and $Q = \{\leftarrow a, b., \leftarrow not\, c, d.\}$.

$m \in Mod(Q)$	$X = Fal(P, m)$	$AS((P \setminus X) \cup Q)$	$CR(P, Q, m)$
$\emptyset$	$\{r_2, r_3, r_4, r_5, r_6\}$	$\{\emptyset\}$	$\{Fal(P, \emptyset)\}$
$\{a\}$	$\{r_3, r_5, r_6\}$	$\{\{a\}\}$	$\{Fal(P, \{a\})\}$
$\{b\}$	$\{r_1, r_2, r_4, r_5\}$	$\{\{b\}\}$	$\{Fal(P, \{b\})\}$
$\{c, d\}$	$\{r_2, r_3\}$	$\{\{c, d\}\}$	$\{Fal(P, \{c, d\})\}$
$\{a, c, d\}$	$\{r_3\}$	$\emptyset$	$\emptyset$
$\{b, c, d\}$	$\{r_1, r_2\}$	$\{\{b, c, d\}\}$	$\{Fal(P, \{b, c, d\})\}$
$\{a, c\}$	$\{r_3, r_5, r_6\}$	$\{\{a\}\}$	$\emptyset$
$\{b, c\}$	$\{r_1, r_2, r_5\}$	$\{\{b, c\}\}$	$\{Fal(P, \{b, c\})\}$
$\{c\}$	$\{r_2, r_3, r_5, r_6\}$	$\{\{c\}\}$	$\{Fal(P, \{c\})\}$

The last column of the table gives the set of canonical removed sets corresponding to a classical model of Q given in the first column of the table. Hence, if we restrict our attention to minimal canonical removed sets w.r.t. inclusion, we have $Min(CR(P, Q), \subseteq) = \{\{r_2, r_3\}, \{r_1, r_2\}, \{r_3, r_5, r_6\}\}$ and, if we consider minimality w.r.t. cardinality, we have $Min(CR(P, Q), \leq) = \{\{r_2, r_3\}, \{r_1, r_2\}\}$.

The following theorems give the equivalence between synatactic (potential) removed sets and semantic canonical removed sets.

Theorem 1. *Let P and Q be logic programs.* $\mathcal{PR}(P, Q) = Min(CR(P, Q), \subseteq)$.

Proof (sketch). The proof is based on the fact that the rules of a potential removed set X are exactly the rules of P falsified by the answer sets of $(P \setminus X) \cup Q$.

The following theorem is a direct consequence of Definition 2 and Theorem 1.

Theorem 2. *Let P and Q be logic programs.* $\mathcal{R}(P, Q) = Min(CR(P, Q), \leq)$.

We introduce a preference relation between interpretations, denoted by $<_{R(P)}$ as follows. Let m and m' be two interpretations, $m <_{R(P)} m'$ means that $|Fal(P, m)| < |Fal(P, m')|$.

The following result directly follows from Theorem 2 and Definition 3. It provides a semantic characterization of ASR revision operators for logic programs.

Theorem 3. *Let P and Q be two logic programs.*
   Let $M = \{m \in Mod(Q) \ s.t. \ CA(P,Q,m) \neq \emptyset\}$. (i) For each selection function f, if $m \in AS(P\star_{RSR(f)}Q)$ then $m \in Min(M, \leq_{R(P)})$. (ii) If $m \in Min(M, \leq_{R(P)})$ then there exists a selection function f s.t. $m \in AS(P\star_{RSR(f)}Q)$.

Example 4 (Example 1 continued). $P = \{r_1 : a \leftarrow b., r_2 : a., r_3 : b., r_4 : c \leftarrow not\,a., r_5 : d., r_6 : d \leftarrow not\,b.\}$ and $Q = \{\leftarrow a, b., \leftarrow not\,c, d.\}$. From the table in Example 3 we have $\mathcal{PR}(P,Q) = Min(CR(P,Q), \subseteq)$, $\mathcal{R}(P,Q) = Min(CR(P,Q), \leq)$ and $Min(M, \leq_{R(P)}) = \{\{c,d\}, \{b,c,d\}\}$. Let f_1 and f_2 be the functions that select respectively $\{r_2, r_3\}$ and $\{r_1, r_2\}$ the respective revised logic programs are $P\star_{RSR(f_1)}Q = P\backslash\{r_2, r_3\}\cup Q$ and $P\star_{RSR(f_2)}Q = P\backslash\{r_1, r_2\}\cup Q$ with $AS(P\star_{RSR(f_1)}Q) = \{\{c,d\}\}$ and $AS(P\star_{RSR(f_2)}Q) = \{\{b,c,d\}\}$.

4   ASP Base Revision by Addition

This section is dedicated to ASP base revision by addition. Let P and Q be logic programs, this revision strategy stems from the addition of rules to P when $P \cup Q$ is inconsistent. This strategy relies on the non-monotonicity of the ASP framework. Indeed, adding a new rule may block a rule which contributes to inconsistency. We allow P and Q to be inconsistent. Note that, with this strategy, revision is not always possible, even if P and Q are consistent. Moreover, since the addition must block an existing rule, we restrict the addition to the vocabulary of P and Q. Revising by addition allows for adding any kind of rules but adding rules is equivalent to adding a set of facts which makes the revision process easier.

4.1   Rule-Based Revision by Addition

The strategy of Added Set Revision (ASR) focuses on the minimal number of new rules to add in order to restore consistency. We first introduce the notion of *potential added set*.

Definition 5 (potential added set). *Let P and Q be two logic programs, let Y be a set of rules made from the vocabulary of P and Q. A potential added set Y is such that: (i) $(P \cup Y) \cup Q$ is consistent. (ii) For each $Y' \subset Y$, $(P \cup Y') \cup Q$ is inconsistent.*

$\mathcal{PA}(P,Q)$ denotes the set of potential added sets for P and Q. Note that rules from a potential added set Y cannot already belong to $P \cup Q$. Indeed, if $(P \cup Y) \cup Q$ is consistent and some rule r from Y already belongs to $P \cup Q$, then there exists some $Y' = Y \backslash \{r\}$ such that $Y' \subseteq Y$ and $(P \cup Y') \cup Q = (P \cup Y) \cup Q$ is consistent. According to the definition if $P \cup Q$ is consistent then $\mathcal{PA}(P,Q) = \{\emptyset\}$.

Example 5. Let P and Q be two logic programs such that $P = \{a \leftarrow not\,b.\}$ and $Q = \{\leftarrow a, not\,c., \leftarrow a, not\,d.\}$. If we restrict ourselves to the addition of facts, we have $\mathcal{PA}(P,Q) = \{\{c., d.\}, \{b.\}\}$.

Note that revision by addition is not always feasible, even if P and Q are consistent.

Example 6. Let P and Q be two logic programs such that $P = \{a.\}$ and $Q = \{\leftarrow a.\}$. We have $\mathcal{PA}(P, Q) = \emptyset$. This is because, even if the constraint $\leftarrow a$. can be read as $h \leftarrow a, not\ h.$, we do not allow using the implicit atom h for adding rules and thus the constraint $\leftarrow a$. cannot be blocked by addition.

For ASR, the minimality criterion is cardinality, we introduce the notion of *added set* by selecting the potential added sets minimal w.r.t. cardinality.

Definition 6 (added set). *Let P and Q be two logic programs, let Y be a set of rules. An added set Y is such that: (i) Y is a potential added set. (ii) There is no potential added set Z such that $Z < Y$.*

$\mathcal{A}(P, Q)$ denotes the set of added sets for P and Q. According to the definition $\mathcal{A}(P, Q) = Min(\mathcal{PA}(P, Q), \leq)$ and if $P \cup Q$ is consistent then $\mathcal{A}(P, Q) = \{\emptyset\}$.

Example 7 (Example 5 continued). For these programs, $\mathcal{A}(P, Q) = \{\{b.\}\}$ is reduced to only one added set.

Example 8. Let P and Q be two logic programs such that $P = \{a \leftarrow b., b \leftarrow a., c.\}$ and $Q = \{\leftarrow c, not\ a.\}$. We have $\mathcal{PA}(P, Q) = \mathcal{A}(P, Q) = \{\{a.\}, \{b.\}\}$.

We now define the *Added Set Revision* family of operators.

Definition 7 (ASR operators). *Let P and Q be two logic programs, $\mathcal{A}(P, Q)$ the set of added sets and f a selection function. The revision operator denoted by $\star_{ASR(f)}$ is a function from $\mathcal{P} \times \mathcal{P}$ to $\mathcal{P}$ such that $P \star_{ASR(f)} Q = (P \cup f(\mathcal{A}(P, Q))) \cup Q$.*

Note that if $\mathcal{A}(P, Q) = \emptyset$, $f(\mathcal{A}(P, Q))$ is not defined. That means that the program cannot be revised by addition.

4.2 Semantic Characterization of ASP Base Revision by Addition

We now present the semantic counterparts of the potential added set and added set notions. We first introduce the notion of *canonical added set*. Given P and Q two logic programs, a *canonical added set* is a set of facts corresponding to the least subset (w.r.t. inclusion) of atoms to add to $P \cup Q$ so that a model of $P \cup Q$ becomes an answer set.

Definition 8 (canonical added set). *Let P and Q be two logic programs and m a model of $P \cup Q$. A canonical added set Y is such that: (i) $Y \subseteq Nded(m, P \cup Q)$. (ii) $m \in AS(P \cup Q \cup Y)$. (iii) $\forall Y' \subset Y, m \notin AS(P \cup Q \cup Y')$.*

$CA(P, Q, m)$ denotes the set of all canonical added sets for m and $CA(P, Q) = \bigcup_{m \in Mod(P \cup Q)} CA(P, Q, m)$. Note that $CA(P, Q, m) = Min(\{Y \mid Y \subseteq Nded(m, P \cup Q)$ and $m \in AS(P \cup Q \cup Y)\}, \subseteq)$.

Example 9. Let $P = \{r_1 : a \leftarrow b, not\ c., r_2 : c \leftarrow d, e, not\ a., r_3 : b \leftarrow d., r_4 : d \leftarrow b., r_5 : e.\}$ and $Q = \{r_6 :\leftarrow not\ a, not\ c., r_7 :\leftarrow a, not\ b, not\ c., r_8 :\leftarrow c, not\ d, not\ a.\}$.

$m \in$ $Mod(P \cup Q)$	$GR(P \cup Q, m)$	$Nded(m, P \cup Q)$	Y	$AS(P \cup Q \cup Y)$	$CA(P, Q, m)$
$\{a, c, e\}$	$\{r_5\}$	$\{a., c.\}$	$\{a., c.\}$	$\{\{a, c, e\}\}$	$\{\{a., c.\}\}$
$\{a, b, d, e\}$	$\{r_1, r_3, r_4, r_5\}$	$\{a., b., d.\}$	$\{b.\}$	$\{\{a, b, d, e\}, \{b, c, d, e\}\}$	
			$\{d.\}$	$\{\{a, b, d, e\}, \{b, c, d, e\}\}$	$\{\{b.\}, \{d.\}\}$
$\{b, c, d, e\}$	$\{r_2, r_3, r_4, r_5\}$	$\{b., c., d.\}$	$\{b.\}$	$\{\{a, b, d, e\}, \{b, c, d, e\}\}$	
			$\{d.\}$	$\{\{a, b, d, e\}, \{b, c, d, e\}\}$	$\{\{b.\}, \{d.\}\}$
$\{a, b, c, d, e\}$	$\{r_3, r_4, r_5\}$	$\{a., b., c., d.\}$	$\{a., b., c.\}$ $\{a., c., d.\}$	$\{\{a, b, c, d, e\}\}$ $\{\{a, b, c, d, e\}\}$	$\{\{a., b., c.\}, \{a., c., d.\}\}$

The last column of the table gives the set of canonical added sets corresponding to a classical model of $P \cup Q$ given in the first column of the table. Hence $Min(CA(P, Q), \subseteq) = \{\{a., c.\}, \{b.\}, \{d.\}\}$ and $Min(CA(P, Q), \leq) = \{\{b.\}, \{d.\}\}$.

Note that canonical added sets only consist of facts. Thus the semantic characterization of ASR operators is limited to ASR operators that require the addition of facts (not the addition of general rules).

Theorem 4. *Let P and Q be logic programs. $\mathcal{PA}(P, Q) \doteq Min(CA(P, Q), \subseteq)$.*

Proof (sketch). The proof is based on the fact that the rules (facts) of a potential added set Y are a subset of the (facts corresponding to the) atoms from the answer sets of $P \cup Y \cup Q$ that can not be deduced from $P \cup Q$.

The following theorem is a direct consequence of Theorem 4 and Definition 6.

Theorem 5. *Let P and Q be logic programs. $\mathcal{A}(P, Q) = Min(CA(P, Q), \leq)$.*

We introduce a preference relation between interpretations, denoted by $<_{A(P,Q)}$ as follows. Let m and m' be two interpretations, $m <_{A(P,Q)} m'$ means that $Min(CA(P, Q, m), \leq) < Min(CA(P, Q, m'), \leq)$.

The following result directly follows from Theorem 2 and Definition 7. It provides a semantic characterization of ASR revision operators for logic programs.

Theorem 6. *Let P and Q be two logic programs.*
Let $M = \{m \in Mod(Q)$ s.t. $CA(P, Q, m) \neq \emptyset\}$. (i) For each f, if $m \in AS(P \star_{ASR(f)} Q)$ then $m \in Min(M, \leq_{A(P,Q)})$. (ii) If $m \in Min(M, \leq_{A(P,Q)})$ then there exists f s.t. $m \in AS(P \star_{ASR(f)} Q)$.

Example 10 (Example 9 continued). Let $P = \{a \leftarrow b, not\ c., c \leftarrow d, e, not\ a., b \leftarrow d., d \leftarrow b., e.\}$ and $Q = \{\leftarrow not\ a, not\ c., \leftarrow a, not\ b, not\ c., \leftarrow c, not\ d, not\ a.\}$. From the table in Example 9 we have

$\mathcal{PA}(P,Q) = Min(CA(P,Q), \subseteq)$, $\mathcal{A}(P,Q) = Min(CA(P,Q), \leq) = \{\{b.\}, \{d.\}\}$. $Min(M, \leq_{A(P,Q)}) = \{\{a,b,d,e\}, \{b,c,d,e\}\}$. Let f_1 and f_2 be the functions that select respectively $\{b.\}$ and $\{d.\}$ the respective revised logic programs are $P \star_{ASR(f_1)} Q = P \cup Q \cup \{b.\}$ and $P \star_{ASR(f_2)} Q = P \cup Q \cup \{d.\}$ with $AS(P \star_{ASR(f_1)} Q) = AS(P \star_{ASR(f_2)} Q) = \{\{a,b,d,e\}, \{b,c,d,e\}\}$.

5   ASP Base Revision by Modification

This section now focuses on ASP base revision by modification. Modification strategy means combining the removal strategy and the addition one. Let P and Q be logic programs, removing some rules from P and in the same time adding some new rules to P allows one to construct a new logic program which is consistent with Q and differs the least from P. Indeed, revision by removal and revision by addition can be viewed as particular cases of revision by modification.

5.1   Rule-Based Revision by Modification

The strategy of Modified Set Revision (MSR) focuses on the minimal number of rules to remove and/or to add in order to restore consistency. A (potential) modified set is a pair of sets of rules, where the first component is the set of rules to remove and the second one is the set of new rules to add. We first define preference relations between pairs of sets of rules w.r.t. set inclusion and w.r.t. cardinality as follows.

Definition 9. *Let* X, Y, X', Y' *be sets of rules.* $(X', Y') \subset (X, Y)$ *if* $X' \subset X$ *and* $Y' \subseteq Y$, *or* $X' \subseteq X$ *and* $Y' \subset Y$. $(X', Y') \leq (X, Y)$ *if* $|X' \cup Y'| \leq |X \cup Y|$.

We now introduce the notion of *potential modified set*.

Definition 10 (potential modified set). *Let* P *and* Q *be two logic programs, let* (X, Y) *be a pair of sets of rules. A potential modified set* (X, Y) *is such that:* (i) $X \subseteq P$. (ii) $(P \backslash X) \cup Y \cup Q$ *is consistent.* (iii) *For each* (X', Y') *such that* $(X', Y') \subset (X, Y)$, $(P \backslash X') \cup Y' \cup Q$ *is inconsistent.*

$\mathcal{PM}(P,Q)$ denotes the set of potential modified sets for P and Q. According to the definition if $P \cup Q$ is consistent then $\mathcal{PM}(P,Q) = \{(\emptyset, \emptyset)\}$.

Example 11. Let $P = \{r_1 : a \leftarrow not\ b., r_2 : c \leftarrow not\ b., r_3 :\leftarrow f.\}$ and $Q = \{\leftarrow a., \leftarrow c., f \leftarrow not\ g., f \leftarrow not\ h.\}$. $\mathcal{PM}(P,Q) = \{(\{r_1, r_2, r_3\}, \emptyset), (\emptyset, \{b., g., h.\}), (\{r_3\}, \{b.\}), (\{r_1, r_2\}, \{g., h.\})\}$.

As for RSR and ASR, the minimality criterion for MSR is cardinality, we introduce the notion of *modified set* by selecting the potential modified sets minimal w.r.t. cardinality.

Definition 11 (modified set). *Let* P *and* Q *be two logic programs, let* X *and* Y *be two sets of rules. A modified set* (X, Y) *is such that:* (i) (X, Y) *is a potential modified set.* (ii) *There is no potential modified set* (X', Y') *such that* $(X', Y') < (X, Y)$.

We denote by $\mathcal{M}(P,Q)$ the set of modified sets. According to the definition $\mathcal{M}(P,Q) = Min(\mathcal{PM}(P,Q),\leq)$ and if $P \cup Q$ is consistent then $\mathcal{M}(P,Q) = \{(\emptyset,\emptyset)\}$.

Example 12 (Example 11 continued). $\mathcal{M}(P,Q) = \{(\{r_3\},\{b.\})\}$.

We now define the *Modified Set Revision* family of operators.

Definition 12 (MSR operators). *Let P and Q be two logic programs, $\mathcal{M}(P,Q)$ the set of modified sets and f a selection function. The revision operator denoted by $\star_{MSR(f)}$ is a function from $\mathcal{P} \times \mathcal{P}$ to $\mathcal{P}$ such that $P \star_{MSR(f)} Q = (P \setminus X) \cup Y \cup Q$ where $(X,Y) = f(\mathcal{M}(P,Q))$.*

5.2 Semantic Characterization of ASP Base Revision by Modification

We now present the semantic counterpart of potential modified set and modified set notions. We first introduce the notion of *canonical modified set*. Given P and Q two logic programs, a canonical modified set is a pair of sets (X,Y) where X is the set of rules from P falsified by a model of Q and Y is the set of facts corresponding to least subsets of atoms of a model of Q not deduced from $(P \setminus X) \cup Q$.

Definition 13 (canonical modified set). *Let P and Q be two logic programs and m be a model of Q. A canonical modified set (X,Y) is such that: (i) $X = Fal(P,m)$ (ii) $Y \subseteq Nded(m,(P \setminus X) \cup Q)$ (iii) $m \in AS((P \setminus X) \cup Q \cup Y)$ (iv) $\forall(X',Y') \subset (X,Y), m \notin AS((P \setminus X') \cup Q \cup Y')$.*

$CM(P,Q,m) = Min(\{(X,Y) \mid X = Fal(P,m), Y \subseteq Nded(m,(P \setminus X) \cup Q)$ and $m \in AS((P \setminus X) \cup Q \cup Y)\},\subseteq)$ denotes the set of all canonical modified sets for m and $CM(P,Q) = \bigcup_{m \in Mod(Q)} CM(P,Q,m)$.

Example 13 (Example 11 continued). $P = \{r_1 : a \leftarrow not\ b., r_2 : c \leftarrow not\ b., r_3 :\leftarrow f.\}$ and $Q = \{\leftarrow a., \leftarrow c., f \leftarrow not\ g., f \leftarrow not\ h.\}$.

$m \in$ $Mod(Q)$	$X =$ $Fal(P,m)$	$Y \subseteq$ $Nded((P \setminus X) \cup Q, m)$	$AS((P \setminus X) \cup Q \cup Y)$
$\{f\}$	$\{r_1,r_2,r_3\}$	$\emptyset$	$\{\{f\}\}$
$\{b,f\}$	$\{r_3\}$	$\{b.\}$	$\{\{b,f\}\}$
$\{f,g\}$	$\{r_1,r_2,r_3\}$	$\{g.\}$	$\{\{f,g\}\}$
$\{f,h\}$	$\{r_1,r_2,r_3\}$	$\{h.\}$	$\{\{f,h\}\}$
$\{b,f,g\}$	$\{r_3\}$	$\{b.,g.\}$	$\{\{b,f,g\}\}$
$\{b,f,h\}$	$\{r_3\}$	$\{b.,h.\}$	$\{\{b,f,h\}\}$
$\{g,h\}$	$\{r_1,r_2\}$	$\{g.,h.\}$	$\{\{g,h\}\}$
$\{f,g,h\}$	$\{r_1,r_2,r_3\}$	$\{f.,g.,h.\}$	$\{\{f,g,h\}\}$
$\{b,g,h\}$	$\emptyset$	$\{b.,g.,h.\}$	$\{\{b,g,h\}\}$
$\{b,f,g,h\}$	$\emptyset$	$\{b.,f.,g.,h.\}$	$\{\{b,f,g,h\}\}$

The second and the third column of the table give the first and the second component respectively of the canonical modified set corresponding to a classical model of Q given in the first column of the table. Hence $Min(CM(P,Q), \subseteq) = \{((\{r_1, r_2, r_3\}, \emptyset), \quad (\{r_3\}, \{b.\}), \quad (\{r_1, r_2\}, \{g., h.\}), \quad (\emptyset, \{b., g., h.\}))\}$ and $Min(CM(P,Q), \leq) = \{((\{r_3\}, \{b.\}))\}$.

Theorem 7. *Let P and Q be programs. $\mathcal{PM}(P,Q) = Min(CM(P,Q), \subseteq)$.*

Proof (sketch). The proof is based on the fact that if (X, Y) is a potential modified set of P and Q, then X is a potential removed set of P and $Q \cup Y$, and Y is a potential added set of $P \setminus X$ and Q.

The following theorem is a direct consequence of Theorem 7 and Definition 11.

Theorem 8. *Let P and Q be logic programs. $\mathcal{M}(P,Q) = Min(CM(P,Q), \leq)$.*

We introduce a preference relation between interpretations, denoted by $<_{M(P,Q)}$. Let m and m' be interpretations, $m <_{M(P,Q)} m'$ means that $Min(CM(P,Q,m), \leq) < Min(CM(P,Q,m'), \leq)$. The following result directly follows from Theorem 8 and Definition 12. It provides a semantic characterization of MSR revision operators.

Theorem 9. *Let P and Q be logic programs.*
Let $M = \{m \in Mod(Q)$ s.t. $CM(P,Q,m) \neq \emptyset\}$. (i) For each f, if $m \in AS(P \star_{MSR(f)} P)$ then $m \in Min(M, \leq_{M(P,Q)})$. (ii) If $m \in Min(M, \leq_{M(P,Q)})$ then there exists f s.t. $m \in AS(P \star_{MSR(f)} P)$.

Example 14 (Example 11 continued). $P = \{r_1 : a \leftarrow not\ b., r_2 : c \leftarrow not\ b., r_3 :\leftarrow f.\}$ and $Q = \{\leftarrow a., \leftarrow c., f \leftarrow not\ g., f \leftarrow not\ h.\}$. From the table in Example 13 we have $\mathcal{PM}(P,Q) = Min(CM(P,Q), \subseteq)$, $\mathcal{M}(P,Q) = Min(CM(P,Q), \leq) = \{((\{r_3\}, \{b.\}))\}$ and $M = \{\{b, f\}\}$. There is only one modified set thus f selects $(\{r_3\}, \{b.\})$ and $P \star_{MSR(f)} Q = \{r_1, r_2\} \cup \{b.\} \cup Q$ and $AS(P \star_{MSR(f)} Q) = \{\{b, f\}\}$.

6 Concluding Discussion

Belief base revision has first been extended to ASP in [19] with the "remainder Sets" approach and in [16,17] with the "removed Sets" approach. These two approaches rely on the removal of rules. More recently, another approach called SLP revision [32,33] has been proposed. The strategy stems from the removal or the addition and/or removal of rules. Let P be the initial logic program and Q be the new one. The removal (respectively addition and addition and/or removal) strategies stem from the construction of "s-removal" (respectively "s-expansion" and "s-compatible") logic programs which are subsets of P consistent with Q maximal w.r.t. set inclusion (respectively sets of rules containing P consistent with Q and minimal w.r.t. set inclusion, and a combination of both for the third strategy). They are the dual sets of potential removed sets, potential added sets and potential modified sets respectively.

Note that the families of revision operators proposed in this paper differ from SLP revision operators since the maximality criterion is set inclusion for SLP whereas the minimality criterion for the RSR, ASR and MSR revision operators is cardinality.

Moreover, in this paper we go a step further since we provide a semantic characterization in terms of answer sets for the RSR, ASR and MSR revision families of operators. This is an important contribution since it allows one to go from the evolution of a syntactic rule-based revision operator to the evolution of its semantic content.

Future work will be dedicated to the study of logical properties of the proposed revision operators in terms of the satisfaction of Hansson's postulates for base revision adapted to the ASP framework. We also plan to implement the proposed families of revision operators and to conduct an experimental study. Another issue is the study of their computational complexity.

References

1. Alchourrón, C.E., Makinson, D.: On the logic of theory change: safe contraction. Stud. Log. **44**(4), 14–37 (1985)
2. Alferes, J.J., Leite, J.A., Pereira, L.M., Przymusinska, H., Przymusinski, T.C.: Dynamic updates of non-monotonic knowledge bases. J. Log. Prog. **45**(1–3), 43–70 (2000)
3. Baral, C.: Knowledge Representation, Reasoning and Declarative Problem Solving. Cambridge University Press, Cambridge (2003)
4. Benferhat, S., Bennaim, J., Papini, O., Würbel, E.: Answer set programming encoding of prioritized removed sets revision: application to GIS. Appl. Intell. **32**, 60–87 (2010)
5. Benferhat, S., Cayrol, C., Dubois, D., Lang, J., Prade, H.: Inconsistency management and prioritized syntax-based entailment. In: Proceedings of IJCAI 1993, pp. 640–645 (1993)
6. Delgrande, J., Peppas, P., Woltran, S.: AGM-style belief revision of logic programs under answer set semantics. In: Cabalar, P., Son, T.C. (eds.) LPNMR 2013. LNCS, vol. 8148, pp. 264–276. Springer, Heidelberg (2013). doi:10.1007/978-3-642-40564-8_27
7. Delgrande, J.P., Schaub, T., Tompits, H., Woltran, S.: Belief revision of logic programs under answer set semantics. In: Proceedings of KR 2008, pp. 411–421 (2008)
8. Delgrande, J., Schaub, T., Tompits, H., Woltran, S.: Merging logic programs under answer set semantics. In: Hill, P.M., Warren, D.S. (eds.) ICLP 2009. LNCS, vol. 5649, pp. 160–174. Springer, Heidelberg (2009). doi:10.1007/978-3-642-02846-5_17
9. Delgrande, J.P., Schaub, T., Tompits, H., Woltran, S.: A model-theoretic approach to belief change in answer set programming. ACM Trans. Comput. Log. **14**(2), 14:1–14:46 (2013)
10. Eiter, T., Fink, M., Sabbatini, G., Tompits, H.: On properties of update sequences based on causal rejection. TPLP **2**(6), 711–767 (2002)
11. Gebser, M., Kaufmann, B., Schaub, T.: Conflict-driven answer set solving: from theory to practice. Artif. Intell. **187**, 52–89 (2012)
12. Gelfond, M., Lifschitz, V.: The stable model semantics for logic programming. In: Proceedings of ICLP 1988, pp. 1070–1080 (1988)

13. Hansson, S.O.: Semi-revision. J. Appl. Non-class. Log. **7**, 151–175 (1997)
14. Hansson, S.O.: A Text of Belief Dynamics: Theory Change and Database Updating. Springer, Netherlands (1999)
15. Hué, J., Papini, O., Würbel, E.: Removed sets fusion: performing off the shelf. In: Proceedings of ECAI 2008, pp. 94–98 (2008)
16. Hué, J., Papini, O., Würbel, E.: Merging belief bases represented by logic programs. In: Sossai, C., Chemello, G. (eds.) ECSQARU 2009. LNCS, vol. 5590, pp. 371–382. Springer, Heidelberg (2009). doi:10.1007/978-3-642-02906-6_33
17. Hué, J., Papini, O., Würbel, E.: Extending belief base change to logic programs with ASP. In: Trends in Belief Revision and Argumentation Dynamics. Studies in Logic, December 2013
18. Katsuno, H., Mendelzon, A.O.: Propositional knowledge base revision and minimal change. Artif. Intell. **52**(3), 263–294 (1991)
19. Krümpelmann, P., Kern-Isberner, G.: Belief base change operations for answer set programming. In: del Cerro, L.F., Herzig, A., Mengin, J. (eds.) JELIA 2012. LNCS, vol. 7519, pp. 294–306. Springer, Heidelberg (2012). doi:10.1007/978-3-642-33353-8_23
20. Lehmann, D.: Belief revision, revised. In: Proceedings of IJCAI 1995, pp. 1534–1540 (1995)
21. Leone, N., Pfeifer, G., Faber, W., Eiter, T., Gottlob, G., Perri, S., Scarcello, F.: The DLV system for knowledge representation and reasoning. ACM Trans. Comput. Log. **7**(3), 499–562 (2006)
22. Niemelä, I.: Logic programs with stable model semantics as a constraint programming paradigm. AMAI **25**(3–4), 241–273 (1999)
23. Papini, O.: A complete revision function in propositional calculus. In: Neumann, B. (ed.) Proceedings of ECAI 1992, pp. 339–343. Wiley (1992)
24. Reiter, R.: A logic for default reasoning. Artif. Intell. **13**(1–2), 81–132 (1980)
25. Sakama, C., Inoue, K.: Updating extended logic programs through abduction. In: Gelfond, M., Leone, N., Pfeifer, G. (eds.) LPNMR 1999. LNCS, vol. 1730, pp. 147–161. Springer, Heidelberg (1999). doi:10.1007/3-540-46767-X_11
26. Schaub, T.: Here's the beef: answer set programming!. In: Proceedings of ICLP 2008, pp. 93–98 (2008)
27. Slota, M., Leite, J.: On semantic update operators for answer-set programs. In: Proceedings of ECAI 2010, pp. 957–962 (2010)
28. Slota, M., Leite, J.: The rise and fall of semantic rule updates based on se-models. TPLP **14**(6), 869–907 (2014)
29. Turner, H.: Strong equivalence made easy: nested expressions and weight constraints. TPLP **3**, 609–622 (2003)
30. Würbel, E., Jeansoulin, R., Papini, O.: Revision: an application in the framework of GIS. In: Proceedings of KR 2000, pp. 505–518 (2000)
31. Zhang, Y., Foo, N.Y.: Updating logic programs. In: Proceedings of ECAI 1998, pp. 403–407 (1998)
32. Zhuang, Z., Delgrande, J.P., Nayak, A.C., Sattar, A.: A new approach for revising logic programs. In: Proceedings of NMR 2016, pp. 171–176 (2016)
33. Zhuang, Z., Delgrande, J.P., Nayak, A.C., Sattar, A.: Reconsidering AGM-style belief revision in the context of logic programs. In: Proceedings of ECAI 2016, pp. 671–679 (2016)

On Iterated Contraction: Syntactic Characterization, Representation Theorem and Limitations of the Levi Identity

Sébastien Konieczny[1]([⊠]) and Ramón Pino Pérez[2]

[1] CRIL - CNRS, Université d'Artois, Lens, France
konieczny@cril.fr
[2] Facultad de Ciencias, Universidad de Los Andes, Mérida, Venezuela
pino@ula.ve

Abstract. In this paper we study iterated contraction in the epistemic state framework, offering a counterpart of the work of Darwiche and Pearl for iterated revision. We provide pure syntactical postulates for iterated contraction, that is, the postulates are expressed only in terms of the contraction operator. We establish a representation theorem for these operators. Our results allow to highlight the relationships between iterated contraction and iterated revision. In particular we show that iterated revision operators form a larger class than that of iterated contraction operators. As a consequence of this, in the epistemic state framework, the Levi identity has limitations; namely, it doesn't allow to define all iterated revision operators.

1 Introduction

Belief change theory [1,7–9,11,12] aims at modelling the evolution of the logical beliefs of an agent according to new inputs the agent receives.

The two main classes of operators are revision operators, which allow to correct some wrong beliefs of the agent, and contraction operators, which allow to remove some pieces of beliefs from the beliefs of the agent.

Contraction and revision, though being different processes, are closely linked. Two *identities* allow to define contraction from revision and vice-versa. One can define a revision operator from a contraction operator by the *Levi identity*, which states that, in order to define a revision by α, one can first perform a contraction by $\neg\alpha$ and then an expansion[1] by α [15]. Conversely, one can define a contraction operator from a revision operator by using the *Harper identity*: what is true after contraction by α is what is true in the current state and in the result of the revision by $\neg\alpha$ [10]. To give the formal definition of these identities, let us denote by K a theory (a deductively closed set of logical sentences), and let α be a formula. Let us denote $\star$ a revision operator, $\div$ a contraction operator, and $\oplus$ the expansion:

---

[1] See [8] for exact definition, but one can safely identify expansion with conjunction/union in most cases.

S. Moral et al. (Eds.): SUM 2017, LNAI 10564, pp. 348–362, 2017.
DOI: 10.1007/978-3-319-67582-4_25

| Levi identity | $K \star \alpha = (K \div \neg \alpha) \oplus \alpha$ |
| Harper identity | $K \div \alpha = K \cap (K \star \neg \alpha)$ |

The connection obtained through these identities is very strong, since one obtains in fact a bijection between the set of revision operators and the set of contraction operators [8]. So, in the AGM framework these two classes of operators are two sides of the same coin, and one can study either revision or contraction, depending on which operator is chosen as more basic/natural.

Although intrinsically a dynamic process, initial works on belief change only address the (static) one-step change [1,8,11], and were not able to cope with iterated change.

After many unsuccessful attempts, a solution for modelling iterated revision was provided by Darwiche and Pearl [6]. They provide additional postulates to govern iterated change. These additional constraints for iteration cannot use simple logical theories for belief representation, as used in the one-step case. One has to shift to a more powerful representation, epistemic states, which allow to encode the revision strategy of the agent, and allow to ensure dynamic coherence of changes.

It is interesting to note that the work of Darwiche and Pearl was dedicated to iterated revision. One could expect that a characterization of iterated contraction could be obtained safely from generalizations of the identities. In fact this is not the case. First, until now there is no proposal for postulates nor representation theorem for iterated contraction. There were some works on iterated contraction that we will discuss in the related work section, but no real counterpart of the work of Darwiche and Pearl for contraction existed so far. This is what we propose in this paper.

The class of iterated contraction operators obtained is very interesting in different respects. It allows to obtain a better understanding of belief change theory. Actually, some of the consequences of our representation theorem are quite surprising.

First, when comparing the two representation theorems (the one for iterated contraction and the one for iterated revision), it is clear that they share the same class of (faithful) assignments. This means that the difference between iterated contraction and iterated revision is not a matter of nature, but a matter of degree (revision being a bigger change than contraction, not a different kind of change). This also means that there is a deep relationship between iterated contraction and iterated revision via the representation theorems and the assignments. One could think that the generalization of the Levi and Harper identities could be easily attained thanks to these theorems. Nevertheless, it is not the case.

More surprisingly, we prove that there are more iterated revision operators than contraction operators (oppositely to the bijection obtained in the classical - AGM - framework). As a consequence, one can not expect to have generalizations of the Levi identity for the iterated case: some iterated revision operators are out of reach from iterated contraction ones. This seems to suggest that iterated belief revision operators are more basic than iterated belief contraction ones.

In the next Section we will provide the formal preliminaries for this paper. In Sect. 3 we give the logical postulates for modelling iterated contraction. In Sect. 4 we provide a representation theorem for these contractions. In Sect. 5 we study the links between iterated contraction and iterated revision. In Sect. 6 we discuss some related works and we conclude in Sect. 7.

2 Preliminaries

We consider a propositional language $\mathcal{L}$ defined from a finite set of propositional variables $\mathcal{P}$ and the standard connectives. Let $\mathcal{L}^*$ denote the set of consistent formulae of $\mathcal{L}$.

An interpretation ω is a total function from $\mathcal{P}$ to $\{0,1\}$. The set of all interpretations is denoted by $\mathcal{W}$. An interpretation ω is a model of a formula $\phi \in \mathcal{L}$ if and only if it makes it true in the usual truth functional way. $[\![\alpha]\!]$ denotes the set of models of the formula α, i.e., $[\![\alpha]\!] = \{\omega \in \mathcal{W} \mid \omega \models \alpha\}$. $\vdash$ denotes implication between formulae, i.e. $\alpha \vdash \beta$ means $[\![\alpha]\!] \subseteq [\![\beta]\!]$.

$\leq$ denotes a pre-order on $\mathcal{W}$ (i.e., a reflexive and transitive relation), and $<$ denotes the associated strict order defined by $\omega < \omega'$ if and only if $\omega \leq \omega'$ and $\omega' \not\leq \omega$. A pre-order is *total* if for all $\omega, \omega' \in \mathcal{W}$, $\omega \leq \omega'$ or $\omega' \leq \omega$. If $A \subseteq \mathcal{W}$, then the set of minimal elements of A with respect to a total pre-order $\leq$, denoted by $\min(A, \leq)$, is defined by $\min(A, \leq) = \{\omega \in A \mid \nexists \omega' \in A \text{ such that } \omega' < \omega\}$.

We will use epistemic states to represent the beliefs of the agent, as usual in iterated belief revision [6]. An epistemic state Ψ represents the current beliefs of the agent, but also additional conditional information guiding the revision process (usually represented by a pre-order on interpretations, a set of conditionals, a sequence of formulae, etc.). Let $\mathcal{E}$ denote the set of all epistemic states. A projection function $B : \mathcal{E} \longrightarrow \mathcal{L}^*$ associates to each epistemic state Ψ a consistent formula $B(\Psi)$, that represents the current beliefs of the agent in the epistemic state Ψ. We will call models of the epistemic state the models of its beliefs, i.e. $[\![\Psi]\!] = [\![B(\Psi)]\!]$.

A concrete and very useful representation of epistemic states are total pre-orders over interpretations. In this representation, if $\Psi = \leq$, $B(\Psi)$ is a propositional formula which satisfies $[\![B(\Psi)]\!] = \min(\mathcal{W}, \leq)$. We call this concrete representation of epistemic states the *canonical representation*.

For simplicity purpose we will only consider in this paper consistent epistemic states and consistent new information. Thus, we consider change operators as functions $\circ$ mapping an epistemic state and a consistent formula into a new epistemic state, i.e. in symbols, $\circ : \mathcal{E} \times \mathcal{L}^* \longrightarrow \mathcal{E}$. The image of a pair (Ψ, α) under $\circ$ will be denoted by $\Psi \circ \alpha$.

Let us now recall Darwiche and Pearl proposal for iterated revision [6]. Darwiche and Pearl modified the list of KM postulates [11] to work in the more general framework of epistemic states:

(R*1) $B(\Psi \star \alpha) \vdash \alpha$
(R*2) If $B(\Psi) \wedge \alpha \nvdash \bot$ then $B(\Psi \star \alpha) \equiv \varphi \wedge \alpha$
(R*3) If $\alpha \nvdash \bot$ then $B(\Psi \star \alpha) \nvdash \bot$

(R*4) If $\Psi_1 = \Psi_2$ and $\alpha_1 \equiv \alpha_2$ then $B(\Psi_1 \star \alpha_1) \equiv B(\Psi_2 \star \alpha_2)$
(R*5) $B(\Psi \star \alpha) \wedge \psi \vdash B(\Psi \star (\alpha \wedge \psi))$
(R*6) If $B(\Psi \star \alpha) \wedge \psi \nvdash \bot$ then $B(\Psi \star (\alpha \wedge \psi)) \vdash B(\Psi \star \alpha) \wedge \psi$

For the most part, the DP list is obtained from the KM list by replacing each φ by $B(\Psi)$ and each $\varphi \star \alpha$ by $B(\Psi \star \alpha)$. The only exception to this is (R*4), which is stronger than its simple translation.

In addition to this set of basic postulates, Darwiche and Pearl proposed a set of postulates devoted to iteration:

(DP1) If $\alpha \vdash \mu$ then $B((\Psi \star \mu) \star \alpha) \equiv B(\Psi \star \alpha)$
(DP2) If $\alpha \vdash \neg\mu$ then $B((\Psi \star \mu) \star \alpha) \equiv B(\Psi \star \alpha)$
(DP3) If $B((\Psi \star \alpha) \vdash \mu$ then $B((\Psi \star \mu) \star \alpha) \vdash \mu$
(DP4) If $B((\Psi \star \alpha) \nvdash \neg\mu$ then $B((\Psi \star \mu) \star \alpha) \nvdash \neg\mu$

And then they give a representation theorem in terms of pre-orders on interpretations:

Definition 1. *A faithful assignment is a mapping that associates to any epistemic state Ψ a total pre-order $\leq_\Psi$ on interpretations such that:*

1. *If $\omega \models B(\Psi)$ and $\omega' \models B(\Psi)$, then $\omega \simeq_\Psi \omega'$*
2. *If $\omega \models B(\Psi)$ and $\omega' \nvDash B(\Psi)$, then $\omega <_\Psi \omega'$*
3. *If $\Psi_1 = \Psi_2$, then $\leq_{\Psi_1} = \leq_{\Psi_2}$*

Theorem 1 ([6]). *An operator $\star$ satisfies (R*1)–(R*6) if and only if there is a faithful assignment that maps each epistemic state Ψ to a total pre-order on interpretations $\leq_\Psi$ such that:*

$$[\![\Psi \star \mu]\!] = \min([\![\mu]\!], \leq_\Psi)$$

Theorem 2 ([6]). *Let $\star$ be a revision operator that satisfies (R*1)–(R*6). This operator satisfies (DP1)–(DP4) if and only if this operator and its faithful assignment satisfies:*

(CR1) *If $\omega \models \mu$ and $\omega' \models \mu$, then $\omega \leq_\Psi \omega' \Leftrightarrow \omega \leq_{\Psi \star \mu} \omega'$*
(CR2) *If $\omega \models \neg\mu$ and $\omega' \models \neg\mu$, then $\omega \leq_\Psi \omega' \Leftrightarrow \omega \leq_{\Psi \star \mu} \omega'$*
(CR3) *If $\omega \models \mu$ and $\omega' \models \neg\mu$, then $\omega <_\Psi \omega' \Rightarrow \omega <_{\Psi \star \mu} \omega'$*
(CR4) *If $\omega \models \mu$ and $\omega' \models \neg\mu$, then $\omega \leq_\Psi \omega' \Rightarrow \omega \leq_{\Psi \star \mu} \omega'$*

The first aim of this paper is to provide a similar direct characterization of iterated contraction. Then we will study the links between iterated revision and iterated contraction operators.

3   Iterated Contraction

Let us first give the basic postulates for contraction of epistemic states. We use the contraction postulates given for propositional logic formulas in [4], that are equivalent to the original AGM ones [1], that we only adapt for epistemic states:

(C1) $B(\Psi) \vdash B(\Psi - \alpha)$

(C2) If $B(\Psi) \nvdash \alpha$, then $B(\Psi - \alpha) \vdash B(\Psi)$

(C3) If $B(\Psi - \alpha) \vdash \alpha$, then $\vdash \alpha$

(C4) $B(\Psi - \alpha) \wedge \alpha \vdash B(\Psi)$

(C5) If $\alpha_1 \equiv \alpha_2$ then $B(\Psi - \alpha_1) \equiv B(\Psi - \alpha_2)$

(C6) $B(\Psi - (\alpha \wedge \beta)) \vdash B(\Psi - \alpha) \vee B(\Psi - \beta)$

(C7) If $B(\Psi - (\alpha \wedge \beta)) \nvdash \alpha$, then $B(\Psi - \alpha) \vdash B(\Psi - (\alpha \wedge \beta))$

(C1) states that contraction can just remove some information, so the beliefs of the posterior epistemic state are weaker than the beliefs of the prior one. (C2) says that if the epistemic state does not imply the formula by which one wants to contract, then the posterior epistemic state will have the same beliefs as the prior one. (C3) is the success postulate, it states that the only case where contraction fails to remove a formula from the beliefs of the agent is when this formula is a tautology. (C4) is the recovery postulate, it states that if we do the contraction by a formula followed by a conjunction by this formula, then we will recover the initial beliefs. This ensures that no unnecessary information is discarded during the contraction. (C5) is the irrelevance of syntax postulate, that says that the syntax does not have any impact on the result of the contraction. (C6) says that the contraction by a conjunction implies the disjunction of the contractions by the conjuncts. (C7) says that if α is not removed during the contraction by the conjunction by $\alpha \wedge \beta$, then the contraction by α implies the contraction by the conjunction.

Now let us introduce the postulates for iterated contraction:

(C8) If $\neg\alpha \vdash \gamma$ then $B(\Psi - (\alpha \vee \beta)) \vdash B(\Psi - \alpha) \Leftrightarrow B(\Psi - \gamma - (\alpha \vee \beta)) \vdash B(\Psi - \gamma - \alpha)$

(C9) If $\gamma \vdash \alpha$ then $B(\Psi - (\alpha \vee \beta)) \vdash B(\Psi - \alpha) \Leftrightarrow B(\Psi - \gamma - (\alpha \vee \beta)) \vdash B(\Psi - \gamma - \alpha)$

(C10) If $\neg\beta \vdash \gamma$ then $B(\Psi - \gamma - (\alpha \vee \beta)) \vdash B(\Psi - \gamma - \alpha) \Rightarrow B(\Psi - (\alpha \vee \beta)) \vdash B(\Psi - \alpha)$

(C11) If $\gamma \vdash \beta$ then $B(\Psi - \gamma - (\alpha \vee \beta)) \vdash B(\Psi - \gamma - \alpha) \Rightarrow B(\Psi - (\alpha \vee \beta)) \vdash B(\Psi - \alpha)$

(C8) expresses the fact that if a contraction by a disjunction implies the contraction by one of the disjuncts, then it will be the same if we first contract by a formula that is a consequence of the negation of that disjunct. (C9) captures the fact that if a contraction by a disjunction implies the contraction by one of the disjuncts, then it will be the same if we first contract by a formula that implies this disjunct. (C10) expresses the fact that if a contraction by a disjunction implies the contraction by one of the disjunct after a contraction by a formula that is a consequence of the negation of the other disjunct, then it was already the case before this contraction. (C11) captures the fact that if a contraction by a disjunction implies the contraction by one of the disjunct after a contraction by a formula that implies the other disjunct, then it was already the case before this contraction.

The operators satisfying (C1)–(C7) will be called contraction operators. The operators satisfying (C1)–(C11) will be called iterated contraction operators.

4 Representation Theorem

Let us now provide a representation theorem for iterated contraction operators in terms of faithful assignments. These assignments associate to each epistemic state a total pre-order on interpretations, this total pre-order representing the relative plausibility of each interpretation, and the current beliefs of the agent being the most plausible ones.

And let us now state the basic theorem for contraction postulates in the epistemic state framework.

Theorem 3. *An operator $-$ satisfies the postulates (C1)–(C7) if and only if there exists a faithful assignment that associates to each epistemic state Ψ a total pre-order $\leq_\Psi$ on interpretations such that*

$$[\![\Psi - \alpha]\!] = [\![\Psi]\!] \cup \min([\![\neg\alpha]\!], \leq_\Psi)$$

We will say that the faithful assignment given by the previous theorem *represents* the operator $-$. The proof of this theorem follows the same lines as the proof for the representation theorem in the propositional case [4]. For space reason it will not be included here. We concentrate our effort in proving a representation theorem for iterated contraction operators.

Let us now state the representation theorem for iterated contraction.

Theorem 4. *Let $-$ be a contraction operator that satisfies (C1)–(C7). This operator satisfies (C8)–(C11) if and only if this operator and its faithful assignment satisfies:*

4. *If $\omega, \omega' \in [\![\gamma]\!]$ then $\omega \leq_\Psi \omega' \Leftrightarrow \omega \leq_{\Psi - \gamma} \omega'$*
5. *If $\omega, \omega' \in [\![\neg\gamma]\!]$ then $\omega \leq_\Psi \omega' \Leftrightarrow \omega \leq_{\Psi - \gamma} \omega'$*
6. *If $\omega \in [\![\neg\gamma]\!]$ and $\omega' \in [\![\gamma]\!]$ then $\omega <_\Psi \omega' \Rightarrow \omega <_{\Psi - \gamma} \omega'$*
7. *If $\omega \in [\![\neg\gamma]\!]$ and $\omega' \in [\![\gamma]\!]$ then $\omega \leq_\Psi \omega' \Rightarrow \omega \leq_{\Psi - \gamma} \omega'$*

We will call a faithful assignment that satisfies properties 4 to 7 an iterated faithful assignment.

Condition 4 captures the fact that the plausibility between the models of γ is exactly the same before the contraction and after the contraction by γ. Condition 5 captures the fact that the plausibility between the models of $\neg\gamma$ is exactly the same before the contraction and after the contraction by γ. Conditions 4 and 5 will be called *rigidity conditions* (called also ordered preservation conditions in [18]). Condition 6 captures the fact that if a model of $\neg\gamma$ is strictly more plausible than a model of γ before the contraction by γ then it will be the case after the contraction by γ. Condition 7 captures the fact that the plausibility of the models of $\neg\gamma$ does not decrease with respect the models of γ after contraction by γ. More precisely, if a model of $\neg\gamma$ is at least as plausible as a model of γ before the contraction it will be the case after the contraction by γ. Conditions 6 and 7 will be called *non-worsening conditions*.

Please note that the iterated faithful assignments are directly related to the ones for iterated revision (cf. Theorem 2), there are only some inversions in conditions 6 and 7 (compared to CR3 and CR4), that are due to the fact that one can see a contraction by $\neg\alpha$ as a softer change (improvement [14]) than a revision by α (see discussion at the beginning of Sect. 6).

For space reasons we only give a sketch of the proof of Theorem 4. An iterated contraction operator is indeed a contraction operator. Thus, we know by Theorem 3 that there is a faithful assignment representing it. Thus the proof of Theorem 4 will consist in proving that under the assumption of the basic contraction postulates, the iteration postulates entail conditions 4–7 for the faithful assignment representing the operator. And reciprocally, that if the iterated faithful assignment represents the operator the postulates of the iteration are satisfied. Thus, from now on, in this section we suppose that $-$ is a contraction operator and $\Psi \mapsto \leq_\Psi$ is the faithful assignment representing it, that is the equation in Theorem 3 holds. Actually, we prove the following facts which are enough to conclude:

(i) The assignment satisfies condition 4 if and only if postulate (C8) holds.
(ii) The assignment satisfies condition 5 if and only if postulate (C9) holds.
(iii) Suppose the contraction operator satisfies (C8). Then the assignment satisfies condition 6 if and only if postulate (C10) holds.
(iv) The assignment satisfies condition 7 if and only if postulate (C11) holds.

In order to give a flavor of the whole proof we give the proof of the Fact (iii) (which is perhaps a little more complicated than the other three facts). First we have the following observations:

Observation 1. *Suppose that the assignment satisfies condition 4, then for any μ such that $\mu \vdash \gamma$ we have $\min(\llbracket\mu\rrbracket, \leq_\Psi) = \min(\llbracket\mu\rrbracket, \leq_{\Psi-\gamma})$.*

Observation 2. *Condition 6 is equivalent to the following condition:*

6'. If $\omega \in \llbracket\neg\gamma\rrbracket$ and $\omega' \in \llbracket\gamma\rrbracket$ then $\omega' \leq_{\Psi-\gamma} \omega \Rightarrow \omega' \leq_\Psi \omega$

Now we proceed to prove (iii). Note that, since α, β and γ are any formulas, using (C5), the postulate (C10) can be rewritten as follows:

If $\beta \vdash \gamma$ then $B((\Psi - \gamma) - \neg(\alpha \wedge \beta)) \vdash B((\Psi - \gamma) - \neg\alpha) \Rightarrow B(\Psi - \neg(\alpha \wedge \beta)) \vdash B(\Psi - \neg\alpha)$

First we prove that postulate (C10) entails condition 6 of an iterated assignment. By Observation 2, it is enough to prove that (C10) entails 6'. Thus, assume (C10) holds. Suppose $\omega \in \llbracket\neg\gamma\rrbracket$, $\omega' \in \llbracket\gamma\rrbracket$ and $\omega' \leq_{\Psi-\gamma} \omega$. We want to show that $\omega' \leq_\Psi \omega$.

Let α and β be formulas such that $[\![\alpha]\!] = \{\omega, \omega'\}$ and $[\![\beta]\!] = \{\omega'\}$. Note that $\{\omega'\} \subseteq \min([\![\alpha]\!], \leq_{\Psi - \gamma})$ because $\omega' \leq_{\Psi - \gamma} \omega$. We have also $\min([\![\alpha \wedge \beta]\!], \leq_{\Psi - \gamma}) = \{\omega'\}$.

Then, we have $[\![(\Psi - \gamma) - \neg(\alpha \wedge \beta)]\!] = [\![\Psi - \gamma]\!] \cup \min([\![\alpha \wedge \beta]\!], \leq_{\Psi - \gamma})$; the last expression is equal to $[\![\Psi - \gamma]\!] \cup \{\omega'\}$ which is a subset of $[\![\Psi - \gamma]\!] \cup \min([\![\alpha]\!], \leq_{\Psi - \gamma})$. But this last expression is $[\![(\Psi - \gamma) - \neg\alpha]\!]$. So, $[\![(\Psi - \gamma) - \neg(\alpha \wedge \beta)]\!] \subseteq [\![(\Psi - \gamma) - \neg\alpha]\!]$, that is $B((\Psi - \gamma) - \neg(\alpha \wedge \beta)) \vdash B((\Psi - \gamma) - \neg\alpha)$. Then, by (C10), we have $B(\Psi - \neg(\alpha \wedge \beta)) \vdash B(\Psi - \neg\alpha)$, that is $[\![\Psi]\!] \cup \{\omega'\} \subseteq [\![\Psi]\!] \cup \min(\{\omega, \omega'\}, \leq_\Psi)$. Therefore, $\omega' \in [\![\Psi]\!]$ or $\omega' \in \min(\{\omega, \omega'\}, \leq_\Psi)$. In both cases we get $\omega' \leq_\Psi \omega$ (In the first case we use the fact that $\leq_\Psi$ is a faithful assignment, in particular $[\![\Psi]\!] = \min(\mathcal{W}, \leq_\Psi)$).

Now we prove that Condition 6 entails Postulate (C10). Assume that $\beta \vdash \gamma$. We suppose that $\alpha \wedge \beta \nvdash \bot$ (the other case is trivial because the contraction by a tautology doesn't change the beliefs). Suppose $B((\Psi - \gamma) - \neg(\alpha \wedge \beta)) \vdash B((\Psi - \gamma) - \neg\alpha)$, that is

$$[\![(\Psi - \gamma) - \neg(\alpha \wedge \beta)]\!] \subseteq [\![(\Psi - \gamma) - \neg\alpha]\!] \tag{1}$$

We want to show that $B(\Psi - \neg(\alpha \wedge \beta)) \vdash B(\Psi - \neg\alpha)$, that is to say

$$[\![\Psi - \neg(\alpha \wedge \beta)]\!] \subseteq [\![\Psi - \neg\alpha]\!] \tag{2}$$

By Theorem 3, Eqs. (1) and (2) can be rewritten, respectively, as

$$[\![\Psi - \gamma]\!] \cup \min([\![\alpha \wedge \beta]\!], \leq_{\Psi - \gamma}) \subseteq [\![\Psi - \gamma]\!] \cup \min([\![\alpha]\!], \leq_{\Psi - \gamma}) \tag{3}$$

and

$$[\![\Psi]\!] \cup \min([\![\alpha \wedge \beta]\!], \leq_\Psi) \subseteq [\![\Psi]\!] \cup \min([\![\alpha]\!], \leq_\Psi) \tag{4}$$

First we are going to prove that (3) entails

$$\min([\![\alpha \wedge \beta]\!], \leq_{\Psi - \gamma}) \subseteq \min([\![\alpha]\!], \leq_{\Psi - \gamma}) \tag{5}$$

In order to see that, take $\omega \in \min([\![\alpha \wedge \beta]\!], \leq_{\Psi - \gamma})$. Then, by Eq. (3), we have either $\omega \in [\![\Psi - \gamma]\!]$ or $\omega \in \min([\![\alpha]\!], \leq_{\Psi - \gamma})$. In the first case, by the fact of having a faithful assignment, $\omega \in \min(\mathcal{W}, \leq_{\Psi - \gamma})$ and therefore $\omega \in \min([\![\alpha]\!], \leq_{\Psi - \gamma})$. In the second case is trivial. Thus, in any case we have 5.

Now, towards a contradiction, suppose that (4) doesn't hold. Thus, there exists $\omega \in \min([\![\alpha \wedge \beta]\!], \leq_\Psi)$ such that $\omega \notin \min([\![\alpha]\!], \leq_\Psi)$. So there is $\omega' \in \min([\![\alpha]\!], \leq_\Psi)$ such that

$$\omega' <_\Psi \omega \tag{6}$$

Note that $\omega \models \beta$, and, by hypothesis, $\beta \vdash \gamma$, thus $\omega \models \gamma$. We are going to consider the following two cases: $\omega' \models \gamma$ and $\omega' \models \neg\gamma$

Suppose we are in the first case, i.e. $\omega' \models \gamma$. Then, by Condition 4 (equivalent to our assumption of Postulate (C8)), $\omega' <_{\Psi - \gamma} \omega$. Now, suppose we are in the second case, i.e. $\omega' \models \neg\gamma$. Then, because $\omega \models \gamma$ and (6), by Condition 6, $\omega' <_{\Psi - \gamma} \omega$. Thus in any case we have

$$\omega' <_{\Psi - \gamma} \omega \tag{7}$$

Since $\alpha \wedge \beta \vdash \gamma$, by Observation 1, $\omega \in \min([\![\alpha \wedge \beta]\!], \leq_{\Psi - \gamma})$. Then, by (5), $\omega \in \min([\![\alpha]\!], \leq_{\Psi - \gamma})$. But this is a contradiction with (7).

5   Iterated Contraction Vs Iterated Revision

We would like to investigate now the relationship between iterated contraction and iterated revision.

A natural tendency would be to try to generalize Levi and Harper Identity to the iterated case. In fact some related works followed this path [2,5,17].

In the following we will first argue and show that it is not so simple. We will also show that there are some problems when one follows this way for connecting iterated contraction and iterated revision. Actually, we will show that in the iterated case, they are not two sides of a same coin (i.e. two classes of operators linked by a bijection), but that they rather are two instances of a same kind of change operators, and that the link and difference is just a matter of degree of change.

5.1   Identities in the General Case

Let us first recall the Levy and Harper Identities:

Levi identity $\Psi \star \alpha = (\Psi \div \neg\alpha) \oplus \alpha$
Harper identity $\Psi \div \alpha = \Psi \sqcap (\Psi \star \neg\alpha)$

Let us note the problems of using these identities for iterated contraction and revision in the epistemic state framework. First, in the AGM case, these two identities are definitional, that means that, for instance, using Levi identity one can obtain the revision operator $\star$ that defines the theory $\Psi \star \alpha$ from the right side of the identity using the contraction and expansion operators. But in our general framework, epistemic states are abstract objects, which can only be apprehended at the logical level from the projection function B. Thus, in this general framework, we do not fully know what $\Psi \div \neg\alpha$ is, and so we can not use it to define what should be $\Psi \star \alpha$. So using a definitional equality here is difficult. The only way to proceed seems to be choosing a particular representation of epistemic states to work with (such as total preorders over the interpretations), but then the results are given on this representation and not in the general case. Second, whereas $\oplus$ and $\sqcap$ have a clear meaning in the AGM framework, one has to figure out a definition in the epistemic state framework. That is by itself a non-trivial task (studying the possible definitions of $\sqcap$ is one of the main aims of [2]).

A possibility would be to restrict these identities to the beliefs of the epistemic states, as:

Belief Levi equivalence $B(\Psi \star \alpha) \equiv B(\Psi \div \neg\alpha) \oplus \alpha$
Belief Harper equivalence $B(\Psi \div \alpha) \equiv B(\Psi) \vee B(\Psi \star \neg\alpha)$

But we do not have identities anymore, but only equivalences that are not definitional. So one has to first identify two operators $\star$ and $\div$ and check that they are linked through these equivalences.

5.2   Identities Under the Canonical Representation

So to go further we have to commit to a particular representation of epistemic states. The canonical one, using total preorders (described in Sect. 2), can be used together with the faithful assignment, to define completely the new epistemic state after contraction (or revision). Thus, suppose that we have a faithful assignment $\Psi \mapsto \leq_\Psi$. We identify Ψ with $\leq_\Psi$, and we define $\leq_{\Psi-\gamma}$ satisfying the properties (4–7) of Theorem 4. Then, by Theorem 4, the operator defined by $\Psi - \gamma = \leq_{\Psi-\gamma}$ is an iterated contraction operator. In such a case we say that the operator $-$ is *given by* the assignment. We can proceed, in the same way when the assignment satisfies the requirements of an iterated assignment (for revision [6]) and then the operator defined by $\Psi \star \alpha = \leq_{\Psi\star\alpha}$ is an iterated revision operator.

Thus, one can restate the identities on the total pre-orders associated to the epistemic states by the operators[2]:

Tpo Levi identity $\leq_{\Psi\star\alpha} = \leq_{\Psi\div\neg\alpha} \oplus \alpha$
Tpo Harper identity $\leq_{\Psi\div\alpha} = \leq_\Psi \sqcap \leq_{\Psi\star\neg\alpha}$

So now we can define these pre-orders using the identities and check that we correctly obtain operators with the expected properties. The only remaining problem is to define the operators $\oplus$ and $\sqcap$ in this setting.

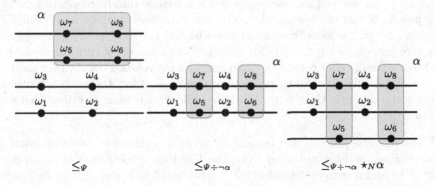

Fig. 1. From contraction to revision

As for $\oplus$ let us show that using Boutilier natural revision operator $\star_N$ [3] is a correct option, in the sense that using this operator as $\oplus$ we obtain a DP (Darwiche and Pearl [6]) revision operator (see Proposition 1).

Let us recall the definition of this operator on total pre-orders, that amounts to look at the most plausible models of the new piece of information and define them as the new most plausible interpretations while nothing else changes: Let $\leq_\Psi$ be the pre-order associated to the epistemic state Ψ by the faithful assignment, and let α be the new piece of information, then $\leq_{\Psi\star_N\alpha}$ (we will also use the equivalent notation $\leq_\Psi \star_N\alpha$) is defined as:

---

[2] Tpo means Total pre-order.

- If $\omega \models \min([[\alpha]], \leq_\Psi)$ and $\omega' \not\models \min([[\alpha]], \leq_\Psi)$, then $\omega <_{\Psi \star_N \alpha} \omega'$
- In all the other cases $\omega \leq_{\Psi \star_N \alpha} \omega'$ iff $\omega \leq_\Psi \omega'$

Then one can show that the Tpo Levi identity holds for the iterated case:

Proposition 1. *Let $\div$ be an iterated contraction operator given by its assignment $\Psi \mapsto \leq_\Psi$. Then the assignment defined by $\leq_{\Psi \star \alpha} = \leq_{\Psi \div \neg \alpha} \star_N \alpha$ satisfies properties (CR1)–(CR4), and can be used to define a Darwiche and Pearl iterated revision operator in the framework of the canonical representation of epistemic states.*

This proposition implies in particular that to each iterated contraction operator one can associate a corresponding iterated revision operator. So, this means that the cardinality of the class of iterated revision operators obtained via the Tpo Levi identity is at least equal to the cardinality of the class of iterated contraction operators. Note that this observation does not depend on the interpretation of the symbol $\oplus$ utilized.

The following example illustrates the use of our concrete Tpo Levi identity.

Example 1. Let us consider the total pre-order $\leq_\Psi$ represented in Fig. 1. So in that figure $[[\leq_\Psi]] = \{\omega_1, \omega_2\}$ and $[[\alpha]] = \{\omega_5, \omega_6, \omega_7, \omega_8\}$. In this Figure the lower an interpretation is, the more plausible it is. For instance in $\leq_\Psi$ we have that $\omega_1 <_\Psi \omega_3$. An iterated contraction by $\neg\alpha$ is a change that increases (improves) the plausibility of the models of α, with the condition that the most plausible models of α in $\leq_\Psi$ joins (become as plausible as) the most plausible models of $\leq_\Psi$ (We give on such possibility for $\leq_{\Psi \div \neg \alpha}$). The relation between the models of α doesn't change after contraction and nor does the relation between the models of $\neg\alpha$. From this, to define a revision, one can just select these most plausible models of α and take them as the most plausible models using Boutilier's natural revision ($\leq_{\Psi \div \neg \alpha} \star_N \alpha$).

The converse process, that is, defining iterated contraction operators starting from iterated revision operators using the Tpo Harper identity, requires in particular to find a correct definition for $\sqcap$. This problem is investigated by Booth and Chandler [2], where they show that there is not a single, canonical way to proceed.

One can see this as an additional richness of the iterated framework. However, this richness of the epistemic state representation has its counterparts. In fact in the iterated case there are more revision operators than contraction ones. More precisely we have the following result:

Theorem 5. *There are more iterated revision operators than iterated contraction operators. In particular, this entails that it is impossible to find an interpretation of the expansion $\oplus$ in the Tpo Levi identity in order to obtain all the iterated revision operators via this identity.*

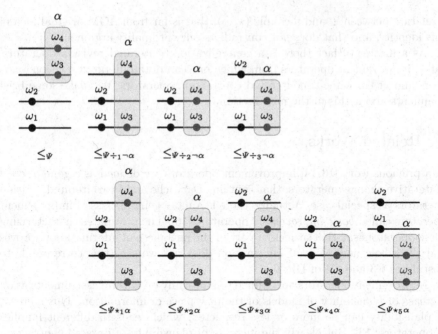

Fig. 2. All possible contractions by $\neg\alpha$ and revisions by α from $\leq_\Psi$

Proof. Let $\leq_\Psi$ and α be as illustrated in Fig. 2. Due to rigidity conditions for the iterated contraction, there are only three possible different outputs as results of contraction of $\leq_\Psi$ by $\neg\alpha$ ($\leq_{\Psi\div_1\neg\alpha}$, $\leq_{\Psi\div_2\neg\alpha}$ and $\leq_{\Psi\div_3\neg\alpha}$). Contrastingly there are five different possible output for revision. That is due to the rigidity postulates for iterated revision. Three of these possible revision outputs can be obtained from contraction outputs using natural revision as in the previous example ($\leq_{\Psi\star_1\alpha}$, $\leq_{\Psi\star_2\alpha}$, $\leq_{\Psi\star_3\alpha}$). The other two ones are $\leq_{\Psi\star_4\alpha}$ and $\leq_{\Psi\star_5\alpha}$, that are not related to any contraction result using the identity.

The previous theorem is important because it tells us that in the iterated case the Levy identity has limitations. This theorem is also important since it gives us a true distinction between classical AGM framework and the iterated framework. In the classical AGM framework there is a bijection between revision and contraction operators. Contraction is often considered as a more fundamental operator since a revision can be defined, through Levi identity, as a contraction followed by a conjunction (expansion). In the iterated case there are more iterated revision than iterated contraction operators, so the more general change operator seems to be revision.

One can object that $\leq_{\Psi\star_4\alpha}$ and $\leq_{\Psi\star_5\alpha}$ could maybe be obtained through Levi identity by using another definition of $\oplus$ than Boutilier's natural revision operator, as we used here. But this does not change the fact that there is only three possible contraction results versus five possible revision results, and then with any alternative, there is still no way to define a bijection. We can just define a relation between $\div$ and a couple $(\star, \oplus)$, that is far from AGM original idea of this identity, and that does not contradicts our cardinality argument.

As a matter of fact there is a generalization of iterated revision operators, called improvement operators from which one can obtain iterated revision operators and at the same time iterated contraction operators. We make some brief comments about this in the next section.

6   Related Works

In a previous work [13,14] improvement operators are defined as a general class of iterative change operators, that contains Darwiche and Pearl iterated revision operators as special case. Actually, there is a more general class of improvement operators called *basic improvement* operators [16]. The postulates characterizing these operators say that at least a part of the new piece of information improves and the whole new piece of information does not worsen (this corresponds to postulates C3 and C4 of DP [6]).

Improvement operators are defined semantically on faithful assignments as an increase of plausibility of models of the new piece of information. This increase of plausibility can be more or less restricted, which leads to different families of operators [13]. But clearly the increase of plausibility of iterated contraction operators is limited due to the fact that the most plausible models of the new piece of information can not become more plausible than the models of the previous beliefs of the agent. Whereas for revision there is no such constraint, and so much more freedom is granted for improvement.

The following proposition says that our iterated contraction operators are also weak improvement operators (by the negation of the input).

Proposition 2. *Let $\div$ be an iterated contraction operator, then the operator $\hat{\div}$ defined as $\Psi \hat{\div} \alpha = \Psi \div \neg\alpha$ is a weak improvement operator [14], moreover it is a basic improvement operator.*

Thus, the previous proposition and the fact that iterated revision operators are also basic improvement operators seems to mean that this class of operator as the most primitive operators in iterated belief change.

Chopra et al. [5] also give postulates for iterated contraction, but they did it using iterated revision operators in their postulates, so the iterated contraction operators are not defined independently, but as a byproduct of iterated revision ones. Actually their starting point is a couple of given operators $*$ and $-$ that satisfy the Levi and Harper equivalences. Then, they characterize the four iterated semantic properties of Definition 4 in terms of syntactical postulates mixing the operators $*$ and $-$. In this work we propose a direct characterization of iterated contraction operators (not depending of any iterated revision operator).

Booth and Chandler [2] investigate the problem of the definition of iterated contraction through the Harper Identity for the concrete case of pre-orders on interpretations. The paper shows the richness of the question. In this paper we explain this richness by the fact that there are much more iterated revision operators than iterated contraction operators, so this means that several different iterated revision operators correspond, via the Harper identity, to the same iterated contraction operator.

The work of Ramachadran et al. [18] is very interesting. It concerns the characterization of three iterated contraction operator in the framework of the canonical representation of epistemic states. They give a pure syntactical characterization of these three operators. However they don't characterize the full class of iterated contraction operators.

7 Conclusion

To sum up, in this paper we proposed the first direct logical characterization of the class of iterated contraction operators having an iterated behavior similar to the one proposed by Darwiche and Pearl for iterated revision operators. We stated a representation theorem in terms of total pre-orders on interpretations. We discussed the fact that there is no easy way to generalize the Levi and Harper identity in the iterated case, but more importantly, that this is not a primordial issue, since, conversely to the classical (AGM) case, these two classes of operators are not linked by a bijection in the iterated case. There are more iterated revision operators than iterated contraction operators, and both are special cases of the more general class of improvement operators, where iterated contractions produce a smaller change than iterated revision operators. So, these two classes of change operators are not different in nature, but in degree of change.

Acknowledgments. The authors would like to thank the reviewers for their helpful comments.

References

1. Alchourrón, C.E., Gärdenfors, P., Makinson, D.: On the logic of theory change: partial meet contraction and revision functions. J. Symbolic Logic **50**, 510–530 (1985)
2. Booth, R., Chandler, J.: Extending the harper identity to iterated belief change. In: Kambhampati, S., (ed.) Proceedings of the Twenty-Fifth International Joint Conference on Artificial Intelligence, IJCAI 2016, New York, NY, USA, 9–15 July 2016, pp. 987–993. IJCAI/AAAI Press (2016)
3. Boutilier, C.: Iterated revision and minimal change of conditional beliefs. J. Philos. Logic **25**(3), 262–305 (1996)
4. Caridroit, T., Konieczny, S., Marquis, P.: Contraction in propositional logic. In: Destercke, S., Denoeux, T. (eds.) ECSQARU 2015. LNCS (LNAI), vol. 9161, pp. 186–196. Springer, Cham (2015). doi:10.1007/978-3-319-20807-7_17
5. Chopra, S., Ghose, A., Meyer, T.A., Wong, K.-S.: Iterated belief change and the recovery axiom. J. Philos. Logic **37**(5), 501–520 (2008)
6. Darwiche, A., Pearl, J.: On the logic of iterated belief revision. Artif. Intell. **89**, 1–29 (1997)

7. Fermé, E., Hansson, S.O.: AGM 25 years. J. Philos. Logic **40**(2), 295–331 (2011)
8. Gärdenfors, P.: Knowledge in Flux. MIT Press, Cambridge (1988)
9. Hansson, S.O.: A Textbook of Belief Dynamics. Theory Change and Database Updating. Kluwer, Dordrecht (1999)
10. Harper, W.L.: Rational conceptual change. In: 1976 Proceedings of the Biennial Meeting of the Philosophy of Science Association, PSA, vol. 2, pp. 462–494. Philosophy of Science Association, East Lansing, Mich (1977)
11. Katsuno, H., Mendelzon, A.O.: Propositional knowledge base revision and minimal change. Artif. Intell. **52**, 263–294 (1991)
12. Katsuno, H., Mendelzon, A.O.: On the difference between updating a knowledge base and revising it. In: Belief Revision, pp. 183–203. Cambridge University Press (1992)
13. Konieczny, S., Medina Grespan, M., Pino Pérez, R.: Taxonomy of improvement operators and the problem of minimal change. In: Proceedings of the Twelfth International Conference on Principles of Knowledge Representation And Reasoning (KR 2010), pp. 161–170 (2010)
14. Konieczny, S., Pino Pérez, R.: Improvement operators. In: Proceedings of the Eleventh International Conference on Principles of Knowledge Representation and Reasoning (KR 2008), pp. 177–187 (2008)
15. Levi, I.: Subjunctives, dispositions and chances. Synthese **34**, 423–455 (1977)
16. Grespan, M.M., Pino Pérez, R.: Representation of basic improvement operators. In: Trends in Belief Revision and Argumentation Dynamics, pp. 195–227. College Publications (2013)
17. Nayak, A., Goebel, R., Orgun, M., Pham, T.: Taking LEVI IDENTITY seriously: a plea for iterated belief contraction. In: Lang, J., Lin, F., Wang, J. (eds.) KSEM 2006. LNCS (LNAI), vol. 4092, pp. 305–317. Springer, Heidelberg (2006). doi:10.1007/11811220_26
18. Ramachandran, R., Nayak, A.C., Orgun, M.A.: Three approaches to iterated belief contraction. J. Philos. Logic **41**(1), 115–142 (2012)

Handling Topical Metadata Regarding the Validity and Completeness of Multiple-Source Information: A Possibilistic Approach

Célia da Costa Pereira[1], Didier Dubois[2], Henri Prade[2], and Andrea G.B. Tettamanzi[1(✉)]

[1] Université Côte d'Azur, CNRS, I3S, Sophia Antipolis, France
{celia.pereira,andrea.tettamanzi}@unice.fr
[2] IRIT – CNRS, 118, route de Narbonne, Toulouse, France
{dubois,prade}@irit.fr

Abstract. We study the problem of aggregating metadata about the validity and/or completeness, with respect to given topics, of information provided by multiple sources. For a given topic, the validity level reflects the certainty that the information stored is true. The completeness level of a source on a given topic reflects the certainty that a piece of information that is not stored is false. We propose a modeling based on possibility theory which allows the fusion of such multi-source information into a graded belief base.

1 Introduction and Related Work

The relation between *beliefs* and *knowledge* plays a central role in epistemology. Much of epistemology revolves around questions about when and how our beliefs are justified or qualify as knowledge [19]. Without taking a position in this debate, in this paper, we will use the term *knowledge* when referring to information provided by an information source, but we will use the term *beliefs* to refer to a (possibly partial, incomplete, or uncertain) representation of reality obtained by combining information provided by one or more sources with metadata about its validity and completeness.

The problem of representing validity and completeness of information stored in databases has started drawing attention many years ago. For example, we can consider the model of database integrity proposed by Motro [17] and the work by Demolombe [8] who used modal logic for representing information stored in relational databases. Our aim is to consider validity and completeness in more general knowledge bases (KBs) in which the closed world assumption is not made. Therefore, a mechanism for representing uncertainty in the beliefs induced by KBs fed by sources which can provide invalid and/or incomplete pieces of information is needed.

Cholvy [6] uses the theory of evidence for proposing an interesting way to compute the extent to which an agent should believe a new piece of information

© Springer International Publishing AG 2017
S. Moral et al. (Eds.): SUM 2017, LNAI 10564, pp. 363–376, 2017.
DOI: 10.1007/978-3-319-67582-4_26

provided by an imperfect information source. A difference with respect to our work is that we explicitly associate these metadata concerning validity and completeness to topics and this allows us to describe these metadata for a source at a finer grain.

Bacchus *et al.* [3] proposed the "random worlds" method, an approach for inducing degrees of belief from KBs fed with different types of information like statistical correlations, physical laws, default rules, etc. They apply the principle of indifference and, therefore, all the possible worlds derived from the agent's KB are equally probable. The uncertainty about information is directly represented in the KB (as statistical information, defeasible information and so on), not as metadata.

We consider a possibilistic representation of beliefs to take uncertainty into account. We assume the beliefs of an agent come from various information sources, which may be more or less reliable (this has to do with information *validity*) and more or less exhaustive (this has to do with *completeness*). The validity level reflects the certainty that the information an agent stores on a given topic is true, while the completeness level reflects the certainty that, on a given topic, a missing piece of information is false.

The goal of our model is to support inferences, thus to answer queries, by providing a weighted summary of the different (and possibly conflicting) opinions of the available sources. An important point in our framework is that we provide the user (requestor) with the different answers that can be obtained from the information system in case of conflict. The user must then be aware of which sources give which answer to his/her query, and with which certainty degree.

We adapt and extend the formalism by Dubois and Prade [12] for completeness and validity of databases, to reason about the beliefs (opinions) of a source. We give a possibilistic reasoning algorithm for those beliefs, whose complexity is in the same class as reasoning on a crisp KB and less expensive than reasoning on a general possibilistic belief base. Furthermore, we combine this solution with a multi-source generalization of possibilistic logic [9] to summarize and reason about the different (and possibly conflicting) beliefs of the sources.

The paper is organized as follows: the next section states the problem we study; Sect. 3 gives then some background about the formal tools we use. Section 4 explains how a gradual set of beliefs can be constructed from validity and completeness metadata. Section 5 exploits multi-source possibilistic logic to merge beliefs from multiple sources. Section 7 concludes the paper.

2   Problem Statement

The problem we study can be schematically depicted as in Fig. 1. We are given n KBs $K_1, \ldots, K_n$, fed by n imperfect, independent information sources $s_1, \ldots, s_n$, about which two kinds of metadata are known: for each topic, on the one hand, we know to what degree a source s_i provides *valid* information. On the other hand, we know to what degree information provided by source s_i is *complete*. Here, we use the term *knowledge base* to mean a (possibly noisy and incomplete)

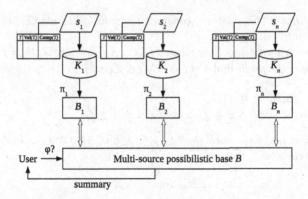

Fig. 1. A schematic illustration of an abstract information system, consisting of n knowledge bases K_i fed by n independent information sources s_i, with metadata about their validity and completeness, whence a possibility distribution π_i and a corresponding possibilistic belief base B_i are constructed and used to answer queries

collection of *facts*, for which the open world assumption (OWA) holds. We answer two important questions:

1. How can the facts contained in each KB K_i be combined with metadata about the validity and completeness of its source s_i to construct a gradual belief base B_i taking the uncertainty of K_i (due to its imperfection) into account?
2. How can the n belief bases be used to answer queries while merging (possibly conflicting) information coming from the n sources?

The former is a problem of metadata aggregation, whereas the latter is a problem of information (or, more properly, belief) fusion. We argue that possibility theory provides suitable tools to solve both problems.

To illustrate our proposal, we will use simple examples inspired, much like in [8], by an air travel planning application. One might suppose that some of the KBs in the system would be fed with flight information directly by an airline and some by an airport, each source being more authoritative about information which falls directly under its control and less for other information.

3 Background

3.1 Knowledge Representation Language

For the sake of simplicity, we base our treatment on a decidable fragment of pure (i.e., without function symbols and identity) first-order predicate logic, namely the Schönfinkel-Bernays class of first-order formulas [5].

Definition 1 (Language $\mathcal{L}$). *Let $\mathcal{L}_{QF}$ be the set of quantifier-free formulas inductively defined as follows:*

– *a term is either a variable or a constant (including literals denoting numbers, times, character strings, etc.);*
– *if P is an n-ary predicate symbol and $t_1, \ldots, t_n$ are terms, then $P(t_1, \ldots, t_n)$ is an (atomic) formula and $P(t_1, \ldots, t_n) \in \mathcal{L}_{QF}$;*
– $\bot, \top \in \mathcal{L}_{QF}$
– *if $\phi \in \mathcal{L}_{QF}$, then $\neg\phi \in \mathcal{L}_{QF}$;*
– *if $\phi, \psi \in \mathcal{L}_{QF}$ then $\phi \wedge \psi \in \mathcal{L}_{QF}$ and $\phi \vee \psi \in \mathcal{L}_{QF}$.*

$\mathcal{L}$ is the smallest language such that $\mathcal{L}_{QF} \subseteq \mathcal{L}$ and, if $\phi \in \mathcal{L}_{QF}$ and $x_1, \ldots, x_m$, $y_1, \ldots, y_n$ are variables, then

$$\exists x_1 \ldots \exists x_m \forall y_1 \ldots \forall y_n \phi \in \mathcal{L}.$$

A variable x is free *in formula ϕ if it is not quantified; otherwise it is* bound. *A formula without free variables is* closed. *A formula with free variables is* open. *A formula not containing variables is* ground.

The semantics of $\mathcal{L}$ can be defined as follows:

Definition 2. *The Herbrand base of $\mathcal{L}$ is the set $H_{\mathcal{L}}$ of all ground atoms in $\mathcal{L}$. An interpretation (or model) is a function $I : H_{\mathcal{L}} \to \{0, 1\}$, which can also be viewed as a subset of the Herbrand base, $I \subseteq H_{\mathcal{L}}$ (the set of all atoms ϕ such that $\phi^I = 1$). We denote $\Omega = 2^{H_{\mathcal{L}}}$ the set of all interpretations.*

We observe that $H_{\mathcal{L}}$ is finite, because there are no function symbols in $\mathcal{L}$.

Definition 3 (Satisfaction). *Let P be an n-ary predicate, $\phi, \psi \in \mathcal{L}$ closed formulas and I an interpretation of $H_{\mathcal{L}}$:*

– $\models_I \top$ *and* $\not\models_I \bot$;
– $\models_I P(t_1, \ldots, t_n)$ *if and only if $P(t_1, \ldots, t_n) \in I$;*
– $\models_I \neg\phi$ *if and only if* $\not\models_I \phi$;
– $\models_I \phi \wedge \psi$ *if and only if $\models_I \phi$ and $\models_I \psi$;*
– $\models_I \phi \vee \psi$ *if and only if $\models_I \phi$ or $\models_I \psi$;*
– $\models_I \forall x \phi(x)$ *if and only if $\models_I \phi(c)$ for all constant c;*
– $\models_I \exists x \phi(x)$ *if and only if $\models_I \phi(c)$ for some constant c.*

An open formula $\phi(x_1, \ldots, x_n)$ is satisfied by I iff $\models_I \forall x_1 \ldots \forall x_n \phi(x_1, \ldots, x_n)$.

It is a well-known result that the satisfiability of the formulas of $\mathcal{L}$, besides being decidable, is in the NEXPTIME-complete complexity class [15].

We impose the restriction that only ground formulas of $\mathcal{L}_{QF}$ without negation and disjunction (which we shall call *facts*) can be stored in a KB (one does not usually say, when stating facts, things like "Tom is not from NY" or "Tom is from NY or LA"). We denote such restricted language $\mathcal{L}_{\text{fact}}$.

Notice that, by Definition 2, the three languages $\mathcal{L}_{\text{fact}} \subset \mathcal{L}_{QF} \subset \mathcal{L}$ share the same identical Herbrand base $H_{\mathcal{L}}$.

Definition 4. *Let $\phi, \psi \in \mathcal{L}$: $\phi \models \psi$ if and only if, for all $I \subseteq H_{\mathcal{L}}$, if $\models_I \phi$, then also $\models_I \psi$.*

$\mathcal{L}$ may be viewed as an abstraction of popular ways to encode information used in state-of-the-art technologies, such as relational databases, datalog, and RDF + OWL.

Example 1. The set of facts $S \subset \mathcal{L}_{\text{fact}}$,

$$S = \{ \quad \text{Flight}(\text{AF1680}), \text{Origin}(\text{AF1680}, \text{CDG}), \text{Dest}(\text{AF1680}, \text{LHR}),$$
$$\text{Depart}(\text{AF1680}, 07{:}25), \text{Arrival}(\text{AF1680}, 07{:}45), \text{Airline}(\text{AF1680}, \text{AF}) \}$$

describes a morning flight connecting Paris Charles de Gaulle to London Heathrow. Formula $\phi = \exists x (\text{Flight}(x) \wedge \text{Origin}(x, \text{CDG}) \wedge \text{Dest}(x, \text{LHR}))$ states that there is a flight connecting those two airports.

3.2 Possibility Theory and Possibilistic Logic

Possibility theory [11] is a mathematical theory of uncertainty that relies upon fuzzy set theory [20], in that the (fuzzy) set of possible values for a variable of interest is used to describe the uncertainty as to its precise value. At the semantic level, the membership function of such set, π, is called a *possibility distribution* and its range is $[0, 1]$. By convention, $\pi(\mathcal{I}) = 1$ means that it is totally possible for $\mathcal{I}$ to be the real world, $0 < \pi(\mathcal{I}) < 1$ means that $\mathcal{I}$ is only somehow possible, while $\pi(\mathcal{I}) = 0$ means that $\mathcal{I}$ is ruled out. A possibility distribution π is said to be normalized if there exists at least one interpretation $\mathcal{I}_0$ such that $\pi(\mathcal{I}_0) = 1$.

Definition 5. *(Possibility and Necessity Measures) A possibility distribution π induces a possibility measure and its dual necessity measure, denoted by Π and N respectively. Both measures apply to a classical set of interpretation $S \subseteq \Omega$ and are defined as follows:*

$$\Pi(S) = \max_{\mathcal{I} \in S} \pi(\mathcal{I}); \tag{1}$$

$$N(S) = 1 - \Pi(\bar{S}) = \min_{\mathcal{I} \in \bar{S}} \{1 - \pi(\mathcal{I})\}. \tag{2}$$

In words, $\Pi(S)$ expresses to what extent S is consistent with the available knowledge. Conversely, $N(S)$ expresses to what extent S is entailed by the available knowledge. Among the properties of Π and N induced by a normalized possibility distribution on a finite universe of discourse Ω, we can mention, for all subsets $S \subseteq \Omega$:

1. $\Pi(S) = 1 - N(\bar{S})$ (duality);
2. $N(S) > 0 \Rightarrow \Pi(S) = 1; \quad \Pi(S) < 1 \Rightarrow N(S) = 0$.

Possibilistic logic [10] has been originally motivated by the need to manipulate syntactic expressions of the form (ϕ, α), where ϕ is a classical logic formula, and α is a certainty level, with the intended semantics that $N(\phi) \geq \alpha$, where N is a necessity measure.

A possibilistic belief base B is a set $\{(\phi_i, \alpha_i)\}_{i=1,\ldots,m}$ of possibilistic logic formulas. Clearly, B can be layered into a set of nested classical bases $B_\alpha = \{\phi_i \mid$

$(\phi_i, \alpha_i) \in B$ and $\alpha_i \geq \alpha\}$ such that $B_\alpha \subseteq B_\beta$ if $\alpha \geq \beta$. Proving syntactically $B \vdash (\phi, \alpha)$ amounts to proceeding by refutation and proving $B \cup \{(\neg\phi, 1)\} \vdash (\bot, \alpha)$ by repeated application of the resolution rule $(\neg\phi \vee \psi, \alpha), (\phi \vee \nu, \beta) \vdash (\psi \vee \nu, \min(\alpha, \beta))$. Moreover, $B \vdash (\phi, \alpha)$ if and only if $B_\alpha \vdash \phi$ and $\alpha > inc(B)$, where $inc(B)$ is the inconsistency level of B defined as $inc(B) = \max\{\alpha \mid B \vdash (\bot, \alpha)\}$. It can be shown that $inc(B) = 0$ iff B_0 is consistent, with $B_0 = \{\phi_i \mid (\phi_i, \alpha_i) \in B\}$. Thus reasoning from a possibilistic base just amounts to reasoning classically with subparts of the base whose formulas are strictly above the certainty level.

A possibilistic belief base $B = \{(\phi_i, \alpha_i)\}_{i=1,\ldots,m}$ encodes the constraints $N(\phi_i) \geq \alpha_i$. B is thus semantically associated with a possibility distribution [10]

$$\pi_B(\mathcal{I}) = \min_{i=1,\ldots,m} \max(\phi_i^{\mathcal{I}}, 1 - \alpha_i),$$

where $\phi_i^{\mathcal{I}} = 1$ if $\mathcal{I}$ is a model of ϕ_i, and $\phi_i^{\mathcal{I}} = 0$ otherwise; π_B is the largest possibility distribution, i.e., the least specific distribution assigning the largest possibility levels in agreement with the constraints $N(\phi_i) \geq \alpha_i$ for $i = 1, \ldots, m$. The distribution π_B rank-orders the interpretations $\mathcal{I}$ of the language induced by the ϕ_i's according to their plausibility on the basis of the strength of the pieces of information in B. If the set of formulas B_0 is consistent, then the distribution π_B is normalized. The semantic entailment is defined by $B \models (\phi, \alpha)$ iff $\forall \mathcal{I}, \pi_B(\mathcal{I}) \leq \pi_{\{(\phi, \alpha)\}}(\mathcal{I})$. Reasoning by refutation in propositional possibilistic logic is sound and complete, applying the syntactic resolution rule. Namely, it can be shown that $B \models (\phi, \alpha)$ iff $B \vdash (\phi, \alpha)$ and $inc(B) = 1 - \max_{\mathcal{I}} \pi_B(\mathcal{I})$.

Algorithms for reasoning in possibilistic logic and an analysis of their complexity, which is similar to the one of classical logic, multiplied by the logarithm of the number of levels used in the necessity scale, can be found in [14].

4   Representing and Reasoning with Validity and Completeness

When dealing with relational databases, only the statements explicitly present in the database are considered as true (valid). The others are considered as false—the closed world assumption (CWA). When dealing with more general *knowledge* bases, i.e., sets of logical formulas, from which other formulas can be deduced, the true statements are those explicitly represented in the KB, plus those which can be inferred thanks to a reasoner. However, due to the OWA, we cannot suppose that statements that cannot be inferred are false—the truth status of some statements may be unknown in case of incomplete knowledge. In fact, insofar as for any formula ϕ we have a tool to decide if ϕ can be inferred and if $\neg\phi$ can be inferred, CWA makes no sense since when neither ϕ nor $\neg\phi$ can be inferred, CWA would lead to a contradiction, unless we put syntatic restrictions on ϕ, e.g., $\neg\phi$ cannot be expressed in the language.

In this section, we recall the notions of validity and completeness for dealing with relational databases [8,12] and adapt them to the more general case of a KB. We treat validity and completeness of information at the fine grain of a *topic*, defined as follows.

Definition 6. *(Topic) Given a formula $\phi \in \mathcal{L}_{QF}$ without negation, the topic $T(\phi)$ is the set of all the ground formulas that can be obtained by substituting all the free variables in ϕ with all possible constants.*

Example 2. The topic of "all flights departing from Heathrow" may be described by the open formula Origin(x, LHR).

Let $\mathcal{T}$ be the set of topics and let K be a KB of formulas in $\mathcal{L}_{\text{fact}}$. In practice, K is a conjunction of ground atoms in $H_{\mathcal{L}}$.

Unlike for databases, in the case of a general KB, the OWA holds and logical inferences can be performed. Therefore, we must think in terms of logical entailment of formulas.

Example 3. Assume the following KB is given:
K = {Flight(AF1680), Origin(AF1680, CDG), Dest(AF1680, LHR), Airline (AF1680, AF)} then $K \models \exists x(\text{Flight}(x) \wedge \text{Airline}(x, \text{AF}))$ (there is a flight operated by AF), but $K \not\models \exists x \forall y(\neg\text{Flight}(y) \vee \text{Airline}(y, x))$ (all flights are operated by one airline), because one cannot logically rule out other facts not contained in K (K is not complete), such as, for instance, Flight(BA303) and Airline(BA303, BA).

In absolute terms, the notions of *validity* and *completeness* of a KB K with respect to a topic may be defined as follows:

- K is *valid* with respect to a topic iff, for every formula ϕ in that topic, $K \models \phi$ implies that ϕ is indeed true;
- K is *complete* with respect to a topic iff, for every formula ψ in that topic, $K \not\models \psi$ implies that ψ is false.

A formula may be believed to different degrees. We suppose that these degrees depend on both the degree of completeness of the set of facts contained in K and on their validity, which depends on the reliability (or trustworthiness) or even safety [7] of their information source. For example, information related to an Air France flight should be complete if the source is the Air France carrier itself. However, the completeness could be lesser if the source is a private travel agency with a partial coverage about the current flights from the different companies including those of Air France. Similarly, the degree of trust to be associated with information fed by a clerk should be less than the one to be associated with information fed by a supervisor. Still, we would like to emphasize that the way in which such degrees are obtained is out of the scope of this paper. A good source in the literature about trust can be, for example, [16], for a computational view of trust.

We assume that metadata about validity and completeness of information stored in K is given in the form of two functions, **Val** and **Comp**, which associate a degree of validity and completenes, respectively, to each topic.

Definition 7. *Let* **Val** $: \mathcal{T} \rightarrow [0, 1]$ *be such that, for all $T \in \mathcal{T}$,* **Val**(T) *is the degree to which K contains valid information about topic T, which means, for all formulas ϕ such that $K \models \phi$ and $\phi \in T$, $N(\phi) \geq$ **Val**(T).*

Definition 8. *Let* **Comp** $: \mathcal{T} \rightarrow [0, 1]$ *be such that, for all $T \in \mathcal{T}$,* **Comp**(T) *is the degree to which K contains complete information about topic T, which means, for all formulas ψ such that $K \not\models \psi$ and $\psi \in T$, $\Pi(\psi) \leq 1 -$ **Comp**(T).*

In practice, the **Val** and **Comp** functions may be implemented efficiently by a hash table having the formulas representing the topics as keys; a missing key would imply a degree of zero. Now, K plus the metadata provided by **Val** and **Comp** allow us to compute the degree of possibility and necessity for any arbitrary formulas ϕ and ψ as follows:

$$\Pi^-(\phi) = \min_{T:\phi\in T} 1 - \mathbf{Comp}(T), \qquad \text{if } K \not\models \phi; \qquad (3)$$

$$N^+(\psi) = \max_{T:\psi\in T} \mathbf{Val}(T), \qquad \text{if } K \models \psi. \qquad (4)$$

Notice that Π^- and N^+ are associated to two distinct possibility distributions π^+ (the least specific distribution induced by the necessity measure of Eq. 4) and π^- (the least specific distribution induced by the possibility measure of Eq. 3). We now show that if K is consistent, intersecting π^+ and π^- yields a normalized possibility distribution π, for all models I, of the form $\pi(I) = \min\{\pi^+(I), \pi^-(I)\}$, such that there is a single model I^* with $\pi(I^*) = 1$. We recall that normalization is the equivalent, within possibilistic logic, of consistency in crisp logic.

Let B a hypothetical possibilistic belief base corresponding to it. We now prove that such a possibility distribution exists and is normalized.

Let $H_{\mathcal{L}}$ be the Herbrand base constructed over $\mathcal{L}$ and $\Omega = 2^{H_{\mathcal{L}}}$ be the set of all interpretations. A possibilistic data base K^+ will be a collection of pairs (g_i, ν_i) made of ground atoms $g_i \in H_K \subset H_{\mathcal{L}}$, and necessity levels obtained from validity degrees as per Eq. 4. The uncertain completeness assumptions comes down to the assumption of another (virtual) data base K^- containing a collection of pairs $(\neg g_j, \kappa_j)$ made of all ground atoms $g_j \in H$ that do not appear in K^+, and necessity levels obtained from completeness degrees as per Eq. 3.

Theorem 1. *There exists a normalized possibility distribution $\pi : \Omega \to [0,1]$ of the form $\pi(I) = \min\{\pi^+(I), \pi^-(I)\}$, such that there is a single model I^* with $\pi(I^*) = 1$, inducing the possibility and necessity measures of Eqs. 3 and 4.*

Proof. As K^+ contains only positive ground atoms $g_i \in H_K$, it is consistent. So the possibility distribution π^+ induced by K^+ is normalised. Let K_α^+ be a cut of K^+. Its set of models is rectangular in the sense that it is of the form $\wedge_{\nu_i \geq \alpha} g_i$. The set of models of possibility 1 corresponds to the largest conjunction. Likewise we can consider K^- that contains only negative ground atoms $\neg g_j, g_j \notin H_K$, and is thus consistent as well. Let K_α^- be a cut of K^-. Its set of models is rectangular in the sense that it is of the form $\wedge_{\kappa_j \geq \alpha} \neg g_j$. It is clear that everything behaves as if the actual base were $K^+ \cup K^-$. As it contains all literals in the negative or positive form only once, there is a model with positive necessity, namely, $\wedge_{g_i \in K^+} g_i \bigwedge \wedge_{g_j \in K^-} \neg g_j$ with necessity at least $\min(\min_{g_i \in K^+} \nu_i \min_{g_j \in K^-} \neg \kappa_j$. Hence the possibility of this model is 1, and is unique since there can be at most one model with positive necessity. The least specific possibility distribution induced by $K^+ \cup K^-$ obviously enforces the original necessity degrees as all formulas in $K^+ \cup K^-$ are logically independent from one another.

Given that such π exists, it is not important to know it or to represent one of its corresponding possibilistic bases B explicitly, since K, its associated metadata

Val and **Comp**, together with a classical reasoner are enough to compute any possibilistic inference, as shown by the following algorithm:

Algorithm 1 (Inference from B).

Input: $K \subset \mathcal{L}_{\text{fact}}$: a KB; $\phi \in \mathcal{L}$: a formula;

Output: $N(\phi)$.

```
 1: α ← 0
 2: if K ⊨ φ then
 3:     for T ∈ 𝒯 do
 4:         if φ ∈ T and α < Val(T) then
 5:             α ← Val(T)
 6: else if K ⊭ ¬φ then
 7:     for T ∈ 𝒯 do
 8:         if ¬φ ∈ T and α < Comp(T) then
 9:             α ← Comp(T)
10: return α.
```

Property 1. *Algorithm 1 is correct (i.e., it computes $N(\phi)$).*

Proof. If $K \models \phi$, Eq. 4 is applied; otherwise, Eq. 3 together with duality: $N(\phi) = 1 - \Pi(\neg\phi)$.

Property 2. *The cost of Algorithm 1 is two classical inferences.*

Proof. Algorithm 1 needs to execute at most two classical inferences: the one in Line 2 and, in case $K \not\models \phi$, the one in Line 6. Checking whether a formula belongs in a topic can be done in a purely syntactic fashion (linear in the length of ϕ) and its cost is thus negligible.

Example 4. Let K be the same as in the previous example, with the following metadata:

T	$\mathbf{Val}(T)$	$\mathbf{Comp}(T)$
Origin(x, y)	α	β
Airline(x, AF)	γ	δ

There are four constants in K (AF, AF1680, CDG, and LHR) and four predicates: Flight$(\cdot)$, Airline$(\cdot, \cdot)$, Dest$(\cdot, \cdot)$, and Origin$(\cdot, \cdot)$. Since there is no typing of the constants in $\mathcal{L}$, we thus construct the Herbrand base

$$H_K = \{ \quad \text{Flight(AF)}, \quad \ldots, \quad \text{Flight(LHR)},$$
$$\text{Airline(AF, AF)}, \ldots, \text{Airline(LHR, LHR)},$$
$$\text{Dest(AF, AF)}, \quad \ldots, \quad \text{Dest(LHR, LHR)},$$
$$\text{Origin(AF, AF)}, \ldots, \text{Origin(LHR, LHR)} \},$$

with $\|H_K\| = 52$, which gives $\|\Omega\| = \|2^{H_K}\| = 2^{52} \approx 4.5 \cdot 10^{15}$ interpretations. However, we do not need to explicitly construct π over such an impossibly huge domain. By applying Algorithm 1, we can easily compute, for instance:

$$N(\text{Origin}(\text{AF1680}, \text{CDG})) = \alpha,$$
$$N(\neg\text{Origin}(\text{AF1680}, \text{CDG})) = 0,$$
$$N(\text{Airline}(\text{AF1680}, \text{AF})) = \gamma,$$
$$N(\neg\text{Airline}(\text{AF1680}, \text{AF})) = 0,$$
$$N(\exists x(\text{Flight}(x) \wedge \text{Origin}(x, \text{LHR}))) = 0,$$
$$N(\forall x(\neg\text{Flight}(x) \vee \neg\text{Origin}(x, \text{LHR}))) = \beta.$$

5   Merging Beliefs from Multiple Sources

Information is provided by different sources. So we need not only to keep track of the certainty levels of the pieces of information, but also of their sources [9]. Keeping track of sources is especially important, in case of conflicting information, to be able to report which sources support what opinions and thus give the user the elements required for a choice. This is why we need a multi-source generalization of possibilistic logic, like the one proposed in [9] and further developed in [4], to combine and reason about the belief bases obtained, as explained in the previous section, by taking the validity and completeness metadata of the source into account.

We shall denote the set of all the sources in the system by S.

A multi-source possibilistic logic formula is a pair (ϕ, F), where ϕ is a logical formula, and $F \subseteq S$ is a *fuzzy* subset of the set of the sources in the system, i.e., F belongs to the complete distributive lattice $L = [0, 1]^S$, equipped with the max-based union $\cup$, min-based intersection $\cap$, and, if we consider another fuzzy set $G \subseteq S$, the inclusion $F \subseteq G \Leftrightarrow \forall a \in S, F(a) \leq G(a)$.

The intended meaning of a formula (ϕ, F) is that formula ϕ is believed by a source a at least to degree $F(a)$. Each source believing ϕ somehow belongs to the fuzzy set F. The certainty of ϕ, say $C(\phi)$, is then given by the maximal degree of belief in ϕ associated to the sources in F, which believe ϕ to some extent, and, for any source $a \in S$, we have that $C(\phi) \geq F(a)$ (a believes that ϕ is true at least at degree $F(a)$). Formulas of the form $(\phi, \emptyset)$ are not written (the system only considers the formulas which are somehow believed by at least one source).

Example 5. Assume there are three sources, $S = \{a, b, c\}$, where a is Air France, b is British Airways, and c is the Charles de Gaulle airport. Now, let their belief bases be:

ϕ	$F(a)$	$F(b)$	$F(c)$
Dest(AF1680, LHR)	1	0	0.8
Depart(AF1680, 06:25)	0	0.5	0
Depart(AF1680, 07:25)	1	0	1
Arrival(AF1680, 07:45)	1	0.5	0

Let us consider the particular fuzzy sets of sources of the form $F = \alpha/A$, defined as

$$(\alpha/A)(a) = \begin{cases} \alpha \in (0,1], & \text{if } a \in A; \\ 0, & \text{if } a \in \overline{A}. \end{cases}$$

They correspond to a subset A of sources having the same lower bound α on the certainty level of some considered formula. The following equivalence holds between possibilistic logic bases:

$$\{(\phi, \alpha/A), (\phi, \beta/B)\} \equiv \{(\phi, (\alpha/A) \cup (\beta/B))\}. \tag{5}$$

Example 6. We will thus have:

(Dest(AF1680, LHR), $(1/\{a\})$ ∪ $(0.8/\{c\})$)), Depart(AF1680, 06:25), $0.5/\{b\})$,
Depart(AF1680, 07:25), $1/\{a, c\})$, Arrival(AF1680, 07:45), $(1/\{a\})$ ∪ $(0.5/\{b\})$)),
Arrival(AF1680, 08:45), $0.8/\{c\})$

A multi-source possibilistic base (which, in the context of this paper, represents a summary of the opinions of multiple sources) is defined as a finite set (i.e., a conjunction) of multi-source possibilistic formulas.

Inference in multi-source possibilistic logic proceeds by refutation, as in standard possibilistic logic: given a base $B = \{(\phi_i, \alpha_i/A_i)\}_{i=1,\dots,m}$, proving $B \vdash (\phi, F)$ amounts to proving $B \cup \{(\neg\phi, S)\} \vdash (\bot, F)$ by repeated application of the equivalence of Eq. 5 and of the resolution rule

$$\frac{(\neg P \vee Q, \alpha/A), (P \vee R, \beta/B)}{(Q \vee R, \min(\alpha, \beta)/(A \cap B))}. \tag{6}$$

The semantics of the multi-source possibilistic logic may be given in terms of a generalization of possibility theory based on a fuzzy-set-valued possibility distribution $\pi : \Omega \to [0,1]^S$. In the context of this work, $\Omega = 2^{H_\mathcal{L}}$. The fuzzy-set-valued possibility distribution π associates to every interpretation $\mathcal{I}$ a fuzzy set of sources for which $\mathcal{I}$ is possible; $(\pi(\mathcal{I}))(a)$ is the degree to which source a deems $\mathcal{I}$ possible. Distribution π is normalized if $\exists \mathcal{I}_0 \in \Omega : \pi(\mathcal{I}_0) = S$. This means that the sources are *collectively* consistent since there exists at least one interpretation that all sources find fully possible. There exists another, weaker form of normalization for such a distribution, which only expresses that the sources are *individually* consistent, namely: $\bigcup_{\mathcal{I} \in \Omega} \pi(\mathcal{I}) = S$. For instance, the multi-source possibilistic base $B = \{(\phi, 1/A), (\neg\phi, 1/\overline{A})\}$, where $\overline{A} = S \setminus A$, is clearly not collectively consistent, but it is individually consistent. Indeed here there is partition of the sources into two subsets, those in A that support ϕ and those in $\overline{A}$ that support $\neg\phi$.

The relevant possibility and necessity measures may be defined as follows: for all formulas ϕ,

$$\Pi(\phi) = \bigcup_{\substack{\mathcal{I} \in \Omega \\ \mathcal{I} \models \phi}} \pi(\mathcal{I}), \quad N(\phi) = \bigcap_{\substack{\mathcal{I} \in \Omega \\ \mathcal{I} \not\models \phi}} \overline{\pi(\mathcal{I})}. \tag{7}$$

The distribution associated with base $B = \{(\phi_i, \alpha_i/A_i)\}_{i=1,\ldots,m}$ is

$$\pi_B(\mathcal{I}) = \begin{cases} \mathcal{S}, & \text{if } \mathcal{I} \models \phi_1 \wedge \ldots \wedge \phi_m; \\ \bigcap_{i:\mathcal{I} \not\models \phi_i} (1-\alpha_i)/A_i \cup \overline{A_i}, & \text{otherwise.} \end{cases}$$

This reflects the fact that if a source in A_i believes with certainty α_i that ϕ_i is true, such a source can find possible an interpretation that violates ϕ_i only at a level that is upper bounded by $1 - \alpha_i$. Multiple source possibilistic logic is sound and complete for refutation, with respect to the above semantics [9].

We have now all the formal tools needed to solve the belief fusion problem of providing a coherent answer to queries in presence of possibly conflicting beliefs. The model we propose can process queries which take the form of a formula $\phi \in \mathcal{L}$. If ϕ is closed, then the expected answer is just the fuzzy set of sources according to which ϕ holds. If ϕ is open, the expected answer is a list of substitutions of its free variables, annotated with the fuzzy set of the sources that support it.

To answer a query, the answers provided by the n belief bases are aggregated in a multi-source possibilistic base $B = \{(\phi_i, \alpha_i/A_i)\}_{i=1,\ldots,m}$, which is then used to compile the answer.

Example 7. Continuing the previous example, the result of query $\text{Dest}(x, \text{LHR}) \wedge \text{Depart}(x, y) \wedge \text{Arrival}(x, z)$ requesting all flights with destination London Heathrow, together with their departure and arrival times, would be

x	y	z	$F(a)$	$F(b)$	$F(c)$
AF1680	7:25	7:45	1	0	0
AF1680	7:25	8:45	0	0	0.8

or, in a more synthetic form,

x	y	z	F
AF1680	7:25	7:45	$1/\{a\}$
AF1680	7:25	8:45	$0.8/\{c\}$

The result of query $\exists x \text{Dest}(x, \text{LHR}) \wedge \text{Arrival}(x, 8{:}45)$ asking whether a flight exists with destination London Heathrow arriving at 8:45, would be, in synthetic form, $0.8/\{c\}$.

6   Related Work

Our proposal fills a gap at the intersection of two fields of in- vestigation, namely distributed information systems and possibilistic logic.

The problem of reasoning about validity and complete- ness in relational databases was first addressed by Demolombe [8] in the setting of modal logic.

Recent work on collaborative access control in distributed datalog [1] shares some common intuitions and concerns with our model. However, this approach, which is based on provenance calculus [13], does not handle uncertainty (although probabibistic c-tables are also encompassed by provenance calculus). The approach proposed in the present paper is anyway more in the spirit of possibilistic c-tables, which have been recently introduced in [18].

Finally, the idea of associating subsets of sources as supporting arguments to answers has been suggested in [2] in the context of numerical information fusion.

7   Conclusion

We have presented a solution to construct a possibilistic belief base from a crisp knowledge base using topical validity and completeness metadata. The main result is that possibilistic inferences from such belief base can be performed at the cost of two classical inferences, which is less than the cost of inference on a general possibilistic belief base. Furthermore, our solution can be straightforwardly adapted to KB representation standards like datalog and RDF + OWL.

We have also shown how to exploit the expressive power of multi-source possibilistic logic to provide the user with a comprehensive logical summary of the different opinions held by the sources. Nevertheless, it is likely that a user might be happier with receiving less detailed information in response to her queries. We see basically two directions that might be followed to alleviate the cognitive load for the end user:

- give the user the option of specifying a maximum number k of answers, to be used to select only the k most certain answers according to their supporting sources, so that each answer be simply annotated with a crisp set of sources that support it;
- if a taxonomy of sources is available (e.g., based on their sector, geographical location, etc.), the sets of sources supporting an answer could be "linguistically synthesized" (in the sense of Zadeh's [21]) by categorical labels, like "all the airlines based in the UK" or "most airport operators", which are certainly easier to understand and process than extensive lists of sources.

References

1. Abiteboul, S., Bourhis, P., Vianu, V.: A formal study of collaborative access control in distributed datalog. In: ICDT. LIPIcs, vol. 48, pp. 10:1–10:17. Schloss Dagstuhl - Leibniz-Zentrum fuer Informatik (2016)
2. Assaghir, Z., Napoli, A., Kaytoue, M., Dubois, D., Prade, H.: Numerical information fusion: lattice of answers with supporting arguments. In: ICTAI, pp. 621–628. IEEE Computer Society (2011)
3. Bacchus, F., Grove, A.J., Halpern, J.Y., Koller, D.: From statistical knowledge bases to degrees of belief. Artif. Intell. **87**(1–2), 75–143 (1996)
4. Belhadi, A., Dubois, D., Khellaf-Haned, F., Prade, H.: Multiple agent possibilistic logic. J. Appl. Non-Class. Logics **23**(4), 299–320 (2013)

5. Bernays, P., Schönfinkel, M.: Zum Entscheidungsproblem der mathematischen Logik. Math. Ann. **99**, 342–372 (1928)
6. Cholvy, L.: Collecting information reported by imperfect information sources. In: Greco, S., Bouchon-Meunier, B., Coletti, G., Fedrizzi, M., Matarazzo, B., Yager, R.R. (eds.) IPMU 2012. CCIS, vol. 299, pp. 501–510. Springer, Heidelberg (2012). doi:10.1007/978-3-642-31718-7_52
7. Cholvy, L., Demolombe, R., Jones, A.: Reasoning about the safety of information: from logical formalization to operational definition. In: Raś, Z.W., Zemankova, M. (eds.) ISMIS 1994. LNCS, vol. 869, pp. 488–499. Springer, Heidelberg (1994). doi:10.1007/3-540-58495-1_49
8. Demolombe, R.: Answering queries about validity and completeness of data: from modal logic to relational algebra. In: FQAS. Datalogiske Skrifter (Writings on Computer Science), vol. 62, pp. 265–276. Roskilde University (1996)
9. Dubois, D., Lang, J., Prade, H.: Dealing with multi-source information in possibilistic logic. In: ECAI, pp. 38–42 (1992)
10. Dubois, D., Lang, J., Prade, H.: Possibilistic logic. In: Handbook of Logic in Artificial Intelligence and Logic Programming (vol. 3): Nonmonotonic Reasoning and Uncertain Reasoning, NY, USA, pp. 439–513. Oxford University Press, New York (1994)
11. Dubois, D., Prade, H.: Possibility Theory–An Approach to Computerized Processing of Uncertainty. Plenum Press, New York (1988)
12. Dubois, D., Prade, H.: Valid or complete information in databases —a possibility theory-based analysis—. In: Hameurlain, A., Tjoa, A.M. (eds.) DEXA 1997. LNCS, vol. 1308, pp. 603–612. Springer, Heidelberg (1997). doi:10.1007/BFb0022068
13. Green, T.J., Karvounarakis, G., Tannen, V.: Provenance semirings. In: Libkin, L. (ed.) Proceedings of 26th ACM SIGACT-SIGMOD-SIGART Symposium on Principles of Database Systems, Beijing, 11–13 June, pp. 31–40. ACM (2007)
14. Lang, J.: Possibilistic logic: complexity and algorithms. In: Kohlas, J., Moral, S. (eds.) Algorithms for Uncertainty and Defeasible Reasoning, pp. 179–220. Gabbay, D.M., Smets, Ph. (eds.) vol. 5. Handbook of Defeasible Reasoning and Uncertainty Management Systems. Kluwer Acad. Publ., Dordrecht (2001)
15. Lewis, H.R.: Complexity results for classes of quantificational formulas. J. Comput. Syst. Sci. **21**, 317–353 (1980)
16. Marsh, S.P.: Formalising Trust as a Computational Concept. Ph.D. thesis, Department of Computing Science and Mathematics University of Stirling (1994)
17. Motro, A.: Integrity = validity + completeness. ACM Trans. Database Syst. **14**(4), 480–502 (1989)
18. Pivert, O., Prade, H.: Possibilistic conditional tables. In: Gyssens, M., Simari, G. (eds.) FoIKS 2016. LNCS, vol. 9616, pp. 42–61. Springer, Cham (2016). doi:10.1007/978-3-319-30024-5_3
19. Schwitzgebel, E.: Belief. In: Zalta, E.N. (ed.) The Stanford Encyclopedia of Philosophy. Stanford University (Fall 2008), http://plato.stanford.edu/archives/fall2008/entries/belief/
20. Zadeh, L.A.: Fuzzy sets. Inf. Control **8**, 338–353 (1965)
21. Zadeh, L.A.: A theory of approximate reasoning. In: Hayes, J.E., Mitchie, D., Mikulich, L.I. (eds.) Machine intelligence, vol. 9, pp. 149–194. Halstead Press, New York (1979)

Aggregation of Preferences on Criteria Importance Expressed on Various Subsets by Several Decision Makers

Christophe Labreuche[(✉)]

Thales Research and Technology, Palaiseau, France
christophe.labreuche@thalesgroup.com

Abstract. We are interested in the aggregation of preference information provided by several decision makers regarding the relative importance of criteria. When a large number of decision makers are involved, they have specific areas of expertise and express their preferences on possibly different subsets of criteria. The standard aggregation methods do not apply to this situation where the preference information of the decision makers have different support. We consider four possible types of preference information provided by the decision makers: binary preference relations comparing the relative importance of criteria, quaternary relations comparing differences of relative importance between pairs of criteria, classification of the difference of importance between any two criteria in predefined categories, and numerical values of the criteria weights. In the three first cases, we use an extension of the relational analysis.

1 Introduction

Group Decision Making arises in many domains such as public policy making, crisis management or complex system engineering. Each Decision Maker (DM) expresses his preference on multiple criteria. We are in this context interested in constructing a unique multi-criteria model representing a good compromise among the individual preferences of all stakeholders. Let us consider the following example.

Example 1. In crisis management, the assessment of the impact of mitigation actions is complex as they have a potential impact on many sectors of activity. At short term, evacuation and rescue efficiencies are critical. At mid-term, the satisfaction of vital needs of citizens, efficiency of public services, and working status of transport and telecom infrastructures are crutial. At long term, consequences on economy and environment are important. The impact assessment of mitigation actions depends thus on multiple criteria. The elicitation of such multi-criteria model naturally involves several Decision Makers (DMs), such as the Regional Operational Leader (DM1) in charge of crisis mitigation, the first responder ambulance (DM2) and the road infrastructure operator (DM3). The DMs do not have the same role. They have specific areas of expertise and express their preferences on possibly different subsets of criteria, as illustrated in Table 1. For instance, DM2 does not say anything on criteria $c3, c4$.

© Springer International Publishing AG 2017
S. Moral et al. (Eds.): SUM 2017, LNAI 10564, pp. 377–388, 2017.
DOI: 10.1007/978-3-319-67582-4_27

Table 1. In this table, crosses are criteria on which a DM is not familiar with. DM2 (resp. DM3) provides preference information only on criteria c_1 and c_2 (resp. c_2, c_3, c_4). The numerical values are the values of the weights for the three DMs

	c_1: victim dispatch centre	c_2: health care efficiency	c_3: road transport efficiency	c_4: recovery costs at long term
DM1: Regional Operational Leader	$\frac{1}{6}$	$\frac{2}{6}$	$\frac{2}{6}$	$\frac{1}{6}$
DM2: First responder ambulance	$\frac{1}{3}$	$\frac{2}{3}$	$\times$	$\times$
DM3: Road infrastructure operator	$\times$	$\frac{2}{5}$	$\frac{2}{5}$	$\frac{1}{5}$

When numerical weights are available, we denote by w_i^k the weight alloted by DM k on criterion i. It is not easy to combine these weights w^k to provide a weight vector w representing a good consensus among DMs' preferences. As these weight vectors have different support, their aggregation through, for example, an arithmetic or geometric mean does not make sense.

Weights w^k and w are used in a quantitative multi-criteria model, such as a weighted sum. Hence the weights shall correspond to an interval scale and even a ratio scale. This means that the following quantities $\frac{w_i^k - w_j^k}{w_l^k - w_h^k}$ and $\frac{w_i^k}{w_j^k}$ shall be well-defined. This allows for instance DM k to say that criterion c_2 is twice as more important than criterion c_1 when $\frac{w_2^k}{w_1^k} = 2$. We propose to construct the consensual weights w given the w^k's, not by comparing or aggregating directly the values of the weights w^k, but by comparing the previous ratios (Sect. 6). In Example 1, DM 2 says that criterion 2 is twice as more important than criterion 1, which is what DM 1 also tells us. Moreover DMs 1 and 3 both say that criteria 2 and 3 are twice more important than criterion 4. Hence, the weights provided by DM 1 are completely consistent with the preference information of DMs 2 and 3, so that weights w^1 can be taken as the consensual weights w.

We have presented so far a situation in which the input information is directly the criteria weights of the DMs. In practice, these numerical weights can be constructed thanks to several techniques. The weakest assumption is that the DMs provide a binary preference relation that simply compares the relative importance of criteria. One looks then for a consensual binary relation (Sect. 3). Numerical weights are obtained from an interval scale, which can be classically constructed from a quaternary relation, saying that the difference of importance between two criteria is at least as large as that for two other criteria. One can thus assume that all DMs provide a quaternary relation, and we look for a consensual quaternary relation (Sect. 4). A convenient way to derive these quaternary relations, is to ask each DM to classify the difference of importance between any two criteria in predefined categories, as in the MACBETH approach [3]. We look for a consensual assignment of pairs of criteria to categories (Sect. 5). In these three cases, we use an extension of the relational analysis [21,22]. It aims at minimizing the distance between the consensual relation and that provided by the DMs.

The novelty of the paper lies in the aggregation of partial preference information regarding the relative importance of criteria, where these partial preferences are expressed in ways that often used in multi-criteria decision making - namely assignment of pairs of criteria to categories and numerical values. The MACBETH approach is a very efficient and widely used approach to construct numerical weights from decision makers [3]. It has been applied to many real applications. The second case, where we assume we have directly the numerical values of criteria, typically arises when each DM has used his preferred elicitation method to construct his weights (MACBETH, AHP,...). In this case, we directly consider the numerical weights. The two other preference models (namely binary relation over criteria, and a quaternary relation) can be seen as intermediate steps. We start with the simplest and most classical situation, where each DM provides only a binary relation over criteria. The relational analysis is already known in this context [21,22]. The only novelty here is to rewrite the problem allowing incomplete binary relations. The extension of the relational analysis to the quaternary relation, and to assignment of pairs of criteria to categories is new. Our work is mostly based on the relational analysis as it is based in the Condorcet criteria. Finally, when DMs provide numerical weights, we cannot use anymore the relational analysis as we do not explicitly have (binary, quaternary or categorization) relations. We propose a novel approach based on the comparison of ratios.

2   Notation

The set of DMs and criteria are denoted by $M = \{1, \ldots, m\}$ and $N = \{1, \ldots, n\}$ respectively. The preferences of the DMs are given over the parameters $P = \{1, \ldots, p\}$ of a multi-criteria model on N. The values of the parameters are described by a vector $w = (w_1, \ldots, w_p)$ satisfying some normalization and monotonicity constraints. The parameters can be classically weights alloted to the criteria, where $P = N$. In this latter case, the set of parameters fulfilling the constraints is:

$$\mathcal{W}(P) = \Big\{ w \in \mathbb{R}^P : \forall i \in P \ w_i \geq 0 \text{ and } \sum_{i \in P} w_i = 1 \Big\}.$$

We will not explore in this paper situations where P contains more parameters – as for instance for capacities.

As we have seen in Example 1, each DM has his own expertise and does not express his preferences on all parameters. DM k provides his preferences only on the subset $P^k \subseteq P$ of parameters. In Example 1, $P^1 = \{1,2,3,4\} = N$, $P^2 = \{1,2\}$ and $P^3 = \{2,3,4\}$. The numbers in the table are the values of

w^1, w^2, w^3. The opinion of DM k is represented by a weight vector $w^k \in \mathcal{W}(P^k)$, and the consensual weights are $w \in \mathcal{W}(P)$.

3   Case of Binary Preference Relations

The simplest situation arises when the DMs provide a binary relation on the parameters in P. Each DM $k \in M$ expresses his preferences only on the subset P^k of parameters, by a binary relation $\succsim^k$ on P^k, where $i \succsim^k j$ (for $i, j \in P^k$) means that DM k find i at least as important as j. We wish to aggregate these binary relations in order to produce an order relation $\succsim$ on P.

The problem of aggregating binary relations is very classical in social choice. The well-known Arrow theorem shows that there is no *ideal* voting procedure. This explains the diversity of existing voting procedures: Condorcet method, Borda count, Smith set, Copeland tournament, plurality with a runoff, KemenyYoung, ... Each of them satisfies a different set of properties.

The fact that each DM provides a binary relation only on a subset of P, makes the application of the voting procedures not possible or not relevant. There is a recent literature in social choice and AI interested in voting procedures with partial information. One is in particular interested in knowing whether a given candidate is a necessary/possible winner or loser despite the missing information from the voters [17]. A candidate is a necessary winner if he is a winner according to the chosen voting rule, whatever the completion of the incomplete voters' preferences. If there is no necessary winner, one can wonder what is the minimal number of queries to be asked to the voters in order to know for sure the winner [8].

These concepts are not relevant in our case, as we cannot interpret the fact that each DM expresses his preferences only on a subset of P as some missing information. In Example 1, the fact that DM2 does not express preference information for criteria $c2$ and $c3$ is not a missing information, and we do not look for some completion of DM2' preferences on criteria $c2$ and $c3$. DM2 has no legitimacy on these two criteria and we do not expect any preferences on them. Hence we are not looking for a necessary winner. It does not make sense to try all completions of $\succsim^k$ given on P^k extended to P.

The binary preference relations are described as Boolean relations:

$$\forall i, j \in P, \quad R_{ij} = \begin{cases} 1 & \text{if } i \succsim j \\ 0 & \text{otherwise} \end{cases} \qquad \forall i, j \in P^k, \quad R_{ij}^k = \begin{cases} 1 & \text{if } i \succsim^k j \\ 0 & \text{otherwise} \end{cases}$$

We will use a method based on the Condorcet principle (i.e. counting the number of candidates beaten by candidate i minus the number of candidates beating i), but without the possibility of having cycles (as in the Condorcet method). The relational analysis realizes such a compromise [21, 22]. It looks for a median weak order $\succsim$ on P [16] which is the nearest to the partial preferences $\succsim^k$ in the sense of the L^1 distance: $\sum_{k \in M} \sum_{i,j \in P^k} |R_{ij} - R_{ij}^k|$. As the relations are Boolean, it is equivalent to minimizing the sum of the squares of the errors:

$$\sum_{k \in M} \sum_{i,j \in P^k} \left(R_{ij} - R_{ij}^k\right)^2 = \sum_{k \in M} \sum_{i,j \in P^k} \left((R_{ij})^2 - 2R_{ij}R_{ij}^k + (R_{ij}^k)^2\right)$$
$$= \sum_{k \in M} \sum_{i,j \in P^k} \left(R_{ij} - 2R_{ij}R_{ij}^k + R_{ij}^k\right)$$

where the unknowns are terms R_{ij}. Hence minimizing the previous functional is equivalent to minimizing $\sum_{k \in M} \sum_{i,j \in P^k} \left(1 - 2R_{ij}^k\right) R_{ij}$. We look for a weak order R, and thus R satisfies reflexivity, completeness and transitivity. These three conditions can be written as linear constraints [21, 22]:

$$R_{ii} = 1 \quad \forall i \in P \qquad \qquad \text{(Reflexivity)} \qquad (1)$$
$$R_{ij} + R_{ji} \geq 1 \quad \forall i,j \in P \qquad \qquad \text{(Completeness)} \qquad (2)$$
$$R_{ih} \geq R_{ij} + R_{jh} - 1 \quad \forall i,j,h \in N \qquad \text{(transitivity)} \qquad (3)$$

The binary relation R over P is thus solution to the following Mixed Integer Programming (MIP) [21, 22]:

$$\text{Minimize} \sum_{k \in M} \sum_{i,j \in P^k} \left(1 - 2R_{ij}^k\right) R_{ij}$$
$$\text{under} \begin{vmatrix} \forall i,j \in P \quad R_{ij} \in \{0,1\} \\ (1), (2), (3) \end{vmatrix} \qquad (4)$$

There exists a heuristics to solve problem (4) [22]. It can be noted that minimizing the functional in (4) is equivalent to minimizing the Condorcet criterion. Indeed, the complement of relations R and R^k are $\overline{R}_{ij} = 1 - R_{ij}$ and $\overline{R}_{ij}^k = 1 - R_{ij}^k$. We note that $R_{ij}^k R_{ij} + \overline{R}_{ij}^k \overline{R}_{ij} = (2R_{ij}^k - 1)R_{ij} - R_{ij}^k + 1$. Hence minimizing (4) is equivalent to maximizing $\sum_{k \in M} \sum_{i,j \in P^k} \left(R_{ij}^k R_{ij} + \overline{R}_{ij}^k \overline{R}_{ij}\right)$. One looks thus at maximizing the similarity between R and the R^k's, and that between the complement of R and the complements of the R^k's. This corresponds to the Condorcet principle.

Constructing $\succsim$ from the $\succsim^k$'s is not sufficient to identify numerical weights w. In this case, the DMs have to provide more preference information, e.g. quaternary relations (see next section).

4   Case of Quaternary Preference Relations

As already said, the parameters in P are typically numerical weights w_i $(i \in P)$ used in a weighted sum. Hence the weights correspond to interval scales. Such a scale is typically constructed given a binary relation and a quaternary relation. For DM $k \in M$, a quaternary relation is relation $\succsim^{*,k}$ on $P^k \times P^k$ such that $i,j \succsim^{*,k} l,h$ (for $i,j,l,h \in P^k$) means that the difference of importance between i and j is at least as large as that between l and h, i.e. $w_i^k - w_j^k \geq w_l^k - w_h^k$. Each DM $k \in M$ is supposed to provide a binary relation $\succsim^k$ on P^k and a quaternary relation $\succsim^{*,k}$ on $P^k \times P^k$.

We wish to find a binary relation $\succsim$ on P, and a quaternary relation $\succsim^\star$ on $P \times P$. We first aggregate the $\succsim^k$'s to construct $\succsim$, as in Sect. 3. We thus need to determine $\succsim^\star$ given $\succsim$ and the $\succsim^{\star,k}$'s.

Let us write the quaternary relations as Boolean relations: $R^\star_{ijlh} = 1$ (for $i,j,l,h \in P$) if $i,j \succsim^\star l,h$ and $R^\star_{ijlh} = 0$ otherwise; $R^{\star,k}_{ijlh} = 1$ (for $i,j,l,h \in P^k$) if $i,j \succsim^{\star,k} l,h$ and $R^{\star,k}_{ijlh} = 0$ otherwise. Following Sect. 3, we extend the relational analysis to quaternary relations. To this end, let us consider:

$$\sum_{k \in M} \sum_{i,j,l,h \in P^k} \left(R^\star_{ijlh} - R^{\star,k}_{ijlh} \right)^2 = \sum_{k \in M} \sum_{i,j,l,h \in P^k} \left(R^\star_{ijlh} - 2R^\star_{ijlh} R^{\star,k}_{ijlh} + R^{\star,k}_{ijlh} \right)$$

Hence minimizing this functional is equivalent to minimizing $\sum_{k \in M} \sum_{i,j,l,h \in P^k} \left(1 - 2R^{\star,k}_{ijlh} \right) R^\star_{ijlh}$. A quaternary relation $R^\star$ is representable by an interval scale when there exist a weight vector w such that

$$R^\star_{ijlh} = 1 \quad \Longleftrightarrow \quad w_i - w_j \geq w_l - w_h. \tag{5}$$

According to [18, Theorem 2, Sect. 4.4], relation (5) holds when $\langle N \times N, R^\star \rangle$ is an *algebraic-difference structure*, i.e. it satisfies the following relations:

$$\langle N \times N, R^\star \rangle \text{ is a weak order} \tag{6}$$

$$\text{If } R^\star_{ijlh} = 1, \text{ then } R^\star_{hlji} = 1 \tag{7}$$

$$\text{If } R^\star_{iji'j'} = 1 \text{ and } R^\star_{jhj'h'} = 1 \text{ then } R^\star_{ihi'h'} = 1 \tag{8}$$

and two last technical (archimidean) conditions that are not relevant in the case of discrete scales. Relation (6) means that $R^\star$ is reflexive ($R^\star_{ijij} = 1$ for all $i,j \in N$), complete (for all $i,j,l,h \in N$, we have $R^\star_{ijlh} = 1$ or $R^\star_{lhij} = 1$ [or both]) and transitive (if $R^\star_{iji'j'} = 1$ and $R^\star_{i'j'i''j''} = 1$ then $R^\star_{iji''j''} = 1$). These three conditions can be written as linear constraints:

$$R^\star_{ijij} = 1 \quad \forall i,j \in N \tag{Reflexive} \quad (9)$$

$$R^\star_{ijlh} + R^\star_{lhij} \geq 1 \quad \forall i,j,l,h \in N \tag{Complete} \quad (10)$$

$$R^\star_{iji''j''} \geq R^\star_{iji'j'} + R^\star_{i'j'i''j''} - 1 \quad \forall i,j,i',j',i'',j'' \in N \tag{transitive} \quad (11)$$

Relation (7) can be written in the linear form

$$R^\star_{hlji} \geq R^\star_{ijlh} \quad \forall i,j,l,h \in N \tag{12}$$

Moreover, (8) can be put in the form

$$R^\star_{ihi'h'} \geq R^\star_{iji'j'} + R^\star_{jhj'h'} - 1 \quad \forall i,j,h,i',j',h' \in N \tag{13}$$

Finally $R^\star$ shall be consistent with R (i.e. $\succsim$). These conditions are given in [18, Definition 1 and Theorem 1, Sect. 4.4]:

$$\left[i \succsim j \text{ and } j \succsim h \right] \implies \left[R^\star_{ihij} = 1 \text{ and } R^\star_{ihjh} = 1 \right] \tag{14}$$

$$\left[i \succsim j, \ j \succsim h, \ i' \succsim j', \ j' \succsim h', \ R^\star_{iji'j'} = 1 \text{ and } R^\star_{jhj'h'} = 1 \right] \implies R^\star_{ihi'h'} = 1 \tag{15}$$

Relation (15) is already included in (8).

We introduce the following linear optimization problem:

$$\text{Minimize} \sum_{k \in M} \sum_{i,j,l,h \in P^k} \left(1 - 2R^{\star,k}_{ijlh}\right) R^{\star}_{ijlh}$$

$$\text{under} \begin{vmatrix} \forall i,j,l,h & R^{\star}_{ijlh} \in \{0,1\} \\ (9),(10),(11),(12),(13),(14) \end{vmatrix} \tag{16}$$

5 Case of Quaternary Preference Relations Expressed by Categories

A convenient way to express a quaternary relation is to use categories, such as in the MACBETH method [3]. We assume that we have t categories $\{1,\ldots,t\}$ of intensity of preferences ranging from category 1 (very small preference) to t (extreme preference). Each DM $k \in M$ is supposed to provide Boolean relations $C^{k,\geq s}$ and $C^{k,\leq s}$ such that $C^{k,\geq s}_{ij} = 1$ (resp. $C^{k,\leq s}_{ij} = 1$) if DM k finds that the difference between the importance of i and j belongs to a category greater or equal (resp. lower or equal) to s.

We assume that each DM k provides the binary relations $\succsim^k$, $C^{k,\geq s}$ and $C^{k,\leq s}$ ($s \in \{1,\ldots,t\}$) on P^k. We use the approach of Sect. 3 to construct $\succsim$ from the $\succsim^k$'s. We now wish to construct at the same time the consensual quaternary relation $\succsim^{\star}$, and the category binary relations $C^{\geq s}$ and $C^{\leq s}$, given $\succsim$, $C^{k,\geq s}$ and $C^{k,\leq s}$. We need to enforce some conditions on $C^{\geq s}$ and $C^{\leq s}$.

We have first some monotonicity conditions. If $C^{\geq s}_{ij} = 1$, then $C^{\geq s-1}_{ij} = 1$. If $C^{\leq s}_{ij} = 1$, then $C^{\leq s+1}_{ij} = 1$. Then

$$\forall i,j \in P \; \forall s \in \{2,\ldots,t\} \quad C^{\geq s}_{ij} \leq C^{\geq s-1}_{ij} \tag{17}$$

$$\forall i,j \in P \; \forall s \in \{1,\ldots,t-1\} \quad C^{\leq s}_{ij} \leq C^{\leq s+1}_{ij} \tag{18}$$

We have the interval consistency condition saying that if $C^{\geq s}_{ij} = 1$ and $s > s'$ then $C^{\leq s'}_{ij} = 0$. Hence

$$\forall i,j \in P \; \forall s,s' \in \{1,\ldots,t\} \text{ with } s > s' \quad C^{\geq s}_{ij} + C^{\leq s'}_{ij} \leq 1 \tag{19}$$

If $C^{\geq s}_{ij} = 1$ and $C^{\leq s-1}_{lh} = 1$, then $R^{\star}_{ijlh} = 1$. Hence

$$\forall i,j,l,h \in P \; \forall s \in \{2,\ldots,t\} \quad R^{\star}_{ijlh} \geq C^{\geq s}_{ij} + C^{\leq s-1}_{lh} - 1 \tag{20}$$

We wish to minimize the following functional

$$\sum_{k \in M} \sum_{i,j \in P} \sum_{s \in \{1\ldots,t\}} \left[\left(C^{\geq s}_{ij} - C^{k,\geq s}_{ij}\right)^2 + \left(C^{\leq s}_{ij} - C^{k,\leq s}_{ij}\right)^2 \right]$$

Hence we need to solve the MIP (where the unknowns are $C_{ij}^{k,\geq s}$, $C_{ij}^{\geq s}$ and $R_{ijlh}^{\star}$)

$$\text{Minimize} \sum_{k \in M} \sum_{i,j \in P} \sum_{s \in \{1,\ldots,t\}} \left[\left(1 - 2C_{ij}^{k,\geq s} \right) C_{ij}^{\geq s} + \left(1 - 2C_{ij}^{k,\leq s} \right) C_{ij}^{\leq s} \right]$$

$$\text{under} \quad \begin{vmatrix} \forall i,j,l,h \quad R_{ijlh}^{\star} \in \{0,1\} \\ \forall i,j \; \forall s \in \{1\ldots,t\} \quad C_{ij}^{\geq s}, C_{ij}^{\leq s} \in \{0,1\} \\ \text{Conditions on } R^{\star} : (9),(10),(11),(12),(13),(14) \\ \text{Conditions on } C^{\geq s} \text{ and } C^{\leq s} : (17),(18),(19),(20) \end{vmatrix} \tag{21}$$

6   Case of Numerical Weights

6.1   Consistency Among Expert Opinions

We assume in this section that the DMs directly provide weights w^k, that need to be aggregated to obtain a synthetic weight vector w on P. One could think of using an aggregation function, such as a simple arithmetic mean or an Ordered Weighted Average [26], to deduce w_i from the w_i^k's. The main difficulty we face is that the support P^k of the expert weights w^k are different from one DM to another one. As the weights sum-up to one ($\sum_{i \in P^k} w_i^k = 1$), their average value for DM k is $\frac{1}{|P^k|}$. Thus the weights w_i^k and $w_i^{k'}$ of two different DMs k and k' at a same parameter i are not necessarily directly comparable. This makes the aggregation of weights w^k not trivial. For instance, taking for w_i the average value $\frac{\sum_{k \in M : i \in P^k} w_i^k}{|\{k \in M : i \in P^k\}|}$ of the weights w_i^k over all DMs k such that $i \in P^k$ is not adequate.

Weights w^k and w are used in a quantitative multi-criteria model, such as a weighted sum. In order to make sense, the weights shall correspond to an interval scale and even a ratio scale. Hence the following quantities $\frac{w_i^k - w_j^k}{w_l^k - w_h^k}$ and $\frac{w_i^k}{w_j^k}$ shall be well-defined and shall make sense. This allows for instance to say that a parameter is twice as more important than another one, when $\frac{w_i^k}{w_j^k} = 2$.

We note that ratio $\frac{w_i^k}{w_j^k}$ can be seen as a special case of ratio $\frac{w_i^k - w_j^k}{w_l^k - w_h^k}$ with $j = h = 0$, by adding a fictitious parameter $w_0^k := 0$. We thus set $\overline{P} = P \cup \{0\}$, $\overline{P}^k = P^k \cup \{0\}$. We also set

$$\mathcal{W}(\overline{P}) = \{w \in \mathbb{R}^{\overline{P}} : w_0 = 0 \text{ and } w_P \in \mathcal{W}(P)\},$$

where w_P is the restriction of w on P.

Definition 1 (Consistency). We say that the DMs' opinions $(w^k)_{k \in M}$ (with $w^k \in \mathcal{W}(\overline{P}^k)$ for every $k \in M$) are *consistent* if

$$\forall i,j,l,h \in \overline{P} \; \forall k,k' \in M \text{ with } i,j,l,h \in \overline{P}^k \cap \overline{P}^{k'} \qquad \frac{w_i^k - w_j^k}{w_l^k - w_h^k} = \frac{w_i^{k'} - w_j^{k'}}{w_l^{k'} - w_h^{k'}} \tag{22}$$

In Example 1, the opinions of experts are consistent as

$$\frac{w_2^1}{w_1^1} = 2 = \frac{w_2^2}{w_1^2} \quad , \quad \frac{w_2^1}{w_3^1} = 1 = \frac{w_2^3}{w_3^3},$$

$$\frac{w_2^1}{w_4^1} = 2 = \frac{w_2^3}{w_4^3} \quad , \quad \frac{w_2^1 - w_4^1}{w_3^1 - w_4^1} = 1 = \frac{w_2^3 - w_4^3}{w_3^3 - w_4^3}.$$

Definition 1 implies the existence of $w \in \mathcal{W}(\overline{P})$ such that

$$\forall k \in M, \; \forall i, j, l, h \in \overline{P}^k \qquad \frac{w_i - w_j}{w_l - w_h} = \frac{w_i^k - w_j^k}{w_l^k - w_h^k}. \tag{23}$$

In Example 1, w^1 is the resulting weight vector.

6.2   Construction of the Parameters' Value

The consistency condition is very strong and unlikely to be fulfilled. We wish to identify how to relax equality (23) when the DMs are not consistent. The first idea is to minimize quantities

$$\left| \frac{w_i - w_j}{w_l - w_h} - \frac{w_i^k - w_j^k}{w_l^k - w_h^k} \right|, \tag{24}$$

where the unknowns are the components of vector w. In order to have a linear relation in w, we can rather use quantity

$$\left| w_i - w_j - \frac{w_i^k - w_j^k}{w_l^k - w_h^k}(w_l - w_h) \right|, \tag{25}$$

or even

$$\left| (w_i - w_j)(w_l^k - w_h^k) - (w_l - w_h)(w_i^k - w_j^k) \right|, \tag{26}$$

in order to avoid having a denominator. Terms (25) and (26) are two different normalizations of (24). We will keep (26) as there is no problem with potential null denominators. We wish thus to minimize the following functional:

$$\sum_{k \in M} \sum_{i,j,l,h \in \overline{P}^k} \left| (w_i - w_j)(w_l^k - w_h^k) - (w_l - w_h)(w_i^k - w_j^k) \right|^b$$

where b is the power parameter. In the case of binary relations, we saw that the results are the same for any power b. In general, the minimization of sum of L^1 distances ($b = 1$) corresponds to the geometric median problem. This problem does not have an analytical solution, but can be approximated by a fixed point algorithm. In our case, we obtain a Linear Problem (LP):

Minimize $\displaystyle\sum_{k \in M} \sum_{i,j,l,h \in \overline{P}^k} \varepsilon_{i,j,l,h}^k$

under $\left| \begin{array}{l} w \in \mathcal{W}(\overline{P}) \\[4pt] \forall k \in M, \ \forall i,j,l,h \in \overline{P}^k \\[2pt] \quad -\varepsilon_{i,j,l,h}^k \le (w_i - w_j)\,(w_l^k - w_h^k) - (w_l - w_h)\,(w_i^k - w_j^k) \le \varepsilon_{i,j,l,h}^k \\[4pt] \forall k \in M, \ \forall i,j,l,h \in \overline{P}^k \quad \varepsilon_{i,j,l,h}^k \ge 0 \end{array} \right.$

$$\tag{27}$$

In general, the minimization of sum of L^2 distances ($b = 2$) has an analytical solution, which is the centroid. In our case, the weights w are thus solutions to the convex optimization problem:

Minimize $\displaystyle\sum_{k \in M} \sum_{i,j,l,h \in \overline{P}^k} \left((w_i - w_j)\,(w_l^k - w_h^k) - (w_l - w_h)\,(w_i^k - w_j^k) \right)^2$

under $\left| \, w \in \mathcal{W}(\overline{P}) \right.$

$$\tag{28}$$

7   Related Works and Future Works

7.1   Related Works

There are several works supporting Group Decision Making based a Multi-Criteria Decision Aid [24]. There are two classes of such approaches. The first one consists directly in supporting the GDM activity, by helping to find the best compromise option given a set of alternatives [13]. The alternatives are evaluated thanks to the DMs' preference models. Voting processes can be applied when no interaction is required [1]. In the opposite situation, negotiation protocols can be recommended [14]. A clustering approach can allow to form coalitions of DMs having relatively similar preferences [19].

The second kind of approaches aims first at synthesizing all DMs' preference models into one unique preference model that will be considered as representative of the whole group. The alternatives are then evaluated by this group model. The group weight can be obtained by simple aggregation functions, such as the geometric mean [4,20], or a centroid function [23]. A compromise weight minimizing the conflict among the DMs' preferences is computed thanks to linear programming in [25]. The case of ELECTRE-TRI model for sorting problems is analysed in [5]. Some works consider the imprecision of the individual preferences to identify the common model. The aggregation of partial rankings provided by the DMs is proposed in [11]. The common model can also be obtained by a robust approach [7]. This allows comparing the individual models to the group opinion [7].

7.2   Future Works

We have focused in this paper the elicitation process of some parameter vector P by a group of decision makers. We have intrepreted the vector of parameters P as weights of criteria, such as in the weighted sum. It would be interesting to extend our approach to other Multi-Criteria Decision Making models. One can think of a utility function associated to an attribute [3]. One can also think of an *additive utility* model [15]. Another possibility is to give weights not only to single criteria but to any subset of criteria, which corresponds to the concept of capacity [6]. Capacities are used in the Choquet integral [6,12]. A last model we can mention is the Generalized Additive Independence (GAI) model [2,9,10]. How could our approach be extended to these models? These models are characterized by a vector of parameters P'. The set $\mathcal{W}(P')$ of admissible parameters' vectors is a convex polytope. Compared to $\mathcal{W}(P)$, we need to add some monotonicity conditions – saying for instance that the capacity is monotone or that the utility function is additive. The linear program (4) (Sect. 3) can be extended, by adding monotonicity conditions on R relation. The same can be done with quaternary relation – see (16) (Sect. 4) –, or with quaternary preference relations expressed by categories – see (21) (Sect. 5). The case where we are given the numerical values of parameters P' is relatively straightforward extension of (28) (Sect. 6), as we basically just need to replace $\mathcal{W}(P)$ by $\mathcal{W}(P')$.

The proposed method to find a consensus among preferences expressed by several DMs is basically an optimization approach, where we are looking for a model minimizing some distance with the provided information. An alternative approach to find a consensus is the so-called axiomatic characterization of the consensus rules (see e.g. [1]). For future works, we will explore which axioms are satisfied by our proposal. This is not an easy task. The relational analysis is the solution that is the median points among the DMs' preferences.

For future works, we will also implement the proposed methods and study their relevance by simulations. There is no current existing proposal addressing points raised in Sects. 4, 5 and 6. The problem is to assess the quality of the aggregation returned by the algorithms. We will start with some prototypical examples where an intuition can be provided on what should be the consensual result.

Acknowledgments. This work has been supported by the European project FP7-SEC-2013-607697, PREDICT "PREparing the Domino effect In crisis siTuations".

References

1. Arrow, K., Sen, A., Suzumura, K.: Handbook of Social Choice and Welfare. Handbooks in Economics. Elsvier, Amsterdam (2002)
2. Bacchus, F., Grove, A.: Graphical models for preference and utility. In: Conference on Uncertainty in Artificial Intelligence (UAI), Montreal, Canada, pp. 3–10, July 1995
3. Bana, C.A., Costa, J., Corte, J.-C.: Vansnick. MACBETH. Int. J. Inf. Technol. Decis. Mak. **11**, 359–387 (2012)
4. Barzilai, J., Lootsma, F.: Power relations and group aggregation in the multiplicative ahp and smart. J. Multi-Criteria Decis. Anal. **6**, 155–165 (1997)
5. Cailloux, O., Mayag, B., Meyer, P., Mousseau, V.: Operational tools to build a multicriteria territorial risk scale with multiple stakeholders. Reliab. Eng. Syst. Saf **120**, 88–97 (2013)

6. Choquet, G.: Theory of capacities. Annales de l'Institut Fourier **5**, 131–295 (1953)
7. Dias, L., Clmaco, J.: Dealing with imprecise information in group multicriteria decisions: a methodology and a GDSS architecture. Eur. J. Oper. Res. **160**(2), 291307 (2005)
8. Ding, N., Lin, F.: Voting with partial information: minimal sets of questions to decide an outcome. In: Proceedings of the Fourth International Workshop on Computational Social Choice (COMSOC 2012), Krakow, Poland (2012)
9. Fishburn, P.: Interdependence and additivity in multivariate, unidimensional expected utility theory. Int. Econ. Rev. **8**, 335–342 (1967)
10. Fishburn, P.: Utility Theory for Decision Making. Wiley, New York (1970)
11. Gonzalez-Pachon, J., Romero, C.: Aggregation of partial ordinal rankings: an interval goal programming approach. Comput. Oper. Res. **28**, 827–834 (2001)
12. Grabisch, M.: The application of fuzzy integrals in multicriteria decision making. Europ. J. Oper. Res. **89**, 445–456 (1996)
13. Hochbaum, D., Levin, A.: Methodologies and algorithms for group-rankings decision. Manage. Sci. **52**(9), 1394–1408 (2006)
14. Jennings, N., Faratin, P., Lomuscio, A.R., Parsons, S., Wooldridge, M., Sierra, C.: Automated negotiation: prospects, methods and challenges. Int. J. Group Decis. Negot. **10**(2), 199–215 (2001)
15. Keeney, R.L., Raiffa, H.: Decision with Multiple Objectives. Wiley, New York (1976)
16. Kemeny, J.: Mathematics without numbers. Daedalus **4**, 577–591 (1959)
17. Konczak, K., Lang, J.: Voting procedures with incomplete preferences. In: IJCAI 2005 Multidisciplinary Workshop on Advances in Preference Handling (2005)
18. Krantz, D., Luce, R., Suppes, P., Tversky, A.: Foundations of Measurement Volume 1: Additive and Polynomial Representations. Academic Press, Cambridge (1971)
19. Liu, B., Shen, S., Chen, X.: A two-layer weight determination method for complex multi-attribute large-group decision-making experts in a linguistic environment. Inf. Fusion **23**, 156–165 (2015)
20. Lootsma, F., Bots, P.: The assignment of scores for output-based research funding. J. Multi-Criteria Decis. Anal. **8**, 44–50 (1999)
21. Marcotorchino, J.: Relational analysis theory as a general approach to data analysis and data fusion. In: COGIS 2006 - COGnitive systems with Interactive Sensors (2006)
22. Marcotorchino, J., Michaud, P.: Optimisation en analyse ordinale des données. Masson, Paris (1978)
23. Mateos, A., Jimneza, A., Ros-Insuaa, S.: Monte carlo simulation techniques for group decision making with incomplete information. Europ. J. Oper. Res. **174**, 1842–1864 (2006)
24. Matsatsinis, N., Samaras, A.P.: Mcda and preference disaggregation in group decision support. Eur. J. Oper. Res. **130**, 414–429 (2001)
25. Wei, Q., Ya, H., Ma, J., Fan, Z.: A compromise weight for multi-criteria group decision making with individual preference. J. Oper. Res. Soc. **51**, 625–634 (2000)
26. Yager, R.R.: On ordered weighted averaging aggregation operators in multicriteria decision making. IEEE Trans. Systems, Man Cybern. **18**, 183–190 (1988)

Short Papers

Probabilistic Local Link Prediction in Complex Networks

Víctor Martínez[✉], Fernando Berzal, and Juan-Carlos Cubero

Department of Computer Science and Artificial Intelligence
and Research Center for Information and Communications Technologies (CITIC),
University of Granada, Granada, Spain
{victormg,berzal}@acm.org, jc.cubero@decsai.ugr.es

Abstract. Link prediction is the problem of inferring future or missing relationships between nodes in a given network. This problem has attracted great attention since it has a large number of applications. In this problem, there is always some degree of uncertainty because the absence of a link between a pair of nodes may be the result of the non-existence of the link or the result of it not being observed but actually existing. In this paper, we propose a local link prediction technique that aggregates the observed evidence to estimate the probability of each possible non-observed link. We also show how our scalable link prediction technique achieves higher precision than other well-established local techniques in several networks from very different domains.

Keywords: Link prediction · Uncertainty · Evidence aggregation

1 Introduction

A problem that commonly appears in a large number of domains is, given a set of observed relationships or interactions between entities, predicting the most likely non-observed links. This task is known as the link prediction problem [1]. It has been applied with great success to a large number of tasks, including the prediction of interactions among proteins [2], the prediction of future author collaborations [3], the suggestion of people we may know in social networks [1], or the recommendation of commercial products [4].

Very different approaches have been proposed to deal with the link prediction problem [5]. In this work, we focus our attention on local techniques, which take into account only local information, leading to highly scalable algorithms. Efficiency is paramount because link prediction is commonly applied to massive networks, where scalability is a crucial requirement. Almost all local techniques consider the shared neighbors between nodes to estimate the likelihood of the existence of a link, weighting the contribution of each shared neighbor according to certain feature (such as the degree of the shared node). However, different networks may require different weightings for each shared neighbor contribution, instead of the fixed weighting most local techniques use. To estimate the contribution of each shared neighbor, we must take into account the uncertainty that

© Springer International Publishing AG 2017
S. Moral et al. (Eds.): SUM 2017, LNAI 10564, pp. 391–396, 2017.
DOI: 10.1007/978-3-319-67582-4_28

is present due to the fact that an unobserved link may indicate the non-existence or the actual existence of the link.

We propose a novel local link prediction technique that takes the features of the current network into account and weighs the contribution of each node according to the expected contribution for nodes of its degree. Despite our technique requiring sampling the whole network to estimate the required parameters, this process can be done efficiently and changes in the network can be handled locally. We propose a probabilistic framework where the degree of belief on the existence of a link is increased as more evidence is accumulated in terms of shared neighbors.

This paper is organized as follows. Our proposal is described in full detail in Sect. 2. An empirical evaluation and the results we obtained are discussed in Sect. 3. Finally, conclusions extracted from our work are presented in Sect. 4.

2   Method

Let us assume that we have access to a snapshot of a complex network, observing links between nodes. L_{xy} denotes the event corresponding to the existence of a link between nodes x and y. Γ_x denotes the set of neighbors of a node x and $\Gamma_{x \cap y}$ denotes the set of shared neighbors of a pair of nodes x and y.

According to Bayes' theorem, we can express the complementary probability of a link between nodes x and y given the set of shared neighbors Γ_{xy} as

$$P(\overline{L_{xy}}|\Gamma_{x \cap y}) = \frac{P(\Gamma_{x \cap y}|\overline{L_{xy}})P(\overline{L_{xy}})}{P(\Gamma_{x \cap y})}.$$

Like most local link prediction techniques, we assume independence among shared neighbors, thus we can rewrite the previous expression as

$$P(\overline{L_{xy}}|\Gamma_{x \cap y}) = \prod_{z \in \Gamma_{x \cap y}} \frac{P(z|\overline{L_{xy}})}{P(z)} P(\overline{L_{xy}}).$$

According to Bayes' theorem, the term $\frac{P(z|\overline{L_{xy}})}{P(z)}$ is equal to the term $\frac{P(\overline{L_{xy}}|z)}{P(\overline{L_{xy}})}$ and, after applying this substitution, we obtain the expression

$$P(\overline{L_{xy}}|\Gamma_{x \cap y}) = \prod_{z \in \Gamma_{x \cap y}} \frac{P(\overline{L_{xy}}|z)}{P(\overline{L_{xy}})} P(\overline{L_{xy}}).$$

According to basic probability axioms, the probabilities of an event and of its complementary event always total 1; thus, we can express the previous equation in terms of the probability of the existence of links as

$$P(L_{xy}|\Gamma_{x \cap y}) = 1 - \prod_{z \in \Gamma_{x \cap y}} \frac{1 - P(L_{xy}|z)}{1 - P(L_{xy})}(1 - P(L_{xy})),$$

where $P(L_{xy}|z)$ is the probability of the existence of a link between x and y given the shared neighbor z. This value can be estimated using different methods. In this work, we propose the estimation of this value as the probability of the existence of a link given a shared neighbor of the same degree than z in the network, which is computed as

$$P(L_{xy}|z) = \frac{1}{|N_{|\Gamma_z|}|} \sum_{k \in N_{|\Gamma_z|}} \frac{\sum_{i \neq j, i, j \in \Gamma_k} P(L_{ij})}{|\Gamma_k|(|\Gamma_k| - 1)},$$

where $N_{|\Gamma_z|}$ is the set of nodes with the same degree than z. It is important to note that this value has to be computed only once for each node degree value, rather than once for each shared neighbor.

For each pair of nodes x and y, we assume $P(L_{xy}) = 1$ if a link is currently observed. Since we are only interested in ranking links, the value of $P(L_{xy})$ when no link is observed is irrelevant for us provided that we keep it smaller than 1. Applications requiring a true probability estimation may need to estimate the prior probabilities of unobserved links.

In order to analyze the computational complexity of the proposed approach, we divide the analysis in two phases. In the first phase, the average probability of a link given a shared neighbor of a particular degree is computed. This phase has a computational complexity $O(vd^2)$, where v refers to the number of nodes in the network and d refers to the degree of these nodes. Once these values are computed, the computational complexity of the second phase, regarding the estimation of the likelihood of a link, is just $O(2d)$, since shared neighbors can be efficiently computed using hash sets. As we can see, our method is highly scalable, since the range of node degrees tends to be much smaller than the number of nodes in a complex network.

3 Evaluation and Results

In order to measure the performance of our proposal, we performed a battery of tests by applying our technique to networks gathered from very different domains. We considered the following networks: a social network from a website called Advogato (ADV, [6]), a protein-protein interaction network in budding yeast (YST, [7]), a network of e-mail exchanges between university members (EML, [8]), the metabolic network of Caenorhabditis elegans (CEG, [9]), a human protein-protein interaction network (HPD, [10]), four citation networks (CGS, LDG, SMG, ZWL, [11]), and a power distribution network (UPG, [12]).

Our experimentation consisted of a 5-fold cross-validation where links in each network were randomly divided into five sets of the same size n. Considering each set as the test set, we used the four remaining sets as the training network to predict the most probable n links. As performance score, we used precision, which is computed as

$$precision = \frac{true\ positives}{true\ positives + false\ positives} = \frac{true\ positives}{n},$$

where a link instance is classified as positive if it is ranked within the top n links, and negative otherwise. Results from each run were averaged to obtain a single performance score for each method and network combination.

We compared our approach to some well-known similarity-based local link prediction techniques. Despite their simplicity, these techniques have shown to obtain good results in practice and they are usually considered to be reasonable choices for link prediction. The techniques included in our comparison are the following ones:

- **Common neighbors (CN):** In this method, the likelihood of the existence of a link between two nodes is proportional to the number of shared neighbors between both nodes [1]. This method was proposed after observing a correlation between the number of shared neighbors of two nodes and the probability that they will collaborate in the future in scientific collaboration networks. It is computed as

$$P(L_{xy}) \propto |\Gamma_x \cap \Gamma_y|.$$

- **The Adamic-Adar index (AA):** This method measures the likelihood of the link between two entities based on the logarithmically-penalized degree of each shared neighbor. This index weighs nodes with low degree much heavier than nodes with high degree, assuming that nodes with few neighbors are more likely to contribute to the formation of a link between a pair of nodes of their neighborhood [13]. It is computed as

$$P(L_{xy}) \propto \sum_{z \in \Gamma_x \cap \Gamma_y} \frac{1}{\log |\Gamma_z|}.$$

- **The Resource Allocation index (RA):** This method is similar to the Adamic-Adar index, yet considering the degree of each shared neighbor without a logarithmic penalization [14]. RA models the resource allocation process that takes place in complex networks, where two unconnected nodes exchange units of resources by equally distributing them among their neighbor nodes. The amount of exchanged resources between a pair of nodes can be viewed as a measure of similarity. It is defined as

$$P(L_{xy}) \propto \sum_{z \in \Gamma_x \cap \Gamma_y} \frac{1}{|\Gamma_z|}.$$

- **Local Naïve Bayes (LNB):** This method estimates the role or degree of influence of each shared neighbor using probability theory [15], computing the probability of the existence of a link between a pair of nodes as

$$P(L_{xy}) \propto \sum_{z \in \Gamma_x \cap \Gamma_y} f(z) \log (oR_z)$$

where o is a constant for the network computed as

$$o = \frac{p_{unconnected}}{p_{connected}} = \frac{\frac{1}{2}|V|(|V| - 1)}{|E|} - 1$$

and R_z is the role or influence of the node computed as

$$R_z = \frac{2|\{e_{x,y} : x,y \in \Gamma_z, e_{x,y} \in E\}| + 1}{2|\{e_{x,y} : x,y \in \Gamma_z, e_{x,y} \notin E\}| + 1}.$$

The function $f(z)$ measures the influence of the shared neighbor. The authors suggest $f(z) = 1$ from common neighbors, $f(z) = \frac{1}{\log |\Gamma_z|}$ from the Adamic-Adar index, or $f(z) = \frac{1}{|\Gamma_z|}$ from the resource allocation method.

The results we obtained using each technique for each dataset are shown in Table 1. Our technique is listed as *PLLP*, an acronym for probabilistic local link prediction. It can be observed that our technique achieves a higher average precision than previous techniques for all the complex networks considered in our experimentation.

Table 1. Precision obtained by each method (rows) for each dataset (columns). The best results are highlighted in bold

	ADV	YST	EML	CEG	CGS	HPD	LDG	SMG	UPG	ZWL	Average
CN	0.1489	0.0876	0.1304	0.1119	0.1810	0.0627	0.1113	0.1357	0.0296	0.1516	0.11507
AA	0.1783	0.1080	0.1923	0.1532	0.3985	0.0828	0.1662	0.1581	0.0165	0.2110	0.16649
RA	0.1821	0.0876	0.1800	0.1448	0.4178	0.0669	0.1633	0.1407	0.0132	0.2075	0.16039
LNB-CN	0.1717	0.1189	0.1912	**0.1569**	0.2507	0.0850	0.1529	0.1583	**0.0425**	0.2005	0.15286
LNB-AA	**0.1906**	**0.1190**	0.1972	0.1555	0.4068	0.0867	0.1723	**0.1597**	0.0176	0.2158	0.17212
LNB-RA	0.1874	0.0954	0.1778	0.1462	0.4186	0.0658	0.1615	0.1471	0.0150	0.2067	0.16215
PLLP	0.1875	0.1141	**0.1982**	0.1536	**0.4273**	**0.0874**	**0.1729**	0.1593	0.0347	**0.2167**	**0.17517**

4 Conclusions

In this paper, we have proposed a novel technique for link prediction. Our method aggregates local evidence to estimate the probability of an uncertain event such as the existence of a link. It works by increasing the degree of belief for each potential link by aggregating the evidence provided by each shared neighbor. Our proposal achieves better average precision results than some well-established local link prediction techniques for several networks from very different domains. Local Naïve Bayes outperforms our technique for certain networks under specific configurations. However, our method performs better in average without the need of testing parameters. Instead of including ad hoc parameters, our method achieves better results reasoning from first principles.

In future work, we will explore and evaluate alternative approaches to the estimation of the probability of a link given a shared neighbor. Models to estimate the prior probability of links will also be studied, since they can be useful for those applications requiring true probability estimations.

Acknowledgments. This work is partially supported by the Spanish Ministry of Economy and the European Regional Development Fund (FEDER), under grant TIN2012-36951 and the program "Ayudas para contratos predoctorales para la formación de doctores 2013" (grant BES-2013-064699).

References

1. Liben-Nowell, D., Kleinberg, J.: The link-prediction problem for social networks. J. Assoc. Inf. Sci. Technol. **58**(7), 1019–1031 (2007)
2. Martínez, V., Cano, C., Blanco, A.: ProphNet: a generic prioritization method through propagation of information. BMC Bioinform. **15**(1), S5 (2014)
3. Pavlov, M., Ichise, R.: Finding experts by link prediction in coauthorship networks. FEWS **290**, 42–55 (2007)
4. Huang, Z., Li, X., Chen, H.: Link prediction approach to collaborative filtering. In: Proceedings of the 5th ACM/IEEE-CS Joint Conference on Digital libraries, pp. 141–142. ACM (2005)
5. Martínez, V., Berzal, F., Cubero, J.C.: A survey of link prediction in complex networks. ACM Comput. Surv. **49**(4), 69 (2016)
6. Massa, P., Salvetti, M., Tomasoni, D.: Bowling alone and trust decline in social network sites. In: Eighth IEEE International Conference on Dependable, Autonomic and Secure Computing, DASC 2009, pp. 658–663. IEEE (2009)
7. Bu, D., Zhao, Y., Cai, L., Xue, H., Zhu, X., Lu, H., et al.: Topological structure analysis of the protein-protein interaction network in budding yeast. Nucleic Acids Res. **31**(9), 2443–2450 (2003)
8. Guimera, R., Danon, L., Diaz-Guilera, A., Giralt, F., Arenas, A.: Self-similar community structure in a network of human interactions. Phys. Rev. E **68**(6), 065103 (2003)
9. Duch, J., Arenas, A.: Community detection in complex networks using extremal optimization. Phys. Rev. E **72**(2), 027104 (2005)
10. Peri, S., Navarro, J.D., Amanchy, R., Kristiansen, T.Z., Jonnalagadda, C.K., Surendranath, V., et al.: Development of human protein reference database as an initial platform for approaching systems biology in humans. Genome Res. **13**(10), 2363–2371 (2003)
11. Pajek datasets. http://vlado.fmf.uni-lj.si/pub/networks/data
12. Watts, D.J., Strogatz, S.H.: Collective dynamics of small-world networks. Nature **393**(6684), 440–442 (1998)
13. Adamic, L.A., Adar, E.: Friends and neighbors on the web. Soc. Netw. **25**(3), 211–230 (2003)
14. Lü, L., Jin, C.H., Zhou, T.: Similarity index based on local paths for link prediction of complex networks. Phys. Rev. E **80**(4), 046122 (2009)
15. Liu, Z., Zhang, Q.M., Lü, L., Zhou, T.: Link prediction in complex networks: a local naïve Bayes model. EPL (Europhys. Lett.) **96**(4), 48007 (2011)

A Fuzzy Ontology-Based System for Gait Recognition Using Kinect Sensor

Fernando Bobillo[1]([✉]), Lacramioara Dranca[2], and Jorge Bernad[1]

[1] Aragon Institute of Engineering Research (I3A),
University of Zaragoza, Zaragoza, Spain
{fbobillo,jbernad}@unizar.es
[2] Defense University Center (CUD), General Military Academy, Zaragoza, Spain
licri@unizar.es

Abstract. Gait recognition involves the automatic classification of human people from sequences of data about their movement patterns. This paper describes our ongoing work in the development of a gait recognition system using Microsoft Kinect data and based on fuzzy ontologies to manage the imprecision of the data and to improve the system scalability.

Keywords: Gait recognition · Fuzzy ontologies · Kinect sensor

1 Introduction

The problem of gait recognition consists of automatically classifying human people by analyzing data about their movement patterns. Gait recognition has many applications, including security (e.g. authentication and surveillance) and medicine (e.g. automatic support for the diagnosis of neurological diseases). Furthermore, it has several advantages with respect to other biometrical measures for human recognition. For example, it is non-intrusive, does not require any collaboration from the subject, and involves less confidential data than other techniques, such as face recognition.

In the last years we have witnessed an increase in the number of low cost sensors to capture pose sequences to compute biometrical measures related to the human gait. An example is Microsoft Kinect, a motion sensing input device originally conceived as a peripheral for video game consoles.

Although there is a notable effort in the gait recognition using Microsoft Kinect (see Sect. 2 for a discussion), existing approaches generate big amounts of data which are difficult to understand by a non-expert or to reuse between different applications. For this reason, we advocate for the combination of Semantic Web technologies to represent human Microsoft Kinect data and the biometrical features for human gait motion analysis computed using them. Due to the intrinsic imprecision of the original data, we propose to use fuzzy ontologies to use fuzzy sets rather than precise crisp values at production stage.

This paper describes our ongoing work. The main objectives of our research and the main contributions so far can be summarized as follows:

© Springer International Publishing AG 2017
S. Moral et al. (Eds.): SUM 2017, LNAI 10564, pp. 397–404, 2017.
DOI: 10.1007/978-3-319-67582-4_29

- Our system uses fuzzy ontologies to represent Microsoft Kinect data and biometrical features for human gait motion analysis.
- Our data segmentation is based on steps rather than on full sequences.
- We reduce the number of variables with respect to previous works.
- At production stage, data corresponding to a new recording can be compared only with a customizable number of candidates from our database.
- We plan to provide a public database with data obtained using Kinect V2.

The remainder is organized as follows. Section 2 overviews some related work. Then, Sect. 3 details the architecture of our system and summarizes some preliminary results. Section 4 ends with conclusions and ideas for future research.

2   Related Work

This section summarizes most of the previous work on gait recognition using the Kinect sensor or on the use of ontologies to represent Kinect data.

Gait recognition and Kinect. After the commercial launch of Kinect in 2011, several research papers have approached the gait analysis for human recognition. One of the first approaches is [9], where the authors observed promising results concerning person recognition using a Naive Bayes classifier and a simple set of features obtained from nine people with a Kinect sensor. A real time approach for Kinect based recognition is presented in [6]. In that case the features are characterized as static (height, length of bones) or dynamic (angles of joints). Several distances where used between these features and finally a nearest neighbor classifier obtained around 80% accuracy for ten people. In [8] a framework for gait-based recognition is proposed, based on a publicly available Kinect dataset with 30 people. The authors extract 16 dimensional vectors from the dataset and use several dissimilarity tests and achieve 93.29% identification rate and 99.11% gender recognition rate. Some steps further are taken in [7] with a new method for fusing information from Riemannian and Euclidean features representation that achieves 95.67% accuracy. Moreover, the authors mention a new dataset for gait recognition captured from 30 people using the more recent Kinect V2 but, unfortunately, it is not currently publicly available.

Ontologies and Kinect. There have also been some previous approaches to represent Kinect-related data using classical ontologies [4] and fuzzy ones [3]. The authors even developed the so-called *Kinect ontology*. However, despite this generic name, their approach is strongly focused on a different application, recognition of human activity, and cannot be reused in our scenario. For example, *Kinect ontology* was not designed to encode directly the information directly obtained from the sensors, and its fuzzy extension does not discuss a fuzzy representation of the relevant features for gait recognition.

3   Architecture of the System

The proposed system has four main components: a data capture phase, a pre-processing phase, an ontology and a decision phase, as shown in Fig. 1.

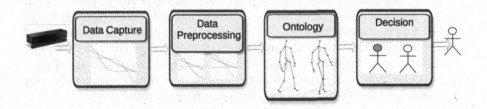

Fig. 1. Architecture of the system

Data Capture. The Data Capture module interacts directly with the sensor and collects raw data from a Kinect sensor. The Kinect sensor actually integrates several sensors (e.g., RGB camera, depth sensor, or infrared sensor) from which several joint points of the human skeleton are obtained. These joint points are retrieved as points in a 3D-space where the coordinate origin is located at the center of the Kinect sensor. As an example, Fig. 1(Data Capture) shows both the feet and the spine base values in the depth axis (z-axis) of a person walking two steps. Due to inaccuracy of the sensor, the coordinates for each joint point are not precise, so it is needed to prune the data that can be incorrectly taken by the sensor. The module explained in the next section is entrusted of such task. This inaccuracy will be also managed by the fuzzy ontology.

Data Preprocessing. Next, the data captured in the previous phase are pre-processed. The data preprocessing module contains several algorithms for steps segmentation, noise reduction, and feature extraction.

Our system uses a step-based identification approach rather than using entire sequences. Sequence means in this context a Kinect register of a person walking towards the camera. Usually, sequences contain 3–4 steps. In order to detect the steps in a sequence, a strategy based on local maximums of the distance between the feet time series is used. For training purpose we used the *UPCVGait* dataset [8], a publicly available dataset acquired using Kinect V1. Figure 1(Data Preprocessing) shows an example of the segmentation done for a person walking two steps (both feet and the spine base data are shown). Some strategies based on length of the bones, height of a person, and variation of the movement direction in a step have been used for noise reduction purposes.

Then, the feature extraction is performed for each of the steps identified previously. We identified three kind of features:

- *Anthropological features:* height, humerus length, forearm length, thigh length and shin length. As these measures may slightly vary from frame to frame, we compute mean, max, min, and standard deviation for each detected step.

- *Step related features:* step length, step width. Since the walking direction of a person may vary and is not always straight to the location of the camera, an angular rotation over the y-axis (vertical axis) has been applied firstly in order to align the direction of walking of each person with the position of the Kinect sensor.
- *Angle related features:* angles of the projections on the xz and yz plane of each of the previously mentioned bones, both left and right ones. Inspired by [8] and before computing these angles, a new angular rotation has been applied in order to align the vertical axis with the inclination of the torso (the line connecting the center of the hips with the center of the shoulders) of each person. As these angles are different from frame to frame, mean, max, min and standard deviation values are computed for each step.

We have used WEKA (namely the WrapperSubsetEval method, that evaluates attribute sets by using a learning scheme) to obtain a selection of the most representative attributes, which in our case were 12 [10]. This way, our system uses a smaller number of attributes than previous works.

Fuzzy Ontology for Gait Recognition. The benefits of using a fuzzy ontology are two-fold: firstly, data representation is more appropriate for human understanding and machine reuse; secondly, we can provide a reduced number of candidates to the Decision Module.

Our fuzzy ontology is able to represent raw Kinect data about the movement of a person but also biometric features computed from them. Fuzzy ontologies extend classical ontologies with ideas of fuzzy logic [11]. While in classical set theory elements either belong to a set or not, in fuzzy set theory elements can belong to some degree, usually ranging in $[0, 1]$. As in the classical case, 0 means no-membership and 1 full membership, but now a value between 0 and 1 represents the extent to which x can be considered as an element of the fuzzy set. When applying these ideas to fuzzy ontologies, it is possible to define fuzzy extensions of the concepts, properties, axioms, and datatypes. In our case, our fuzzy ontology includes *fuzzy datatypes*, replacing crisp values with a more general fuzzy membership function. For example, assume that we want to recognize an individual human001 using some biometrical metrics, such as its maximal height. Rather than representing that the value of the data property maxHeight for human001 is 190 cm, we can take into account the imprecision of the sensor by considering instead a triangular function (see Fig. 2(a)) such that $\pm d$ cm is considered as acceptable. While one could consider using a better sensor, we aim at using low cost devices and thus must deal with such imprecision.

Our ontology has been developed using Protégé [5] ontology editor. Classes, properties, individuals, and most of the axioms are represented as usual. To represent the fuzzy datatypes, we have used Protégé plug-in called *Fuzzy OWL 2* that can be used to create and edit fuzzy ontologies [1]. The idea is to use a classical OWL 2 ontology with OWL 2 annotations to add the fuzzy information that OWL 2 cannot directly encode, the plug-in makes the annotations syntax transparent to the users.

As discussed in Sect. 2, we needed to develop it from scratch. To represent the Kinect data, we consider 4 mutually disjoint classes. For each instance of Human, there are several recordings. Every video obtained using Kinect is represented as an instance of Sequence and each sequence is composed of several instances of Frame. After some preprocessing, we can also divide a sequence in several instance of Step, so that each step is related to a unique sequence. Each step contains several frames, but each frame is associated at most to one step (if the human stops walking at some point of the video, there might be frames not associated to any step). We also have object and data properties with their corresponding some domain, range, and functionality restrictions.

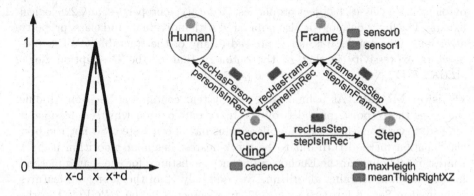

Fig. 2. (a) A triangular membership function; (b) An excerpt of our ontology.

Relationships between classes are modeled using object properties person-IsInRecording, recordingHasFrame, recordingHasStep, and stepIsInFrame, together with their inverses recordingHasPerson, frameIsInRecording, stepIsInRecording, and frameHasStep, respectively. Figure 2(b) shows the classes and their relationships. We use subproperty chains to infer missing information. For example, the chain frameIsInRecording ∘ recordingHasStep is a subproperty of frameHasStep.

Each frame has 25 datatype properties linking it with each of the 25 joints identified by the Kinect V2 sensor. For example, sensor0 related a frame with a xsd:double number. The names of these datatype properties use the common numeration of the joints, buth we also added 25 equivalent datatype properties with more readable names. For example, spineBase is equivalent to sensor0.

Regarding the biometric features, each step has several datatype properties, such as maxHeight (of the person), or meanThighRightXZ (average value of the angles formed by the right thigh). We not only represent those attributes selected by WEKA, but also other ones such as that stepLength, or leg (left or right) that could be interesting for other people. Similarly, our ontology also allows representing biometric features of a sequence (although our Decision Module do not use them) such as the cadence (number of steps per unit of time).

Reasoning with a fuzzy ontology requires using a fuzzy reasoner. There are several implementations, one of them being *fuzzyDL* [2], accessible from the

Protégé plug-in. Firstly, it is possible to check the consistency of a fuzzy ontology. Secondly, it is possible to solve the instance retrieval problem, that is, obtaining the individuals that belong to a fuzzy concept together with their membership degrees, so we can order them and retrieve only the top-k candidates. For example, at production stage we can obtain data from a person using Kinect and then, before trying to identify him/her, retrieve the top-k individuals belonging to the fuzzy concept of people having some particular values of the attributes (e.g., a given height and a given right thigh angle) represented by means of fuzzy datatypes. Only the retrieved individuals will be transferred to the Decision Module for further processing.

Let us add some statistical information. The current schema of our fuzzy ontology has 4 classes, 8 object properties, 79 datatype properties, and 228 logical axioms. The fuzzy ontology is also populated with individuals and class/property assertions. Our ontology is not expressed in any of the tractable OWL 2 profiles: its expressivity is that of the ontology that of the Description Logic $\mathcal{ALCRIF}(\mathbf{D})$.

Decision Module. As a final step, our system contains a Decision Module that uses the response provided by the fuzzy ontology on which a classifier is applied. Several machine learning algorithms have been tested. So far, the best classification method turn out to be the k-nearest neighbor algorithm using 1 nearest neighbour and the Euclidean distance as distance for the search method for this machine learning algorithm. We used only 12 of the aggregate features computed in Sect. 3 for each step (392 steps detected in the *UPCVGait* in the preprocessing phase) and we obtained 89.03% correctly classified instances using this algorithm and 10-fold cross-validation. This result is slightly lower than 93.29%, obtained in [8] for the same dataset. However, our system can be more scalable; in [8], the search space is much bigger since they use all individuals and the entire gait sequence to classify new recordings (we use the candidates given by the fuzzy ontology and we analyze steps). Moreover, we believe that an improvement to our approach would be to introduce a voting-based scheme for the steps of a new sequence, that is, given the steps of a gait sequence, we will use the classification algorithm to classify each step, and so, each step will be a vote for one individual. The individual with more votes will be the final result of the classification.

4   Conclusions and Future Work

In this work we have summarized our ongoing research project about the combination of a gait recognition system based on Microsoft Kinect and fuzzy ontologies. We have designed an architecture for our system, developed a fuzzy ontology that makes it possible to represent Microsoft Kinect V1 and V2 data in a better way (easier to understand by humans and to reuse by intelligent applications), and discussed how to populate it with biometric features. Interestingly, for the sake of scalability, our fuzzy ontology makes it possible to reduce the number of candidates for the recognition algorithm. So far, we have only performed preliminary experiments with Microsoft Kinect V1 data, with the main differences that we use a smaller number of variables (obtained after an attribute selection to compute the most discriminant ones) and step segmentation rather than a sequence-based approach, showing promising results.

Future work will include the recording of a significant number of video sequences to develop a complete benchmark with Microsoft Kinect V2 data, making both our ontology and dataset publicly available. More experiments are also needed to evaluate *(i)* the scalability, *(ii)* the interpretability of the knowledge by humans, *(iii)* the performance of different machine learning algorithms in the framework of gait recognition systems, and *(iv)* tuning some parameters such as the width of the fuzzy triangular functions or the number of top-k candidates retrieved by the fuzzy ontology.

Acknowledgement. We were funded by DGA/FEDER and projects UZCUD 2016-TEC-02 (University of Zaragoza and Defense University Center), TIN2013-46238-C4 and TIN2016-78011-C4 (Ministerio de Economía y Competitividad).

References

1. Bobillo, F., Straccia, U.: Fuzzy ontology representation using OWL 2. Int. J. Approx. Reason. **52**(7), 1073–1094 (2011)
2. Bobillo, F., Straccia, U.: The fuzzy ontology reasoner fuzzyDL. Knowl.-Based Syst. **95**, 12–34 (2016)
3. Diaz-Rodriguez, N., Pegalajar-Cuellar, M., Lilius, J., Delgado, M.: A fuzzy ontology for semantic modelling and recognition of human behaviour. Knowl.-Based Syst. **66**, 46–60 (2014)
4. Díaz Rodríguez, N., Wikström, R., Lilius, J., Cuéllar, M.P., Delgado Calvo Flores, M.: Understanding movement and interaction: an ontology for kinect-based 3D depth sensors. In: Urzaiz, G., Ochoa, S.F., Bravo, J., Chen, L.L., Oliveira, J. (eds.) UCAmI 2013. LNCS, vol. 8276, pp. 254–261. Springer, Cham (2013). doi:10.1007/978-3-319-03176-7_33
5. Horridge, M., Tudorache, T., Nyulas, C., Vendetti, J., Noy, N.F., Musen, M.A.: WebProtege: a collaborative web-based platform for editing biomedical ontologies. Semant. Web **4**(1), 89–99 (2013)
6. Jiang, S., Wang, Y., Zhang, Y., Sun, J.: Real time gait recognition system based on Kinect skeleton feature. In: Jawahar, C.V., Shan, S. (eds.) ACCV 2014. LNCS, vol. 9008, pp. 46–57. Springer, Cham (2015). doi:10.1007/978-3-319-16628-5_4
7. Kastaniotis, D., Theodorakopoulos, I., Economou, G., Fotopoulos, S.: Gait based recognition via fusing information from euclidean and riemannian manifolds. Pattern Recogn. Lett. **84**, 245–251 (2016)
8. Kastaniotis, D., Theodorakopoulos, I., Theoharatos, C., Economou, G., Fotopoulos, S.: A framework for gait-based recognition using Kinect. Pattern Recogn. Lett. **68**(Part 2), 327–335 (2015)
9. Preis, J., Kessel, M., Werner, M., Linnhoff-Popien, C.: Gait recognition with Kinect. In: Proceedings of the 1st International Workshop on Kinect in Pervasive Computing, pp. P1–P4 (2012)

10. Smith, T.C., Frank, E.: Introducing machine learning concepts with WEKA. In: Mathé, E., Davis, S. (eds.) Statistical Genomics. MMB, vol. 1418, pp. 353–378. Springer, New York (2016). doi:10.1007/978-1-4939-3578-9_17
11. Straccia, U.: Foundations of Fuzzy Logic and Semantic Web Languages. CRC Studies in Informatics Series. Chapman & Hall, New York (2013)

Evidential Joint Calibration of Binary SVM Classifiers Using Logistic Regression

Pauline Minary[1,2]([⊠]), Frédéric Pichon[1], David Mercier[1], Eric Lefevre[1], and Benjamin Droit[2]

[1] Laboratoire de Génie Informatique et d'Automatique de l'Artois (LGI2A), Univ. Artois, EA 3926, F-62400 Béthune, France
{frederic.pichon,david.mercier,eric.lefevre}@univ-artois.fr
[2] Département des Télécommunications, SNCF Réseau, La Plaine Saint Denis, France
{pauline.minary,benjamin.droit}@reseau.sncf.fr

Abstract. In a context of multiple classifiers, a calibration step based on logistic regression is usually used to independently transform each classifier output into a probability distribution, to be then able to combine them. This calibration has been recently refined, using the evidence theory, to better handle uncertainties. In this paper, we propose to use this logistic-based calibration in a multivariable scenario, *i.e.*, to consider jointly all the outputs returned by the classifiers, and to extend this approach to the evidential framework. Our evidential approach was tested on generated and real datasets and presents several advantages over the probabilistic version.

Keywords: Belief functions · Information fusion · Evidential calibration

1 Introduction

Using several classifiers to obtain different information on a given object and combining their outputs is a means to obtain better classification performance. These classifiers may be trained with different data or may not rely on the same training models. Thus, their outputs may not be of the same type or not scaled with each other. To be able to combine them, they first have to be made comparable: a technique called calibration is usually applied, enabling to transform a classifier output into a probability. One of the most commonly used calibration is based on logistic regression [8].

Recently, Xu et al. [11] proposed a refinement of this calibration within a framework for reasoning under uncertainty called evidence theory [9,10]. This theory models more precisely the uncertainties inherent to such calibration process and thus enables to prevent an over-fitting issue that may appear, especially when few training data are available. Thus, given a single classifier returning a confidence score after observing a given object, Xu et al.'s approach transforms this score into a belief function.

© Springer International Publishing AG 2017
S. Moral et al. (Eds.): SUM 2017, LNAI 10564, pp. 405–411, 2017.
DOI: 10.1007/978-3-319-67582-4_30

There exists a multivariable version of the logistic regression, called the multiple logistic regression [5], where the technique is defined with more than one input. If we apply this approach to the vector of scores returned by the classifiers for a given object, we can obtain a joint calibration, which returns a probability. Yet, this technique is also prone to the uncertainty problem. Within this scope, we propose to use the evidential extension of calibration proposed by Xu *et al.*, and to apply it to the calibration based on the multiple logistic regression. Thus, for a given object, our proposed approach transforms the vector of scores returned by the classifiers into a belief function.

This paper is organized as follows. First, Sect. 2 recalls the necessary background on evidence theory. Then, Sect. 3 exposes the probabilistic calibration based on the multiple logistic regression and the extension to the evidential framework that we propose. In Sect. 4, the proposed approach and its probabilistic version are compared. Finally, conclusion and perspectives are given in Sect. 5.

2   Evidence Theory

In this section, basic notions of the evidence theory are first exposed in Sect. 2.1. Applications of this theory to inference and prediction, which are useful to define calibration in the evidential framework, are addressed in Sect. 2.2.

2.1 Basic Notions

The theory of evidence is a framework for reasoning under uncertainty. Let Ω be a finite set called the frame of discernment, which contains all the possible answers to a given question of interest Q. In this theory, uncertainty with respect to the answer to Q is represented using a *Mass Function* (MF) defined as a mapping $m^{\Omega} : 2^{\Omega} \rightarrow [0, 1]$ that satisfies $\sum_{A \subseteq \Omega} m^{\Omega}(A) = 1$ and $m^{\Omega}(\emptyset) = 0$. The quantity $m^{\Omega}(A)$ corresponds to the share of belief that supports the claim that the answer is contained in $A \subseteq \Omega$ and nothing more specific. Any subset A of Ω such that $m^{\Omega}(A) > 0$ is called a focal set of m^{Ω}. When the focal sets are nested, m^{Ω} is said to be consonant. A mass function can be equivalently represented by the belief and plausibility functions, respectively defined by

$$Bel^{\Omega}(A) = \sum_{B \subseteq A} m^{\Omega}(B), \quad Pl^{\Omega}(A) = \sum_{B \cap A \neq \emptyset} m^{\Omega}(B), \quad \forall A \subseteq \Omega. \quad (1)$$

The plausibility function restricted to singletons is called the contour function, denoted pl^{Ω} and defined by $pl^{\Omega}(\omega) = Pl^{\Omega}(\{\omega\}), \forall \omega \in \Omega$. When a mass function is consonant, the plausibility function can be recovered from its contour function with $Pl^{\Omega}(A) = \sup_{\omega \in A} pl^{\Omega}(\omega), \quad \forall A \subseteq \Omega$.

Different decision strategies exist to make a decision about the true answer to Q, given a MF m^{Ω} on this answer [4]. In particular, the answer having the smallest so-called *upper* or *lower expected costs* may be selected. When the set of focal elements is reduced to singletons and Ω, and when the costs are taken equal to 0 if the answer is correct and 1 otherwise, the upper and lower expected costs of some answer $\omega \in \Omega$ are respectively defined as $R^*(\omega) = 1 - m^\Omega(\{\omega\})$ and $R_*(\omega) = 1 - m^\Omega(\{\omega\}) - m^\Omega(\Omega)$. Choosing the answer minimizing the lower (resp. upper) expected costs is called the optimistic (resp. pessimistic) strategy. To avoid making risky decisions, when the expected costs are high, a reject decision can be introduced: we define $R_{rej} \in [0,1]$, and the reject decision is made when R_{rej} is lower than the other expected costs.

2.2   Statistical Inference and Forecasting

The evidence theory can be used for inference and forecasting. Consider $\theta \in \Theta$ an unknown parameter, $x \in \mathbb{X}$ some observed data and $f_\theta(x)$ the density function generating the data. Statistical inference consists in making statements about θ after observing the data x. Shafer [9] proposed to represent the knowledge about θ by a consonant belief function Bel_x^Θ based on the likelihood function $L_x : \theta \to f_\theta(x)$, whose contour function is the normalized likelihood function:

$$pl_x^\Theta(\theta) = \frac{L_x(\theta)}{\sup\limits_{\theta' \in \Theta} L_x(\theta')}, \qquad \forall \theta \in \Theta. \tag{2}$$

Suppose now that we have some knowledge about θ after observing some data x, in the form of a contour function pl_x^Θ. The aim of forecasting is to make statements about a not yet observed data $Y \in \mathbb{Y}$, whose conditional distribution given $X = x$ depends on θ. A solution consists in using the sampling model of Dempster [3] to deduce a belief function on $\mathbb{Y}$ [6,7]. This model proposes to express Y as a function of the parameter θ and some unobserved variable, whose distribution is independent of θ.

Let us consider an important particular case. Assume that $Y \in \mathbb{Y} = \{0,1\}$ is a random variable with a Bernoulli distribution. In that case, Xu et al. [11] showed, by applying inference and forecasting, that we have

$$Bel_x^\mathbb{Y}(\{1\}) = \hat{\theta} - \int_0^{\hat{\theta}} pl_x^\Theta(u)du, \quad Pl_x^\mathbb{Y}(\{1\}) = \hat{\theta} + \int_{\hat{\theta}}^1 pl_x^\Theta(u)du, \tag{3}$$

where $\hat{\theta}$ maximizes pl_x^Θ.

3   An Evidential Joint Calibration Approach

Assume that after observing an object which belongs either to class 0 or 1, a SVM classifier returns a confidence score $s \in \mathbb{R}$. To learn how to interpret what this score represents with respect to the true label $y \in \mathbb{Y} = \{0,1\}$ of the object, a step called calibration may be performed. In particular, the one based on logistic regression is commonly used [8]. It aims to estimate the probability distribution $p^\mathbb{Y}(\cdot|s)$ and relies on a training set. Yet, the less training samples are available, the more the estimated probabilities are uncertain. To manage these uncertainties, Xu *et al.* proposed to refine this calibration using the theory of evidence [11].

We propose to use the multiple version of the logistic regression [5] and to apply it to the outputs of multiple classifiers, *i.e.*, to perform a joint calibration of the scores provided by J binary SVM classifiers. It relies on a training set defined by $\mathcal{X} = \{(S_{11}, ..., S_{J1}, Y_1), ..., (S_{1n}, ..., S_{Jn}, Y_n)\}$, where S_{ji} corresponds to the score given by the jth classifier for the ith test sample, and Y_i its true label. Given a vector of scores $\mathbf{s} = (s_1, ..., s_J)$, with s_j the score returned by the jth classifier, the calibration based on the multiple logistic regression can be defined by

$$P^{\mathbb{Y}}(y = 1|\mathbf{s}) \approx h_{\mathbf{s}}(\sigma) = \frac{1}{1 + \exp(\sigma_0 + \sigma_1 s_1 + \sigma_2 s_2 + ... + \sigma_J s_J)}, \quad (4)$$

where the parameter $\sigma = \{\sigma_0, ..., \sigma_J\} \in \mathbb{R}^{J+1}$ is obtained by maximizing the likelihood function L, defined by

$$L(\sigma) = \prod_{i=1}^{n} p_i^{Y_i}(1 - p_i)^{1-Y_i}, \text{ with } p_i = \frac{1}{1 + \exp(\sigma_0 + \sigma_1 S_1 + ... + \sigma_J S_J)}. \quad (5)$$

To better handle the uncertainties, we propose to extend this approach to the evidential framework by following the same likelihood-based reasoning as in [11]. Calibration of a given vector of scores $\mathbf{s}$ based on logistic regression can be seen as a prediction problem of a Bernoulli variable Y with parameter θ, where $\theta = h_{\mathbf{s}}(\sigma)$. A belief function $Bel^{\mathbb{Y}}(\cdot|\mathbf{s})$ can be derived from the contour function $pl_{\mathcal{X}}^{\Theta}(\cdot|\mathbf{s})$ using Eq. (3). Following Xu *et al.* [11], this contour function can be computed from $Pl_{\mathcal{X}}^{\Sigma}$, which is the plausibility function of $pl_{\mathcal{X}}^{\Sigma}$ defined by

$$pl_{\mathcal{X}}^{\Sigma}(\sigma) = \frac{L(\sigma)}{L(\hat{\sigma})}, \quad \forall \sigma \in \Sigma, \quad (6)$$

with $\hat{\sigma} = (\hat{\sigma}_0, ..., \hat{\sigma}_J)$ the Maximum Likelihood Estimate (MLE) of σ and L the likelihood defined in Eq. (5). As $\theta = h_{\mathbf{s}}(\sigma)$, we have

$$pl_{\mathcal{X}}^{\Theta}(\theta|\mathbf{s}) = \begin{cases} 0 & \text{if } \theta \in \{0, 1\}, \\ Pl_{\mathcal{X}}^{\Sigma}(h_{\mathbf{s}}^{-1}(\theta)) & \text{otherwise,} \end{cases} \quad (7)$$

with

$$h_{\mathbf{s}}^{-1}(\theta) = \left\{ (\sigma_0, \sigma_1, ..., \sigma_J) \in \Sigma \Big| \frac{1}{1 + \exp(\sigma_0 + \sigma_1 s_1 + ... + \sigma_J s_J)} = \theta \right\}, \quad (8)$$

$$= \left\{ (\sigma_0, \sigma_1, ..., \sigma_J) \in \Sigma | \sigma_0 = \ln(\theta^{-1} - 1) - \sigma_1 s_1 - ... - \sigma_J s_J \right\}. \quad (9)$$

Thus, Eqs. (7) and (9) yield the following contour function

$$pl_{\mathcal{X}}^{\Theta}(\theta|\mathbf{s}) = \sup_{\sigma_1, ..., \sigma_J \in \mathbb{R}} pl_{\mathcal{X}'}^{\Sigma}(\ln(\theta^{-1} - 1) - \sigma_1 s_1 - \sigma_2 s_2 - ... - \sigma_J s_J, \sigma_1, ..., \sigma_J), \quad (10)$$

for all $\theta \in [0, 1]$. The vector of parameters $(\sigma_1, \sigma_2, ..., \sigma_J)$ which maximizes $pl_{\mathcal{X}}^{\Sigma}$ can be approximated using an iterative maximization algorithm. The computational complexity of such algorithm is $O(nJ)$ per iteration.

4   Experiments

We simulated a binary dataset composed of randomly generated instance vectors from a multivariate normal distribution, composed of two features, with means $\mu_0 = (-1, 0)$ in class 0 and $\mu_1 = (1, 1)$ in class 1, and with a covariance matrix equals to $\begin{bmatrix} 1 & 0.5 \\ 0.5 & 1 \end{bmatrix}$ for both classes. The possibility of deciding to reject a test sample was introduced, and we used both pessimistic and optimistic strategies for the evidential approach. We generated a set of 290 training samples: three SVM classifiers were trained, using the LIBSVM library [2], with three non-overlapping subset of 30 training samples of this set, and our evidential joint calibration was trained with the remaining 200 samples. Then, the same experiment was performed but with 15 examples to train the approach. The decision frontiers in both cases are illustrated in Fig. 1, for $R_{rej} = 0.2$. As it can be seen, the approach based on the optimistic strategy tends to decide more, hence to reject less, the test samples than the two others and it is the exact opposite for the pessimistic strategy. Furthermore, the frontiers are a lot more distant from each other when there are less examples to train the approach (Fig. 1b), *i.e.*, when there are more uncertainties. The probabilistic calibration only yields one frontier so the impact of the uncertainties is not visible. Thus, evidential joint approaches better reflect the uncertainties than the probabilistic one, and using an evidential approach enables to choose between a strategy which decide more often and reject less test samples, or the opposite.

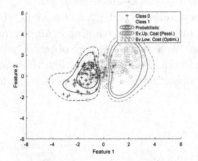
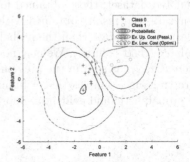

(a) Approach trained with 200 examples   (b) Approach trained with 15 examples

Fig. 1. Decision frontiers in feature space of the joint calibration trained with 200 (1a) and 15 (1b) training examples, for $R_{rej} = 0.2$.

With the same set repartition, we calculated the error rate and accuracy rates for 100 test samples and $R_{rej} = 0.2$. Accuracy rate corresponds to the number of correctly classified objects over the number of classified objects, *i.e.*, not over the total number of test samples as some of them are rejected. The process was repeated for 100 rounds of random partitioning. The obtained points are more distant from each other when few training examples are available (Fig. 2).

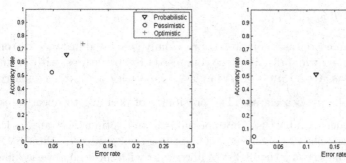

(a) Approach trained with 200 examples   (b) Approach trained with 15 examples

Fig. 2. Error and accuracy rates for $R_{rej} = 0.2$ and with 200 (2a) and 15 (2b) training examples.

This interval reflects the uncertainties as it is larger when they are more important. This information cannot be obtained with the probabilistic approach, which is represented by only one point.

Furthermore, we performed the same experiment with R_{rej} varying from 0 to 1, on four datasets (*Australian, Diabetes, Heart, Ionosphere*) of UCI repository [1] and on the simulated dataset. The classifiers were still trained on non-overlapping subsets of 30 examples, either for simulated or real data. Our joint calibration was trained with 45 then 15 samples. Figure 3 shows the results obtained for the simulated dataset; those obtained for the real datasets are similar. For a given error rate, the results obtained with the pessimistic strategy has a higher (or equal) accuracy rate than the probabilistic one when few training examples are available (right column). We may notice that these two points are obtained with different R_{rej}, as seen in the previous experiment. Furthermore, when there are more training examples (left column), the obtained results become similar for the probabilistic and evidential approaches.

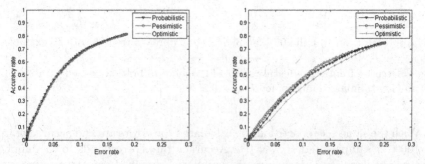

(a) Simulated data – 45 training examples (b) Simulated data – 15 training examples

Fig. 3. Error and accuracy rates with 45 (left) and 15 training examples (right).

Finally, we compared our evidential joint approach to Xu *et al.*'s approach [11], which independently calibrate the scores given by single classifier and combine them with Dempster's rule [9]. We performed the same experiment as the first one detailed in [11], where the training set size for the third classifier was varying. The training of our joint calibration was performed by concatenating the calibration training subsets of the three classifiers. The joint proposed approach presents lower error rates than Xu *et al.*'s approach on the simulated dataset as well as on the real data (results cannot be shown due to space limitations).

5   Conclusion

In this paper, an evidential joint calibration based on logistic regression was proposed. Logistic regression is commonly used to calibrate the scores of a single classifier and we used its multiple version to take into account together the scores returned by multiple classifiers for an object. The application of evidence theory enables to better handle the process uncertainties than the probabilistic version.

We only studied the calibration using logistic regression but the same reasoning can be applied to other calibration techniques. Finally, an extension of our approach to multiclass problem could also be considered in future works.

References

1. Bache, K., Lichman, M.: UCI machine learning repository (2013). http://archive. ics.uci.edu/ml
2. Chang, C-C., Lin, C-J.: LIBSVM: A library for support vector machines. Trans. Intel. Syst. Technol. **2**, 27:1–27:27 (2011). Software, http://www.csie.ntu.edu.tw/~cjlin/libsvm
3. Dempster, A.P.: New methods for reasoning towards posterior distributions based on sample data. Ann. Math. Stat. **37**(2), 355–374 (1966)
4. Denœux, T.: Analysis of evidence-theoretic decision rules for pattern classification. Pattern Recogn. **30**(7), 1095–1107 (1997)
5. Hosmer, D.W., Lemeshow, S.: Applied Logistic Regression. Wiley, New York (2004)
6. Kanjanatarakul, O., Denœux, T., Sriboonchitta, S.: Prediction of future observations using belief functions: a likelihood-based approach. Int. J. Approximate Reasoning **72**, 71–94 (2015)
7. Kanjanatarakul, O., Sriboonchitta, S., Denœux, T.: Forecasting using belief functions: an application to marketing econometrics. Int. J. Approximate Reasoning **55**(5), 1113–1128 (2014)
8. Platt, J.C.: Probabilistic outputs for support vector machines and comparisons to regularized likelihood methods. Adv. in large margin classifiers **10**(3), 61–74 (1999)
9. Shafer, G.: A Mathematical Theory Of Evidence. Princeton University Press, Princeton (1976)
10. Smets, P., Kennes, R.: The transferable belief model. Artif. Intell. **66**, 191–243 (1994)
11. Xu, P., Davoine, F., Zha, H., Denœux, T.: Evidential calibration of binary SVM classifiers. Int. J. Approximate Reasoning **72**, 55–70 (2016)

Ad Hoc Metric for Correspondence Analysis Between Fuzzy Partitions

Carlos Molina[1], María D. Ruiz[2], Daniel Sánchez[3], and José M. Serrano[1(✉)]

[1] University of Jaén, Jaén, Spain
{carlosmo,jschica}@ujaen.es
[2] University of Cádiz, Cádiz, Spain
mariadolores.ruiz@uca.es
[3] University of Granada, Granada, Spain
daniel@decsai.ugr.es

Abstract. Correspondence analysis is a very common and renowned statistical technique, with applications in data summarization, classification, regression, etc. One particular approach is that of comparing different partitions over the same set of objects. Moreover, it can be interesting to analyze correspondences at different detail levels, not only between partitions, but between classes in these partitions. In addition, the case of fuzzy partitions over data is still a researching milestone in development. In this work we propose a novel measure following a previous definition of an alternate methodology in terms of data mining tools, in order to overcome some limitations of the former one for the case of considering partial and global correspondences between fuzzy partitions.

Keywords: Fuzzy correspondence analysis · Fuzzy partitions comparison · Ad hoc metrics

1 Introduction

Correspondence analysis [3] is a well-known statistical technique that can be commonly applied to obtain and describe existing relations between two categorical variables. It is a helpful tool for data dimensionality reduction, as an initial step before more complex processes such as classification, regression, discriminant analysis, etc. Further extensions and applications of this technique can be found throughout the literature [9, 12].

Nevertheless, since it is based on distances and graphical representations, the interpretation can be subjective and sometimes confusing. As a way to overcome this, an alternative to classical correspondence analysis based on data mining techniques was introduced in [19]. This approach allows to obtain local, partial, and global correspondences, according to the required detail level. In contrast to the usual graphical interpretation of distances, correspondences are expressed in terms of data mining tools such as association rules and approximate dependencies, and as a consequence, we can apply the same metrics to interpret and measure the original correspondences.

© Springer International Publishing AG 2017
S. Moral et al. (Eds.): SUM 2017, LNAI 10564, pp. 412–419, 2017.
DOI: 10.1007/978-3-319-67582-4_31

Furthermore, it must be taken into account the fact that in most of real world problems, unclear boundaries between partitions can be found, as some particular elements, due to their nature, may belong to more than one class, with different degrees, inside a same partition. Fuzzy logic allows us to extend existing techniques such as classification, clustering, etc., in order to cope with this issue. As a result, techniques for comparing sets of partitions have been extended in the same way. Renowned metrics as the Rand [17] or Jaccard indices [14] meet their counterparts in fuzzy contexts as, for example, approaches as those of Campello [8], Frigui et al. [11], Brouwer [6], Hüllermeier and Rifqi [13] and Anderson et al. [2]. In [1] the reader may find a more extensive comparison of the cited indices.

Similarly, in [7] the mentioned methodology for correspondence analysis in terms of data mining tools is extended to the fuzzy case, and in [16] an initial comparison with some of the previous measures is discussed. Nevertheless, as it is discussed in [7], some restrictions apply in the original definition of fuzzy partial and global correspondences, as non-atomic values (i.e., elements belonging to more than one partition) are not fully allowed. This paper is intended to continue this research line, introducing an ad hoc measure, in order to overcome the cited drawback. The document is structured as follows. After this introduction, the original proposal for (fuzzy) correspondence analysis in terms of data mining tools is recalled. Following this, we define our new index, and some examples of use are discussed. Concluding remarks as well as future works proposals end the paper.

2 Correspondences as Data Mining Tools

Correspondence analysis is usually applied as an early stage for integration or fusion of different classifications over a same set of objects. In classical correspondence analysis, partitions are displayed by means of a contingency table. Instead, we represent partitions by means of a relational table. For sake of brevity, we will refer directly to the fuzzy case, since the crisp case is easy to particularize from the former one.

Let O be a finite set of objects, and $\widetilde{\mathcal{P}} = \{\widetilde{P}_1, \ldots, \widetilde{P}_p\}$ and $\widetilde{\mathcal{Q}} = \{\widetilde{Q}_1, \ldots, \widetilde{Q}_q\}$ be two fuzzy partitions over O. Let $\widetilde{T}_{\widetilde{\mathcal{P}}\widetilde{\mathcal{Q}}}$ be the fuzzy transactional table associated to O, where each transaction represents an object, that is, $|\widetilde{T}_{\widetilde{\mathcal{P}}\widetilde{\mathcal{Q}}}| = |O|$. Table 1 shows an example of representation (let us remark that, in this particular example, partitions are not in Ruspini form). Given $o \in O$, $\widetilde{P}_i \in \widetilde{\mathcal{P}}$ and $\widetilde{Q}_j \in \widetilde{\mathcal{Q}}$, we noted for $\widetilde{P}_i(o)$ (respectively, $\widetilde{Q}_j(o)$) the membership degree of o in $\widetilde{P}_i$ (respectively, $\widetilde{Q}_j$). Each object must belong to at least one class of each partition, that is, $\forall o \in O, \exists \widetilde{P}_i \in \widetilde{\mathcal{P}}/\widetilde{P}_i(o) > 0$, and each class must contain at least one object, that is, $\widetilde{P}_i, \widetilde{Q}_j \neq \emptyset$. Let us note that, for sake of simplicity, each class in $\widetilde{\mathcal{P}}$ (resp. $\widetilde{\mathcal{Q}}$) can be associated to a single column. Without loss of generality, we can say that columns $\widetilde{P}_1 \ldots \widetilde{P}_p$ (resp. $\widetilde{Q}_1 \ldots \widetilde{Q}_q$) represent the set of possible classes in $\widetilde{\mathcal{P}}$ (resp. $\widetilde{\mathcal{Q}}$).

Table 1. Example of fuzzy transactional table $\widetilde{T}_{\widetilde{\mathcal{P}}\widetilde{\mathcal{Q}}}$

O	$\widetilde{\mathcal{P}}$			$\widetilde{\mathcal{Q}}$		
	$\widetilde{P}_1$	$\widetilde{P}_2$	$\widetilde{P}_3$	$\widetilde{Q}_1$	$\widetilde{Q}_2$	$\widetilde{Q}_3$
o_1	0.81	0	0	0.47	0.63	0
o_2	0.72	0.35	0	0	0.93	0
o_3	0.41	0.65	0	0	1.0	0
o_4	0.09	0.9	0	0	1.0	0.02
o_5	0	0.69	0.1	0	0.78	0.51
o_6	0	0	0.7	0	0.52	0.89
o_7	0	0	0.89	0	0.02	0.63

Let us remark that this approach allows us to consider not only perfect correspondences, but also those with possible exceptions. Hence, we are concerned with measuring the accuracy of correspondences between partitions.

2.1   Local, Partial, and Global Correspondences

Due to space restriction issues, we will recall only the definitions regarding the fuzzy case. A complete discussion about crisp correspondence analysis by means of data mining tools can be found in [19]. One of the advantages of this approach is that correspondences can be measured with the same metrics as those of data mining tools. In particular, certainty factor [20] returns a value between -1 (perfect, negative correspondence) and 1 (perfect, positive correspondence).

This methodology was later extended in order to manage correspondences between fuzzy partitions in [7]. Representing fuzzy partitions as in Table 1, the following types of fuzzy correspondences can be defined.

Definition 1 ([7]) *Fuzzy local correspondence.* Let $\widetilde{P}_i \in \widetilde{\mathcal{P}}$ *and* $\widetilde{Q}_j \in \widetilde{\mathcal{Q}}$. *There exists a fuzzy local correspondence from* $\widetilde{P}_i$ *to* $\widetilde{Q}_j$, *noted* $\widetilde{P}_i \Rightarrow \widetilde{Q}_j$, *if* $\widetilde{P}_i \subseteq \widetilde{Q}_j$, *that is,* $\forall o \in O$, $\widetilde{P}_i(o) \leq \widetilde{Q}_j(o)$.

Fuzzy local correspondences can be obtained in terms of fuzzy association rules (e.g., following the formal model proposed in [10]). Fuzzy partial and global correspondences were defined as well, following the model for fuzzy approximate dependencies introduced in [5]. But, as it is addressed in [7], in these cases, we must manage not classes, but partitions. It would be necessary to define an overall membership degree of an object regarding a whole partition, that is, $\widetilde{\mathcal{A}}(o)$. This issue introduced a multidimensionality problem and, hence, objects were limited to belong to only one class in every partition, for example, that one with the highest membership degree.

Definition 2 ([7]) *Fuzzy partial correspondence.* There exists a fuzzy partial correspondence from $\widetilde{\mathcal{P}}$ *to* $\widetilde{\mathcal{Q}}$, *noted* $\widetilde{\mathcal{P}} \Rrightarrow \widetilde{\mathcal{Q}}$, *when* $\forall \widetilde{P}_i \in \widetilde{\mathcal{P}} \ \exists \widetilde{Q}_j \in \widetilde{\mathcal{Q}}$ *such that* $\widetilde{P}_i \subseteq \widetilde{Q}_j$, *that is,* $\forall o \in O/t_o[\widetilde{\mathcal{P}}] = \widetilde{P}_i$ *implies* $t_o[\widetilde{\mathcal{Q}}] = \widetilde{Q}_j$ *and* $\widetilde{\mathcal{P}}(o) \dot{\leq} \widetilde{\mathcal{Q}}(o)$.

$\leq$ defines a vectorial order relation that, for this particular case, corresponds to a classic order relation. Finally, the step from fuzzy partial correspondences to fuzzy global correspondences is straightforward.

Definition 3 ([7]) *Fuzzy global correspondence. There exists a fuzzy global correspondence between $\widetilde{\mathcal{P}}$ and $\widetilde{\mathcal{Q}}$, noted $\widetilde{\mathcal{P}} \equiv \widetilde{\mathcal{Q}}$, when $\widetilde{\mathcal{P}} \Rightarrow \widetilde{\mathcal{Q}}$ and $\widetilde{\mathcal{Q}} \Rightarrow \widetilde{\mathcal{P}}$.*

In order to continue and complete this approach, in the following section we propose a new index, specifically intended for measuring fuzzy partial (and global) correspondences between fuzzy partitions.

3   Ad Hoc Index for Fuzzy Partial Correspondences

According to Definition 2, there is a fuzzy partial correspondence between two partitions, when we find that classes from the first partition are included, to some extent, in classes from the second partition. Hence, if we are capable of measure these inclusions for each pair of classes and aggregate the obtained values into a general index, we could measure a partial (and later, global) correspondence between these two partitions. With this idea in mind, we define our index as follows:

Definition 4. *Let $O = \{o_1, \ldots, o_n\}$, be again a set of objects, with $\widetilde{\mathcal{P}} = \{\widetilde{P}_1, \ldots, \widetilde{P}_p\}$ and $\widetilde{\mathcal{Q}} = \{\widetilde{Q}_1, \ldots, \widetilde{Q}_q\}$, two fuzzy partitions over O. There is a partial correspondence from $\widetilde{\mathcal{P}}$ to $\widetilde{\mathcal{Q}}$ when all classes from partition $\widetilde{\mathcal{P}}$ are included in classes from $\widetilde{\mathcal{P}}$, to some extent, which we measure by means of the following index:*

$$adhoc(\widetilde{\mathcal{P}}, \widetilde{\mathcal{Q}}) = AGGR_{i=1}^{p}\left(\bigoplus_{j=1}^{q}\left(AVG_{k=1}^{n}\left(\widetilde{\mathcal{P}}_i(o_k) \otimes \widetilde{\mathcal{Q}}_j(o_k)\right)\right)\right) \qquad (1)$$

where $\otimes$ is a t-norm, $\oplus$ a t-conorm, AGGR is an aggregation operator, and AVG is an averaging operator.

The reasoning behind this definition is that, for each pair $\widetilde{\mathcal{P}}_i \in \widetilde{\mathcal{P}}, \widetilde{\mathcal{Q}}_j \in \widetilde{\mathcal{Q}}$, we check to what extent is the former one included in the latter one according to all objects in O, by means of the t-norm $\otimes$. In our experiments we have considered $a \otimes b = min(a, b)$. Next, by means of an averaging operator (in our case, an average mean), we aggregate all these values for each $\widetilde{\mathcal{Q}}_j \in \widetilde{\mathcal{Q}}$ in order to obtain an estimated inclusion degree. Among all these degrees, we select the most representative one for each $\widetilde{\mathcal{P}}_i \in \widetilde{\mathcal{P}}$ (we took $\oplus = $ max). Finally, we obtain our index as an aggregation ($AGGR = sum$, in our case) of the previous values. The closer the value to 1, the more similar the partitions are. In fact, $adhoc(\widetilde{\mathcal{P}}, \widetilde{\mathcal{Q}}) = 1$, if $\widetilde{\mathcal{P}} = \widetilde{\mathcal{Q}}$. Algorithm 1 describes the process in a more formal way.

It must be remarked that, reviewing the literature, a similar index has been already proposed by Beringer and Hüllermeier in [4], where similarities between classes within partitions, instead of objects, are taken into account.

4   Experiments

As an initial but illustrative example, let us remember the example shown in Table 1. Following the original approach for fuzzy partial correspondences introduced in [7], a certainty factor $CF = 0.80$ (resp., 0.20) was returned for the fuzzy partial correspondence $\widetilde{\mathcal{P}} \Rrightarrow \widetilde{\mathcal{Q}}$ (resp., $\widetilde{\mathcal{Q}} \Rrightarrow \widetilde{\mathcal{P}}$). Our index returned a value of 0.839 (resp., 0.641). Apart from this, we have compared different set of partitions. Starting from randomly generated values, we compare a 5-classes fuzzy partition with a 7-classes one over an hypothetical set of 400 objects. Let $\widetilde{\mathcal{A}}_5$ be the former one, and $\widetilde{\mathcal{A}}_7$, the latter one. We measured fuzzy partial correspondence $\widetilde{\mathcal{A}}_5 \Rrightarrow \widetilde{\mathcal{A}}_7$ (resp. $\widetilde{\mathcal{A}}_7 \Rrightarrow \widetilde{\mathcal{A}}_5$) with a value of our index of $adhoc(\widetilde{\mathcal{A}}_5, \widetilde{\mathcal{A}}_7) = 0.571$ (resp. 0.787). This first experimental instance was mainly intended to test the behavior of the metric.

Table 2. Fuzzy partitions computed over wiki4HE dataset

	$\widetilde{\mathcal{W}}_1$	$\widetilde{\mathcal{W}}_2$	$\widetilde{\mathcal{W}}_3$	$\widetilde{\mathcal{W}}_4$
Distance	Euclidean		Manhattan	
Clusters	19	11	19	11
Error	1.3715	3.2144	0.4853	0.8590

In second place, we took *wiki4HE* Dataset [15] from UCI Machine Learning Repository, and applied different FCM (R package *e1071*) executions in order to generate different partitions (Table 2). Two different metrics (Euclidean and Manhattan) were applied, and for each one, two possible partitions were computed, with different number of classes. It is expected that, since both metrics are relatively similar, our index should reflect this with a high value. Moreover, high values for fuzzy partial correspondences are expected from more detailed (higher number of classes) partitions to more general (lower number of classes) ones, and vice versa.

Our index, together with the proposed one in [4], were computed between those partitions, in order to measure the fuzzy partial correspondences between them. The results are summarized in Table 3, the first value being that of our index, and the second one, Beringer and Hüllermeier's.

Table 3. Fuzzy partial correspondences between partitions (row $\Rrightarrow$ column)

	$\widetilde{\mathcal{W}}_1$	$\widetilde{\mathcal{W}}_2$	$\widetilde{\mathcal{W}}_3$	$\widetilde{\mathcal{W}}_4$
$\widetilde{\mathcal{W}}_1$	1.000/1.000	0.998/0.974	0.819/0.974	0.988/0.974
$\widetilde{\mathcal{W}}_2$	0.468/0.943	1.000/1.000	0.476/0.943	0.763/0.943
$\widetilde{\mathcal{W}}_3$	0.853/0.971	0.955/0.971	1.000/1.000	0.955/0.971
$\widetilde{\mathcal{W}}_4$	0.483/0.947	0.903/0.947	0.521/0.947	1.000/1.000

It must be noticed how fuzzy partial correspondences $\widetilde{\mathcal{W}_1} \Rrightarrow \widetilde{\mathcal{W}_2}$ and $\widetilde{\mathcal{W}_3} \Rrightarrow \widetilde{\mathcal{W}_4}$ are strong (index value close to 1), since the latter ones are summarizations of the former ones. That is, a reduction in the number of clusters induces that the former clusters are included, to some degree, in the latter ones. The opposite correspondences have a lower index value, which, according to the previous reasoning, seems logical. Since this issue is not detected in Beringer and Hüllermeier's proposal, whose index shows similar values for each pair of partitions, a deeper study should be conducted in order to explain it.

Finally, we also computed our index over the same partitions considered in [7], and found an interesting issue; our ad hoc index returned a value higher than 1. This could be due to the fact that one of the partitions was not in Ruspini [18] form. This situation may suggest that fuzzy operators in Eq. 1 needs to be properly adjusted.

Algorithm 1. Algorithm AdHoc

Input : $O = \{o_1, \ldots, o_n\}$, a set of objects, $\widetilde{\mathcal{P}} = \{\widetilde{P}_1, \ldots, \widetilde{P}_p\}$ and
$\widetilde{\mathcal{Q}} = \{\widetilde{Q}_1, \ldots, \widetilde{Q}_q\}$, two fuzzy partitions over O.
Output: $adhoc(\widetilde{\mathcal{P}}, \widetilde{\mathcal{Q}})$, measure of the fuzzy partial correspondence from $\widetilde{\mathcal{P}}$ to
$\widetilde{\mathcal{Q}}$, $\widetilde{\mathcal{P}} \Rrightarrow \widetilde{\mathcal{Q}}$.

1 $V_P \leftarrow \emptyset$
2 **foreach** $\widetilde{P}_i \in \widetilde{\mathcal{P}}$ **do**
3      $V_Q \leftarrow \emptyset$
4   **foreach** $\widetilde{Q}_j \in \widetilde{\mathcal{Q}}$ **do**
        /* Consider how $\widetilde{P}_i$ is included in every $\widetilde{Q}_j$ according to O */
5          $V_Q[j] \leftarrow AVG_{o \in O}\left(\left(\widetilde{P}_i(o) \otimes \widetilde{Q}_j(o)\right)\right)$
6   **end**
    /* For each $\widetilde{P}_i$, select the most representative value in V_q */
7      $V_P[i] \leftarrow \bigoplus_1^q (V_Q[j])$
8 **end**
  /* Finally, aggregate all values in V_P */
9 $adhoc(\widetilde{\mathcal{P}}, \widetilde{\mathcal{Q}}) \leftarrow AGGR_{i=1}^p (V_P[i])$

---

5 Concluding Remarks and Further Works

In this work our intention has been to continue a previous methodology for fuzzy correspondence analysis. To this purpose, we have proposed a new *ad hoc* index (in absence of a better name) to measure fuzzy partial, and global, correspondences between two fuzzy partitions, based on the extent to which the classes of a partition are included in the classes of the second partition, according to every object in a collection. First experiments suggest that the obtained results seem reasonable (values close to 1 where expected, and vice versa), although a deeper analysis, interesting properties study, and comparison with existing indices is still pending in order to validate and refine our proposal. They will be properly addressed in a future extension of this paper.

Acknowledgements. This work has been partially supported by the Spanish Ministry of Economy and Competitiveness and the European Regional Development Fund (FEDER) under projects TIN2015-64776-C3-1-R and TIN2014-58227-P, and by the Energy IN TIME project funded from the European Union in the Seventh Framework Programme under grant agreement No. 608981.

References

1. Anderson, D.T., Bezdek, J.C., Keller, J.M., Popescu, M.: A comparison of five fuzzy rand indices. In: Hüllermeier, E., Kruse, R., Hoffmann, F. (eds.) IPMU 2010. CCIS, vol. 80, pp. 446–454. Springer, Heidelberg (2010). doi:10.1007/978-3-642-14055-6_46
2. Anderson, D.T., Bezdek, J.C., Popescu, M., Keller, J.M.: Comparing fuzzy, probabilistic, and possibilistic partitions. IEEE Trans. Fuzzy Syst. **18**(5), 906–918 (2010)
3. Benzécri, J.: Cours de Linguistique Mathématique. Faculté des Sciences, Université de Rennes (1964)
4. Beringer, J., Hüllermeier, E.: Fuzzy clustering of parallel data streams. In: Advances in Fuzzy Clustering and Its Application, pp. 333–352 (2007)
5. Berzal, F., Blanco, I., Sánchez, D., Serrano, J., Vila, M.: A definition for fuzzy approximate dependencies. Fuzzy Sets Syst. **149**(1), 105–129 (2005). Fuzzy Sets in Knowledge Discovery
6. Brouwer, R.K.: Extending the rand, adjusted rand and jaccard indices to fuzzy partitions. J. Intell. Inform. Syst. **32**(3), 213–235 (2009)
7. Calero, G., Delgado, J., Serrano, J., Sánchez, D., Vila, M.: A proposal of fuzzy correspondence analysis based on flexible data mining techniques. In: Soft Methodology and Random Information Systems, pp. 447–454. Springer, Heidelberg (2004)
8. Campello, R.J.: A fuzzy extension of the rand index and other related indexes for clustering and classification assessment. Pattern Recogn. Lett. **28**(7), 833–841 (2007)
9. Cox, M.A.A., Cox, T.F.: Multidimensional Scaling, pp. 315–347. Springer, Heidelberg (2008)
10. Delgado, M., Ruiz, M., Sánchez, D., Serrano, J.: A formal model for mining fuzzy rules using the RL representation theory. Inform. Sci. **181**(23), 5194–5213 (2011)
11. Frigui, H., Hwang, C., Rhee, F.C.-H.: Clustering and aggregation of relational data with applications to image database categorization. Pattern Recogn. **40**(11), 3053–3068 (2007)
12. Greenacre, M.: Correspondence Analysis in Practice, 3rd edn. CRC Press, Boca Raton (2016)
13. Hullermeier, E., Rifqi, M., Henzgen, S., Senge, R.: Comparing fuzzy partitions: a generalization of the rand index and related measures. IEEE Trans. Fuzzy Syst. **20**(3), 546–556 (2012)
14. Jaccard, P.: Étude comparative de la distribution florale dans une portion des alpes et des jura. Bulletin del la Société Vaudoise des Sciences Naturelles **37**, 547–579 (1901)
15. Meseguer Artola, A., Aibar Puentes, E., Lladós Masllorens, J., Minguillón Alfonso, J., Lerga Felip, M.: Factors that influence the teaching use of wikipedia in higher education (2014)

16. Molina, C., Prados, B., Ruiz, M.-D., Sánchez, D., Serrano, J.-M.: Comparing partitions by means of fuzzy data mining tools. In: Hüllermeier, E., Link, S., Fober, T., Seeger, B. (eds.) SUM 2012. LNCS (LNAI), vol. 7520, pp. 337–350. Springer, Heidelberg (2012). doi:10.1007/978-3-642-33362-0_26
17. Rand, W.M.: Objective criteria for the evaluation of clustering methods. J. Am. Stat. Assoc. **66**(336), 846–850 (1971)
18. Ruspini, E.H.: A new approach to clustering. Inform. Control **15**(1), 22–32 (1969)
19. Sánchez, D., Serrano, J.M., Vila, M.A., Aranda, V., Calero, J., Delgado, G.: Using data mining techniques to analyze correspondences between user and scientific knowledge in an agricultural environment. In: Piattini, M., Filipe, J., Braz, J. (eds.) Enterprise Information Systems IV, pp. 75–89. Kluwer Academic Publishers, Hingham (2003)
20. Shortliffe, E., Buchanan, B.: A model of inexact reasoning in medicine. Math. Biosci. **23**, 351–379 (1975)

On Similarity-Based Unfolding

Ginés Moreno[1]([✉]), Jaime Penabad[2], and José Antonio Riaza[1]

[1] Department of Computing Systems, UCLM, 02071 Albacete, Spain
{Gines.Moreno,JoseAntonio.Riaza}@uclm.es
[2] Department of Mathematics, UCLM, 02071 Albacete, Spain
Jaime.Penabad@uclm.es

Abstract. The unfolding transformation has been widely used in many declarative frameworks for improving the efficiency and scalability of programs after applying computational steps on their rules. Inspired by our previous experiences in fuzzy logic languages not dealing with similarity relations, in this work we adapt such operation to the so-called FASILL language (acronym of "Fuzzy Aggregators and Similarity Into a Logic Language") which has been recently designed and implemented in our research group for coping with implicit/explicit truth degree annotations, a great variety of connectives and unification by similarity.

Keywords: Fuzzy logic programming · Similarity relations · Unfolding

1 Introduction

The challenging research area of *Fuzzy Logic Programming* is devoted to introduce *fuzzy logic* concepts into *logic programming* in order to explicitly deal with vagueness and uncertainty in a natural way. It has provided an extensive variety of Prolog dialects along the last three decades. *Fuzzy logic languages* can be classified (among other criteria) according to the emphasis they assign to fuzzifying the original unification/resolution mechanisms of Prolog. Whereas some approaches are able to cope with similarity/proximity relations at unification time [5,9], other ones extend their operational principles (maintaining syntactic unification) for managing a wide variety of fuzzy connectives and truth degrees on rules/goals beyond the simpler case of *true* or *false* [6,7]. As in many other fuzzy languages, in this paper we use the lattice of real numbers $([0,1], \leq)$ for modeling truth degrees and, in particular, we also assume the presence of fuzzy connectives such as the arithmetical average whose truth function is defined as $@_{aver}(x,y) = \frac{x+y}{2}$, and the unary connective $@_{very}(x) = x^2$, as well as the classical *Gödel's conjunction* defined as $x \wedge y = min(x,y)$. Our unifying approach, where lattices of truth degrees cohabit with similarity relations, is represented by the design of the FASILL language [3,4], for which we have developed the

This work has been partially supported by the EU (FEDER), the State Research Agency (AEI) and the Spanish *Ministerio de Economía y Competitividad* under grant TIN2016-76843-C4-2-R (AEI/FEDER, UE).

© Springer International Publishing AG 2017
S. Moral et al. (Eds.): SUM 2017, LNAI 10564, pp. 420–426, 2017.
DOI: 10.1007/978-3-319-67582-4_32

FLOPER system that has been used for coding real-word applications (see [1]) and can be also executed on-line via http://dectau.uclm.es/floper/?q=sim.

Definition 1 (Similarity relation). *Given a domain* $\mathcal{U}$ *and a lattice L with a fixed t-norm* $\wedge$ *(usually the Gödel's conjunction), a similarity relation* $\mathcal{R}$ *is a fuzzy binary relation on* $\mathcal{U}$, *that is, a fuzzy subset on* $\mathcal{U} \times \mathcal{U}$ *(namely, a mapping* $\mathcal{R} : \mathcal{U} \times \mathcal{U} \rightarrow L$), *fulfilling the reflexive, symmetric and transitive (that is,* $\mathcal{R}(x,z) \geq \mathcal{R}(x,y) \wedge \mathcal{R}(y,z), \forall x, y, z \in \mathcal{U}$) *properties.*

Definition 2 (Rule and Program). *A rule has the form* $A \leftarrow \mathcal{B}$, *where A is an atomic formula called head and* $\mathcal{B}$, *called body, is a well-formed formula (ultimately built from atomic formulas* $B_1, \ldots, B_n$, *truth values of L and connectives). A* FASILL *program (or simply program) is a tuple* $\langle \Pi, \mathcal{R}, L \rangle$ *where* Π *is a set of rules,* $\mathcal{R}$ *is a similarity relation between the elements of the signature* Σ *of* Π, *and L is a complete lattice.*

Example 1. In this paper we will deal with the following program $\mathcal{P} = \langle \Pi, \mathcal{R}, L \rangle$ based on lattice $L = ([0,1], \leq)$ and the set of rules Π, and similarity relation $\mathcal{R}$ (expressed as a matrix on $\mathcal{U} = \{vanguardist, elegant, metro, taxi, bus\}$) below:

$$\Pi = \begin{cases} R_1 : vanguardist(hydropolis) & \leftarrow 0.9 \\ R_2 : elegant(ritz) & \leftarrow 0.8 \\ R_3 : close(hydropolis, taxi) & \leftarrow 0.7 \\ R_4 : good\_hotel(x) & \leftarrow @_{aver}(elegant(x), @_{very}(close(x, metro))) \end{cases}$$

$\mathcal{R}$	vanguardist	elegant	metro	taxi	bus
vanguardist	1	0.6	0	0	0
elegant	0.6	1	0	0	0
metro	0	0	1	0.4	0.5
taxi	0	0	0.4	1	0.4
bus	0	0	0.5	0.4	1

It is easy to check that $\mathcal{R}$ fulfills the reflexive, symmetric and transitive properties. Particularly, we have that: $\mathcal{R}(taxi, metro) \geq \mathcal{R}(metro, bus) \wedge \mathcal{R}(bus, taxi) = min(0.5, 0.4) = 0.4$.

Furthermore, the natural extension of $\mathcal{R}$ from symbols to terms, denoted as $\hat{\mathcal{R}}$, determines that $elegant(taxi)$ and $vanguardist(metro)$ are similar terms, since: $\hat{\mathcal{R}}(elegant(taxi), vanguardist(metro)) = \mathcal{R}(elegant, vanguardist) \wedge \hat{\mathcal{R}}(taxi, metro) = 0.6 \wedge \mathcal{R}(taxi, metro) = 0.6 \wedge 0.4 = min(0.6, 0.4) = 0.4$.

Instead of syntactic unification, and similarly to other fuzzy languages, FASILL uses weak unification for coping with similarity relations [5,9]. In essence, the *weak most general unifier* of two terms t and s, say $wmgu(t, s) = \langle \sigma, r \rangle$, is the simplest substitution σ together with value $r \in L$ verifying $r = \hat{\mathcal{R}}(t\sigma, s\sigma)$. So, w.r.t. the previous example we have that $wmgu(elegant(taxi), vanguardist(metro)) = \langle id, 0.4 \rangle$, being id the empty substitution, whereas $wmgu(vanguardist(x), elegant(taxi)) = \langle \{x/taxi\}, 0.6 \rangle$.

In order to describe the procedural semantics of the FASILL language, in the following we denote by $\mathcal{C}[A]$ a formula where A is a sub-expression (usually an atom) which occurs in the –possibly empty– context $\mathcal{C}[]$ whereas $\mathcal{C}[A/A']$ means the replacement of A by A' in the context $\mathcal{C}[]$. Moreover, $Var(s)$ denotes the set of distinct variables occurring in the syntactic object s and $\theta[\mathcal{V}ar(s)]$ refers to the substitution obtained from θ by restricting its domain to $\mathcal{V}ar(s)$. In the next definition, we always consider that A is the selected atom in a goal $\mathcal{Q}$ and L is the complete lattice associated to Π.

Definition 3 (Computational Step). *Let $\mathcal{Q}$ be a goal and let σ be a substitution. The pair $\langle \mathcal{Q}; \sigma \rangle$ is a state. Given a program $\langle \Pi, \mathcal{R}, L \rangle$ and a t-norm $\wedge$ in L, a computation is formalized as a state transition system, whose transition relation $\rightsquigarrow$ is the smallest relation satisfying these rules:*

(1) Successful step (denoted as$\overset{SS}{\rightsquigarrow}$) :

$$\frac{\langle \mathcal{Q}[A], \sigma \rangle \quad A' \leftarrow \mathcal{B} \in \Pi \quad wmgu(A, A') = \langle \theta, r \rangle}{\langle \mathcal{Q}[A/\mathcal{B} \wedge r]\theta, \sigma\theta \rangle} \; SS$$

(2) Failure step (denoted as$\overset{FS}{\rightsquigarrow}$) :

$$\frac{\langle \mathcal{Q}[A], \sigma \rangle \quad \nexists A' \leftarrow \mathcal{B} \in \Pi : wmgu(A, A') = \langle \theta, r \rangle, r > \bot}{\langle \mathcal{Q}[A/\bot], \sigma \rangle} \; FS$$

(3) Interpretive step (denoted as$\overset{IS}{\rightsquigarrow}$) :

$$\frac{\langle \mathcal{Q}[@(r_1, \ldots, r_n)]; \sigma \rangle \quad \overset{.}{@}(r_1, \ldots, r_n) = r_{n+1}}{\langle \mathcal{Q}[@(r_1, \ldots, r_n)/r_{n+1}]; \sigma \rangle} \; IS$$

A *derivation* is a sequence of arbitrary length $\langle \mathcal{Q}; id \rangle \rightsquigarrow^* \langle \mathcal{Q}'; \sigma \rangle$. As usual, rules are renamed apart. When $\mathcal{Q}' = r \in L$, the state $\langle r; \sigma \rangle$ is called a *fuzzy computed answer* (f.c.a.) for that derivation.

Example 2. Let $\mathcal{P} = \langle \Pi, \mathcal{R}, L \rangle$ be the program from Example 1. It is possible to perform this derivation with fuzzy computed answer $\langle 0, 4, \{x/ritz\} \rangle$ for $\mathcal{P}$ and goal $\mathcal{Q} = good\_hotel(x)$:

$$D1 : \langle good\_hotel(x), id \rangle \overset{SS^{R4}}{\rightsquigarrow}$$

$$\langle @_{aver}(elegant(x), @_{very}(close(x, metro))), \{x_1/x\} \rangle \overset{SS^{R2}}{\rightsquigarrow}$$

$$\langle @_{aver}(0.8, @_{very}(close(ritz, metro))), \{x_1/ritz, x/ritz\} \rangle \overset{FS}{\rightsquigarrow}$$

$$\langle @_{aver}(0.8, @_{very}(0)), \{x_1/ritz, x/ritz\} \rangle \overset{IS}{\rightsquigarrow}$$

$$\langle @_{aver}(0.8, 0), \{x_1/ritz, x/ritz\} \rangle \overset{IS}{\rightsquigarrow}$$

$$\langle 0.4, \{x_1/ritz, x/ritz\} \rangle$$

Apart from this derivation, there exists a second one ending with the alternative f.c.a. $\langle 0.38, \{x/hydropolis\} \rangle$ associated to the same goal.

As we will see in the following section, the application of computational steps according Definition 3 on the body of FASILL program rules, is the basis of our similarity-based unfolding transformation.

2   Unfolding FASILL programs

Unfolding is a well-known, widely used, semantics-preserving program transformation rule which is able to improve programs, generating more efficient code. The unfolding transformation traditionally considered in pure logic programming consists in the replacement of a program clause C by the set of clauses obtained after applying a symbolic computation step in all its possible forms on the body of C [8,10]. Although in [2] we successfully adapted such operation to fuzzy logic programs dealing with lattices of truth degrees, there are not precedents coping with similarity relations, which motivates the present work.

Definition 4 (Similarity-based Unfolding). *Let* $\mathcal{P} = \langle \Pi, \mathcal{R}, L \rangle$ *be a* FASILL *program and let* $R : A \leftarrow B \in \Pi$ *be a program rule with no empty body. Then, the similarity-based unfolding of rule* R *in program* $\mathcal{P}$ *is the new program* $\mathcal{P}' = (\mathcal{P} - \{R\}) \cup \mathcal{U}$ *where* $\mathcal{U} = \{A\sigma \leftarrow B' \mid \langle B; id \rangle \rightsquigarrow \langle B'; \sigma \rangle\}$.

From now on we consider that the selection function (also called "computation rule") used when applying computational steps on a given goal, only applies $\overset{\text{IS}}{\rightsquigarrow}$ steps (always from left to right) whenever there are no atoms to exploit (again from left to right) with $\overset{\text{SS}}{\rightsquigarrow}$ and/or $\overset{\text{FS}}{\rightsquigarrow}$ steps.

Example 3. Let us built a transformation sequence where each FASILL program in the sequence is obtained from the immediately preceding one by applying fuzzy unfolding, except the initial one $\mathcal{P}_0$, which in our case is the one illustrated in Example 1, that is:

$R_1 : vanguardist(hydropolis) \leftarrow 0.9$
$R_2 : elegant(ritz) \qquad\qquad \leftarrow 0.8$
$R_3 : close(hydropolis, taxi) \quad \leftarrow 0.7$
$R_4 : good\_hotel(x) \qquad\qquad \leftarrow @_{aver}(elegant(x), @_{very}(close(x, metro)))$

Program $\mathcal{P}_1$ is obtained after unfolding rule R_4 (with selected atom $elegant(x)$) by applying a $\overset{\text{SS}}{\rightsquigarrow}$ step with rules R_1 and R_2:

$R_1 : \quad vanguardist(hydropolis) \leftarrow 0.9$
$R_2 : \quad elegant(ritz) \qquad\qquad \leftarrow 0.8$
$R_3 : \quad close(hydropolis, taxi) \quad \leftarrow 0.7$
$R_{4-1} : good\_hotel(hydropolis) \quad \leftarrow @_{aver}(0.9 \wedge 0.6,$
$\qquad\qquad\qquad\qquad\qquad\qquad\qquad @_{very}(close(hydropolis, metro)))$
$R_{4-2} : good\_hotel(ritz) \qquad\qquad \leftarrow @_{aver}(0.8 \wedge 1,$
$\qquad\qquad\qquad\qquad\qquad\qquad\qquad @_{very}(close(ritz, metro)))$

After unfolding rule R_{4-1} (with selected atom $close(hydropolis, metro)$) by applying a $\overset{\text{SS}}{\rightsquigarrow}$ step with rule R_3, we obtain program $\mathcal{P}_2$:

$R_1 : \qquad vanguardist(hydropolis) \leftarrow 0.9$
$R_2 : \qquad elegant(ritz) \qquad\qquad \leftarrow 0.8$
$R_3 : \qquad close(hydropolis, taxi) \quad \leftarrow 0.7$
$R_{4-1-3} : good\_hotel(hydropolis) \quad \leftarrow @_{aver}(0.9 \wedge 0.6,$
$\qquad\qquad\qquad\qquad\qquad\qquad\qquad @_{very}(0.7 \wedge 0.4))$
$R_{4-2} : \quad good\_hotel(ritz) \qquad\qquad \leftarrow @_{aver}(0.8 \wedge 1,$
$\qquad\qquad\qquad\qquad\qquad\qquad\qquad @_{very}(close(ritz, metro)))$

Program $\mathcal{P}_3$ is obtained by unfolding rule R_{4-1-3} (with selected expression $0.9 \wedge 0.6$) after applying a $\overset{IS}{\rightsquigarrow}$ step:

$R_1 :$ $vanguardist(hydropolis) \leftarrow 0.9$

$R_2 :$ $elegant(ritz)$ $\leftarrow 0.8$

$R_3 :$ $close(hydropolis, taxi)$ $\leftarrow 0.7$

$R_{4-1-3I} :$ $good\_hotel(hydropolis)$ $\leftarrow @_{aver}(0.6, @_{very}(0.7 \wedge 0.4))$

$R_{4-2} :$ $good\_hotel(ritz)$ $\leftarrow @_{aver}(0.8 \wedge 1,$
                                        $@_{very}(close(ritz, metro)))$

When unfolding rule R_{4-1-3I} (with selected expression $0.7 \wedge 0.4$) by applying a $\overset{IS}{\rightsquigarrow}$ step we reach program $\mathcal{P}_4$:

$R_1 :$ $vanguardist(hydropolis) \leftarrow 0.9$

$R_2 :$ $elegant(ritz)$ $\leftarrow 0.8$

$R_3 :$ $close(hydropolis, taxi)$ $\leftarrow 0.7$

$R_{4-1-3II} :$ $good\_hotel(hydropolis)$ $\leftarrow @_{aver}(0.6, @_{very}(0.4))$

$R_{4-2} :$ $good\_hotel(ritz)$ $\leftarrow @_{aver}(0.8 \wedge 1,$
                                        $@_{very}(close(ritz, metro)))$

Program $\mathcal{P}_5$ is obtained by unfolding rule $R_{4-1-3II}$ (with selected expression $@_{very}(0.4)$) after applying a $\overset{IS}{\rightsquigarrow}$ step:

$R_1 :$ $vanguardist(hydropolis) \leftarrow 0.9$

$R_2 :$ $elegant(ritz)$ $\leftarrow 0.8$

$R_3 :$ $close(hydropolis, taxi)$ $\leftarrow 0.7$

$R_{4-1-3III} :$ $good\_hotel(hydropolis)$ $\leftarrow @_{aver}(0.6, 0.16)$

$R_{4-2} :$ $good\_hotel(ritz)$ $\leftarrow @_{aver}(0.8 \wedge 1,$
                                        $@_{very}(close(ritz, metro)))$

Now, by unfolding rule $R_{4-1-3III}$ (with selected expression $@_{aver}(0.6, 0.16)$) after applying a $\overset{IS}{\rightsquigarrow}$ step, we obtain program $\mathcal{P}_6$:

$R_1 :$ $vanguardist(hydropolis) \leftarrow 0.9$

$R_2 :$ $elegant(ritz)$ $\leftarrow 0.8$

$R_3 :$ $close(hydropolis, taxi)$ $\leftarrow 0.7$

$R_{4-1-3IIII} :$ $good\_hotel(hydropolis)$ $\leftarrow 0.38$

$R_{4-2} :$ $good\_hotel(ritz)$ $\leftarrow @_{aver}(0.8 \wedge 1,$
                                        $@_{very}(close(ritz, metro)))$

Next, after unfolding rule R_{4-2} (with selected atom $close(ritz, metro)$) by applying a $\overset{FS}{\rightsquigarrow}$ step, we reach program $\mathcal{P}_7$:

$R_1 :$ $vanguardist(hydropolis) \leftarrow 0.9$

$R_2 :$ $elegant(ritz)$ $\leftarrow 0.8$

$R_3 :$ $close(hydropolis, taxi)$ $\leftarrow 0.7$

$R_{4-1-3IIII} :$ $good\_hotel(hydropolis)$ $\leftarrow 0.38$

$R_{4-2F} :$ $good\_hotel(ritz)$ $\leftarrow @_{aver}(0.8 \wedge 1, @_{very}(0))$

Finally, after 3 unfolding (based on $\overset{IS}{\rightsquigarrow}$) steps on rule R_{4-2F}, we reach the final program $\mathcal{P}_{10}$:

$$R_1: \qquad vanguardist(hydropolis) \leftarrow 0.9$$
$$R_2: \qquad elegant(ritz) \qquad\qquad \leftarrow 0.8$$
$$R_3: \qquad close(hydropolis, taxi) \leftarrow 0.7$$
$$R_{4-1-3IIII}: good\_hotel(hydropolis) \quad \leftarrow 0.38$$
$$R_{4-2FIII}: \quad good\_hotel(ritz) \qquad \leftarrow 0.4$$

In the previous example it is easy to see that each program in the sequence produces the same set of f.c.a.'s for a given goal but reducing the length of derivations. For instance, the derivation performed w.r.t. the original program $\mathcal{P}_0$ illustrated in Example 2 can be emulated in the final program $\mathcal{P}_{10}$ with just one computational step (instead of five) as: $\langle good\_hotel(x), id\rangle \overset{SS^{R_{4-2FIII}}}{\rightsquigarrow} \langle 0.4, \{x/ritz\}\rangle$. However, we have found out that some constraints must be imposed when unfolding FASILL programs since the transformation is not safe in general, as we are going to illustrate in the following couple of examples.

Example 4. Consider that the similarity degree between the predicate symbols q and r is $\mathcal{R}(q,r) = 0.8$ in the following original and unfolded programs:

$$\mathcal{P} = \begin{cases} R_1: p(x) \leftarrow @_{aver}(q(x), 1) \\ R_2: r(a) \leftarrow 0.6 \end{cases} \quad \mathcal{P}' = \begin{cases} R_{1-2}: p(a) \leftarrow @_{aver}(0.6 \wedge 0.8, 1) \\ R_2: \quad r(a) \leftarrow 0.6 \end{cases}$$

And now observe that the following derivations developed with each program return different f.c.a.'s for the same goal, thus indicating that, in general, unfolding does not preserve those f.c.a.'s produced on derivations using $\overset{FS}{\rightsquigarrow}$ steps.

$$D_1: \langle p(b); id\rangle \overset{SS^{R_1}}{\rightsquigarrow} \qquad\qquad D_1': \langle p(b); id\rangle \overset{FS}{\rightsquigarrow}$$
$$\langle @_{aver}(q(b), 1); \{x_1/b\}\rangle \overset{FS}{\rightsquigarrow} \qquad \langle 0; id\rangle$$
$$\langle @_{aver}(0, 1); \{x_1/b\}\rangle \overset{IS}{\rightsquigarrow}$$
$$\langle 0.5; \{x_1/b\}\rangle$$

Example 5. In the following programs and derivations we consider now a similarity relation establishing that $\mathcal{R}(a, b) = 0.4$ and $\mathcal{R}(q, r) = 0.5$:

$$\mathcal{P} = \begin{cases} R_1: p(x) \leftarrow @_{very}(q(x)) \\ R_2: r(b) \end{cases} \quad \mathcal{P}' = \begin{cases} R_{1-2}: p(b) \leftarrow @_{very}(0.5) \\ R_2: \quad r(b) \end{cases}$$

$$D_2: \langle p(a); id\rangle \overset{SS^{R_1}}{\rightsquigarrow} \qquad\qquad D_2': \langle p(a); id\rangle \overset{SS^{R_{1-2}}}{\rightsquigarrow}$$
$$\langle @_{very}(q(a)); \{x_1/a\}\rangle \overset{SS^{R_2}}{\rightsquigarrow} \qquad \langle 0.4 \wedge @_{very}(0.5); id\rangle \overset{IS}{\rightsquigarrow}$$
$$\langle @_{very}(0.4 \wedge 0.5); \{x_1/a\}\rangle \overset{IS}{\rightsquigarrow} \qquad \langle 0.4 \wedge 0.25; id\rangle \overset{IS}{\rightsquigarrow}$$
$$\langle @_{very}(0.4); \{x_1/a\}\rangle \overset{IS}{\rightsquigarrow} \qquad \langle 0.25; id\rangle$$
$$\langle 0.16; \{x_1/a\}\rangle$$

Which once again shows that the correctness of the unfolding transformation does not hold because, in this particular case, we have that $@_{very}(0.4 \wedge 0.5) \neq 0.4 \wedge @_{very}(0.5)$. We are nowadays trying to formalize a condition to be required on any connective @ used on the body of unfolded rules which could look as $@(t_1, ..., (t_i \wedge t_{i+1}), ..., t_n) = t_i \wedge @(t_1, ..., t_{i+1}, ..., t_n), \forall i, n \in N, i \leq n$, being $\wedge$ the t-norm fixed in Definitions 1 and 3 for propagating similarities.

3   Conclusions and Future Work

FASILL is a fuzzy logic programming language with implicit/explicit truth degree annotations, a great variety of connectives and unification by similarity. In [3,4] we have recently provided the syntax, operational/declarative semantics, and implementation issues of this language which properly manages similarity and truth degrees in a single framework. In this work we have focused on a preliminary formulation of an unfolding transformation for optimizing FASILL programs. We have pointed out that, in contrast with other precedent languages, the treatment of similarities introduces several risks for preserving the correctness of the transformation, for which we are nowadays identifying a set of sufficient conditions allowing us to prove its soundness and completeness properties.

References

1. Almendros-Jiménez, J.M., Bofill, M., Luna-Tedesqui, A., Moreno, G., Vázquez, C., Villaret, M.: Fuzzy XPath for the Automatic Search of Fuzzy Formulae Models. In: Beierle, C., Dekhtyar, A. (eds.) SUM 2015. LNCS, vol. 9310, pp. 385–398. Springer, Cham (2015). doi:10.1007/978-3-319-23540-0_26
2. Julián-Iranzo, P., Moreno, G., Penabad, J.: On fuzzy unfolding: a multi-adjoint approach. Fuzzy Sets Syst. **154**, 16–33 (2005)
3. Julián-Iranzo, P., Moreno, G., Penabad, J., Vázquez, C.: A fuzzy logic programming environment for managing similarity and truth degrees. In: Escobar, S. (ed.), Proceedings of XIV Jornadas sobre Programación y Lenguajes, PROLE 2014, Cádiz, Spain, vol. 173. EPTCS, pp. 71–86 (2015). http://dx.doi.org/10.4204/EPTCS.173.6
4. Julián-Iranzo, P., Moreno, G., Penabad, J., Vázquez, C.: A Declarative Semantics for a Fuzzy Logic Language Managing Similarities and Truth Degrees. In: Alferes, J.J.J., Bertossi, L., Governatori, G., Fodor, P., Roman, D. (eds.) RuleML 2016. LNCS, vol. 9718, pp. 68–82. Springer, Cham (2016). doi:10.1007/978-3-319-42019-6_5
5. Julián-Iranzo, P., Rubio-Manzano, C.: An efficient fuzzy unification method and its implementation into the bousi prolog system. In: Proceedings of the IEEE International Conference on Fuzzy Systems, Barcelona, Spain, pp. 1–8. IEEE (2010). http://dx.doi.org/10.1109/FUZZY.2010.5584193
6. Kifer, M., Subrahmanian, V.S.: Theory of generalized annotated logic programming and its applications. J. Logic Program. **12**, 335–367 (1992)
7. Medina, J., Ojeda-Aciego, M., Vojtáš, P.: Similarity-based unification: a multiadjoint approach. Fuzzy Sets Syst. **146**, 43–62 (2004)
8. Pettorossi, A., Proietti, M.: Rules and strategies for transforming functional and logic programs. ACM Comput. Surv. **28**(2), 360–414 (1996)
9. Sessa, M.I.: Approximate reasoning by similarity-based SLD resolution. Theoret. Comput. Sci. **275**(1–2), 389–426 (2002)
10. Tamaki, H., Sato, T.: Unfold/Fold transformations of logic programs. In: Tärnlund, S. (ed.), Proceedings of Second International Conference on Logic Programming, pp. 127–139 (1984)

The Causality/Repair Connection in Databases: Causality-Programs

Leopoldo Bertossi[(✉)]

School of Computer Science, Carleton University, Ottawa, Canada
bertossi@scs.carleton.ca

Abstract. In this work, answer-set programs that specify repairs of databases are used as a basis for solving computational and reasoning problems about causes for query answers from databases.

1 Introduction

Causality appears at the foundations of many scientific disciplines. In data and knowledge management, the need to represent and compute *causes* may be related to some form of *uncertainty* about the information at hand. More specifically in data management, we need to understand why certain results, e.g. query answers, are obtained or not. Or why certain natural semantic conditions are not satisfied. These tasks become more prominent and difficult when dealing with large volumes of data. One would expect the database to provide *explanations*, to understand, explore and make sense of the data, or to reconsider queries and integrity constraints (ICs). Causes for data phenomena can be seen as a kind of explanations.

Seminal work on *causality in DBs* introduced in [17], and building on work on causality as found in artificial intelligence, appeals to the notions of counterfactuals, interventions and structural models [15]. Actually, [17] introduces the notions of: (a) a DB tuple as an *actual cause* for a query result, (b) a *contingency set* for a cause, as a set of tuples that must accompany the cause for it to be such, and (c) the *responsibility* of a cause as a numerical measure of its strength (building on [11]).

Most of our research on causality in DBs has been motivated by an attempt to understand causality from different angles of data and knowledge management. In [6], precise reductions between causality in DBs, DB repairs, and consistency-based diagnosis were established; and the relationships where investigated and exploited. In [7], causality in DBs was related to view-based DB updates and abductive diagnosis. These are all interesting and fruitful connections among several forms of non-monotonic reasoning; each of them reflecting some form of uncertainty about the information at hand. In the case of DB repairs [3], it is about the uncertainty due the non-satisfaction of given ICs, which is represented by presence of possibly multiple intended repairs of the inconsistent DB.

L. Bertossi–Research supported by NSERC Discovery Grant #06148.

S. Moral et al. (Eds.): SUM 2017, LNAI 10564, pp. 427–435, 2017.
DOI: 10.1007/978-3-319-67582-4_33

DB repairs can be specified by means of *answer-set programs* (or *disjunctive logic programs with stable model semantics*) [14], the so-called *repair-programs*. Cf. [3,10] for repair-programs and additional references. In this work we exploit the reduction of DB causality to DB repairs established in [6], by taking advantage of repair programs for specifying and computing causes, their contingency sets, and their responsibility degrees. We show that the resulting *causality-programs* have the necessary and sufficient expressive power to capture and compute not only causes, which can be done with less expressive programs [17], but specially minimal contingency sets and responsibilities (which can not). Causality programs can also be used for reasoning about causes. Finally, we briefly show how causality-programs can be adapted to give an account of other forms of causality in DBs.

2   Background

Relational DBs. A relational schema $\mathcal{R}$ contains a domain, $\mathcal{C}$, of constants and a set, $\mathcal{P}$, of predicates of finite arities. $\mathcal{R}$ gives rise to a language $\mathfrak{L}(\mathcal{R})$ of first-order (FO) predicate logic with built-in equality, $=$. Variables are usually denoted by $x, y, z, ...$, and sequences thereof by $\bar{x}, ...$; and constants with $a, b, c, ...$, etc. An *atom* is of the form $P(t_1, \ldots, t_n)$, with n-ary $P \in \mathcal{P}$ and $t_1, \ldots, t_n$ *terms*, i.e. constants, or variables. An atom is *ground* (aka. a tuple) if it contains no variables. A DB instance, D, for $\mathcal{R}$ is a finite set of ground atoms; and it serves as an interpretation structure for $\mathfrak{L}(\mathcal{R})$.

A *conjunctive query* (CQ) is a FO formula, $\mathcal{Q}(\bar{x})$, of the form $\exists \bar{y} \, (P_1(\bar{x}_1) \wedge \cdots \wedge P_m(\bar{x}_m))$, with $P_i \in \mathcal{P}$, and (distinct) free variables $\bar{x} := (\bigcup \bar{x}_i) \smallsetminus \bar{y}$. If $\mathcal{Q}$ has n (free) variables, $\bar{c} \in \mathcal{C}^n$ is an *answer* to $\mathcal{Q}$ from D if $D \models \mathcal{Q}[\bar{c}]$, i.e. $Q[\bar{c}]$ is true in D when the variables in $\bar{x}$ are componentwise replaced by the values in $\bar{c}$. $\mathcal{Q}(D)$ denotes the set of answers to $\mathcal{Q}$ from D D. $\mathcal{Q}$ is a *boolean conjunctive query* (BCQ) when $\bar{x}$ is empty; and when *true* in D, $\mathcal{Q}(D) := \{true\}$. Otherwise, it is *false*, and $\mathcal{Q}(D) := \emptyset$.

In this work we consider integrity constraints (ICs), i.e. sentences of $\mathfrak{L}(\mathcal{R})$, that are: (a) *denial constraints* (DCs), i.e. of the form $\kappa : \neg \exists \bar{x}(P_1(\bar{x}_1) \wedge \cdots \wedge P_m(\bar{x}_m))$, where $P_i \in \mathcal{P}$, and $\bar{x} = \bigcup \bar{x}_i$; and (b) *functional dependencies* (FDs), i.e. of the form $\varphi : \neg \exists \bar{x}(P(\bar{v}, \bar{y}_1, z_1) \wedge P(\bar{v}, \bar{y}_2, z_2) \wedge z_1 \neq z_2)$. Here, $\bar{x} = \bar{y}_1 \cup \bar{y}_2 \cup \bar{v} \cup \{z_1, z_2\}$, and $z_1 \neq z_2$ is an abbreviation for $\neg z_1 = z_2$.[1] A *key constraint* (KC) is a conjunction of FDs: $\bigwedge_{j=1}^{k} \neg \exists \bar{x}(P(\bar{v}, \bar{y}_1) \wedge P(\bar{v}, \bar{y}_2) \wedge y_1^j \neq y_2^j)$, with $k = |\bar{y}_1| = |\bar{y}_2|$. A given schema may come with its set of ICs, and its instances are expected to satisfy them. If this is not the case, we say the instance is *inconsistent*.

Causality in DBs. A notion of *cause* as an explanation for a query result was introduced in [17], as follows. For a relational instance $D = D^n \cup D^x$, where D^n and D^x denote the mutually exclusive sets of endogenous and exogenous tuples, a tuple $\tau \in D^n$ is called a *counterfactual cause* for a BCQ $\mathcal{Q}$, if $D \models \mathcal{Q}$ and

---

[1] The variables in the atoms do not have to occur in the indicated order, but their positions should be in correspondence in the two atoms.

$D \smallsetminus \{\tau\} \not\models \mathcal{Q}$. Now, $\tau \in D^n$ is an *actual cause* for $\mathcal{Q}$ if there exists $\Gamma \subseteq D^n$, called a *contingency set* for τ, such that τ is a counterfactual cause for $\mathcal{Q}$ in $D \smallsetminus \Gamma$. This definition is based on [15].

The notion of *responsibility* reflects the relative degree of causality of a tuple for a query result [17] (based on [11]). The responsibility of an actual cause τ for $\mathcal{Q}$, is $\rho(\tau) := \frac{1}{|\Gamma|+1}$, where $|\Gamma|$ is the size of a smallest contingency set for τ. If τ is not an actual cause, $\rho(\tau) := 0$. Tuples with higher responsibility are stronger explanations.

In the following we will assume all the tuples in a DB instance are endogenous. (Cf. [6] for the general case.) The notion of cause as defined above can be applied to monotonic queries, i.e. whose sets of answers may only grow when the DB grows [6].[2] *In this work we concentrate only on conjunctive queries, possibly with $\neq$.*

Example 1. Consider the relational DB $D = \{R(a_4, a_3), R(a_2, a_1), R(a_3, a_3), S(a_4), S(a_2), S(a_3)\}$, and the query $\mathcal{Q} \colon \exists x \exists y (S(x) \wedge R(x, y) \wedge S(y))$. It holds, $D \models \mathcal{Q}$.

$S(a_3)$ is a counterfactual cause for $\mathcal{Q}$: if $S(a_3)$ is removed from D, $\mathcal{Q}$ is no longer true. Its responsibility is 1. So, it is an actual cause with empty contingency set. $R(a_4, a_3)$ is an actual cause for $\mathcal{Q}$ with contingency set $\{R(a_3, a_3)\}$: if $R(a_3, a_3)$ is removed from D, $\mathcal{Q}$ is still true, but further removing $R(a_4, a_3)$ makes $\mathcal{Q}$ false. The responsibility of $R(a_4, a_3)$ is $\frac{1}{2}$. $R(a_3, a_3)$ and $S(a_4)$ are actual causes, with responsibility $\frac{1}{2}$. $\qquad\qquad\qquad\qquad\qquad\square$

Database repairs. Cf. [3] for a survey on DB repairs and consistent query answering in DBs. We introduce the main ideas by means of an example. The ICs we consider in this work can be enforced only by deleting tuples from the DB (as opposed to inserting tuples). Repairing the DB by changing attribute values is also possible [3–5], [6, sec. 7.4], but until further notice we will not consider this kind of repairs.

Example 2. The DB $D = \{P(a), P(e), Q(a, b), R(a, c)\}$ is inconsistent with respect to the (set of) *denial constraints* (DCs) $\kappa_1 \colon \neg \exists x \exists y (P(x) \wedge Q(x, y))$, and $\kappa_2 \colon \neg \exists x \exists y (P(x) \wedge R(x, y))$. It holds $D \not\models \{\kappa_1, \kappa_2\}$.

A *subset-repair*, in short an *S-repair*, of D wrt. the set of DCs is a $\subseteq$-maximal subset of D that is consistent, i.e. no proper superset is consistent. The following are S-repairs: $D_1 = \{P(e), Q(a, b), R(a, b)\}$ and $D_2 = \{P(e), P(a)\}$. A *cardinality-repair*, in short a *C-repair*, of D wrt. the set of DCs is a maximum-cardinality, consistent subset of D, i.e. no subset of D with larger cardinality is consistent. D_1 is the only C-repair. $\qquad\qquad\qquad\qquad\qquad\square$

For an instance D and a set Σ of DCs, the sets of S-repairs and C-repairs are denoted with $Srep(D, \Sigma)$ and $Crep(D, \Sigma)$, resp.

---

[2] E.g. CQs, unions of CQs (UCQs), Datalog queries are monotonic.

3   Causality Answer Set Programs

Causes from repairs. In [6] it was shown that causes for queries can be obtained from DB repairs. Consider the BCQ $\mathcal{Q}: \exists \bar{x}(P_1(\bar{x}_1) \wedge \cdots \wedge P_m(\bar{x}_m))$ that is (possibly unexpectedly) true in D: $D \models \mathcal{Q}$. Actual causes for $\mathcal{Q}$, their contingency sets, and responsibilities can be obtained from DB repairs. First, $\neg \mathcal{Q}$ is logically equivalent to the DC:

$$\kappa(\mathcal{Q}): \neg \exists \bar{x}(P_1(\bar{x}_1) \wedge \cdots \wedge P_m(\bar{x}_m)). \tag{1}$$

So, if $\mathcal{Q}$ is true in D, D is inconsistent wrt. $\kappa(\mathcal{Q})$, giving rise to repairs of D wrt. $\kappa(\mathcal{Q})$.

Next, we build differences, containing a tuple τ, between D and S- or C-repairs:

(a) $Dif^s(D, \kappa(\mathcal{Q}), \tau) = \{D \smallsetminus D' \mid D' \in Srep(D, \kappa(\mathcal{Q})), \ \tau \in (D \smallsetminus D')\},$ (2)

(b) $Dif^c(D, \kappa(\mathcal{Q}), \tau) = \{D \smallsetminus D' \mid D' \in Crep(D, \kappa(\mathcal{Q})), \ \tau \in (D \smallsetminus D')\}.$ (3)

It holds [6]: $\tau \in D$ is an actual cause for $\mathcal{Q}$ iff $Dif^s(D, \kappa(\mathcal{Q}), \tau) \neq \emptyset$. Furthermore, each S-repair D' for which $(D \smallsetminus D') \in Dif^s(D, \kappa(\mathcal{Q}), \tau)$ gives us $(D \smallsetminus (D' \cup \{\tau\}))$ as a subset-minimal contingency set for τ. Also, if $Dif^s(D, \kappa(\mathcal{Q}), \tau) = \emptyset$, then $\rho(\tau) = 0$. Otherwise, $\rho(\tau) = \frac{1}{|s|}$, where $s \in Dif^s(D, \kappa(\mathcal{Q}), \tau)$ and there is no $s' \in Dif^s(D, \kappa(\mathcal{Q}), \tau)$ with $|s'| < |s|$. As a consequence we obtain that τ is a most responsible actual cause for $\mathcal{Q}$ iff $Dif^c(D, \kappa(\mathcal{Q}), \tau) \neq \emptyset$.

Example 3. (ex. 1 cont.) With the same instance D and query $\mathcal{Q}$, we consider the DC $\kappa(\mathcal{Q}): \neg \exists x \exists y(S(x) \wedge R(x, y) \wedge S(y))$, which is not satisfied by D. Here, $Srep(D, \kappa(\mathcal{Q})) = \{D_1, D_2, D_3\}$ and $Crep(D, \kappa(\mathcal{Q})) = \{D_1\}$, with $D_1 = \{R(a_4, a_3), \ R(a_2, a_1), R(a_3, a_3), S(a_4), S(a_2)\}$, $D_2 = \{R(a_2, a_1), S(a_4), S(a_2), S(a_3)\}$, $D_3 = \{R(a_4, a_3), R(a_2, a_1), S(a_2), S(a_3)\}$.

For tuple $R(a_4, a_3)$, $Dif^s(D, \kappa(\mathcal{Q}), R(a_4, a_3)) = \{D \smallsetminus D_2\} = \{\{R(a_4, a_3), R(a_3, a_3)\}\}$. So, $R(a_4, a_3)$ is an actual cause, with responsibility $\frac{1}{2}$. Similarly, $R(a_3, a_3)$ is an actual cause, with responsibility $\frac{1}{2}$. For tuple $S(a_3)$, $Dif^c(D, \kappa(\mathcal{Q}), S(a_3)) = \{D \smallsetminus D_1\} = \{S(a_3)\}$. So, $S(a_3)$ is an actual cause, with responsibility 1, i.e. a most responsible cause.   □

It is also possible, the other way around, to characterize repairs in terms of causes and their contingency sets. Actually this connection can be used to obtain complexity results for causality problems from repair-related computational problems [6]. Most computational problems related to repairs, specially C-repairs, which are related to most responsible causes, are provably hard. This is reflected in a high complexity for responsibility [6] (see below for some more details).

Answer-set programs for repairs. Given a DB D and a set of ICs, Σ, it is possible to specify the repairs of D wrt. Σ by means of an answer-set program (ASP)

$\Pi(D, \Sigma)$, in the sense that the set, $Mod(\Pi(D, \Sigma))$, of its stable models is in one-to-one correspondence with $Srep(D, \Sigma)$ [2,10] (cf. [3] for more references). In the following we consider a single denial constraint $\kappa \colon \neg\exists\bar{x}(P_1(\bar{x}_1) \wedge \cdots \wedge P_m(\bar{x}_m)).$[3]

Although not necessary for repair purposes, it may be useful on the causality side having global unique tuple identifiers (tids), i.e. every tuple $R(\bar{c})$ in D is represented as $R(t, \bar{c})$ for some integer t that is not used by any other tuple in D. For the repair program we introduce a nickname predicate R' for every predicate $R \in \mathcal{R}$ that has an extra, final attribute to hold an annotation from the set $\{\mathsf{d}, \mathsf{s}\}$, for "delete" and "stays", resp. Nickname predicates are used to represent and compute repairs.

The *repair-ASP*, $\Pi(D, \kappa)$, for D and κ contains all the tuples in D as facts (with tids), plus the following rules:

$$P'_1(t_1, \bar{x}_1, \mathsf{d}) \vee \cdots \vee P'_m(t_n, \bar{x}_m, \mathsf{d}) \leftarrow P_1(t_1, \bar{x}_1), \ldots, P_m(t_m, \bar{x}_m),$$
$$P'_i(t_i, \bar{x}_i, \mathsf{s}) \leftarrow P_i(t_i, \bar{x}_i), \; not \; P'_i(t_i, \bar{x}_i, \mathsf{d}), \; i = 1, \cdots, m.$$

A stable model M of the program determines a repair D' of D: $D' := \{P(\bar{c}) \mid P'(t, \bar{c}, \mathsf{s}) \in M\}$, and every repair can be obtained in this way [10]. For an FD, say $\varphi \colon \neg\exists x y z_1 z_2 v w(R(x, y, z_1, v) \wedge R(x, y, z_2, w) \wedge z_1 \neq z_2)$, which makes the third attribute functionally depend upon the first two, the repair program contains the rules:

$$R'(t_1, x, y, z_1, v, \mathsf{d}) \vee R'(t_2, x, y, z_2, w, \mathsf{d}) \leftarrow R(t_1, x, y, z_1, v), R(t_2, x, y, z_2, w), z_1 \neq z_2.$$
$$R'(t, x, y, z, v, \mathsf{s}) \leftarrow R(t, x, y, z, v), \; not \; R'(t, x, y, z, v, \mathsf{d}).$$

For DCs and FDs, the repair program can be made non-disjunctive by moving all the disjuncts but one, in turns, in negated form to the body of the rule [2,10]. For example, the rule $P(a) \vee R(b) \leftarrow Body$, can be written as the two rules $P(a) \leftarrow Body, not R(b)$ and $R(b) \leftarrow Body, not P(a)$. Still the resulting program can be *non-stratified* if there is recursion via negation [14], as in the case of FDs and DCs with self-joins.

Example 4. (ex. 3 cont.) For the DC $\kappa(\mathcal{Q}) \colon \neg\exists x \exists y(S(x) \wedge R(x, y) \wedge S(y))$, the repair-ASP contains the facts (with tids) $R(1, a_4, a_3), R(2, a_2, a_1), R(3, a_3, a_3),$ $S(4, a_4), S(5, a_2), S(6, a_3)$, and the rules:

$$S'(t_1, x, \mathsf{d}) \vee R'(t_2, x, y, \mathsf{d}) \vee S'(t_3, y, \mathsf{d}) \leftarrow S(t_1, x), R(t_2, x, y), S(t_3, y),$$
$$S'(t, x, \mathsf{s}) \leftarrow S(t, x), \; not \; S'(t, x, \mathsf{d}). \qquad \text{etc.}$$

Repair D_1 is represented by the stable model M_1 containing $R'(1, a_4, a_3, \mathsf{s})$, $R'(2, a_2, a_1, \mathsf{s}), R'(3, a_3, a_3, \mathsf{s}), S'(4, a_4, \mathsf{s}), S'(5, a_2, \mathsf{s})$, and $S'(6, a_3, \mathsf{d})$.   □

Specifying causes with repair-ASPs. According to (2), we concentrate on the differences between the D and its repairs, now represented by $\{P(\bar{c}) \mid P(t, \bar{c}, \mathsf{d}) \in M\}$,

---

[3] It is possible to consider a combination of several DCs and FDs, corresponding to UCQs (possibly with $\neq$), on the causality side [6].

for M a stable model of the repair-program. They are used to compute actual causes and their $\subseteq$-minimal contingency sets, both identified by tids. So, given the repair-ASP for a DC $\kappa(\mathcal{Q})$, a binary predicate $Cause(\cdot, \cdot)$ will contain a tid for cause in its first argument, and a tid for a tuple belonging to its contingency set. Intuitively, $Cause(t, t')$ says that t is an actual cause, and t' accompanies t as a member of the former's contingency set (as captured by the repair at hand or, equivalently, by the corresponding stable model). More precisely, for each pair of predicates P_i, P_j in the DC $\kappa(\mathcal{Q})$ as in (1) (they could be the same if it has self-joins), introduce the rule $Cause(t, t') \leftarrow P_i'(t, \bar{x}_i, \mathsf{d}), P_j'(t', \bar{x}_j, \mathsf{d}), t \neq t'$, with the inequality condition only when P_i and P_j are the same.

Example 5. (ex. 3 and 4 cont.) The causes for the query, represented by their tids, can be obtained by posing simple queries to the program under the *uncertain or brave* semantics that makes true what is true in *some* model of the repair-ASP.[4] In this case, $\Pi(D, \kappa(\mathcal{Q})) \models_{brave} Ans(t)$, where the auxiliary predicate is defined on top of $\Pi(D, \kappa(\mathcal{Q}))$ by the rules: $Ans(t) \leftarrow R'(t, x, y, \mathsf{d})$ and $Ans(t) \leftarrow S'(t, x, \mathsf{d})$.

The repair-ASP can be extended with the following rules to compute causes with contingency sets:
$Cause(t, t') \leftarrow S'(t, x, \mathsf{d}), R'(t', u, v, \mathsf{d})$,
$Cause(t, t') \leftarrow S'(t, x, \mathsf{d}), S'(t', u, \mathsf{d}), t \neq t'$,
$Cause(t, t') \leftarrow R'(t, x, y, \mathsf{d}), S'(t', u, \mathsf{d})$.

For the stable model M_2 corresponding to repair D_2, we obtain $Cause(1, 3)$ and $Cause(3, 1)$, from the repair difference $D \smallsetminus D_2 = \{R(a_4, a_3), R(a_3, a_3)\}$. $\square$

We can use the DLV system [16] to build the contingency set associated to a cause, by means of its extension, DLV-Complex [9], that supports set building, membership and union, as built-ins. For every atom $Cause(t, t')$, we introduce the atom $Con(t, \{t'\})$, and the rule that computes the union of (partial) contingency sets as long as they differ by some element:

$$Con(T, \#union(C_1, C_2)) \leftarrow Con(T, C_1), Con(T, C_2), \#member(M, C_1),$$
$$not\ \#member(M, C_2).$$

The responsibility for an actual cause τ, with tid t, as associated to a given repair D' (with $\tau \notin D'$), and then to a given model M' of the extended repair-ASP, can be computed by counting the number of t's for which $Cause(t, t') \in M'$. This responsibility will be maximum within a repair (or model): $\rho(t, M') := 1/(1 + |d(t, M')|)$, where $d(t, M') := \{Cause(t, t') \in M'\}$. This value can be computed by means of the *count* function, supported by DLV [13], as follows: pre-rho$(T, N) \leftarrow \#count\{T' : Con(T, T')\} = N$, followed by the rule computing the responsibility: $rho(T, M) \leftarrow M * (pre\text{-}rho(T, M) + 1) = 1$. Or equivalently, via $1/|d(M)|$, with $d(M') := \{P(t', \bar{c}, \mathsf{d}) \mid P(t', \bar{c}, \mathsf{d}) \in M'\}$.

---

[4] As opposed to the *skeptical or cautious* semantics that sanctions as true what is true in *all* models. Both semantics as supported by the DLV system [16], to which we refer below.

Each model M of the program so far will return, for a given tuple (id) that is an actual cause, a *maximal-responsibility contingency set within that model*: no proper subset is a contingency set for the given cause. However, its cardinality may not correspond to the (global) *maximum* responsibility for that tuple. For that we need to compute only maximum-cardinality repairs, i.e. C-repairs.

C-repairs can be specified by means of repair-ASPs [1] that contain *weak-program constraints* [8,13]. In this case, we want repairs that minimize the number of deleted tuples. For each DB predicate P, we introduce the weak-constraint[5] $\Leftarrow P(t, \bar{x}), P'(t, \bar{x}, \mathsf{d})$. In a model M the body can be satisfied, and then the program constraint violated, but the number of violations is kept to a minimum (among the models of the program without the weak-constraints). A repair-ASP with these weak constraints specifies repairs that minimize the number of deleted tuples; and *minimum-cardinality* contingency sets and maximum responsibilities can be computed, as above.

Complexity. Computing causes for CQs can be done in polynomial time in data [17], which was extended to UCQs in [6]. As has been established in [6,17], the computational problems associated to contingency sets and responsibility are in the second level of the polynomial hierarchy (PH), in data complexity [12]. On the other side, our causality-ASPs can be transformed into non-disjunctive, unstratified programs, whose reasoning tasks are also in the second level of the PH (in data). It is worth mentioning that the ASP approach to causality via repairs programs could be extended to deal with queries that are more complex than CQs or UCQs. (In [18] causality for queries that are conjunctions of literals was investigated; and in [7] it was established that cause computation for Datalog queries can be in the second level of the PH.)

Causality programs and ICs The original causality setting in [17] does not consider ICs. An extension of causality under ICs was proposed in [7]. Under it, the ICs have to be satisfied by the DBs involved, i.e. the initial one and those obtained by cause- and contingency-set deletions. When the query at hand is monotonic[6], monotonic ICs (e.g. denial constraints and FDs) are not much of an issue since they stay satisfied under deletions associated to causes. So, the most relevant ICs are non-monotonic, such as referential ICs, e.g. $\forall xy(R(x, y) \rightarrow S(x))$ in our running example. These ICs can be represented in a causality-program by means of (strong) program constraints. In the running example, we would have, for example, the constraint: $\leftarrow R'(t, x, y, \mathsf{s}), not\ S'(t', x, \mathsf{s})$.[7]

Preferred causes and repairs. In [6], generalized causes were introduced on the basis of arbitrary repair semantics (i.e. classes of preferred consistent subinstances, commonly under some maximality criterion), basically starting from the characterization in (2) and (3), but using repairs of D wrt. $\kappa(\mathcal{Q})$ in a class,

---

[5] Hard program-constraints, of the form $\leftarrow Body$, eliminate the models where they are violated.

[6] I.e. the set of answers may only grow when the instance grows.

[7] Or better, to make it *safe*, by a rule and a constraint: $aux(x) \leftarrow S'(t', x, \mathsf{s})$ and $\leftarrow R'(t, x, y, \mathsf{s}), not\ aux(x)$.

$Rep(D, \kappa(\mathcal{Q}))$, possibly different from $Srep(D, \kappa(\mathcal{Q}))$ or $Crep(D, \kappa(\mathcal{Q}))$. As a particular case in [6], *causes based on changes of attribute values* (as opposed to tuple deletions) were defined. In that case, admissible updates are replacements of data values by null values, to break joins, in a minimal or minimum way. Those underlying DB repairs were used in [4] to hide sensitive data that could be exposed through CQ answering; and corresponding repair programs were introduced. They could be used, as done earlier in this paper, as a basis to reason about- and compute the new resulting causes (at the tuple or attribute-value level) and their contingency sets.[8]

References

1. Arenas, M., Bertossi, L., Chomicki, J.: Answer sets for consistent query answers. Theor. Pract. Logic Program. **3**(4&5), 393–424 (2003)
2. Barceló, P., Bertossi, L., Bravo, L.: Characterizing and Computing Semantically Correct Answers from Databases with Annotated Logic and Answer Sets. In: Bertossi, L., Katona, G.O.H., Schewe, K.-D., Thalheim, B. (eds.) SiD 2001. LNCS, vol. 2582, pp. 7–33. Springer, Heidelberg (2003). doi:10.1007/3-540-36596-6_2
3. Bertossi, L.: Database Repairing and Consistent Query Answering. Morgan & Claypool, Synthesis Lectures on Data Management (2011)
4. Bertossi, L., Li, L.: Achieving data privacy through secrecy views and null-based virtual updates. IEEE Trans. Knowl. Data Eng. **25**(5), 987–1000 (2013)
5. Bertossi, L., Bravo, L.: Consistency and trust in peer data exchange systems. Theor. Pract. Logic Program. **17**(2), 148–204 (2017)
6. Bertossi, L., Salimi, B.: From causes for database queries to repairs and model-based diagnosis and back. Theor. Comput. Syst. **61**(1), 191–232 (2017)
7. Bertossi, L., Salimi, B.: Causes for query answers from databases: datalog abduction, view-updates, and integrity constraints. To appear in Int. J. Approximate Reasoning. Corr Arxiv Paper cs.DB/1611.01711
8. Buccafurri, F., Leone, N., Rullo, P.: Enhancing disjunctive datalog by constraints. IEEE Trans. Knowl. Data Eng. **12**(5), 845–860 (2000)
9. Calimeri, F., Cozza, S., Ianni, G., Leone, N.: An ASP System with Functions, Lists, and Sets. In: Erdem, E., Lin, F., Schaub, T. (eds.) LPNMR 2009. LNCS, vol. 5753, pp. 483–489. Springer, Heidelberg (2009). doi:10.1007/978-3-642-04238-6_46
10. Caniupan-Marileo, M., Bertossi, L.: The consistency extractor system: answer set programs for consistent query answering in databases. Data Knowl. Eng. **69**(6), 545–572 (2010)
11. Chockler, H., Halpern, J.Y.: Responsibility and blame: a structural-model approach. J. Artif. Intell. Res. **22**, 93–115 (2004)
12. Dantsin, E., Eiter, T., Gottlob, G., Voronkov, A.: Complexity and expressive power of logic programming. ACM Comput. Surv. **33**(3), 374–425 (2001)
13. Faber, W., Pfeifer, G., Leone, N., Dell'Armi, T., Ielpa, G.: Design and implementation of aggregate functions in the DLV system. Theor. Pract. Logic Program. **8**(5-6), 545–580 (2008)
14. Gelfond, M., Kahl, Y.: Knowledge Representation and Reasoning, and the Design of Intelligent Agents. Cambridge University Press, Cambridge (2014)

---

[8] Cf. also [5] for an alternative null-based repair semantics and its repair programs.

15. Halpern, J., Pearl, J.: Causes and explanations: a structural-model approach: part 1. Br. J. Philos. Sci. **56**, 843–887 (2005)
16. Leone, N., Pfeifer, G., Faber, W., Eiter, T., Gottlob, G., Perri, S., Scarcello, F.: The DLV system for knowledge representation and reasoning. ACM Trans. Comput. Logic. **7**(3), 499–562 (2006)
17. Meliou, A., Gatterbauer, W., Moore, K.F., Suciu, D.: The complexity of causality and responsibility for query answers and non-answers. Proc. VLDB. **4**, 34–41 (2010)
18. Salimi, B., Bertossi, L., Suciu, D., Van den Broeck, G.: Quantifying causal effects on query answering in databases. Proceedings of TaPP (2016)

Author Index

Printed in the United States
By Bookmasters